WORLDS IN INTERACTION: SMALL BODIES AND PLANETS OF THE SOLAR SYSTEM

WORLDS IN INTERACTION:
SMALL BODIES AND
PLANETS OF THE SOLAR SYSTEM

*Proceedings of the Meeting "Small Bodies in the Solar System
and their Interactions with the Planets" held in Mariehamn,
Finland, August 8–12, 1994*

Edited by

H. RICKMAN and M. J. VALTONEN

Reprinted from *Earth, Moon, and Planets*,
Volume 72, Nos. 1–3, 1996

KLUWER ACADEMIC PUBLISHERS
DORDRECHT / BOSTON / LONDON

Library of Congress Cataloging-in-Publication Data

ISBN-13: 978-94-010-6578-8   e-ISBN-13: 978-94-009-0209-1
DOI: 10.1007/978-94-009-0209-1

Published by Kluwer Academic Publishers,
P.O. Box 17, 3300 AA Dordrecht, The Netherlands.

Kluwer Academic Publishers incorporates
the publishing programmes of
D. Reidel, Martinus Nijhoff, Dr W. Junk and MTP Press.

Sold and distributed in the U.S.A. and Canada
by Kluwer Academic Publishers,
101 Philip Drive, Norwell, MA 02061, U.S.A.

In all other countries, sold and distributed
by Kluwer Academic Publishers Group,
P.O. Box 322, 3300 AH Dordrecht, The Netherlands.

Printed on acid-free paper

EARTH, MOON, AND PLANETS / *Vol. 72   Nos. 1-3   1996*

WORLDS IN INTERACTION: SMALL BODIES AND PLANETS
OF THE SOLAR SYSTEM

Edited by H. RICKMAN and M.J. VALTONEN

PREFACE

When August comes, the Scandinavian nights get darker. One can go for evening strolls as if in a borderland between near and far, between the flowery meadows and the starry sky. This is a time to realize one of the grandest and most intriguing of all connections. Planet Earth, our benign abode, plays a part in Galactic history. Its atoms, like those in our bodies, were mostly formed as a result of stellar evolution by processes still occurring in stars like Antares. Tiny grains are bringing newly formed organic molecules into the birthplaces of future stars and planets, and this process may once have been essential for rendering our own planet a lively, fertile place. And the Galaxy has kept influencing its earthly creatures. One important way is through gravitational torques applied to our distant reservoir of comets – the Oort cloud. The August Perseid meteors demonstrate the fact that we continually live with cometary orbits crossing that of the Earth. And asteroidal fragments find their ways to us as well. Through the interplay of planets and minor bodies of the Solar System, material is transported and ejected on a massive scale, and planetary surfaces get scarred. These impacts form part of Earth's geologic history as well as its future, and as a consequence, our climate and living conditions can get seriously perturbed. When watching the stars, we probe a sinister environment where processes far beyond our control are taking place over time scales long enough to give us comfort but ultimately setting limits to our existence, continuing to upheave the evolution of terrestrial species.

This book came about as a result of the conference *Small Bodies in the Solar System and their Interactions with the Planets*, held in Mariehamn, Åland, August 8–12 1994. The conference was attended by about 140 scientists from 23 countries worldwide. It had a particular focus on the collisions of the small bodies with the Earth and other planets, and almost as if ordered, the most spetacular collision of comet D/Shoemaker-Levy 9 with Jupiter took place barely a month before the conference. Needless to say, this was one of the hot topics of the meeting, and the event is also reflected in these proceedings.

The importance of our Galactic environment in the discussion of the impacts of small bodies onto the Earth started to be fully realized only in the 1980's, and so the conference on *The Galaxy and the Solar System* was held in Tucson in 1985 to gather together experts on the whole range of topics. The conference *Dynamics and Evolution of Minor Bodies with Galactic and Geological Implications*, held in Kyoto in 1991 as a worldwide extension of a Japanese series of yearly, topical meetings, continued the discussions in the same spirit. Our participation in this meeting gave us the inspiration to organize a follow-up three years later at Mariehamn. Meanwhile, research went on and intensified, and new important developments appeared. The present proceedings describe and summarize the latest research in this very active field.

Fittingly to the topic of the conference, the conference site was close to an ancient impact crater, Lumparn, which is now a roundish bay in the Åland archipelago and formed an interesting destination for a half-day excursion. The town of Mariehamn also happens to lie half-way between the two old university towns, Uppsala and Turku, both of which shared the organizing responsibility. The opening reception was held at Tuorla Observatory of University of Turku, while the closing happened in the gardens of the Astronomical Observatory of the University of Uppsala. In between, most participants took the opportunity to travel by boats of differents sizes to reach from one destination to the next. For the enthusiasts there was a post-conference tour to lake Siljan, one of the largest and most important impact structures known, as well as a prime tourist resort in Sweden.

We would like to take the opportunity here to thank the staffs of the two observatories for the great organizing effort. Especially one should mention Sirpa Reinikainen from Tuorla and Mats Dahlgren from Uppsala for work much beyond duty. In the editorial work we also had very good assistance by Marcus Gunnarsson and Erik Onnela of Uppsala Observatory. The local organization in Mariehamn was in the professional hands of Stig-Björn Ulfsson whose office handled the "small bodies" magnificently. Our thanks are also due to the town of Mariehamn, the two universities, Nordita, the International Science Foundation, Academy of Finland, Swedish Natural Science Research Council, the Swedish Space Board, Turku University foundation and Finland's Academy for Sciences and Letters for sponsoring the meeting.

HANS RICKMAN   MAURI VALTONEN

GUIDE TO THE PHOTO

1 H.J. Schober, 2 E. Lohinger, 3 C.B. Cosmovici, 4 P. Farinella, 5 M. Hoffmann, 6 E. Bowell, 7 H. Rickman, 8 F.L. Whipple, 9 M.J. Valtonen, 10 K.V. Kholshevnikov, 11 K.I. Churyumov, 12 I. Petrovskaya, 13 O. Hernius, 14 Ch. Froeschlé, 15 P.B. Babadzhanov, 16 A.W. Harris, 17 H. Prętka, 18 J.A. Fernández, 19 A. Mylläri, 20 N.O. Kirsanov, 21 M.-F. He, 22 V.V. Emel'yanenko, 23 C. Keay, 24 I. Belskaya, 25 J. Li, 26 D. Jewitt, 27 V. Zappalà, 28 Cl. Froeschlé, 29 D.I. Steel, 30 P.A. Dybczyński, 31 T. Michałowski, 32 J. Henrard, 33 P. Magnusson, 34 G. Andreev, 35 T. Kwiatkowski, 36 J. Borovička, 37 P. Pravec, 38 E. Elst, 39 K. Elst, 40 I.P. Williams, 41 M. Lehtinen, 42 D.P. O'Ceallaigh, 43 A. Lemaître, 44 C.T. Reimann, 45 E.M. Pittich, 46 I.V. Nemtchinov, 47 U. Motschmann, 48 G. Tancredi, 49 C.-I. Lagerkvist, 50 G.J. Flynn, 51 M. Moons, 52 N. Kotsarenko, 53 D.F. Lupishko, 54 B.G. Marsden, 55 A. Milani, 56 M. Elst, 57 S. Elst, 58 J.-Q. Zheng, 59 J.J. Matese, 60 P.K. Seidelmann, 61 A. Morbidelli, 62 M. Lindgren, 63 K. Muinonen, 64 S. Yabushita, 65 Yu.V. Obrubov, 66 A. Brunini, 67 R. Gil-Hutton, 68 L.J. Pesonen, 70 E. Asphaug, 71 E. Kührt, 72 I. Hasegawa, 73 M. Šidlichovský, 74 N. Solovaya, 75 A. Fitzsimmons, 76 W. Bottke, 77 G. Hahn, 78 J. Lagerros, 79 F. López-Garcia, 80 M. Wolf, 81 S. Szutowicz, 82 S. Innanen, 83 K. Innanen, 84 L. Laucenieks, 85 K. Ziołkowski, 86 S. Mikkola, 87 S. Green, 88 W.G. Elford, 89 R. Lorenz, 90 K. Meech, 91 J. Piironen, 92 Å. Claesson, 93 R. Törnberg, 94 A. Salítis, 95 J. Zetzer, 96 W. Waniak, 97 A. Villani, 98 S. Baccili, 99 M. Adolfsson, 100 L. Adolfsson, 101 J. Knollenberg, 102 P.R. Weissman, 103 A.D. Storrs, 104 Yu. Skorov, 105 J.C. Brandt, 106 R. Jedicke, 107 N. McBride, 108 M. English, 109 H. Lehto, 110 M.K. Wallis, 111 S.V.M. Clube, 112 S. Isobe, 113 N. Isobe, 114 K. Isobe, 115 M. Yoshikawa, 116 G.B. Valsecchi, 117 H. Kinoshita, 118 M. Bailey, 119 E. Drobyshevski, 120 M. Terho, 121 H.U. Keller, 122 L. Valtaoja, 123 M. Dahlgren, 124 D.C. Boice, 125 T.-Y. Huang, 126 H. Karttunen, 127 Z. Ceplecha, 128 A. Chernin, 129 S. Reinikainen, 130 J.L. Hilton, 131 M. Banaszkiewicz, 132 W.M. Napier, 133 Ph. De Jager, 134 M.R. Rampino, 135 T. Nakamura, 136 S. Ipatov, 137 N.-B. Svensson

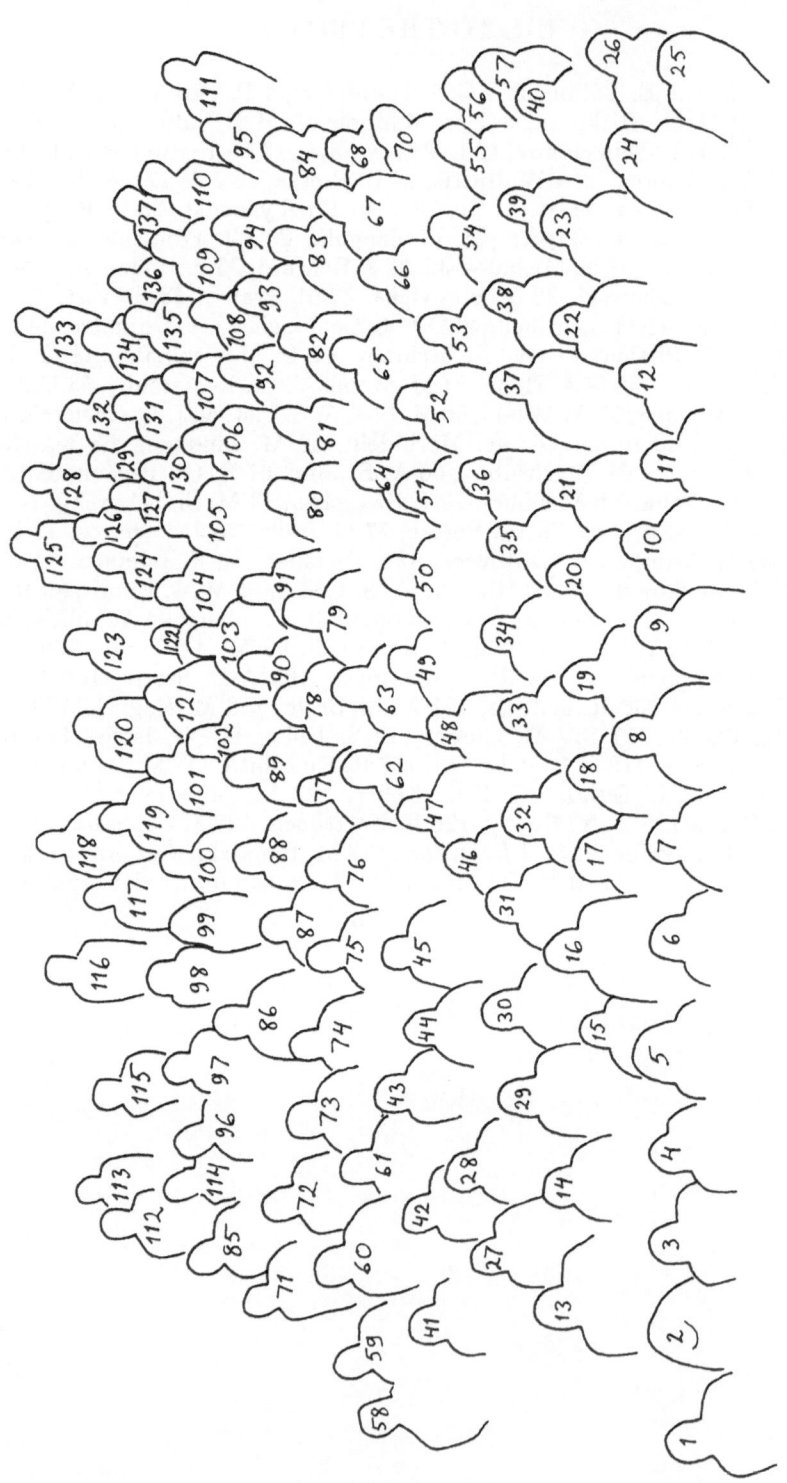

LIST OF PARTICIPANTS

A'HEARN, M.F.
University of Maryland
E-mail: ma@astro.umd.edu

ADOLFSSON, L.
University of Florida
E-mail: adolfsso@astro.ufl.edu

AKSNES, K.
University of Oslo
E-mail: kaare.aksnes@astro.uio.no

ALDAHAN, A.
Uppsala University
E-mail: ala@geosparc.uu.se

ANDREEV, G.V.
Tomsk State University
E-mail: niipmm@urania.tomsk.su

ARTEM'EV, V.I.
Russian Academy of Sciences
E-mail: idg@glas.apc.org

ASPHAUG, E.
NASA–Ames Research Centre
E-mail: asphaug@cosmic.arc.nasa.gov

BABADZHANOV, P.B.
Institute of Astrophysics

BACCILI, S.
University of Pisa
E-mail: baccili@adams.dm.unipi.it

BAILEY, M.E.
Liverpool John Moores University
E-mail: m.bailey%liverpool-john-moores.ac.uk
@hubby.liverpool-john-moores.ac.uk

BANASZKIEWICZ, M.
Space Research Centre
E-mail: marekb@cbk.waw.pl

BELSKAYA, I.
Kharkov University
E-mail: irina@astron.kharkov.ua

BOICE, D.C.
Southwest Research Institute
E-mail: boice@swri.space.swri.edu

BOROVIČKA, J.
Ondřejov Observatory
E-mail: borovic@asu.cas.cz

BOTTKE, W.J.
Lunar & Planetary Laboratory
E-mail: bottke@lpl.arizona.edu

BOWELL, E.
Lowell Observatory
E-mail: elgb@lowell.edu

BRANDT, J.C.
Laboratory of Atmospheric & Space Physics
E-mail: brandt@lyrae.colorado.edu

BRUNINI, A.
Observatorio astronomico de La Plata
E-mail: abrunini@fcaglp.edu.ar

CEPLECHA, Z.
Astronomical Institute, Ondřejov
E-mail: ceplecha@asu.cas.cz

CHERNIN, A.
Moscow State University
E-mail: chernin@sai.msu.su

CHURYUMOV, K.I.
Kiev University
E-mail: aoku@glas.apc.org

CLAESSON, Å.
Uppsala University

CLUBE, S.V.M.
Department of Astrophysics

COLLINI, B.
Uppsala University

Earth, Moon and Planets **72**: xv–xx, 1996.

COSMOVICI, C.
Istituto di Fisica dello Spazio Interplanetario / CNR
E-mail: cosmo@hp.ifsi.fra.cnr.it

DAHLGREN, M.
Uppsala University
E-mail: mats.dahlgren@astro.uu.se

DROBYSHEVSKI, E.M.
Russian Academy of Sciences
E-mail: drob!emdrob%drob.pti.spb.su
@ccpti.ioffe.rssi.ru

DVORAK, R.
Universität Wien
E-mail: dvorak@astro.ast.univie.ac.at

DYBCZYŃSKI, P.A.
Adam Mickiewicz University
E-mail: dybol@phys.amu.edu.pl

ELFORD, W.G.
University of Adelaide

ELST, E.W.
Royal Observatory
E-mail: elst@atmos.oma.be

EMEL'YANENKO, V.V.
Liverpool John Moores University
E-mail: ve@staru1.livjm.ac.uk

ENGLISH, M.
University of Kent
E-mail: mae@ukc.ac.uk

FARINELLA, P.
Observatoire de la Côte d'Azur
E-mail: zappala@rameau.obs-nice.fr

FERNÁNDEZ, J.A.
Departamento de Astronomía
E-mail: julio@fisica.edu.uy

FITZSIMMONS, A.
Queens University
E-mail: a.fitzsimmons@queens-belfast.ac.uk

FLYNN, G.J.
State University of New York

FROESCHLÉ, Ch.
Observatoire de la Côte d'Azur
E-mail: froesch@obs-nice.fr

FROESCHLÉ, Cl.
Observatoire de la Côte d'Azur
E-mail: claude@obs-nice.fr

GAVRILOV, B.G.
Russian Academy of Sciences
E-mail: idg@glas.apc.org

GIL-HUTTON, R.
Universidad Nacional de San Juan
E-mail: rgh@unsjfa.edu.ar

GREEN, S.
University of Kent
E-mail: sfg1@ukc.ac.uk

GRINSPOON, D.
Laboratory of Atmospheric & Space Physics
E-mail: david@sunrd.colorado.edu

HAHN, G.
Institut für Planetenerkendung / DLR
E-mail: hahn@terra.pe.ba.dlr.de

HARRIS, A.W.
Jet Propulsion Laboratory
E-mail: awharris@jpl354.jpl.nasa.gov

HASEGAWA, I.
Otemae Junior College

HE, M.-F.
Shanghai Observatory
E-mail: bmasao@ica.beijing.canet.cn

HENRARD, J.
University of Namur
E-mail: jhenrard@math.fundp.ac.be

HERNIUS, O.
Uppsala University
E-mail: ohe@astro.uu.se

HILTON, J.L.
U.S. Naval Observatory
E-mail: hil@newcomb.usno.navy.mil

HOFFMANN, M.
Observatorium Hoher List
E-mail: hoffmann@astro.uni-bonn.de

HUANG, T.-Y.
Nanjing University
E-mail: bmanju@ica.beijing.canet.cn

INNANEN, K.A.
York University
E-mail: fs300033@sol.yorku.ca

IPATOV, S.I.
Keldysh Institute of Applied Mathematics
E-mail: ipatov@applmat.msk.su

ISOBE, S.
National Astronomical Observatory
E-mail: oisobex@c1.mtk.nao.ac.jp

JEDICKE, R.
Lunar & Planetary Laboratory
E-mail: jedicke@pirl.lpl.arizona.edu

JEWITT, D.
Institute for Astronomy
E-mail: jewitt@galileo.ifa.hawaii.edu

KARTTUNEN, H.
University of Helsinki
E-mail: Hannu.Karttunen@csc.fi

KEAY, C.
University of Newcastle
E-mail: phcslk@cc.newcastle.edu.au

KELLER, H.U.
Max Planck Institut für Aeronomie
E-mail: keller@linax1.mpae.gwdg.de

KHOLSHEVNIKOV, K.V.
St. Petersburg University
E-mail: kvk@aispbu.spb.su

KINOSHITA, H.
National Astronomical Observatory
E-mail: kinoshita@c1.mtk.nao.ac.jp

KIRSANOV, N.O.
Institute for Theoretical Astronomy
E-mail: kirsanov@iipah.spb.su

KNOLLENBERG, J.
Institut für Planetenerkendung / DLR
E-mail: wsat@arzsp7.rz.ba.dlr.de

KONINCKX, C.
Vrije Universiteit Brüssel

KOTSARENKO, N.
Kiev University
E-mail: kotsaren@astron.univ.kiev.uc

KÜHRT, E.
Institut für Weltraumsensorik / DLR
E-mail: ws1s@arzsp7.rz.ba.dlr.de

KWIATKOWSKI, T.
Adam Mickiewicz University
E-mail: tkastr@phys.amu.edu.pl

LAGERKVIST, C.-I.
Uppsala University
E-mail: claes-ingvar.lagerkvist@astro.uu.se

LAGERROS, J.
Uppsala University
E-mail: johan.lagerros@astro.uu.se

LAUCENIEKS, L.
University of Latvia
E-mail: ao@astr.lu.lv

LEHTO, H.
Tuorla Observatory
E-mail: harry.lehto@utu.fi

LEMA^ITRE, A.
University of Namur
E-mail: alemaitr@math.fundp.ac.be

LINDGREN, M.
Uppsala University
E-mail: mats.lindgren@astro.uu.se

LOHINGER, E.
Universität Wien
E-mail: lohinger@astro1.ast.univie.ac.at

LÓPEZ-GARCIA, F.
Universidad Nacional de San Juan
E-mail: quito@unsjfa.edu.ar

LORENZ, R.D.
University of Kent
E-mail: rdl1@ukc.ac.uk

LUPISHKO, D.F.
Kharkov University
E-mail: lupishko@astron.kharkov.ua

MAGNUSSON, P.
Uppsala University
E-mail: per.magnusson@astro.uu.se

MARSDEN, B.G.
Smithsonian Center for Astrophysics
E-mail: marsden@cfa.harvard.edu

MATESE, J.
University of SW Louisiana
E-mail: jjm9638@usl.edu

MCBRIDE, N.
University of Kent
E-mail: n.mcbride@ukc.ac.uk

MEECH, K.J.
Institute for Astronomy
E-mail: meech@uhifa.ifa.hawaii.edu

MICHAŁOWSKI, T.
Adam Mickiewicz University
E-mail: tmich@phys.amu.edu.pl

MIKKOLA, S.
Tuorla Observatory
E-mail: mikkola@utu.fi

MILANI, A.
University of Pisa
E-mail: milani@dm.unipi.it

MOONS, M.
University of Namur
E-mail: mmoons@quick.cc.fundp.ac.be

MORBIDELLI, A.
Observatoire de la Côte d'Azur
E-mail: morby@obs-nice.fr

MOTSCHMANN, U.
University of Braunschweig
E-mail: uwe@geophys.not.tu-bs.de

MUINONEN, K.
University of Helsinki
E-mail: karri@gstar.helsinki.fi

MYLLÄRI, A.
Petrozavodsk State University
E-mail: amul@pgu.karelia.su

NAKAMURA, T.
National Astronomical Observatory
E-mail: nakamura@c1.mtk.nao.ac.jp

NAPIER, W.M.
Department of Astrophysics

NEMTCHINOV, I.V.
Russian Academy of Sciences
E-mail: idg@glas.apc.org

O'CEALLAIGH, D.P.
Queens University
E-mail: d.o'ceallaigh@qub.ac.uk

OBRUBOV, Yu.V.
Agricultural Academy
E-mail: iate@storm.iasnet.com

OWEN, T.
Institute for Astronomy
E-mail: owen@uhifa.ifa.hawaii.edu

PESONEN, L.J.
Geological Survey of Finland

PETROVSKAYA, I.
St. Petersburg University
E-mail: kvk@aispbu.spb.su

PIIRONEN, J.O.
University of Helsinki
E-mail: jpiironen@gstar.helsinki.fi

PITTICH, E.M.
Slovak Academy of Sciences
E-mail: astropit@savba.savba.sk

PRAVEC, P.
Astronomical Institute
E-mail: ppravec@asu.cas.cz

PRĘTKA, H.
Adam Mickiewicz University
E-mail: obsastr@plpuam.bitnet

RAMPINO, M.R.
New York University
E-mail: rampino@acfcluster.nyu.edu

REIMANN, C.T.
Uppsala University
E-mail: scooter@tsl.uu.se

RICKMAN, H.
Uppsala University
E-mail: hans.rickman@astro.uu.se

SALÍTIS, A.
Daugavpils Pedagogical University

SAUER, K.
MPI für Extraterrestrische Physik
E-mail: ks@mpe.fta-berlin.de

SCHOBER, H.J.
Institute für Astronomie
E-mail: schober@bkfug.kfunigraz.ac.at

SEIDELMANN, P.K.
U.S. Naval Observatory
E-mail: omd@ariel.usno.navy.mil

ŠIDLICHOVSKÝ, M.
Astronomical Institute
E-mail: sidli@ig.cas.cz

SKOROV, Yu.
Keldysh Institute of Applied Mathematics
E-mail: lev@ccas.ru

SOLOVAYA, N.A.
Sternberg Astronomical Institute
E-mail: ursa@sai.msk.su

STEEL, D.I.
Anglo-Australian Observatory
E-mail: dis@aaocbn3.aao.gov.au

STORRS, A.D.
Space Telescope Science Institute
E-mail: storrs@stsci.edu

SVENSSON, N.-B.
Uppsala University

SZUTOWICZ, S.
Space Research Centre
E-mail: slawka@cbk.waw.pl

TANCREDI, G.
Departamento de Astronomía
E-mail: gonzalo@fisica.edu.uy

TERHO, M.
Geological Survey of Finland

TÖRNBERG, R.
Stockholm University
E-mail: roger.tornberg@geol.su.se

VALSECCHI, G.B.
I.A.S. / Planetologia
E-mail: giovanni@vm-14s.14s.fra.cnr.it

VALTAOJA, L.
Tuorla Observatory
E-mail: lvaltaoja@sara.cc.utu.fi

VALTONEN, M.J.
Tuorla Observatory
E-mail: mavalto@sara.cc.utu.fi

VANOUPLINES, P.
Vrije Universiteit Brüssel

VILLANI, A.
University of Pisa
E-mail: avillani@dm.unipi.it

WALLIS, M.K.
University of Wales

WANIAK, W.
Jagellonian University
E-mail: waniak@oa.uj.edu.pl

WEISSMAN, P.R.
Jet Propulsion Laboratory
E-mail: pweissman@jpluvs.jpl.nasa.gov

WHIPPLE, F.L.
Smithsonian Center for Astrophysics

WILLIAMS, I.P.
University of London
E-mail: i.p.williams@qmw.ac.uk

WOLF, M.
Astronomical Institute
E-mail: wolf@earn.cvut.cz

YABUSHITA, S.
Kyoto University
E-mail: yabusita@kuamp.kyoto-u.ac.jp

YOSHIKAWA, M.
Communications Research Laboratory
E-mail: makoto@crl.go.jp

ZAPPALÀ, V.
Osservatorio Astronomico
E-mail: zappala@to.astro.it

ZETZER, J.I.
Russian Academy of Sciences
E-mail: idg@glas.apc.org

ZHENG, J.-Q.
Tuorla Observatory
E-mail: zheng@polaris.cc.utu.fi

ZIOŁKOWSKI, K.
Space Research Centre
E-mail: kazet@cbk.waw.pl

THE SOLAR SYSTEM IN THE GALACTIC ENVIRONMENT

K.A. INNANEN

*Dept of Physics & Astronomy, York University, Toronto, Canada
& Tuorla Observatory, University of Turku*

1. Introduction

The Galaxy is a prominent member of the so-called "Local Group", a physical collection of some 30 galaxies, all except a handful of which are relatively modest in size and mass. The Local Group, in turn, is a modest subcluster of galaxies in the outlying parts of the Virgo cluster of galaxies. In addition to the Galaxy, its list of prominent members includes M31, Maffei 1 and IC 342. Recent computer simulations (Byrd et al., 1994) give one the impression that, in contrast to our nearest large neighbour M31, which may have participated in large-scale mergers and interactions, the Galaxy has not. The Galaxy may, of course, have absorbed smaller objects, one example of which is even now evidently in progress (Ibata et al., 1994). Thus there seems to have been a cosmically significant period of time for the Galaxy to have achieved a substantial degree of global equilibrium, apart from some modest perturbations. The Galaxy is, therefore, considered to have a smoothed, axially symmetric mass distribution, and a corresponding gravitational potential, in which the observed stellar and gaseous components have reached a significant degree of statistical-dynamical equilibrium. Most of contemporary Galactic structure and dynamics rests on these assumptions, and the fact that it does so with considerable success (e.g. Kuijken & Tremaine, 1994) is testimony to the quality of the overall foundations of the subject.

Mapping the Galaxy, and performing the census of the Galactic populations from our location in a dusty suburb of the Galactic disk is an enormous challenge. We shall be reminded of this fact later in the review. Some of the observations have recently been interpreted as evidence of global asymmetry (or axisymmetric distortions, (e.g. Kuijken & Tremaine, 1994)). Such distortions could be caused, for example, by a bar-like mass, rotating in the central regions of the Galaxy. The presence of this type of global asymmetry could play a significant role in the past and future of the trajectory of the Solar System in the Galaxy. Apart from noting this possibility, and leaving it for future confirmation, this review will follow the traditional axisymmetric picture.

Earth, Moon, and Planets **72**: 1-6, 1996.

2. Physical parameters of the Galaxy in the Solar neighbourhood

This section is basically an update of earlier conferences on this theme, in particular that edited by Smoluchowski et al. (1986).

The three, basic physical parameters governing study of the Galaxy are: i) the distance R_o of the Solar System from the centre; ii) the circular speed V_o of the local standard of rest around the Galactic centre, and iii) the local total density of matter ρ_o. Alternatively, it is frequently useful to employ Σ_o, the local total surface density in a column of unit cross-section erected perpendicular to the Galactic plane at the Sun.

Despite great efforts during the past generation, there has been little progress in the improvement of the magnitudes of, and the errors in these quantities. A good recent review dealing with i) and ii) can be found in Kuijken & Tremaine (1994). Happily, R_o and V_o are not primary drivers of the Solar Galactic orbit in the present context, although both carry uncertainties of some 15-20%. Unhappily, ρ_o (or Σ_o) is important; each carries a current uncertainty of a factor of about 2. The reasons for this uncertainty in ρ_o are complex, and are the subject of a vigorous, ongoing controversy. Kuijken (1991), and Kuijken & Gilmore (1992) have argued that there is no compelling evidence for unobserved, disk-like matter in the Solar vicinity. They find that ρ_o is $0.10 M_\odot/pc^3$, or alternatively that Σ_o is $50-60 \ M_\odot/pc^2$. On the other hand, Bahcall et al. (1992) argue with equal conviction that ρ_o is $0.20 \ M_\odot/pc^3$, or that Σ_o is about $80 \ M_\odot/pc^2$. A very illuminating summary of this debate can be found in Fuchs (1994), especially his Fig. 2. If the Bahcall et al. point of view is correct, some 50% of the local total density must reside in a very flat, but so far undetected, disk-like, distribution of matter. The stars in the Solar Galactic neighbourhood have a z scale-height of about 300 pc, and the gas has a z scale-height of about 100 pc. The dark matter in the disk, should it exist, would need to have a z scale-height less than that of the gas to keep the dynamics and its physical foundations self-consistent. Maintaining a disk of compact objects of that sheet-like thinness for several Galactic years is a major challenge. In addition, the option list of available dark matter objects has shrunk in recent years. Two remaining possibilities are: (a) large numbers of "Jupiters", or (b) black holes with mass of about $10^4 M_\odot$ (Rix & Lake, 1993). All of these comments are quite independent of the massive, spherical dark-matter halo in which the visible Galaxy is believed to be embedded. That massive halo contributes only about 1% of the total local mass density, and therefore does not affect the Galactic Solar orbit.

3. The Solar orbit in an axisymmetric Galaxy model

Once an axisymmetric gravitational potential for the Galaxy has been adopted, the computation of the Solar orbit, backward or forward in time, is relatively well defined, and uncontroversial. In addition to sharing the assumed circular motion of the local standard of rest, the Solar System has its own peculiar motion amongst the nearby stars. This motion, about 20 km/sec, has to be vectorially added to V_o. As a result, the Solar Galactic orbit can be conveniently considered to be made up of two parts: the motion of revolution about the symmetry axis (ie in the Galactic plane), and the motion of oscillation perpendicular to the Galactic plane. The motion in the plane is approximately elliptical, with a quasi-eccentricity of about 0.1, i.e. rather similar to that of Mars about the Sun. This non-circular (epicyclic) motion has associated with it a radial motion of about 1 kpc from perigalacticon to apogalacticon. Its (epicyclic) period is determined by the rotational behaviour of the Galactic potential near the Sun. The second part of the Solar motion is essentially a harmonic motion perpendicular to the Galactic plane, and where the square root of the local total density is proportional to the "spring constant".

Thus, if K_z is the gravitational acceleration perpendicular to the Galactic plane at the Sun, per unit mass, then, to first order, Poisson's equation is:

$$dK_z/dz = 4\pi G\rho_o \tag{1}$$

and, also to first order, a being a constant,

$$K_z = -a^2 z \tag{2}$$

The period P_z of the harmonic motion will be given by

$$P_z = 2\pi/a \tag{3}$$

and the amplitude of the associated maximum excursion is given by $z_{max} = Z_o/a$, where Z_o is the velocity component of the Sun perpendicular to the Galactic plane (which it is crossing at this epoch), i.e. about 7.5 km/sec.

If one now assumes that $\rho_o = 0.2 M_\odot/pc^3$, then

$$a^2 = 2.67(10)^{-9} \, (AU/yr/yr/pc) = 400 \; m/yr/yr/pc \tag{4}$$

This is a relatively small acceleration, for 400 m/yr is about 10% of the speed of a snail.

Corresponding to the same total density, one also finds that $P_z = 56$ Myr and $z_{max} = 70$ pc. If, on the other hand, one prefers no disk-like dark matter, so that $\rho_o = 0.10 M_\odot/pc^3$, then $P_z = 82$ Myr, and $z_{max} = 140$ pc. A period P_z of 60 Myr therefore requires ρ_o to be 0.175 $M_\odot/pc^3$. The Solar System thus completes 2-3 oscillations perpendicular to the plane during one complete revolution. Normally it is necessary to perform the integration of the orbit of the Solar System in the Galaxy model numerically, and it is customary to present the result in a 2-dimensional (meridional) coordinate frame which rotates with an angular velocity fixed by the angular momentum component of the Sun about the symmetry axis. One such example orbit can be found in Innanen et al. (1978), Fig. 1.

It is a remarkable fact that the Solar System is presently to be found at a statistically improbable place, i.e. simultaneously crossing the Galactic plane and also near perigalacticon. Furthermore, this conclusion does not depend on the Galaxy model.

4. The Galactic tide and the Oort Cloud

The Oort Cloud is usually taken to be a spherical distribution of cometary material of radius approximately 50,000 AU, centred on the Sun. Heisler and Tremaine (1986), and Heisler (1990) have shown that the dominant, systematic, Galactic influence on the Oort Cloud must be the K_z tide. This tide exceeds other effects by a factor of about 2, with a "spheroid of influence" having a semi-axis dimension of some 135,000 AU in the z-direction. The regular, quasi-harmonic motion of the Solar System in the z-direction might therefore be expected to produce a degree of concurrent effects on the Oort Cloud, such as triggering cometary showers. These effects are the subject of ongoing research, some of which is reported in these proceedings. A competitor for the the Galactic tide would be the possibility of a distant, stellar companion to the Sun, ie the Nemesis object, orbiting within the Oort Cloud. Searches for Nemesis have so far proven unsuccessful; a recent review of the relevant theory has been given by Vandervoort and Sather (1993).

5. Other influences in the Galactic environment

From time to time, the Solar System must encounter other objects in its Galactic journey. Generally, these encounters will be gravitational and therefore the Sun's trajectory will be deflected to some extent during each such encounter. The list of objects includes individual stars, star clusters and associations, spiral arms, giant molecular clouds, and perhaps, black holes. Although each kind of possibility is real enough, encounters normally occur randomly in time, so that reconstruction of the past Solar orbit is not deterministically possible. Encounters with neighbouring stars, as one example, are relatively rare. A recent investigation of this situation has been given by Matthews (1994). The Solar System itself must have been born from supernova ejecta. Whether or not the Solar System has subsequently been affected by a nearby supernova event is also a question of random encounters, unless the event occurred relatively recently. To affect the Earth, it seems that a supernova would need to occur closer than 50 pc from the Solar System. Some evidence for the remnants of such an event has recently been found, in which the Solar System is found to be inside of a huge, ultraviolet "bubble", associated with a faint object named Geminga.

Interstellar space must contain some average density of debris from the formation of disks around young stars. This density is not well known, for no convincingly hyperbolic comet trajectory has been observed. The period of observations by humans has, however, been quite short, and the objects may be quite faint. A wealth of additional, valuable review material on this, and other relevant subjects can be found in Weissman (1990), Bailey et al. (1990), Clube et al. (1992), Duncan and Quinn (1993), Hildebrand (1993), Grieve (1993), and Milani et al. (1994). This review in many respects reflects the viewpoints of its author and is not intended to be comprehensive in all its aspects.

Acknowledgements

This review was written while the author enjoyed the hospitality of the Turku University Observatory at Tuorla during the summer of 1994. I wish to express my appreciation to its Acting Director, Dr. Seppo Mikkola and to Professor M.J. Valtonen for making the visit so pleasant, and also for a wide range of technical assistance in the preparation of the review. This research has been supported, in part, by grants from the Academy of Finland, and the Natural Sciences and Engineering Research Council of Canada.

References

Bahcall, J.N., Flynn, C., & Gould, A.: 1992, *Ap. J.* **389**, 234.
Byrd, G., Valtonen, M., McCall, M.. & Innanen, K.: 1994, *Astron. J.* **107**, 2055.
Bailey, M.E., Clube, S.V.M., & Napier, W.M.: 1990, *The Origin of Comets* (Oxford, Pergamon).
Clube, S.V.M., Yabushita, S., & Henrard, J., (eds): 1992, *Dynamics & Evolution of Minor Bodies with Galactic and Geological Implications*, Kluwer (Dordrecht).
Duncan, M.J. & Quinn, T.: 1993, *Ann. Rev. Astron. Astrophys.* **31**, 265.
Fuchs, B.: 1994, *Sterne und Weltraum*, July, 510.
Grieve, R.A.F.: 1993, *Vistas in Astronomy* **36**, 203.
Hildebrand, A.R.: 1993, *J. R. Astr. Soc. Canada* **87**, 77.
Heisler, J.: 1990, *Icarus* **88**, 104.
Heisler, J., & Tremaine, S.D.: 1986, *Icarus* **65**, 13.
Ibata, R.A., Gilmore, G., & Irwin, M.: 1994, *Sky and Telescope* **88**, 14.
Innanen, K.A., Patrick, A.T. & Duley, W.W.: 1978, *Astr. Sp. Sci.* **57**, 511.
Kuijken, K.: 1991, *Ap. J.* **372**, 125.
Kuijken, K. & Gilmore, G.: 1992, *Ap. J.* **367**, L9.

Kuijken, K. & Tremaine, S.D.: 1994, *Ap. J.* **421**, 178.
Matthews, R.A.J.: 1994, *Q. J. R. Astr. Soc.* **35**, 1.
Milani, A., DiMartino, M., & Cellino, A., (eds): 1994, *Asteroids, Comets, Meteors 1993*,
    Kluwer (Dordrecht).
Rix, H-W., & Lake, G.: 1993, *Ap. J.* **417**, L1.
Smoluchowski, R., Bahcall, J.N. & Matthews, M., (eds): 1986, *The Galaxy and the Solar
    System*, Univ. of Ariz. Press (Tucson).
Vandervoort, P.O., & Sather, E.A.: 1993, *Icarus* **105**, 26.
Weissman, P.R.: 1990, *Nature* **344**, 825.

WHY WE STUDY THE GEOLOGICAL RECORD FOR EVIDENCE OF THE SOLAR OSCILLATION ABOUT THE GALACTIC MIDPLANE

JOHN J. MATESE and PATRICK G.WHITMAN
Department of Physics, The University of Southwestern Louisiana, USA

KIMMO A. INNANEN
Department of Physics, York University, Ontario, Canada

and

MAURI J. VALTONEN
Tuorla Observatory, University of Turku, Finland

Abstract. The Solar System oscillates about the plane defined by the disk of matter in our Galaxy. This oscillatory motion gives rise to a substantial modulation in the tidally induced flux of Oort cloud comets. An observational determination of the quasi-periodicity of this motion carries with it significant information about the population, distributions, dynamics and origins of short-period and long-period comets. An additional incentive for emphasizing such a study is the information about dark disk matter that a period can yield. If dark disk matter is completely negligible, the amplitude of the solar motion will be sufficiently large that the peak-to-trough flux ratio will be ≈ 2.5 and the plane-crossing period will exceed 40 Myr. Dark disk matter comparable in mass to bright disk matter and distributed in any manner is inconsistent with K-dwarf distributions and can be rejected as a working hypothesis. But if a modest fraction of the disk matter is dark and distributed like the interstellar medium, as is consistent with limits deduced from K-giant and K-dwarf velocity distributions, the peak-to-trough flux ratio can increase to a factor of 4 even though the solar z amplitude is decreased. In that case the period can be as little as 30 Myr and the implied Oort population is smaller by a factor of 3. We should carefully reconsider the geological record as a potential discriminator of these options.

1. Introduction

The flux of new Jupiter-dominated ($q < 5.8$ AU) Oort cloud comets during the present epoch is largely determined by the tidal field of the relatively smooth distribution of matter in the galactic disk (Heisler *et al.*, 1987; Matese and Whitman, 1989, 1992 and references therein). Episodic sources of the near-parabolic comet flux include stellar impulses which penetrate the inner Oort cloud and create brief comet showers. Substantial stellar-induced showers occur $\approx$ every 100 Myr. Less frequent (but stronger) impulses due to giant molecular clouds can also perturb comets from the inner cloud. These occur on timescales of ≈ 500 Myr. The results described below ignore the smaller impulsive contributions to the Oort flux.

In contrast to these infrequent stochastic shower phenomena is the continuously varying tidal-induced flux due to the galactic disk considered by Matese *et al.*, (1995, hereafter MWIV). As the sun orbits the galactic center

Earth, Moon, and Planets **72**: 7-12, 1996.

it undergoes quasi-harmonic ($T_z \approx$ 60-80 Myr) motion about the galactic midplane and today is within 10 pc of the midplane. This oscillation is superimposed on the small eccentricity, near-Keplerian motion in the plane having epicycle period $\approx$ 170 Myr. In the process the galactic tidal field on the sun/cloud system will vary, causing a modulation of the observable Oort cloud flux. MWIV created a model of the galactic matter distribution as it affects the Solar motion over a time interval ranging from 300 Myr in the past to 100 Myr into the future. The model contained a core, bulge, halo and multi-component disk and is discussed in detail by Innanen (1995). As constraints on the disk's compact dark disk matter component they required consistency with (1) the observed galactic rotation curve, (2) today's energy flux distribution of new comets, and (3) the studies of K-giant (Bahcall, et al., 1992) and K-dwarf (Kuijken and Gilmore, 1989; Kuijken, 1991) stellar velocity distributions. The adiabatically varying galactic tidal torque was then determined and used to predict the time dependence of the flux. MWIV found that a range of model parameters in which $< 1/3$ of the total disk matter is dark was consistent with all these constraints, but only if it was distributed as compactly as the interstellar medium.

2. Results and Conclusions

Because the K-dwarf studies use a population that has a large scale height, the analysis is sensitive to the galactic disk *surface* mass density and requires that galactic halo effects be subtracted out to obtain the total disk contribution, Σ. Kuijken determined that if dark disk matter was proportional to the bright matter, the ratio of the volume dark matter density to the volume bright matter density in the midplane is 0.05 ± 0.17. This assumed $\Sigma_{bright} = 48$ M$_\odot$ pc^{-2}, which itself is uncertain by ± 9 M$_\odot$ pc^{-2}. Kuijken's result is inconsistent with the smaller-scale-height K-giant proportional model analysis of Bahcall et al., $\Sigma = 84^{+29}_{-24}$ M$_\odot$ pc^{-2}. Gould (1990) reanalyzed the work of Kuijken and Gilmore and concluded that their results imply that $\Sigma = 54 \pm 8$ M$_\odot$ pc^{-2} bringing it into marginal agreement with the proportional model results of Bahcall et al..

If the dark disk matter is compact, distributed like the ISM, the two analyses are more consistent since Bahcall et al. found in that case $\Sigma = 70^{+24}_{-15}$ M$_\odot$ pc^{-2} and Kuijken admitted that the K-dwarf analysis poorly constrains very small scale-height dark disk matter. Formal consistency of the two analyses suggests that 55 M$_\odot$ pc^{-2} $< \Sigma <$ 62 M$_\odot$ pc^{-2}. Taking into account the uncertainty in the bright matter content, the dark matter mass fraction may be as large as 1/3 and is unlikely to be much larger. MWIV showed that $\Sigma = 60 \pm 10$ M$_\odot$ pc^{-2} produces consistent galactic rotation curves and new-comet energy-flux distributions. The ranges of values for the z-oscillation period and the flux variability given above correspond to this uncertainty in the disk density.

If as much as 15 $M_\odot$ pc^{-2} of dark disk matter is distributed over ISM-like scale heights of $\approx$ 50 pc, the enhancement in midplane volume density of $\approx$ 0.15 $M_\odot$ pc^{-3} would give rise to a substantial increase in the tidal torque on outer Oort cloud comets being seen at this epoch, with a concurrent increase in the efficiency with which Oort cloud comets could be made observable. MWIV showed that the inferred outer Oort cloud population is *smaller* by a factor of $\approx$ 3 in this case as compared to the "no dark disk" model. Further, the plane crossing period is reduced from $\approx$ 44 Myr to $\approx$ 33 Myr and the Jupiter-dominated Oort flux has a peak-to-trough ratio of $\approx$ 4. Today's flux would be nearly at its peak and would be larger than the long-term tidal-induced average flux by a factor of 2.4. Arguments that we are not presently in a comet shower (Weissman,1995) shed no light on whether we are presently in an enhanced phase of more modest modulation as described here. Small model parameter changes in density or scale height can reduce the plane crossing interval to $\approx$ 30 Myr but a further reduction to the putative extinction period of 26 Myr will require an extreme choice of surface mass density and dark matter scale height (MWIV).

The time-dependent Jupiter-dominated Oort cloud current (Oort comets per year inside 5.8AU) is displayed along with the earth-crossing current (Oort comets per year inside 1AU) in Figure 1. Results are shown for an interval stretching from 100 Myr in the future to 300 Myr in the past. The most recent plane crossing is assumed to have occurred 1 Myr ago with a vertical component of the solar velocity = 7.5 km s^{-1}. We have subjected the results of Fig. 1 to a conventional regression analysis to determine the width of the current peaks. The standard deviations are 4.0-4.5 Myr.

The results indicate that the cycle interval changes, *i.e.*, the "periodic" oscillations have in fact a period variability of $\approx$ ±1 Myr. Further, the phase of the nearest cycle peak is less than 1 Myr in the future with the adopted parameters. The predicted sequence of broad cycle maxima is consistent with the terrestrial cratering record which has peaks at $\approx$ 2, 32, 65 and 99 Myr (Shoemaker and Wolfe, 1986). Evidence from the dating of geological events (Rampino and Stothers, 1986; Shoemaker and Wolfe, 1986; Rampino and Caldeira, 1992) is also fully consistent with a view that the nearest flux maximum may not have occurred as yet. The model predictions are in good agreement with the dating of impact glasses, microspherules, and the iridium anomalies which have peaks at $\approx$1, 35, and 65 Ma (Shoemaker and Wolfe, 1986).

Shoemaker and Wolfe stated that the cratering record appeared to be periodic but considered it to be a statistical fluke as they found no mechanism for the periodicity. They concluded that the case for periodic cratering was not "compelling". In particular they questioned the galactic oscillations

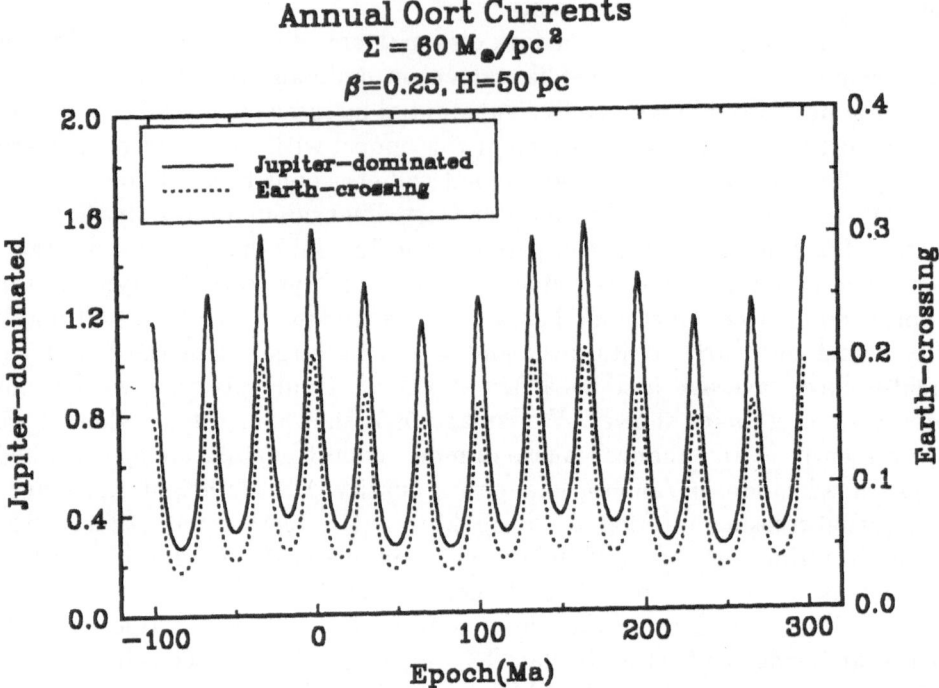

Fig. 1.   Time dependence of the Jupiter-dominated current (comets per year with $q <$ 5.8 AU) from the outer Oort cloud due to the adiabatically changing galactic tide. The earth-crossing rate is also shown. Both results are scaled to 0.2 earth-crossing comets per year at today's epoch (t=0). The epochs range from 100 Myr in the future to 300 Myr in the past.

model of Rampino and Stothers since it invoked the impulsive perturbation of inner Oort cloud comets by molecular clouds. Because the plane crossing epochs are *too close* to the crater peak times to be due to rare stochastic cloud impulses of the inner Oort cloud, the Rampino-Stothers mechanism was rejected as a clock for this phenomena. These objections are obviated in the present model since modulation of outer Oort cloud comet flux is the physical mechanism for the periodicity.Rampino and Haggerty (1994; see also these proceedings) review the case for correlations between large-body impacts and mass extinctions. Along with Yabushita (1995) they discuss the topic of periodicity.

In comparing these results with prior numerical experiments on impact periodicity, we need to estimate the total dispersion expected from a strictly periodic sequence, *i.e.* we should convolve the standard error in the mean cycle interval (≈ 1 Myr) with the standard deviation of the modulation about the cycle peak (≈ 4 Myr) and the dating uncertainties of the most accurately dated craters ($\approx$ 4 Myr). If impact probabilities were modulated as described here with dating uncertainties of this order, we expect a net dispersion about a strictly periodic signal to be $\approx$ 5 – 6 Myr. The steady-state main-belt asteroidal cratering rate would have to be <50% of the total if such a signal were to have a reasonable probability of manifesting itself in the cratering record (Yabushita 1995). This requires that the analysis be restricted to large terrestrial craters.

Valtonen *et al.* (1995) have studied the time evolution of outer Oort cloud comets that have been injected into the Solar System. It was demonstrated there that the time delay for processing new comets into short-period comets may be sufficiently short that earth-crossers can maintain a memory of their time-dependent source. Estimates of crater formation rates from this source amount to a substantial fraction of the observed rates.

Zheng *et al.* (1995) considered the production of short-period orbits from comets captured from the outer Oort cloud. They showed that it may be premature to reject this reservoir as the source of Halley-family and Jupiter-family comets. In conclusion, we emphasize the importance to be attached to observationally determining the numerical value of the real z-oscillation period of the solar motion about the galactic midplane, both for solar-system and galactic studies.

Acknowledgements

JJM and PGW would like to recognize the support of the Louisiana Educational Quality Support Fund under grant RD-A-41. MJV is currently supported as a research professor by the Academy of Finland. KAI acknowledges the support of the Research Council of Canada. KAI has received continuing research support from the Natural Sciences and Engineering Research Council of Canada. He also wishes to acknowledge the hospitality of Professor M.J. Valtonen and Dr. S. Mikkola at Turku University Observatory during the summer of 1994.

References

Bahcall, J. N., Flynn, C., and Gould, A.: 1992, *Astrophys. J.* **389**, 234-250.
Gould, A.: 1990, *MNRAS* **244**, 25-35.
Heisler, J., Tremaine, S., and Alcock, C.: 1987, *Icarus* **70**, 269-288.
Innanen, K. A.: 1995, *these proceedings*.
Kuijken, K.: 1991, *Astrophys. J.* **372**, 125-131.
Kuijken, K., and Gilmore, G.: 1989, *MNRAS* **239**, 650-659.
Matese, J. J., and Whitman, P. G.: 1989, *Icarus* **82**, 389-401.
Matese, J. J., and Whitman, P. G.: 1992, *Celest. Mech. and Dyn. Astron.* **54**, 13-36.
Matese, J. J., Whitman, P. G., Innanen, K. A. and Valtonen, M. J.: 1995, *Icarus* in review.
Rampino, M. R. and Caldeira, K.: 1992, *Celest. Mech. and Dyn. Astron.* **54**, 143-159.

Rampino, M. R. and Stothers, R. B.: 1986, *The Galaxy and the Solar System* (R. Smolu-
    chowski, J. N. Bahcall and M. S. Matthews, Eds.), 241-260. Univ. of Arizona Press,
    Tucson.
Rampino, M. R. and Haggerty, B. M.: 1994, *Hazards due to Comets and Asteroids* (T.
    Gehrels, Ed.), Univ. of Arizona Press, Tucson, *in press*
Shoemaker, E. M. and Wolfe, R. F.: 1986, *The Galaxy and the Solar System* (R. Smolu-
    chowski, J. N. Bahcall and M. S. Matthews, Eds.) 338-386. Univ. of Arizona Press,
    Tucson.
Valtonen, M. J., Zheng, J., Whitman, P. G., and Matese, J. J.: 1995, *Earth, Moon, Planets*
    in review.
Weissman, P. R.: 1995, *these proceedings.*
Yabushita, S.: 1995, *these proceedings.*
Zheng, J., Valtonen, M. J., Korpi, M., Mikkola, S., and Rickman, H. : 1995, *these proceed-
    ings.*

THE STATISTICAL EFFECTS OF GALACTIC TIDES ON THE OORT CLOUD

PIOTR A. DYBCZYŃSKI and HALINA PRĘTKA

Obserwatorium Astronomiczne UAM,
Ul.Słoneczna 36, 60-286 Poznań, Poland
E-mail: dybol@phys.amu.edu.pl, pretka@phys.amu.edu.pl

Abstract. This report is a comment on two papers by Matese and Whitman (1989, 1992). We discuss here the applicability of uniform probability densities for the orbital parameters of the Oort cloud comets.

Key words: Galactic tides – Oort Cloud – comets

1. Introduction

In this work we concentrate on the classical picture of the galactic tides where the local matter density in the galaxy is constant in time. A review of selected papers on galactic tides and a detailed description of the adopted dynamical model can be found in Prętka and Dybczyński (1994, Paper A). The equations of motion in that model are:

$$\ddot{x} = -\frac{\mu}{r^3}x , \qquad \ddot{y} = -\frac{\mu}{r^3}y , \qquad \ddot{z} = -\frac{\mu}{r^3}z - 4\pi G\rho \cdot z \qquad (1)$$

with $\rho = 0.185 M_\odot/pc^3$ (Bahcall 1984), see also Heisler (1990). We also presented there examples of a long-term orbital evolution obtained from numerical integration of those equations, demonstrating the superposition of a long term, strictly periodic variation (with period of order of several hundreds of orbital revolutions) and local short period fluctuations in orbital elements. These short period terms are caused by perturbations in semimajor axis and have a period comparable with orbital one.

The galactic tidal effect can be also studied using secularly averaged Hamiltonian equations. These can be expressed in terms of the Delaunay coordinates and momenta as was done, for example, in two papers by Matese and Whitman (1989, 1992), who provided an analytical solution of the problem.

To test the applicability of the first order secularly averaged Hamiltonian technique to the galactic tide effect, we compare the results obtained with that method to the numerical integration described above. The results are in good agreement even over long time intervals, except for the absence of the short period fluctuations what is obvious result of averaging technique (semimajor axis is constant in this case). The long term periodic variation

Earth, Moon, and Planets **72**: 13-18, 1996.
© 1996 *Kluwer Academic Publishers.*

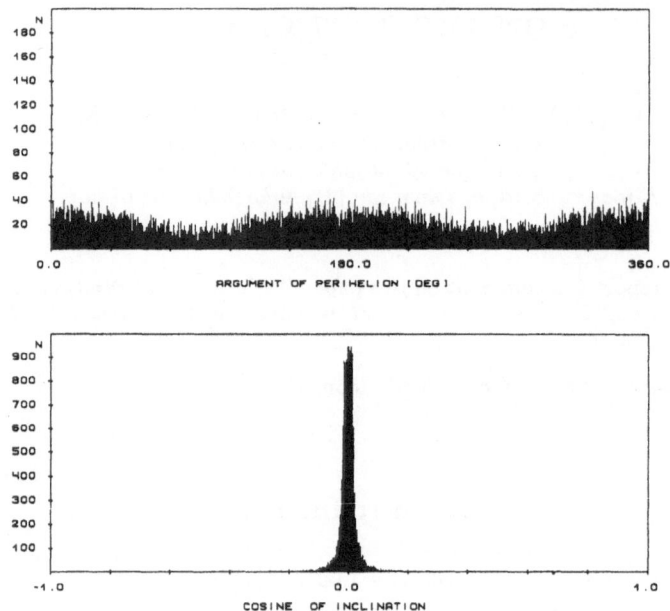

Fig. 1. Orbital elements distributions for potentially observable part of the Oort cloud.

is perfectly reproduced. Analyzing examples of long term orbital evolution
we came to conclusion that the orbits of observable (or almost observable)
comets, with eccentricity greater than 0.99, have short period fluctuations
of negligible amplitude.

We thus decided to use the first order averaged theory in our calculations.
We used both an analytical solution and the numerical integration of Hamiltonian averaged equations. The results of these two approaches were in good
agreement except for certain sets of orbital elements for which the series in
the analytical solution converged extremely slowly. In order to follow the
orbital evolution step by step for long time intervals, dealing with arbitrary
values of elements, numerical integration of the averaged equations appears
to be the simplest and the most effective method.

In their papers, Matese and Whitman used a Monte-Carlo simulation of
the population of the Oort cloud comets, with orbital elements distributed
according to uniform probability densities to produce the observable comets.
After a random selection of all initial parameters for a given comet, they
determinated whether the galactic tidal perturbation was strong enough to
decrease the comet's perihelion distance from initially more than 15 AU to
less than 5 AU after one orbital period. If that was the case, the comet was
marked as 'observable'. In the second paper Matese and Whitman (1992)

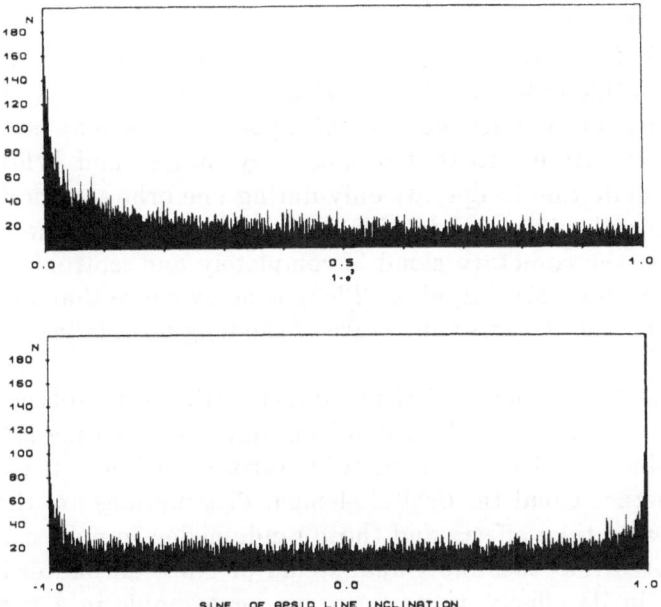

Fig. 2.   Orbital elements distributions for potentially observable part of the Oort cloud.

presented distributions of argument of perihelion, the inclination, and the inclination of line of apsides for the observable comet population, recorded at the end of the one orbital period interval. It means, that this population consists of the majority of comets with q decreasing but there is a significant number of comets with q increasing.

2. Results

Having had the opportunity to observe the individual orbital evolution of arbitrary comets in the cloud, we concluded that treating the orbital elements of comets in the Oort cloud as independent, random variables, with uniform probability densities, may not be the best approximation. Looking at the plots presented in Paper A, it is easy to observe the coincidence of rapid changes in certain pairs of orbital elements during their long term evolution under the influence of the galactic disk tide. For example, in a typical situation the inclination reaches its lowest value when the argument of perihelion equals 90° or 270° exactly. At the same moment the eccentricity reaches its maximum value. Due to the periodic nature of this evolution, a similar situation repeats many times. Moreover, looking at the evolutionary paths of the argument of perihelion one can easily guess that its value

recorded at random moments, will more probably be in the vicinity of 180°
than the vicinity of 90°.

Taking all this into account, we think that the long term perturbation
of the galactic tide induces a deformation in the distributions of orbital elements and their mutual dependence. This point of view may be looked upon
as an opposite extreme to that proposed by Matese and Whitman. They
allow the galactic tide to operate only during one orbital period of a comet
(typically several million years). This treatment assumes that over longer
time intervals the cometary cloud is completely and isotropically randomized by stellar and GMC impulses. There is no evidence that those impulses
are effective enough to erase any trace of the long term influence of galactic
tides.

We present here some statistical characteristics of the observable comet
population in the absence of randomizing impulses for time intervals comparable to the period of the long term variation of orbital elements. For
the real cometary cloud the orbital element distributions are the result of a
balance between tidal effects and those randomizing impulses. Moreover, it
should be noted that each individual stellar or GMC encounter is extremely
nonisotropic in its effects, as one can see for example in a paper by P.A.
Dybczyński presented as a poster at this conference.

As a first step, we observed changes in the distributions of orbital elements
in the cloud over a long time interval. We started with the uniform distributions proposed by Matese and Whitman ($a = 30000$ AU, $1 - e^2 \in (0, 1)$)
and followed the orbital evolution of each comet for 1 Gyr.

The resulting distributions remain almost uniform, showing only small
deviations. Then we determinated the distributions of elements of the observable part of the cloud. During the integration spanning 1 Gyr we marked
as potentially observable all those comets which – at some time – had a
perihelion distance less than the 5 AU threshold. In this manner we divided the Oort cloud into potentially observable and non-observable parts.
Figs. 1(a,b), 2(a,b) present the distributions of ω, $\cos i$, $\sin b$, and $1 - e^2$
for the observable part of the cometary population, recorded at an arbitrary
moment of time. The vast majority of orbits have inclinations close to 90°,
contrary to the observed cometary orbits. This fact can be easily explained.
The inclination of the majority of observable comets follows very similar evolutionary paths. An example of such a variation is shown in Paper A (Fig.
4). The inclination remains almost constant and equal to 90° most of the
time, with the exception of the very short time interval when the perihelion
distance reaches its minimum (and when a comet may be observed).

Finally, we obtained a set of element distributions recorded at the moment
that the comets first crossed the 5 AU observability barrier. As shown in
Fig.3(a,b,c), among the observed comets almost all inclinations are present,
except for the smallest values. However, the decrease in the vicinity of 90°

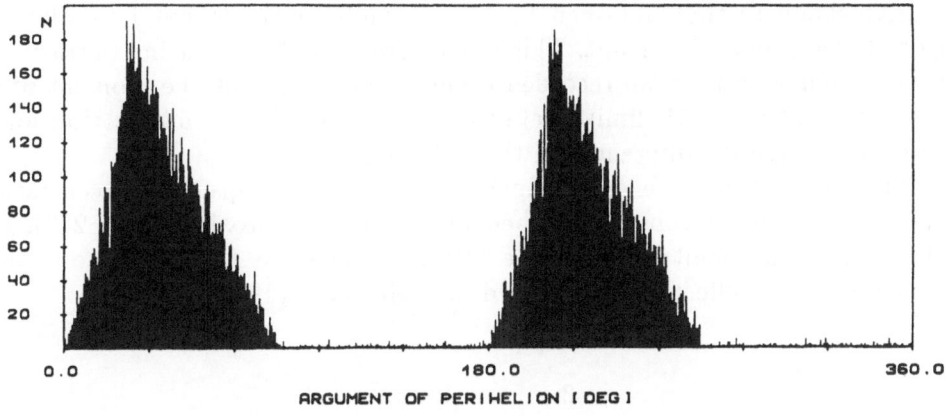

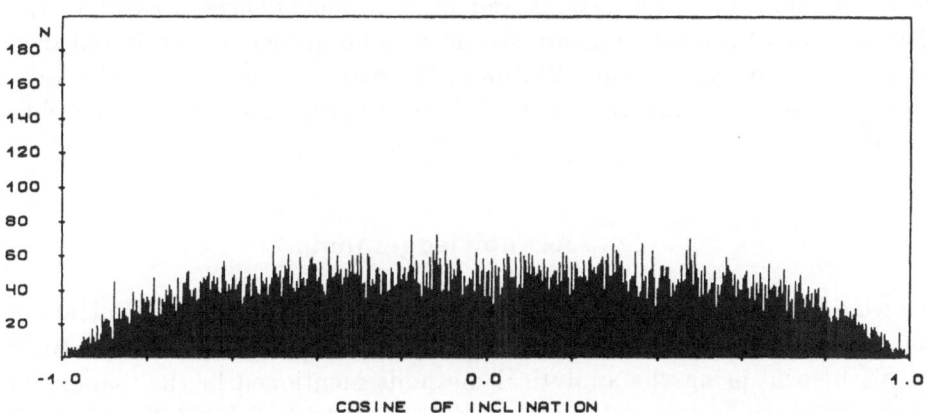

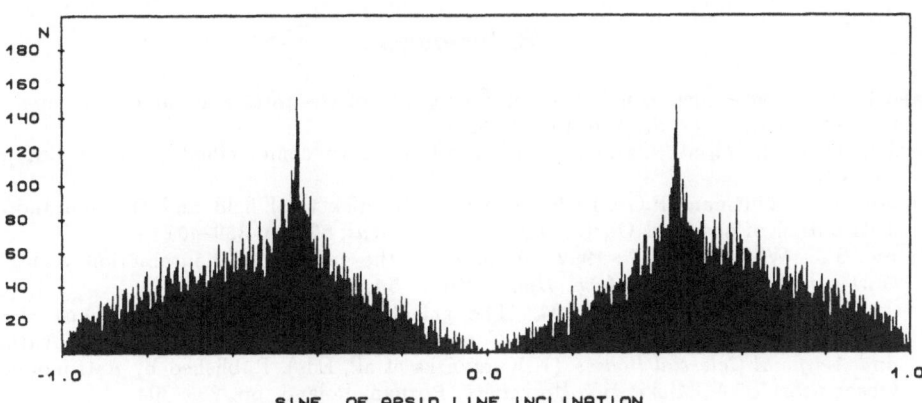

Fig. 3. Orbital elements distributions for the Oort cloud comets at the moment of crossing 5 AU threshold in the perihelion distance.

found by Matese and Whitman is absent here. The ω distribution shows a similar shape to that obtained by Matese and Whitman, but here almost half of the interval is vacant. This results from the fact that in contrary to Matese and Whitman we recorded osculating elements at the moment of q decreasing below 5 AU limit. Detailed comparison of the sin b distributions reveals remarkable differences in their shapes.

In our case the sine of the inclination of line of apsides concentrates around the critical values described in Paper A (approximately $\pm27°$). In the Matese and Whitman case the extreme values are $\pm45°$ because of the strong selection effect as they stated in their paper (1992).

3. Conclusions

In our opinion, all deformations and mutual dependences present in these distributions of orbital elements should also be present in the initial distributions used by Matese and Whitman. We plan to repeat their calculations with this modification. Such a work is in progress and we hope to obtain first results soon.

4. Acknowledgements

We would like to express our thanks to Dr P.R. Weissman for fruitful comments and disscusion. We also thank our colleague Dr Sławomir Breiter for his help in using the analytical methods mentioned in this paper. This work was partially supported by the KBN Grant No. 2 P304 005 07 (Halina Prętka).

References

Bahcall, J.N.: 1984, 'Self-consistent determinations of the total amount of matter near the Sun.', *Astrophys. J.*, **276**, pp. 169–181.
Heisler. J.: 1990, 'Monte Carlo simulations of the Oort comet cloud.', *Icarus*, **88**, pp. 104–121.
Matese, J.J., Whitman, P.G.: 1989, 'The Galactic disk tidal field and the nonrandom distribution of observed Oort cloud comets.', *Icarus*, **82**, pp. 389–401.
Matese, J.J., Whitman, P.G.: 1992, 'A model of the galactic tidal interaction with the Oort comet cloud.', *Cel. Mech. Dyn. Astron.*, **54**, pp. 13.
Prętka, H., Dybczyński, P.A.: 1994, 'The galactic disk influence on the Oort cloud cometary orbits.', *Proceedings of the Conference 'Dynamics and Astrometry of Natural and Artificial Celestial Bodies'* (K.Kurzyńska et al., Eds), Published by Astronomical Observatory of A. Mickiewicz University, Poznań, Poland, pp. 299–304.

ENCOUNTERS OF THE SUN WITH NEARBY STARS IN THE PAST AND FUTURE

A.A. MÜLLÄRI

Department of Mathematics, Petrozavodsk State University, Lenin prospect 33,
Petrozavodsk, 185640 Russia, e-mail: amul@mainpgu.karelia.su

and

V.V. ORLOV

Astronomical Institute, St. Petersburg State University, Bibliotechnaya pl. 2, 198904 St.
Petersburg Peterhof, Russia, e-mail: vor@aispbu.spb.su

Abstract. The relative space motions of the Sun and nearby stars are considered. The coordinates and velocities of the stars are taken from the Catalogue of Nearby Stars by Gliese and Jahreiss (1991). The minimum space separation between the Sun and every star as well as the corresponding moment of time are calculated by two ways. Firstly, the straight line motions are considered. Secondly, the effect of the Galaxy potential is taken into account. The Galaxy model proposed by Kutuzov and Ossipkov (1989) is used. Twenty five stars approaching the Sun closer than two parsecs are selected. The effects of the uncertainties in the observational data are studied. The influence of the encounters to the Oort cloud is discussed.

Key words: Solar neighbourhood, Oort comet cloud

1. Introduction

The large sudden changes of the terrestrial climate could bear evidence of some possible cosmic catastrophes encountered by the Earth. One of the hypothetical reasons for such events is a close passage of a nearby star by the solar system. The encounters could initiate a shower of comets with small perihelia. A collision of the Earth with such a comet may lead to the catastrophic transformation of the climate. The cometary shower forming after a star's passage acts during $\sim 10^6$ years of the passage of the star. Thus it is of interest to trace the mutual trajectories of the nearby stars and the Sun during a short time (e.g. about 10^8 years) to the past and to the future.

2. Observational Data and Results

We consider the nearby stars from the Preliminary Version of the Third Catalogue of the Nearby Stars by Gliese and Jahreiss (1991). The stars with known heliocentric space velocities U, V, W are taken into account (1946 stars). Here the vector U is directed to the galactic center, V in the direction

Earth, Moon, and Planets **72**: 19-23, 1996.

of the galactic rotation, and W to the Northern Galactic Pole. The coordinates and velocities of the stars have been calculated in the galactocentric reference frame.

Firstly, we consider the straight line motions of every star with respect to the Sun. We found the shortest distance r_{min} from the Sun to this line and the corresponding moment of time t_{min}. The stars with $r_{min} < 2$ pc have been selected. The results for these 25 stars are presented in Table I. The values of r_{min} are given in 10^3 astronomical units; the times t_{min} are in 10^3 years. The Sun may have had encounters with three of these stars in the past and can have encounters with another 22 stars in the future.

A similar study was carried out by Revina (1988) who used the data from the previous version of the Catalogue of Nearby Stars (Gliese 1969). She found 25 stars (6 for the past and 19 for the future) having the close (less than 2 pc) encounters with the Sun. Ten stars from her list are the members of our sample. These stars are marked by an asterisk in Table I. Some of the disagreements of values r_{min} and t_{min} could be explained because the data is more precise in the new Catalogue. A similar study was also recently carried out by Matthews (1994). He has considered the stars from the solar neighbourhood with radius 5 pc. For a few stars he used slightly different initial conditions. Our results are in agreement with his results for the same stars.

We have taken into account the effect of the errors in the velocities U, V, W and in the parallaxes π to the values of r_{min} and t_{min}. A Monte Carlo method was applied to estimate the expectations and r.m.s. deviations of r_{min} and t_{min} for 25 stars mentioned above. We varied the additions to the input values U, V, W, and π by a Gaussian distribution with expectation equal to zero and dispersion $\sigma = 3$ km/s for the velocities and corresponding errors from the Catalogue for the parallaxes. The values of the expectations $\langle r_{min} \rangle$ and $\langle t_{min} \rangle$ as well as r.m.s. deviations σ_r and σ_t are also given in Table I.

Secondly, we consider the movements of the stars in the model Galaxy by Kutuzov and Ossipkov (1989). The distance of the Sun from the galactic center is adopted $R_0 = 8.23$ kpc and the height of the Sun upwards the galactic plane is $z_0 = 0.015$ kpc. The circular velocity at the solar distance R_0 is assumed $\Theta_0 = 226$ km/s. The components of the solar motion are $U_0 = +8$, $V_0 = +12$, $W_0 = +7$ km/s. We have integrated the equations of the motion of the Sun and of each star from 1946 stars with known space velocities during 10^8 years forwards and to the past. We neglected the interaction between the stars and the Sun, as well as the influence of the irregular forces. Corresponding values of r_{min} and t_{min} are presented in Table I too.

The two methods are in a good qualitative agreement: the same stars were selected by each of the methods. Also the less is the error of the parallax the

TABLE I

The results for the nearby stars encountering the Sun

N	Name	Lin.	fit	Effect	of errors	Tidal	field
		r_{min}	t_{min}	$r_{min} \pm \sigma_r$	$t_{min} \pm \sigma_t$	r_{min}	t_{min}
82	GJ 2005	154	33.0	156 ± 24	33.2 ± 2.3	154	33.0
305	NN	317	1780	1540 ± 1830	1630 ± 2840	384	1720
456	NN	75	1630	1290 ± 740	1600 ± 560	32	1600
528	Gl 120.1	280	-431	436 ± 216	-435 ± 47	282	-430
943*	Gl 208	341	-530	523 ± 261	-523 ± 83	337	-529
1160	Gl 271	375	985	840 ± 436	963 ± 214	386	990
1718*	Gl 411	291	19.9	291 ± 13	19.9 ± 0.7	291	19.9
1844*	Gl 445	197	43.7	199 ± 28	43.7 ± 1.7	197	43.7
1848	Gl 447	385	70.3	387 ± 43	69.8 ± 6.6	385	70.4
1927*	Gl 459.2	298	418	1290 ± 18000	1200 ± 13200	303	417
1971	Gl 473	59.6	7.5	59.6 ± 4.7	7.5 ± 0.1	59.9	7.5
1973	Gl 474	53.5	427	363 ± 211	452 ± 126	54.4	427
2077	NN	342	1060	2220 ± 9840	2560 ± 11000	373	1050
2290*	Gl 551	218	25.9	217 ± 16	25.6 ± 4.7	218	25.9
2317*	Gl 559	186	27.2	186 ± 17	27.2 ± 3.2	186	27.2
2778	Gl 682	390	64.3	392 ± 42	64.3 ± 3.7	390	64.3
2848*	Gl 699	238	9.8	238 ± 6	9.8 ± 0.3	238	9.8
2853	Gl 700.1	362	427	478 ± 239	435 ± 82	367	429
2891*	Gl 710	259	1030	853 ± 445	999 ± 275	279	1050
2959	Gl 729	393	134	392 ± 83	130 ± 34	393	134
3167*	Gl 783	372	38.2	374 ± 31	38.3 ± 2.3	372	38.2
3536	Gl 860	388	89.0	392 ± 54	88.6 ± 7.9	390	88.7
3706	NN	91	-515	421 ± 230	-523 ± 109	112	-517
3735	GJ 2157	286	427	436 ± 228	432 ± 68	260	425
3742*	Gl 905	195	36.3	196 ± 23	36.4 ± 1.4	195	36.3

better is the quantitative agreement. All $|t_{min}|$ values are less than $2 \cdot 10^6$ years. Therefore our forecast is valid during about 10^6 years.

The minimum separation during this interval will take place with the star 456 ($r_{min} = 32,000\ AU$ and $t_{min} = 1.6 \cdot 10^6$ years). However, as it can be seen from Table I, the uncertainties of $\langle r_{min} \rangle$ and $\langle t_{min} \rangle$ for this star are rather large (mainly due to a big error of the parallax). The most reliable star having a close approach to the Sun is the star 1971 (Gl 473). Corresponding values are $r_{min} \approx 60,000 AU$ and $t_{min} \approx 7,500$ years.

TABLE II

Estimations of influence from the stars to the Oort cloud

N	Name	$M_*, M_\odot$	R_a	R_0	r_{min}	t_{min}
82	GJ 2005	0.18	46	108	154	33.0
305	NN	6.5	276	108	384	1720
456	NN	0.32	11.6	20.5	32	1600
528	Gl 120.1	0.75	131	151	282	-430
943	Gl 208	0.47	137	200	337	-529
1160	Gl 271	2.4	234	152	386	990
1718	Gl 411	0.39	112	179	291	19.9
1844	Gl 445	0.27	67	130	197	43.7
1848	Gl 447	0.24	126	259	385	70.4
1927	Gl 459.2	0.70	138	165	303	417
1971	Gl 473	0.31	21.5	38.4	59.9	7.5
1973	Gl 474	4.0	36.2	18.2	54.4	427
2077	NN	7.0	271	102	373	1050
2290	Gl 551	0.21	69	149	218	25.9
2317	Gl 559	1.8	106	80	186	27.2
2778	Gl 682	0.29	136	254	390	64.3
2848	Gl 699	0.21	75	163	238	9.8
2853	Gl 700.1	1.4	200	167	367	429
2891	Gl 710	0.42	110	169	279	1050
2959	Gl 729	0.23	128	265	393	134
3167	Gl 783	1.0	187	185	372	38.2
3536	Gl 860	0.56	167	223	390	88.7
3706	NN	1.0	56	56	112	-517
3735	GJ 2157	0.78	122	138	260	425
3742	Gl 905	0.40	56	139	195	36.3

3. Discussion

It is interesting to estimate the radius R_a of the action sphere for the stars
with respect to the Sun in the moment of the closest approach. The approximate estimation is as follows:

$$R_a = \frac{r_{min}}{1 + \sqrt{M_\odot/M_*}}, \tag{1}$$

where $M_\odot$ is the solar mass and M_* the mass of the star. We could estimate
the corresponding distances $R_0 = r_{min} - R_a$ from the Sun, where the force
acting to a comet from the Sun is the same as from the star. The masses
of stars and values of R_a and R_0 are given in Table II. The crude mass

estimates are taken from Allen (1973). The values of r_{min} and t_{min} are calculated taking into account the galactic field and they are presented in Table II as well. The star 1973 (Gl 474) will give a maximum effect to the Oort cloud in the near future because its action will exceed the action from the Sun at $r \geq 18,000 AU$ in the direction to the star. The outer parts of the cloud may be essentially deformed by the tidal force from the star.

We note a surprising asymmetry of the numbers of stars encountering the Sun in the past (3 stars) and in the future (22 stars). This asymmetry disappears when the critical distance is increased.

This work uses the most reliable observational data for the nearby stars. Therefore the stars singled out are good candidates for further detailed study of space motions and coordinates.

4. Acknowledgements

The authors are grateful to Mr. Ourusoff for the beautiful computer facilities he provided and for the assistance with operating data.

References

Allen, C.W.: 1973, *Astrophysical quantities*, The Athlone Press, London.
Gliese, W.: 1969, *Veröffent. Astron.-Rechen. Institut Heidelberg* **3**, No. 22.
Gliese, W., Jahreiss, H.: 1991, Unpublished.
Kutuzov, S.A., Ossipkov, L.P.: 1989, *Sov. Astron.*, **66**, 965.
Matthews, R.A.J.: 1994, *Q. J. R. astr. Soc.*, **35**, 1.
Revina, I.A.: 1988, *Analysis of motion of celestial bodies and estimation of accuracy of their observations*, Latvian University, Riga, p. 121.

STAR PASSAGES THROUGH THE OORT CLOUD

PAUL R. WEISSMAN

Jet Propulsion Laboratory, Mail stop 183-601
4800 Oak Grove Drive, Pasadena, CA 91109 USA

Abstract. Stars passing through the Oort cloud eject comets to interstellar space and initiate showers of comets into the planetary region. Monte Carlo simulations of such passages are performed on a representative distribution of cometary orbits. Ejected comets generally lie along a narrow tunnel "drilled" by the star through the cloud. However, shower comets come from the entire cloud, and do not give a strong signature of the star's passage, except in the inverse semimajor axis distribution for the shower comets. The planetary system is likely not experiencing a cometary shower at this time.

Key words: Oort cloud – stellar perturbations – comet showers

1. Introduction

Oort (1950) first recognized that the solar system was surrounded by a cloud of comets, perturbed by random passing stars. Hills (1981) suggested that close stellar passages could excite the cloud and send showers of comets into the planetary region, particularly if there existed a dense inner cloud of comets which are not observed except when such major perturbations occur. Hut et al. (1987) and Fernandez and Ip (1987) modeled the dynamical evolution of such showers, showing that the intense pulse of comets into the planetary region decays in $2 - 3 \times 10^6$ years.

Other dynamical studies have demonstrated the importance of galactic tidal perturbations on the steady-state evolution of the Oort cloud (Byl 1983; Heisler and Tremaine, 1986). However, only penetrating star passages and close approaches by giant molecular clouds can cause the major perturbations which result in cometary showers.

To investigate the effects of penetrating stellar passages, a Monte Carlo simulation model of the Oort cloud was constructed. The model notes the change in orbital elements of comets resulting from the passage of a single star at a given closest approach distance and velocity. The velocity perturbations on the cometary orbits and on the Sun are calculated using the classic impulse approximation, $\Delta V = 2GM_*/DV_*$, where G is the gravitational constant, M_* and V_* are the mass and velocity respectively of the perturbing star, and D is the closest approach distance of the star to the comet or the Sun. Orbital elements for the comets are chosen randomly using the semimajor axis and eccentricity distributions found by Duncan et al. (1987), and assuming random inclinations and mean anomalies. Initial perihelia are restricted to distances > 50 AU and aphelia to distances $< 2 \times 10^5$ AU. A typical Oort cloud constructed in this fashion is shown in Fig. 1.

Earth, Moon, and Planets **72**: 25-30, 1996.

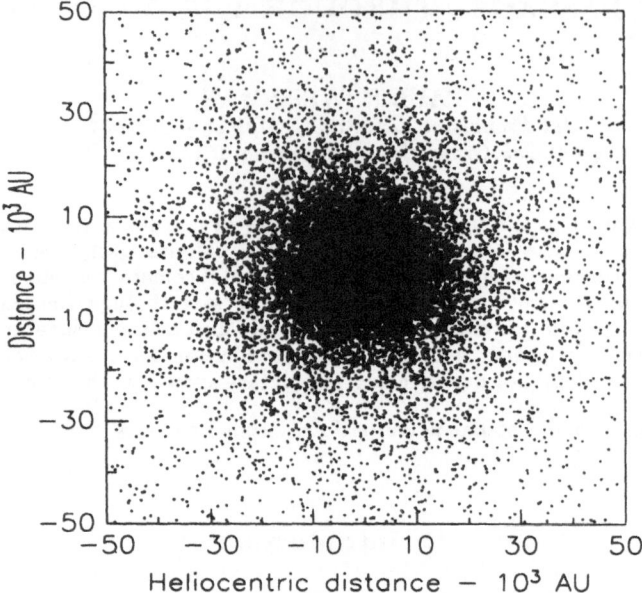

Fig. 1.   Hypothetical Oort cloud constructed for the Monte Carlo simulation, based on the orbital element distributions of Duncan et al. (1987).

2. Results

An example of results for a one solar mass star passing at 30 km s^{-1} through a hypothetical cloud of 2×10^7 comets, at a closest approach distance to the Sun of 10^4 AU, are shown in Fig. 2. The instantaneous location of comets ejected to interstellar space are shown in Fig. 2a, which is a view looking "down" on the path of the star. The ejected comets generally lie close to the star's path, within about 10^3 AU, where the net velocity perturbation exceeds the escape velocity from the solar system, as predicted by earlier studies (Weissman, 1980). This is shown even more clearly in Fig. 2b which is a view along the star's velocity vector. Most of the ejected comets lie along a narrow tunnel "drilled" through the Oort cloud.

The fraction of ejected comets for this case is 8.5×10^{-5} of the total cloud population. The mean hyperbolic velocity of the ejected comets is 0.28 km s^{-1}. Additionally, 2.3×10^{-4} of the cloud population is perturbed to aphelia greater than 2×10^5 AU. These orbits are beyond the Sun's sphere of influence and will likely be lost to interstellar space.

The location of comets perturbed by the same stellar passage to perihelion distances less than 10 AU are shown in Fig. 2c, which is again a view looking "down" on the star's path, and Fig. 2d, which is the view along the stellar velocity vector. In this case, the entire Oort cloud is excited and comets enter the planetary region from all directions. The majority of shower comets

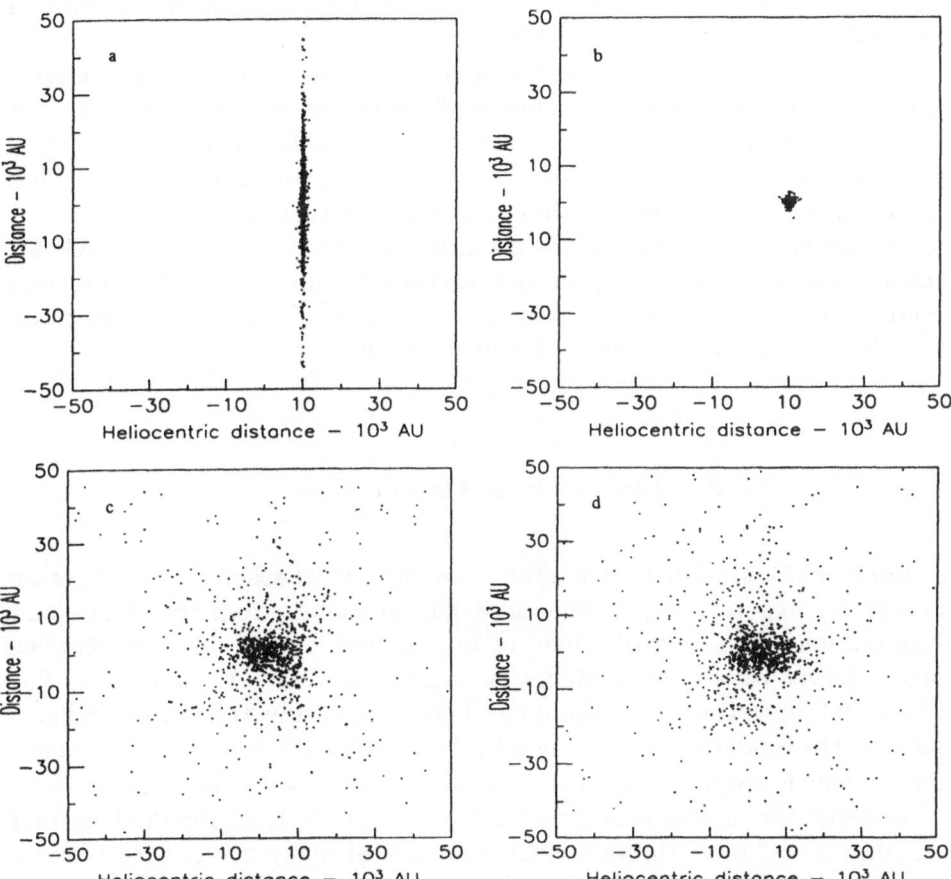

Fig. 2. Location of comets in a hypothetical Oort cloud, lost to various dynamical end-states: a) comets ejected to interstellar space, view from "above" the star's path; b) ejected comets, view along the star's path; c) comets perturbed into the planetary region, $q < 10$ AU, view from above the star's path; d) comets perturbed to $q < 10$ AU, view along the star's path. Ejected comets all come from close to the star's path, whereas comets perturbed to $q < 10$ AU come from the entire Oort cloud.

come from semimajor axes greater than the minimum approach distance of the star to the Sun, indicative of the tidal nature of the perturbation. The dense inner Oort cloud is not easily excited unless a star passes directly through it. The fraction of comets perturbed to $q < 10$ AU for this case is 1.1×10^{-4} of the total cloud population. Most of those comets will be ejected to interstellar space by Jupiter and Saturn within 5 to 10 returns (Weissman, 1979).

The star passage pumps energy and angular momentum into the Oort cloud, as described by Weissman (1991). The total binding energy of the comets is reduced by a factor of 2×10^{-4}. The total angular momentum of the cloud is increased by 1.2×10^{-3}.

Hypothetical cases for different stellar masses, velocities, and closest approach distances were also studied. Results are shown in Fig. 3. The fraction of the cloud population ejected to interstellar space as a function of closest approach distance is shown in Fig. 3a, for two different stellar mass/velocity ratios. The fraction of comets perturbed to perihelia < 10 AU are shown in Fig. 3b, and the fraction lost to aphelia $> 2 \times 10^5$ AU are shown in Fig. 3c. In all cases, the loss fractions are monotonically decreasing functions of encounter distance. Small variations in the curves are the result of statistical noise in the Monte Carlo simulations.

3. Are We In A Comet Shower?

An important question for cometary dynamics is whether the solar system is currently experiencing an enhanced flux of comets from the Oort cloud. Examination of the distributions of orbital elements from a hypothetical comet shower caused by a stellar passage at 5×10^3 AU does not show substantial departures from random. This is especially true when trying to compare the model results to the distributions for the observed long-period comets, which are in limited numbers and are observationally biased.

However, the inverse semimajor axis distribution, $1/a_o$, for the hypothetical cometary shower shows a clear signature of many orbits with semimajor axes less than 2×10^4 AU, which is not seen for the observed long-period comets. The signature is even stronger if one only considers comets arriving in the planetary region in the first 10^6 years after the star's passage. Thus, the solar system does not appear currently to be experiencing a cometary shower, based on the observed distribution of $1/a_o$ for the long-period comets.

Further exploration of the relevant parameter space for star passages through the Oort cloud is currently underway.

4. Acknowledgements

The author thanks Kevin Yau for aid in the preparation of this paper. This work was supported by the NASA Planetary Geology and Geophysics Program and was performed at the Jet Propulsion Laboratory under contract with the National Aeronautics and Space Administration.

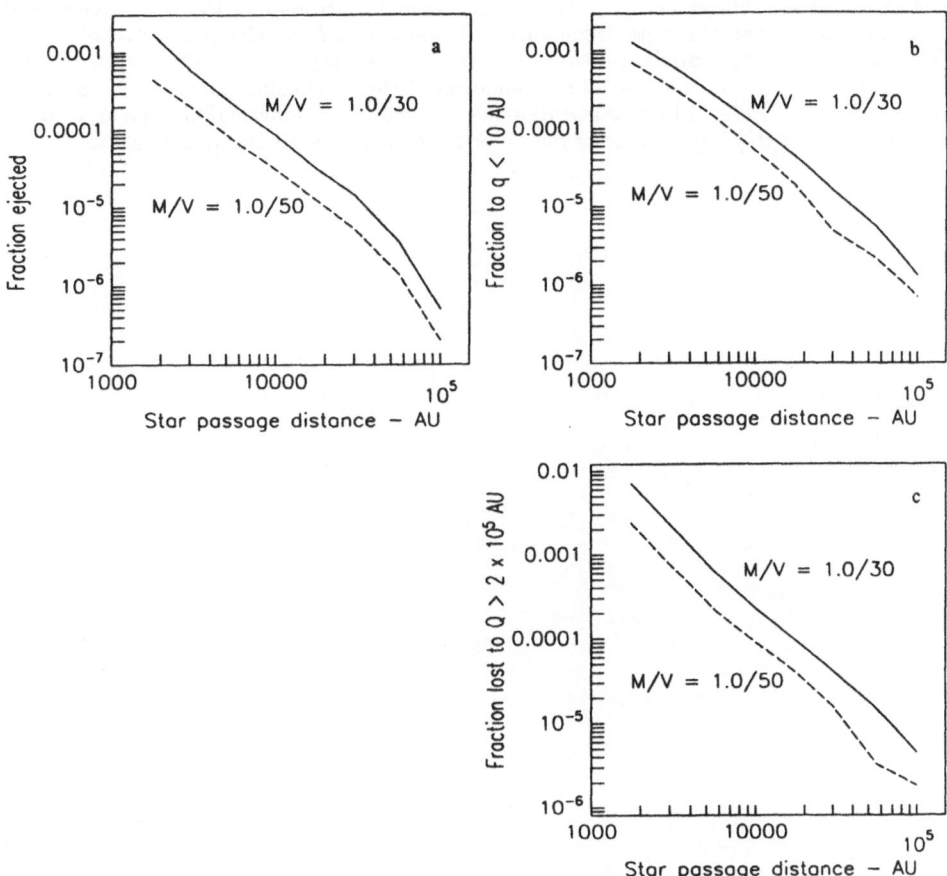

Fig. 3. Fraction of the Oort cloud population lost to various end-states as a function of stellar encounter distance, for two different stellar mass and velocity combinations: a) fraction ejected to interstellar space; b) fraction perturbed into the planetary region, q < 10 AU; c) fraction perturbed to Q > 2 × 10⁵ AU. The units of M/V are solar masses over stellar velocity in km s^{-1}

References

Byl, J.: 1983, 'Galactic perturbations on nearly parabolic cometary orbits', *Moon & Planets* **29**, 121-137

Fernandez, J. A. and Ip, W.-H.: 1987, 'Time dependent injection of Oort cloud comets into Earth-crossing orbits', *Icarus* **71**, 46-56

Heisler, J. and Tremaine, S.: 1986, 'The influence of the galactic tidal field on the Oort comet cloud', *Icarus* **65**, 13-26

Hills, J. G.: 1981, 'Comet showers and the steady-state infall of comets from the Oort cloud', *Astron. J.* **86**, 1730-1740

Hut, P., Alvarez, W., Elder, W. P., Hansen, T., Kauffman, E. G., Keller, G., Shoemaker, E. M. and Weissman, P. R.: 1987, 'Comet showers as possible causes of stepwise mass extinctions', *Nature* **329**, 118-126

Oort, J. H.: 1950, 'The structure of the cloud of comets surrounding the solar system and
    a hypothesis concerning its origin', *Bull. Astron. Inst. Netherlands* **11**, 91-110
Weissman, P. R.: 1980, 'Stellar perturbations of the cometary cloud', *Nature* **288**, 242-243
Weissman, P. R.: 1991, 'The angular momentum of the Oort cloud', *Icarus* **89**, 190-193
Weissman, P. R.: 1979, 'Physical and dynamical evolution of long-period comets' in R. L.
    Duncombe, ed(s)., *Dynamics of the Solar System*, D. Reidel:Dordrecht, 277-282

EVOLUTION OF THE COMETARY CLOUD UNDER THE ACTION OF STELLAR PERTURBATIONS

I.V. PETROVSKAYA

Astronomical Institute of St.Petersburg State University, Bibliotechnaya pl. 2,
St.Petersburg, 198904, Russia

Abstract. The probability of variation of the integrals of the orbit as a result of an encounter was found for a two dimensional system. A method of solution of the Kolmogorov-Feller's equation is obtained using this probability function as a kernel, and it allows us to obtain the distribution of the integrals of the orbit as a function of time. The method is applied to the investigation of the evolution of orbits in the outer cometary cloud under the action of galactic stars. We consider the variations of orbits as a purely discontinuous random process, so we take into account not only distant but also close interactions.

Key words: Oort Cloud, Evolution, Close Encounters

1. Introduction

The importance of close encounters in gravitating systems was indicated by Hénon (1960) who considered the evolution of stellar orbits in a spherical cluster. Under the action of weak interactions the orbits become more elongated and spend more time in the outer regions where the probability of an encounter is small. Therefore the dissipation is a result of strong interaction which must be taken into account. This is true in three dimensional systems with large density gradients, but the time scale increases by factor $\ln N$ as compared with the Fokker-Planck scheme (which is not applicable in this case) if we restrict ourselves only to the processes where close encounters are effective.

If we take into account strong interactions we must consider the scheme of purely discontinuous random process which is described by the Kolmogorov-Feller equation (Petrovskaya, 1969a). This equation has been used for the investigation of the evolution of stellar clusters and to trace the evolution of motions of stars of different masses (Petrovskaya, 1969b, Kaliberda and Petrovskaya, 1972). The stationary solution is achieved at the time $\approx 20t_r$ for the stars with the mass $m = m_1$ and $\approx 10t_r$ for the stars with $m = 0$ where m_1 is the mean mass of field stars, and $t_r \approx 3 \times 10^7$ years for an open cluster.

The kernel of the integro-differential Kolmogorov-Feller equation is the probability of encounter with given changes of the orbital elements of a star (Agekian, 1959). This function, Φ, is multiplied by λ , which takes

Earth, Moon, and Planets **72**: 31-34, 1996.

into account the simultaneous interaction of distant encounters, and $\lambda\Phi \sim (\triangle v)^{-1}$ when the velocity variation $\triangle v \to 0$ (Agekian, 1961). This behavior reflects the nonconvergence of the collision term in three-dimensional systems and special methods are required in order to solve the equation (Petrovskaya, 1969b; Kaliberda and Petrovskaya, 1972).

In the case of a two-dimensional system the calculation of the differential cross section differs from the three-dimensional case. Now the collision term is convergent, and the kernel has a finite limit when $\triangle v \to 0$. Such behavior of $\lambda\Phi$ shows the predominant part of strong interactions in flat systems. A similar behavior takes place in the more complicated case of a layer of finite thickness (Chumak and Chumak 1988, Petrovskaya 1992a).

The method may be applied to the investigation of bodies of small mass which relax in the field of massive bodies. It was shown for the case of the galactic field that the velocity distribution of zero mass stars achieves a stationary state at the time 3×10^9 years (Petrovskaya, 1992b). We may also consider the comets of the Oort Cloud as zero mass bodies when they relax under the action of passing galactic field stars. The first step in a study to this direction is taken in the present paper.

2. Method of analysis

The regular potential is automatically taken into account and the Kolmogorov-Feller equation becomes the kinetic equation if we consider changes of orbit integrals: of energy, H, and of angular momentum, Q,

$$\frac{\partial\phi(t,H,Q)}{\partial t} = -\phi(t,H,Q)\int\int_{-\infty}^{+\infty}\Phi_1(H,Q,h,q)dhdq +$$
$$+ \int_0^{H_0}\int_0^{Q_0}\phi(t,h,q)\Phi_1(h,q,H,Q)dhdq \qquad (1)$$

where $\phi(0,H,Q) = \phi_0(H,Q)$. The integration in the second term of the right hand side is done in region $h \le 0$. The Markovian random process is completely described by the function $\phi(t,H,Q)$ – the density of probability that at time t the energy is in the interval $(H, H + dH)$ and the angular momentum is in the interval $(Q, Q + dQ)$.

The probability $\Phi(H,Q,h,q)dhdq$ has been found in a flat system for a zero mass test body to have an encounter with the variations $h - H, q - Q$ of its orbit integrals H and Q (Petrovskaya, 1994). The multiplicity coefficient λ was obtained for a two dimensional system earlier(Petrovskaya et al. 1984). We take the product $\lambda\Phi = \Phi_1$ as the kernel of equation (1).

The convergence of the collision term ($\lambda\Phi$ is finite when $h - H \to 0$, $q - Q \to 0$) permits to solve (1) by means of integral transformations: Fourier in t and Hankel transformations with finite limits in Q and H (Petrovskaya, 1983). Finally (1) is reduced to the system of linear algebraic equations

$$Y(s, p_i, r_l) \int_0^{H_0} \int_0^{Q_0} [s + \int \int_{-\infty}^{+\infty} \Phi_1(H, Q, h, q)dhdq] \, dH \, dQ =$$

$$\int_0^{H_0} \int_0^{Q_0} \frac{4HQ}{H_0Q_0} J_n(p_iH)J_m(r_lQ) \left\{ \int_0^{H_0} \int_0^{Q_0} dhdq \Phi(h, q, H, Q) \times \right. \quad (2)$$

$$\left. \sum \frac{J_n(p_jh)J_m(r_kq)Y(s, p_j, r_k)}{[J_n'(H_0p_j)J_m'(Q_0r_k)]^2} + \phi_0(H, Q) \right\} dH \, dQ$$

where J_n and J_m are the Bessel functions and p_i and r_l are the roots of equations $J_n(H_0p_i) = 0$ and $J_m(Q_0r_l) = 0$, respectively. The distribution function of H and Q at time t is connected with the inverse Laplace transform of the function $Y(s, p_i, r_l)$ by the relation

$$\phi(t, H, Q) = \sum b_{il}(H, Q)L^{-1}Y(s, p_i, r_l) \quad (3)$$

where

$$b_{il} = \frac{4J_n(p_iH)J_m(r_lQ)}{H_0^2Q_0^2[J_n'(H_0p_i)J_m'(Q_0r_l)]^2} \quad (4)$$

The inverse Laplace transform of Y is easy to find from the peculiar values of determinant of the system (2). Then we can find the distribution function of H and Q at any time t, $\phi(t, H, Q)$, using the linear relation (3) between ϕ and $L^{-1}Y$.

3. Results

The method was applied to the simplest model: a ring of comets at a distance 100 AU from the Sun. It is supposed that all comets have circular orbits at the initial moment. The field stars are supposed to have a Maxwellian two-dimensional distribution function of peculiar velocities and rotate around the galactic centre with the Solar System velocity.

TABLE I

The escape probability for different values of energy.

H_1	0.1	0.2	0.3	0.4	0.5	0.6	0.7	0.8	0.9
γ	0.030	0.012	0.0066	0.0039	0.0023	0.0014	0.00066	0.00034	0.00014

Instead of energy H it was convenient to consider the value $H_1 = -H/U$, where U is the potential. At the initial moment all comets in our model have $H_1 = 0.5$, and then $0 \leq H_1 \leq 1$. The value $H_1 = 0$ corresponds to the escape energy. The eccentricity of orbits $e = 0$ at $t = 0$, and subsequently orbits become noncircular. A stationary state is achieved at $t_0 = 10^7$ years. At $t = t_0$ the mean eccentricity of comets with $H_1 = 0.5$ is ≈ 0.5.

In Table I the ratio

$$\gamma = \frac{\int_0^{H_0} \int_0^{Q_0} \Phi_1 dh dq}{\int \int_{-\infty}^{+\infty} \Phi_1 dh dq} \tag{5}$$

is given which is equal to the probability for a comet to escape as a result of one encounter.

When the stationary state is achieved the rate of dissipation is 14% per 10^7 years. This value may change if we take into account the finite thickness of the system as we propose to do in future. We expect that the rate of evolution will be slowed down in such a more realistic model.

Acknowledgements

This work was supported in part by the International Science Foundation, Grant No. NW4000.

References

Agekian, T.A.: 1959, *Sov. Astron. J.*, **36**, 41.
Agekian, T.A.: 1961, *Sov. Astron. J.*, **38**, 1055.
Chumak, Z.N., Chumak, O.V.: 1988, *Trudy AFI AN KazSSR*, **49**, 126.
Henon, M.: 1960, *Ann. Astrophys.*, **23**, 668.
Kaliberda, V.S., Petrovskaya, I.V.: 1972, *Astrofizika*, **8**, 305.
Petrovskaya, I.V.: 1969a, Sov. Astron. J., **46**, 824.
Petrovskaya, I.V.: 1969b, Sov. Astron. J., **46**, 1220.
Petrovskaya, I.V.: 1983, Vestnik Leningr. Univ., N 13, 94.
Petrovskaya, I.V.: 1992a, *Sov. Astron. J.*, **69**, 408.
Petrovskaya, I.V.: 1992b, *Sel. Mech. and Dyn. Astron.*, **54**, 267.
Petrovskaya, I.V.: 1994, *Astron. and Astroph. Trans.*, **6**, in press.
Petrovskaya, I.V., Chumak, Z.N., Chumak, O.V.: 1984, *Sov. Astron. J.*, **61**, 467.

DYNAMICAL EVOLUTION OF COMETS AND THE PROBLEM OF COMETARY FADING

V.V. EMEL'YANENKO and M.E. BAILEY

*School of Computing and Mathematical Sciences, Liverpool John Moores University,
Byrom Street, Liverpool L3 3AF, U.K.*

Abstract. Possibilities to explain the observed $1/a$-distribution are discussed in the light of improved understanding of the dynamical evolution of long-period comets. It appears that the 'fading problem' applies both to single-injection and continuous-injection models. Although uncertainties due to nongravitational effects do not allow detailed results to be drawn from the observed $1/a$-distribution at small perihelion distance q, that for $q \gtrsim 1.5$ AU shows that a constant fading probability cannot explain all the features of the observed distribution. Assuming that comets can reappear following a period of fading, values for the assumed constant fading and renewal probabilities, and the total cometary flux have been estimated for $q > 1.5$ AU.

Key words: Comets – dynamics – fading problem

1. Introduction and Observed Distribution

The problem of the original $w = 1/a$-distribution, where a is orbital semimajor axis, is long-standing. The peak in the range $0 < w < 10^{-4}$, where $w = 1/a$ in units AU^{-1}, is very narrow in comparison with the dispersion introduced by planetary perturbations, and implies either that new comets coming from the outer solar system are subject to strong fading after perihelion (Oort 1950, Whipple 1953, 1962; Everhart 1979, Weissman 1979) or that the observed shower of comets is rapidly evolving (Kresák 1977, Van Flandern 1978). Yabushita (1979) showed that for a constant fading probability $p \simeq 0.02$, where p denotes the probability per revolution that a comet ceases to be observed, the observed distribution in the interval $0 < w < 5 \times 10^{-4}$ can be explained in a time-dependent model as the result of planetary perturbations. Bailey (1984) estimated the dependence of p on w for a steady-state model and showed that in this case it is necessary to postulate a rapid decrease in p for $w \gtrsim 0.004$.

We return to this problem because first, many new comets have been discovered (Marsden & Williams 1993), in particular comets in orbits of large perihelion distance; and secondly, planetary perturbations, previously modelled as random quantities, can now be treated exactly using an analytical expression for the disturbing function of near-parabolic motion (Emel'yanenko 1991, 1992).

The observed w-distribution depends on perihelion distance and is summarized in Table I for orbits of Classes 1 and 2 (Marsden & Williams 1993).

Earth, Moon, and Planets **72**: 35–40, 1996.

We tabulate the ratio of the observed long-period flux in different intervals of w and for several ranges of q, normalized to that with w in the range $(0, 10^{-4})$. The main feature is that the proportion of comets with large w decreases with increasing q. We also note that the distribution in the range $0 < w < 10^{-4}$ is subject to significant nongravitational effects for small perihelion distances, while the a-distribution for $q > 2.2$ AU has an evident maximum in the range $2 \times 10^4 < a < 3 \times 10^4$ AU.

2. Methods and Models

We consider the motion of long-period comets under the attraction of the sun and four planets (Jupiter, Saturn, Uranus and Neptune), using a mapping of the form $w^{(m+1)} = w^{(m)} + \Delta w_G + \Delta w_N$, where Δw_G and Δw_N are the changes in $1/a$ during the m-th perihelion passage due to gravitational and nongravitational effects respectively (Emel'yanenko 1992). The evolution of the frequency distribution of perihelion passages, or w-distribution, has been investigated for a range of initial values of q, i, ω and a, where i and ω denote the inclination and argument of perihelion respectively. We consider two extreme models: (1) the Single Injection model (SI); and (2) the Continuous Injection model (CI). In the former, the evolution of typically $\sim 10^5$ cometary orbits begins at some initial time $t = 0$ with the planetary phases chosen randomly for each comet. In the CI model, comets are injected at a steady rate of one comet every Δt days, with the initial planetary phases again distributed randomly. Typically $\Delta t \simeq 10^4$–10^5, and the investigation continues for $\sim 10^{10}$ days.

The main features of the dynamical evolution for the SI model have already been described by Yabushita (1979) and Emel'yanenko (1992). The w-distribution does not sensitively depend on the initial semi-major axis a_0 provided $a_0 \gtrsim 10^2$ AU. A maximum in the domain $a > 10^4$ AU moves up in

TABLE I

Normalized frequency distribution of long-period comets versus $w = 1/a$ and q.

	Interval of w			
q range	$(0, 10^{-4})$	$(10^{-4}, 0.02)$	$(0.002, 0.02)$	$(0.01, 0.02)$
0.000–1.100	1	4.60	2.65	1.00
1.100–2.200	1	1.81	0.82	0.19
1.500–2.200	1	1.73	0.53	0.20
≥ 2.200	1	0.89	0.11	0.03

TABLE II

Results for representative SI models at $t = 1.8 \times 10^9$ days.

SI Parameters	Interval of w			
	$(0, 10^{-4})$	$(10^{-4}, 0.02)$	$(0.002, 0.02)$	$(0.01, 0.02)$
$q=0.85$, $p=0.03$, $\Delta w_N = 0$	1	4.57	2.63	0.03
$q=3.00$, $p=0.30$, $\Delta w_N = 0$	1	0.85	0.00	0.00
$q=0.85$, $p=0.03$, $|\Delta w_N| < 0.0002$	1	3.56	1.62	0.34
$q=0.85$, $p=0.03$, $|\Delta w_N| < 0.0004$	1	9.11	5.02	1.26

TABLE III

Results for a representative CI model with $p = 0$ at various times.

t days	Interval of w			
	$(0, 10^{-4})$	$(10^{-4}, 0.02)$	$(0.002, 0.02)$	$(0.01, 0.02)$
0.5×10^8	1	0.94	0.23	0.01
1.0×10^8	1	2.47	1.11	0.05
1.5×10^8	1	4.50	2.66	0.34
1.0×10^9	1	39.57	35.44	18.04

a, and for all initial values of q, i, ω and a_0 the observed maximum (in the range 2×10^4–3×10^4 AU) could arise as a result of planetary perturbations acting for 3–5 Myr.

Despite this, detailed agreement with observations (Table I) cannot be obtained without strong limitations on the cometary lifetime. The small ratio of the number of observed comets with $10^{-4} < w < 0.02$ to that with $0 < w < 10^{-4}$, in the range 1–5, defines the fading problem. We denote this ratio by s. Values of p corresponding to the observed ratio range from 0.03 for $q \sim 1$ AU, to 0.3 for $q \sim 3$ AU. An example of the w-distribution for $q=0.85$ AU, $i=90°$, $\omega=0$, $a_0=4 \times 10^4$ AU and $p=0.03$ is presented in the first line of Table II for $t = 1.8 \times 10^9$ days (~ 5 Myr). The calculated distribution is very close to the observed one for $10^{-4} < w < 0.02$ and $0.002 < w < 0.02$, but there is a substantial discrepancy for the interval $0.01 < w < 0.02$. For larger perihelion distances it is even difficult to find agreement for the region $0.002 < w < 0.02$. Such disagreement occurs for all initial orbital elements.

The CI model also implies a strong limit on the cometary lifetime. Table III gives results for this model, assuming $q=1.85$ AU, $i=90°$, $\omega=0$, $a = 2 \times 10^4$ AU, $p=0$, and $\Delta w_N = 0$. In this case, which assumes no fading, the relative w-distribution agrees with observations in the range $10^{-4} < w < 0.02$ provided the time since injection of comets began is less than 200,000 years,

TABLE IV

Calculated values of p, p^* and g (see text) for SI and CI models.

Model	q (AU)	p	p^*	N	g
SI	(1.500,2.200)	2×10^{-1}	1×10^{-4}	3000	8×10^{-3}
	> 2.200	3×10^{-1}	4×10^{-5}	6000	2×10^{-3}
CI	(1.500,2.200)	2×10^{-1}	3×10^{-3}	3000	6×10^{-2}
	> 2.200	3×10^{-1}	5×10^{-4}	6000	3×10^{-2}

but for larger values of w (e.g. $0.01 < w < 0.02$) the predicted distribution is incorrect. A more general CI model, this time with larger p, can explain the ratio s at larger times, but fails in the same way as the SI model to explain the number of comets at larger w. We conclude that it is impossible to explain the $1/a$-distribution of long-period comets solely with a constant fading probability.

We have also estimated the effects of nongravitational forces. The third and fourth lines of Table II show examples of the w-distribution taking such forces into account. Here values of Δw_N were distributed uniformly in the respective ranges $\pm 0.2 \times 10^{-3}\,\mathrm{AU}^{-1}$ and $\pm 0.4 \times 10^{-3}\,\mathrm{AU}^{-1}$, corresponding to Marsden's parameter $A_2 = \pm 10^{-8}$ and $\pm 2 \times 10^{-8}$ respectively (Marsden & Williams 1993) for $q = 0.85\,\mathrm{AU}$. For each comet Δw_N was assumed to be constant for all perihelion passages. The Table shows that nongravitational forces may have important effects at small perihelion distances.

The peculiarities of the long-term behaviour of nongravitational forces are unknown. In particular, it is difficult to estimate whether they are stable for many revolutions, so a detailed comparison with the observed w-distribution cannot be made for $q \lesssim 1\,\mathrm{AU}$. On the other hand, our simulations show that typical nongravitational forces have little influence on the distribution for $q > 1.5\,\mathrm{AU}$ and so can be ignored.

3. Fading and Renewal Parameters for $q > 1.5\,\mathrm{AU}$

The main features of the w-distribution for $1.5 < q < 2.2\,\mathrm{AU}$ and $q > 2.2\,\mathrm{AU}$ are shown in Table I. These cannot be explained solely by a constant fading probability p, and we have therefore extended the model to include an additional parameter p^*, corresponding to the probability per revolution that a comet is rejuvenated after an unobserved state.

Our calculations have shown that the observed data can be well represented by two parameters p and p^* for both SI and CI models. Representative values are given in Table IV, where we include the dependence on the total cometary lifetime, measured as the number of revolutions N that comets execute (whether as observable or unobservable objects) before their final disappearance. Our studies show that the w-distribution does not depend sensitively on N provided $N > 2 \times 10^3$.

We note that the values of p are approximately the same for different ranges of q for both the SI and CI models. The peculiarities of the w-distribution for different values of q are more closely connected with p^* than with p, suggesting the possible importance of cometary splitting and/or rejuvenation in the interpretation of the $1/a$-distribution. The introduction of p and p^* in the model implies that there may exist many dormant long-period comets, up to now not observed. The ratio of the predicted flux to that for a model with $p = p^* = 0$, for comets with $0 < w < 2 \times 10^{-2}$, may be denoted by the parameter g; the total long-period flux for $q \gtrsim 1.5\,\mathrm{AU}$ thus exceeds the observed flux of comets of Classes 1 and 2 by a factor $g^{-1} \approx 20\text{--}500$.

4. Principal Conclusions

1. A constant fading probability p, whether in an SI or CI model, cannot explain the entire $1/a$-distribution.
2. Nongravitational forces may significantly affect the observed $1/a$-distribution for $q \lesssim 1\,\mathrm{AU}$.
3. Constant fading and renewal probabilities have been found which satisfy the observational data for both SI and CI models, for $q \gtrsim 1.5\,\mathrm{AU}$. Differences in the $1/a$-distribution for different perihelion distances are more closely connected with the renewal probability p^* than the fading probability p.
4. We predict the existence of many unobserved long-period comets.

References

Bailey, M.E.: 1984, *Mon. Not. R. astr. Soc.* **211**, 347.

Emel'yanenko, V.V.: 1991, *Pis'ma Astron. Zh.* **17**, 857.

Emel'yanenko, V.V.: 1992, *Celest. Mech.* **54**, 91.

Everhart, E.: 1979, in *Dynamics of the Solar System*, ed. Duncombe, R.L., Reidel, Dordrecht, p.273.

Kresák, L.: 1977, in *Comets, Asteroids, Meteorites*, ed. Delsemme, A.H., University of Toledo Press, p.97.

Marsden, B.G. and Williams, G.V.: 1993, *Catalogue of Cometary Orbits*, Smithsonian Astrophys. Observatory, Cambridge, Massachusetts.

Oort, J.H.: 1950, *Bull. Astr. Inst. Neth.* **11**, 91.

Van Flandern, T.C.: 1978, *Icarus* **36**, 51.

Weissman, P.R.: 1979, in *Dynamics of the Solar System*, ed. Duncombe, R.L., Reidel, Dordrecht, p.277.

Whipple, F.L.: 1953, *Mem. Soc. R. Sci. Liège (4th Ser.)* **13**, 283.

Whipple, F.L.: 1962, *Astr. J.* **67**, 1.
Yabushita, S.: 1979, *Mon. Not. R. astr. Soc.* **187**, 445.

ORIGIN AND EVOLUTION OF THE LONG-PERIOD COMETS

A. SALĪTIS

Daugavpils Pedagogical University, LV-5400 Daugavpils, Latvia

Abstract.
We consider the changes of cometary perihelion distances as a process of diffusion in the value of q, due to perturbations by stars. We find more comets at large q values than is observed. This suggests that a large number of long-period comets is not observed.

1. Introduction

In this work we have investigated the dynamical evolution of long-period comets by using the diffusion theory of comets.

The basic problem of the diffusion theory is how to explain the process of forming the short-period comets from undoubtedly existing initially long-period ones (with semimajor axis $a \sim 50\,000 - 100\,000$ AU). Many problems connected with the transformation of long-period comets into short-period ones are not yet solved. The distribution of long-period comets according to $z = 1/a$ is not completely explained although the generalization of comet diffusion theory by Salītis (1982) and Salītis and Dīriķis (1990) helps to explain better the observable distribution.

For cometary statistics the transformation of the perihelion distance q is of greater significance than z, because the value of q determines the probability of comet discovery and the rate of physical evolution.

2. Averaged perturbations

For determining the mean value of the change of the perihelion distance $\overline{\Delta q}$ during one revolution of a comet around the Sun we used the formula

$$\Delta q = \frac{2Q \cdot \Delta v_{tr}}{V_q \cdot q}, \tag{1}$$

where V_q is the perihelion velocity of the comet, q is its perihelion distance, Q is its aphelion distance and Δv_{tr} is the change of the transverse velocity. Formula (1) follows from Öpik (1932) where it was derived on the principle of conservation of angular momentum. Assuming the parabolic velocity as V_q in formula (1), which is not very far from reality for near-parabolic orbits, and,

Earth, Moon, and Planets **72**: 41-44, 1996.

Table 1. Perihelion distance change of comets with initial perihelion distances $q = 5$ AU and various values of a during one revolution around the Sun.

a (AU)	$\overline{\Delta q}$ (AU)	$\overline{\Delta q'}$ (AU)
$1.0 \cdot 10^4$	0.20	0.1
$1.5 \cdot 10^4$	0.30	0.6
$2.0 \cdot 10^4$	0.62	2.0
$2.5 \cdot 10^4$	1.08	5.0

for example, examining comets with $q = 8$ AU, we can express formula (1), taking into account that $Q \approx 2a$, in the following way

$$\Delta q = 1.634 \cdot 10^{-9} \cdot a^{5/2} \cdot \Delta v_{tr} \text{ (AU)} \tag{2}$$

From (2) we can derive the change of the perihelion distance during one revolution for comets with various values of semimajor axes a. Results of the calculations are presented in Table 1.

For comparison the results of Rickman (1976) are also presented in the table (marked $\overline{\Delta q'}$). From the table it appears that according to our estimates, with the increase of a the change of the perihelion distance Δq does not increase as fast as according to Rickman. It might be explained by the fact that Rickman has taken into account stellar passages which are too remote. Remote passages are significant over long time intervals which are comparable to the age of the Solar System and which in many cases exaggerate the revolution period of comets with $a \approx 2.5 \cdot 10^4$ AU.

3. Diffusion model

The afore-mentioned examples give us evidence that averaging methods in determining perturbations of perihelion distance Δq give very approximate results. In the present investigation we have refrained from averaging methods. We consider the changes of cometary perihelion distances as a process of diffusion (a slow accumulation of small perturbations of random character, in the value of q). Such an approach is fully admissible for the mutual arrangement of stars and comets, because the orbits of stars and comets at great distance from the Sun are entirely randomly oriented. Making such a presumption we shall compile a balance equation of comet number n with fixed a and q.

$$n(a, q \le q_0) = (1 - k - \kappa) \int_{q_0 - \Delta q}^{q} n(a, a - \Delta q) \, \varphi(a, \Delta q) \, dq, \tag{3}$$

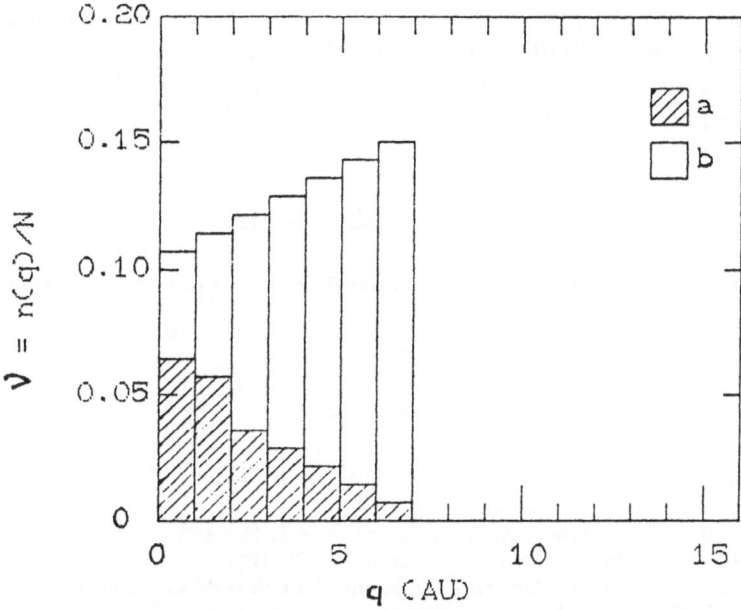

Fig. 1.   Distribution of the long-period comets according to their perihelion distances: a) distribution of 234 observed comets; b) distribution obtained from diffusion equation (3).

Table 2.   Observed long-period comets.

q (AU)	Number of detected long-period comets (%)	
	before 1950	after 1950
3 - 4	27	73
4 - 5	20	80
5 - 6	-	100
6 - 7	-	100

where $n(a, a - \Delta q)$ is the number of comets with fixed a and q in the interval $(a, a - \Delta q)$ and $\varphi(a, \Delta q)$ is the density of probability distribution for perturbations of perihelion distance. The coefficient k determines the fraction of comets which have disappeared from the zone of visibility as a result of physical disintegration and the coefficient κ is the fraction of comets which were lost in the process of dynamical evolution. In order to obtain the solution of this balance equation, it is necessary to know the function $\varphi(a, \Delta q)$ of distribution of perturbations of perihelion distance Δq.

As a first approximation for $\varphi(a, \Delta q)$ we have assumed the normal distribution. Based on this assumption we get the q distribution of long-period comets (Fig. 1). We can see that according to equation (3) we have many

more comets at large q values than is observed (Marsden, Sekanina and Everhart 1978; Marsden 1990). This means that a large number of long-period
comets are not observed. Table 2 gives the statistics of the discovery time
of the long-period comets that confirms our conclusion.

Acknowledgements

I would like to thank the LOC for covering my expenses for attending the
meeting.

References

Marsden, B.G., Sekanina, Z., Everhart, E.: 1978, *Astron.J.* **83**, 64–71.
Marsden, B.G.: 1990, *Astron.J.* **99**, 1971–1973.
Öpik, E.J.: 1932, *Proc. Amer. Acad. Arts and Sc.* **67**, 169–183.
Rickman, H.: 1976, *Bull. Astron. Inst. Czech.* **27**, 92–105.
Salītis, A.: 1982, In *Analiz dvizheniya nebesnyh tel i ih nablyudeniya* (L. Laucenieks, M.
    Dīriķis, E. Kaupusha, Eds.), pp. 10–17.
Salītis, A., Dīriķis, M.: 1990, In *Nordic-Baltic Astronomy Meeting* (C.-I. Lagerkvist,
    D. Kiselman, M. Lindgren, Eds.) pp. 283–286.

ORBITS OF SHORT PERIOD COMETS CAPTURED FROM THE OORT CLOUD

J.Q. ZHENG, M.J. VALTONEN, S. MIKKOLA and M. KORPI

Tuorla Observatory, University of Turku, Finland

and

H. RICKMAN

Astronomical Observatory, Uppsala University, Sweden

Abstract. Oort cloud comets occasionally obtain orbits which take them through the planetary region. The perturbations by the planets are likely to change the orbit of the comet. We model this process by using a Monte Carlo method and cross sections for orbital changes, i.e. changes in energy, inclination and perihelion distance, in a single planet–comet encounter. The influence of all major planets is considered. We study the distributions of orbital parameters of observable comets, i.e. those which have perihelion distance smaller than a given value. We find that enough comets are captured from the Oort cloud in order to explain the present populations of short period comets. The median value of $\cos i$ for the Jupiter family is 0.985 while it is 0.27 for the Halley types. The results may explain the orbital features of short period comets, assuming that the active lifetime of a comet is not much greater than 400 orbital revolutions.

1. Introduction

There is a steady flux of "new" Oort cloud (Oort 1950) comets through the inner Solar System. By integrating the orbits of a large number of such new comets with jovian perturbations upon repeated perihelion passages, Everhart (1972) concluded that short-period comets evolve from new comets by changes in orbit. However, it has been subsequently claimed that the number of short period comets which results from captures from the Oort cloud is not large enough (Joss 1973, Fernández and Ip 1983) or that the distribution of orbital parameters of the observed short period comets does not match with the expected capture population (Duncan *et al.* 1988; Wetherill 1991; Fernández and Gallardo 1993). The situation is still rather unclear (Stagg and Bailey 1989; Valtonen *et al.* 1992).

In this paper we present an improved calculation of the orbital evolution of Oort cloud comets through the Solar System and show that under certain (not implausible) conditions the statistics of short-period comets are well reproduced if they evolve by dynamical means from the Oort cloud.

2. The method of calculation

The most direct way to attack the problem of orbital evolution of Oort cloud comets is to integrate the orbits of large numbers of comets (a million

Earth, Moon, and Planets **72**: 45-50, 1996.

or so) over the relevant time periods, up to the age of the Solar System. Quinn *et al.* (1990) opted for this approach but found that the required time of computation is overwhelming unless the problem is somehow simplified. The simplifications considered were: (1) to have only one planet, Jupiter, to influence the orbits of comets, or (2) to have very massive outer planets in place of Saturn, Uranus and Neptune. Both simplifications change the original problem significantly.

In view of the difficulty with excessive computer time in direct integrations, Wetherill (1991) carried out Monte Carlo calculations using the Öpik (1951) scheme for calculating comet–planet encounters. However, this scheme may not be accurate enough for all the encounters of importance in the current problem (Carusi *et al.* 1990). Instead one should use data from real comet–planet encounters, based on accurate orbit integrations. Following pioneering work by Rickman and Vaghi (1976), Everhart (1977) and Froeschlé and Rickman (1980), Stagg and Bailey (1989) and Fernández and Gallardo (1993) started a push toward this direction by integrating the orbits of comets through the Solar System and using the distribution of the changes of orbital parameters as input to a Monte Carlo scheme. The problem in this previous work is the limited statistical data, which becomes quite obvious upon comparison with the results of Zheng (1994).

The approach used in the present work is outlined by Valtonen *et al.* (1992). By using several millions of accurate orbit integrations, the probability distributions for changes in the cometary orbital elements a (semi-major axis), i (inclination) and q (perihelion distance) are evaluated for orbits of given initial a_0, i_0 and q_0, assuming the remaining orbital elements to be randomly distributed. The planetary orbits are assumed circular. The probability distributions are reported in Zheng (1994), and they are used in a Monte Carlo scheme similar to the approach of Fernández and Gallardo (1994).

We assume that comets arrive from the Oort cloud with a uniform distribution of $\cos i_0$ and with a constant semi-major axis $a_0 = 20\,000$ AU. Several different distributions of q_0 ($\lesssim 30$ AU) were studied. Here we report results based on the Galactic tide model of Matese and Whitman (1989) where the number of new comets increases with q_0 qualitatively like in the stellar perturbation models of Weissman (1985) and Fernández (1982), though differing in the details:

$$N(q_0) = \begin{cases} 1 + 0.014\,q_0^{1.82} & \text{when } q_0 < 13 \text{ AU} \\ 5 & \text{when } 13 \text{ AU} < q_0 < 30 \text{ AU} \end{cases} \tag{1}$$

(in units of the near-parabolic flux at Earth's orbit). For each comet with given initial values of q_0, a_0 and i_0, the probability of scattering by each major planet is calculated and a particular planet is selected according to this weighting. The orbital change caused by the selected planet is then chosen from the relevant, tabulated probability distributions and new orbital elements a, i and q are computed. The comet is then assumed to undergo N revolutions with the new elements until the next scattering, at which time a perturbing planet is again selected and a new orbital change is computed in the same way. This number of revolutions is determined by the inverse of the scattering probabilities for all the planets. This process is continued by selecting new scatterings and new orbits consecutively until the comet escapes from the Solar System or is considered to have become unobservable due to physical decay. As scatterings we only consider strong perturbations, corresponding to $|\Delta 1/a| > 5 \times 10^{-4}\,\mathrm{AU}^{-1}$ for Jupiter, and decreasing proportional to M_p/a_p where M_p is the mass of the planet and a_p its orbital radius, to $|\Delta 1/a| > 5 \times 10^{-6}\,\mathrm{AU}^{-1}$ for Neptune.

We consider a comet to live through N_{max} close encounters with the Sun, where a close encounter is defined by $q < q_{lim}$. We use the values of $N_{max} = 400$, 1000 and 4000 and $q_{lim} = 1.5$ and 2.6 AU. The calculations are continued with a steady influx rate of new comets up to the time limit of T_{max}. We have considered T_{max} ranging from $3 \cdot 10^6$ yr to $300 \cdot 10^6$ yr. Here we report only the case of $T_{max} = 300 \cdot 10^6$ yr, $q_{lim} = 2.6$ AU and $N_{max} = 400$ (see e.g. Delsemme 1973, 1979; Kresák 1981). The full details will be published elsewhere (Zheng et al., in preparation).

3. Captured comets

The model reported here considers the evolution of 10^6 comets, injected at a steady rate of one comet every 300 years. The comets initially have nearly parabolic orbits with semi-major axes $a_0 = 2 \cdot 10^4$ AU, but those considered in the model are the ones that initially suffer scatterings into more tightly bound orbits, leaving them with new semi-major axes less than a_1, where $a_1 = 2000$ AU for Jupiter encounters and $a_1 = 18\,000$ AU for Neptune encounters. Such comets, i.e. those that are perturbed to bound orbits in this sense, constitute approximately 15% of the original near-parabolic flux (see below), so the corresponding near-parabolic flux in the model is one comet every 45 yr with perihelion distance less than 30 AU.

During the simulation any comet which achieves a perihelion distance $<$ q_{lim} is counted as a 'visible' comet; data are reported here for the particular case $q_{lim} = 2.6$ AU. If the orbital period P when the comet is visible is less than 20 yr, the comet is called a Jupiter-family (JF) comet, and each time it passes perihelion in the simulation with such a period and perihelion distance one unit is added to the total number of revolutions as a JF comet in the simulation.

Similarly, if the orbital period P is in the range $20 < P < 200$ yr when the comet is visible, it is called a Halley type (HT) comet, while if $200 < P < 90\,000$ yr (corresponding to $a = 2000$ AU), it is called a long-period (LP) comet. The total number of revolutions of the HT-phenomenon and LP-phenomenon in the simulation is calculated in the same way as for the JF comets. Any comet that undergoes more than N_{max} revolutions as a visible comet is removed from the simulation, being deemed to have been destroyed by physical decay. The results here refer to the particular case $N_{max} = 400$.

We calculate the number of short-period comets both in the Jupiter family $N(JF)$ and the Halley types $N(HT)$ as well as the number of long-period comets $N(LP)$. With the parameters used here we find $N(JF) = 3044$, $N(HT) = 68$ and $N(LP) = 190$. The median values of $\cos i$ for the Jupiter family and for the Halley types are 0.985 and 0.27, respectively. These are close to the observed values. The high value for the Jupiter family is obtained only when $N_{max} = 400$, and the agreement with observations is completely lost when $N_{max} = 4000$.

The numbers of comets $N(JF)$, $N(HT)$ and $N(LP)$ refer to the total number recorded over the calculation period T_{max}. At any point in time only a fraction $N_{max} \cdot \langle P_x \rangle / T_{max}$ of them is observed, so comparison with observed comet numbers can be made after $N(JF)$, $N(HT)$ and $N(LP)$ are multiplied by this fraction. We use $\langle P_{JF} \rangle = 8$ yr for the Jupiter family, $\langle P_{HT} \rangle = 80$ yr for the Halley types and $\langle P_{LP} \rangle = 800$ yr for the long-period comets.

The true influx rate of new comets is obtained as follows. We take the new Earth-crossing comet flux of Bailey and Stagg (1988) of 4.6 comets/AU/yr, which includes a correction for comets missed at the low end of the luminosity function (Everhart 1967; Hughes 1988) to absolute magnitude $H_{10} = 10.8$, and multiply by the fraction of new comets which suffer strong encounters (new $a < a_1$), i.e. by 0.15 (Everhart 1968; Marsden and Williams 1994). In addition, we divide by 1.5 since the present day comet flux may exceed the average over the past few million years by this factor (Matese $et\ al.$ 1995). The resulting long term average flux of new comets which become captured into orbits of $a < a_1$ is therefore about 0.5 comets/AU/yr at 1 AU. But since the comet flux in our model increases with q_0, there is an additional factor of 3.5 which has to be taken into account, as we consider the average Oort cloud flux over the 30 AU interval. Thus the numbers $N(JF)$, $N(HT)$ and $N(LP)$ should be multiplied by 50 /yr $\cdot T_{max}/10^6$ in order that they correspond to the true Oort cloud flux. Since only the fraction $N_{max}\langle P_x \rangle / T_{max}$ is observed at any time, the expected number of comets N_x of type x is

$$N_x = 5 \cdot 10^{-5} \cdot N(x) \cdot N_{max} \cdot \langle P_x \rangle \qquad (2)$$

With our parameters Eq. (1) gives us the expected numbers $N_{JF} = 487$, $N_{HT} = 109$ and $N_{LP} = 3040$.

We should note at this point that our model neglects orbital evolution via distant encounters with planets. We estimate that the inclusion of scattering by small energy steps would increase the HT and LP families by a factor 2–3, while the Jupiter family is less affected by this process. This has to be verified by future work.

Fernández et al. (1992; see also Fig. 2 in Fernández and Gallardo 1994) estimate that the Jupiter family has about 550 members out to $q = 2.5$ AU at the magnitude limit of 10.8. Hughes (1988) estimates the numbers of comets in different dynamical groups on the basis of the cumulative number of discovered comets brighter than H_{10}. Using his Fig. 1, we find 42 Jupiter family comets at $H_{10} = 10.8$ where the discovery rate still appears to be complete. At magnitude $H_{10} = 7.0$ where a break (indicating discovery incompleteness) appears to occur for Halley type comets, these are about 3 times more numerous than the Jupiter family comets. Thus the corrected number of Halley type comets should be ~ 126 at $H_{10} = 10.8$, if the luminosity functions are the same. About one quarter of the long period comets listed by Marsden and Williams (1994) corresponds to our LP-category. At magnitude $H_{10} = 5.8$ they are 50 times more numerous than Jupiter family comets, or there should be ~ 2100 of them at $H_{10} = 10.8$.

The known population of Halley type comets has $q \lesssim 1.5$ AU, and it is not known how to correct the numbers to correspond to the limit of $q = 2.6$ AU. A simple linear extrapolation between the two limits predicts 218 Halley type comets out to $q = 2.6$ AU. Note that some incompleteness of the discovered Halley type comets is due to arise from the orbital periods being comparable to the past interval of efficient comet discovery. For the LP comets we should expect a fourfold increase of the number observed during the last two centuries in order to fit with $\langle P_{LP} \rangle = 800$ yr. Considering the uncertainties (e.g. the proper value of N_{max} in Eq. (1), the Oort Cloud flux at large q, etc.) there is hence no major discrepancy between the (extrapolated) observed comet numbers and the numbers calculated from theory.

Thus the agreement between the Oort cloud capture products and the observed comet families is quite satisfactory, assuming that N_{max} is not very far from 400. This agreement is particularly striking in the inclination distributions which are likely to be better determined observationally than the total population numbers. Within the uncertainties of observational data we may thus state that the short-period comets may derive entirely from the Oort cloud.

Acknowledgements

We acknowledge helpful discussions with Mark Bailey who has given us lots of insight to the comet capture problem during this work. We also acknowledge many useful discussions with John Matese.

References

Bailey, M.E. and Stagg, C.R.: 1988, *Mon. Not. R. astr. Soc.* **235**, 1.
Carusi, A., Valsecchi, G.B. and Greenberg, R.: 1990, *Celest. Mech.* **49**, 111.
Delsemme, A.H.: 1973, *Astron. Astrophys.* **29**, 377.
Delsemme, A.H.: 1979, in *Dynamics of the Solar System*, ed. R.L. Duncombe, *Reidel*, pp. 265–271.
Duncan, M., Quinn, T. and Tremaine, S.: 1988, *Astrophys. J.* **328**, L69.
Everhart, E.: 1967, *Astron. J.* **72**, 1002.
Everhart, E.: 1968, *Astron. J.* **73**, 1039.
Everhart, E.: 1972, *Astrophys. Lett* **10**, 131.
Everhart, E.: 1977, in *Comets, Asteroids, Meteorites*, ed. A.H. Delsemme, *Univ. Toledo*, pp. 99–104.
Fernández, J.A.: 1982, *Astron. J.* **87**, 1318.
Fernández, J.A. and Gallardo, T.: 1993, *Astron. Astrophys.* **281**, 911.
Fernández, J.A. and Ip, W.-H.: 1983, in *Asteroids, Comets, Meteors*, eds. C.-I. Lagerkvist et al., *Uppsala Univ.*, pp. 387–390.
Fernández, J.A., Rickman, H. and Kamél, L.: 1992, in *Proc. Int. Workshop on Periodic Comets*, eds. J.A. Fernández and H. Rickman, *U. de la Republica, Montevideo*, pp. 143–157.
Froeschlé, Cl. and Rickman, H.: 1980, *Astron.Astrophys.* **82**,183.
Hughes, D.W.: 1988, *Icarus* **73**, 149.
Joss, P.C.: 1973, *Astron. Astrophys.* **25**, 271.
Kresák, L.: 1981, *Bull. Astron. Inst. Czech.* **32**, 321.
Marsden, B.G. and Williams, G.V.: 1994, *Catalogue of Cometary Orbits*, IAU Central Bureau for Astron. Telegrams.
Matese, J.J. and Whitman, P.G.: 1989, *Icarus* **82**, 389.
Matese, J.J., Whitman, P.G., Innanen, K.A. and Valtonen, M.J.: 1995, *Icarus*, in press.
Oort, J.H.: 1950, *Bull. Astron. Inst. Netherlands* **11**, 91.
Öpik, E.J.: 1951, *Proc. Roy. Irish Acad.* **54A**, 165.
Quinn, T., Tremaine, S. and Duncan, M.: 1990, *Astrophys. J.* **355**, 667.
Stagg, C.R. and Bailey, M.E.: 1989, *Mon. Not. R. astr. Soc.* **241**, 507.
Rickman, H. and Vaghi, S.: 1976, *Astron. Astrophys.* **51**, 327.
Valtonen, M.J., Zheng, J.-Q. and Mikkola, S.: 1992, *Celest. Mech.* **54**, 37.
Weissman, P.R.: 1985, in *Dynamics of Comets: Their Origin and Evolution*, eds. A. Carusi and G.B. Valsecchi, *Reidel*, pp. 87–96.
Wetherill, G.W.: 1991, in *Comets in the Post-Halley Era* Vol. 1, eds. R.L. Newburn, Jr. et al., *Kluwer*, pp. 537–556.
Zheng, J.-Q.: 1994, *Astron. Astrophys. Suppl.* **108**, 253.

POLYNOMIAL APPROXIMATION OF POINCARÉ MAPS FOR HAMILTONIAN SYSTEMS

CLAUDE FROESCHLÉ AND ELENA LEGA

Observatoire de Nice, B.P.229, F-06304 Nice cedex 4, France

Abstract. Different methods are proposed and tested for transforming a nonlinear differential system, and more particularly a hamiltonian one, into a map without having to integrate the whole orbit as in the well known Poincaré map technique. We construct piecewise polynomial maps by coarse-graining the phase surface of section into parallelograms using values of the Poincaré maps at the vertices to define a polynomial approximation within each cell. The numerical experiments are in good agreement with the standard map taken as a model problem. The agreement is better when the number of vertices and the order of the polynomial fit increase. The synthetic mapping obtained is not symplectic even if at vertices there is an exact interpolation. We introduce a second new method based on a global fitting . The polynomials are obtained using at once all the vertices and fitting by least square polynomes but in such a way that the symplectic character is not lost.

1. A local exact fitting

Poincaré maps are now of common use for studying the qualitative behaviour of differential equations (see Hénon 1981). Morover, in order to study stability problems, many authors have sought explicit algebraic mappings which approximate, at least qualitatively, the Poincaré maps obtained from the original Newton equations. Froeschlé and Petit (1990) have reviewed some of these mappings and showed that they are reliable only as long as one remains within the domain of validity of the approximations made in order to isolate either - in the case of deterministic mappings - an integrable part and some instantaneous perturbations, or - for stochastic mappings - a source of endogeneous/exogeneus stochasticity (see Froeschlé & Rickman, 1988). All these mapping are *ad hoc* and reliable only in some region of the phase space and for some specific purpose. In Froeschlé and Petit (1990) we built a mapping valid everywhere in the phase space, following an idea already used by Varosi et al. (1987) but in the framework of non-Hamiltonian systems (i.e. systems where attractors do exist). The method consists of coarse-graining the phase-space surface of section and then interpolating the value of the image points. Linear interpolation requires a rather fine graining of the phase space, hence it is necessary to compute a lot of points on the grid. However, Taylor expansion of order 3 and 5 can provide very good results as long as symmetrical interpolation formulae have been tested, but their accuracy was found to be inferior. Therefore Petit and Froeschlé (1994) have developed another type of interpolation, where the information, including that

Earth, Moon, and Planets **72**: 51-56, 1996.

on the gradients, is stored to the same level of accuracy only for the nearest-neighbouring vertices. Thus, not only images of vertices are computed, but also tangential mapping at each vertex.

There are in any case two key parameters: the number of bins in each direction $N =$(total number of cells)$^{1/D}$, where D is the dimension of the surface of section, and the order M of the Taylor expansion. In order to explore the validity for the synthetic approach we have applied our method for an algebraic area-preserving mapping for which the computation of orbits is very fast. This allows one to follow a large number of orbits and to carry out enough iterations for a meaningful comparison. In this case we have used the well known standard mapping (Froeschlé 1970, Lichtenberg and Lieberman, 1983):

$$\begin{cases} x_{i+1} = x_i + a\sin(x_i + y_i) \\ y_{i+1} = y_i + x_i \qquad (\mod 2\pi) \end{cases} \tag{1}$$

Fig. 1a shows orbits of the standard mapping for $a = -1.3$. Indeed such a mapping exhibits all the well-known typical features of problems with two degrees of freedom, such as invariant curves, "islands", and stochastic zones where the points wander in a chaotic way. Figs. 1b and 1c are magnifications of the small boxes indicated in fig.1a. At this magnification level, details like second-order islands become evident and the approximation levels of the synthetic maps are easily visualized. Figs. 1d, 1e, 1f correspond to the same orbits and the same magnifications as Figs. 1a-c but using the Taylor interpolation mapping of order $M = 5$ (T5) and a number of bins $N = 20$. In order to get a finer description of the non-area preserving behaviour of T5 for an invariant curve, we have computed at regular time intervals (see Laskar et al. 1992) the rotation number using a linear interpolation method (M. Hénon, private communication). We have displayed (Froeschlé & Petit 1993) a contracting or expanding behaviour of T5 depending on the total number of cells and the starting point. This phenomena are a consequence of the non area preserving character of the local mapping. We present in the next section a new method based on a global area preserving mapping.

2. A global fitting: the least square interpolation

In this new approach we look for the mapping $T^\star$

$$T^\star = \begin{cases} x_{i+1} = P_m(x_i, y_i) \\ y_{i+1} = Q_n(y_i, x_i) \end{cases} \tag{2}$$

where P_m and Q_n are polynomials of order m and n which fit the initial Poincaré map T (here the standard mapping) at the best (in the least square meaning) for the images of a regular grid. Let us emphasize that with this

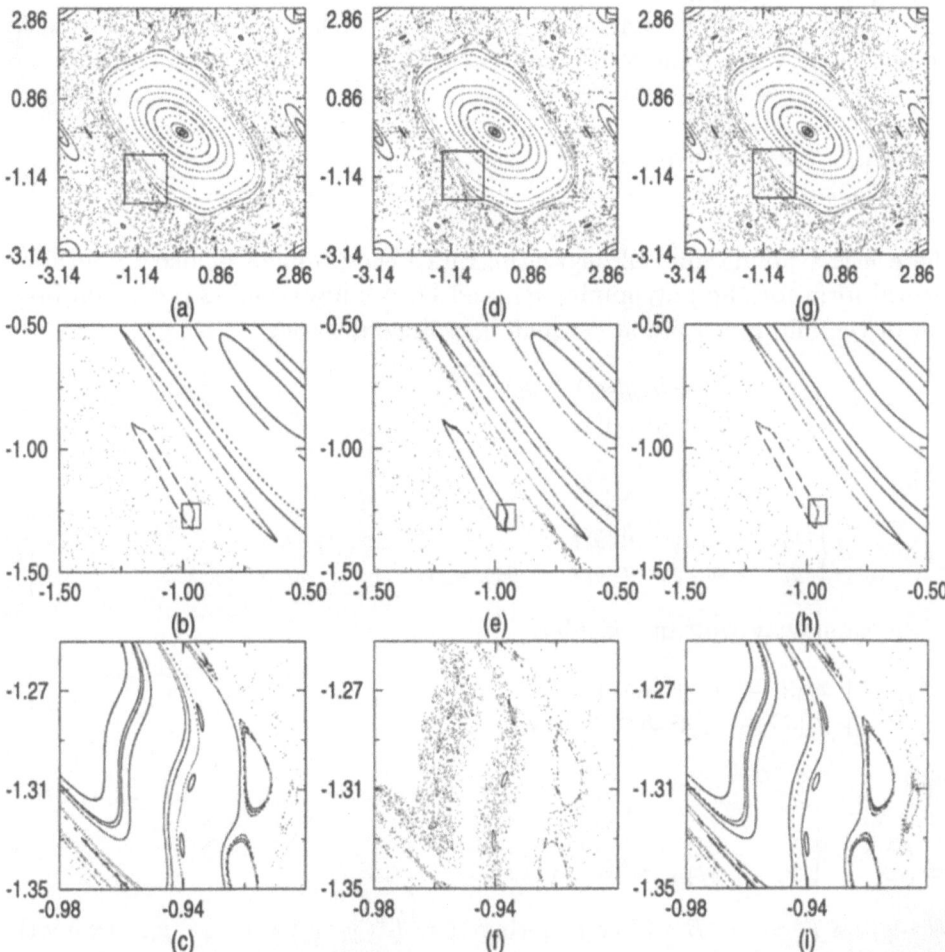

Fig. 1. (a) Plot of the standard map for a=-1.3, (b) and (c) are enlargements of the small boxes shown respectively in (a) and (b); (d), (e) and (f) are the same as (a-b-c) but using a Taylor approximation of order 5. (g), (h) and (i) same as (a-b-c) but using the global method explained in section 2.

approach we lose the exact fitting at the vertices. It is the price we pay in order to obtain a symplectic mapping. In the two-dimensional case of $T^\star$ this means an area preserving mapping i.e. the determinant of the jacobian matrix is one. The problem of getting such polynomial expression is not a trivial one and we used as a basic synthetic mapping the following one:

$$\begin{cases} x_{i+1} = x_i \\ y_{i+1} = y_i + g(x_i) \end{cases} \tag{3}$$

where g is a polynomial of degree n obtained from a generating function by $g(x) = \frac{-\partial S}{\partial x}$. This is a shear along the y axis. Such a mapping is obviously symplectic as well as the following one:

$$\begin{cases} x_{i+1} = x_i + h(y_i) \\ y_{i+1} = y_i \end{cases} \tag{4}$$

where h is a polynomial of degree m. Since the problem is also to obtain a general form for the polynomial P_m and Q_n we have considered a composition of the mappings 3 and 4 which is still symplectic:

$$A = \begin{cases} x_{i+1} = x_i + h(g(x_i) + y_i) \\ y_{i+1} = y_i + g(x_i) \end{cases} \tag{5}$$

with

$$A^{-1} = \begin{cases} x_i = x_{i+1} - h(y_{i+1}) \\ y_i = y_{i+1} - g(x_{i+1} - h(y_{i+1})) \end{cases} \tag{6}$$

In the same way we can consider:

$$B = \begin{cases} x_{i+1} = x_i + \alpha(y_i) \\ y_{i+1} = y_i + \beta(x_i + \alpha(y_i)) \end{cases} \tag{7}$$

with

$$B^{-1} = \begin{cases} x_i = x_{i+1} - \alpha(y_{i+1} - \beta(x_{i+1})) \\ y_i = y_{i+1} - \beta(x_{i+1}) \end{cases} \tag{8}$$

We define $T^\star = A \circ B$ and considering that $T^\star(x_i, y_i) = (x_{i+1}, y_{i+1})$ we write $B(x_i, y_i) = A^{-1}(x_{i+1}, y_{i+1})$ in order to obtain the coefficients of the polynomes g, h, α, β using the least square fitting. Using this trick the computation is straight and does not require any algebric manipulation. Of course we can also use $T^\star = B \circ A$ which will give different values for the coefficients. Fig.2 shows the variation of χ^2 (which measures the goodness of the fit) as a function of the polynomial degree (m and n) and of the number of fitted points ($N \times N$). We notice that the precision does not improve for $N \geq 20$ and we get minimum values respectively for $m = 13$ for the couple h, α and $n = 1$ for the couple g, β. The results on the mapping are shown in fig1 (third column). Let us remark the very good agreement of fig.1c with fig.1i. In order to obtain the same result with the local fitting it is necessary to increase of a factor 4 the total number of points $N \times N$. We have therefore to pay a higher price on computation time in comparaison with the global method. Despite its global character, the global method ($T^\star$) ensures a better fit because of its area preserving properties.

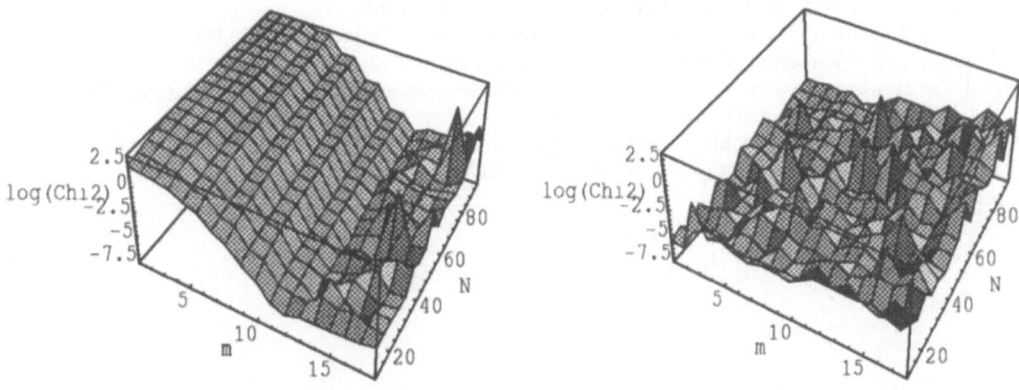

Fig. 2. Variation of χ^2 as a function of the polynomial degree m and of the number of fitted points $N \times N$. Left figure corresponds to the polynomial P, right figure corresponds to the polynomial Q.

3. Conclusion

Synthetic maps appear to be valuable tools for celestial mechanics. We have presented here only some partial results. For instance another important development concerns problems with more than two degrees of freedom, for which the situation is less straightforward than described above. As far as the first method is concerned the number of operations required for the Taylor approximation increases drastically with the dimension of the surface of section. Of course a lower-order map can be used by decreasing the grid size, but a further difficulty lies in the task of storing and recalling the values of the computed images at the vertices. This is the reason why we have used a hash function when dealing with problems with three degree of freedom (Petit and Froeschlé 1994). The restauration of simplecticity for P and Q appears to be promising since for the same accuracy we obtain at once a general fitting.

Acknowledgements

We thank J.M. Petit for his help on computations.

References

C. Froeschlé. *Astron.Astrphys.*, 9:15, 1970.
C. Froeschlé and J.M. Petit. *Astron.Astrphys.*, 238:413, 1990.
C. Froeschlé and J.M. Petit. *Primo convegno nazionale di Meccanica Celeste*. l'Aquila, Italy, 1993.

C. Froeschlé and H Rickman. *Celestial Mech.*, 43:265, 1988.

M. Henon. *Cours des Houches XXXVI*. Noth Holland, Amsterdam 57, 1981.

J. Laskar, C. Froeschlé, and A. Celletti. *Physica D*, 56:253, 1992.

A.J. Lichtenberg and M.A. Lieberman. *Regular and Stochastic motion*. Springer, Berlin, Heidelberg, New York, 1983.

J.M. Petit and C. Froeschlé. *Astron.Astrphys.*, 282:291, 1994.

F. Varosi, C. Grebogi, and J.A. Yorke. *Phys. Lett. A*, 124:59, 1987.

THE PROVENANCE AND EVOLUTION OF COMETS

M.E. BAILEY

School of Computing and Mathematical Sciences, Liverpool John Moores University, Byrom Street, Liverpool L3 3AF, United Kingdom

Abstract. The process of comet formation through the hierarchical aggregation of originally submicron-sized interstellar grains to form micron-sized particles and then larger bodies in the protoplanetary disc, culminating in the formation of planetesimals in the disc extending from Jupiter to beyond Neptune, is briefly reviewed. The 'planetesimal' theory for the origin of comets implies the existence of distinct cometary reservoirs, with implications for the immediate provenance of observed comets (both long-period and short-period) and their evolution as a result of planetary perturbations and physical decay, for example splitting and sublimation. The principal mode of cometary decay and collisional interaction with the terrestrial planets is through the formation and evolution of streams of cometary debris and hitherto undiscovered 'families' of cometary asteroids. Recent dynamical results, in particular the sungrazing and sun-colliding end-state for short-period comet and asteroid orbits, are briefly discussed.

Key words: Comets – interstellar dust – dynamics – solar neutrino problem

1. Introduction

Cometary nuclei contain the most primitive material accessible to direct observations and in situ studies in the solar system. The dust component arises primarily in stellar winds and other outflow phenomena associated with the late stages of stellar evolution, and is incorporated following evolution in the interstellar medium, into cometary building blocks during the earliest stages of accretion of the sun and planetary system. The processes are as yet only poorly understood, and no fully consistent theory exists which explains both the origin of comets and the detailed formation of the solar system. Here we outline the main steps and present results on the dynamical evolution of comets with particular emphasis on the interactions of comets and cometary debris with the planets.

2. Grain Growth

Several distinct phases in the likely presolar evolution of cometary dust grains have been reviewed by Bailey (1994). Briefly, interstellar grains, each carrying isotopic signatures of their formation sites and processing in the interstellar medium (Clayton 1982, Greenberg 1982, Greenberg & Hage 1990), coagulate in dense interstellar clouds to produce loosely bound aggregates with radii ranging up to several microns (e.g. Bailey 1987, 1991). These

Earth, Moon, and Planets **72**: 57–68, 1996.

'giant' interstellar grains, which remain very cold (Mathis 1994), have a porous structure (Daniels & Hughes 1981, Donn & Hughes 1986, Greenberg 1986, Donn 1991) likely to be filled with interstellar ices produced by condensation of molecules from the surrounding cloud. Grain surface chemistry, accretion of molecules from the gas phase, and effects due to irradiation in the interstellar radiation field produce complex organic molecules both on and within these porous grains (cf. Greenberg 1988, Clayton et al. 1989), while grain growth is expected to continue throughout the collapse phase of the protosolar nebula. By the time the cloud has evolved to the point of forming a flattened, disc-like structure the collapsing nebula should already contain a significant number of fundamentally 'interstellar' dust-and-ice particles with sizes in the approximate range $1-100\,\mu m$, while the protoplanetary disc will subsequently generate even larger particles but with internal structures and chemistry reflecting the prior history and evolution of the grains in the interstellar medium. Their diverse isotopic composition provides information on the original sites of dust formation in the Galaxy (Ott 1993, Amari et al. 1993).

Four main processes contribute to the production of comets in the protoplanetary disc: grain-grain collisions in the presence of gas; further grain-grain collisions in the absence of gas (or involving particles too big to be coupled to the gas through turbulence); gravitational instability leading to 'planetesimals' of much larger size; and planetesimal-planetesimal collisions that produce still larger particles and eventually bodies of planetary size.

These processes occur at a rate which decreases with increasing heliocentric distance. We therefore expect bodies such as comets and planetesimals, and their respective building blocks, to differ significantly in respect of factors such as size distribution, chemical composition, dust-to-ice ratio etc., depending on where in the nebula they formed. One of the goals of modelling is to quantify such differences, and to identify tests that will allow theories of cometary origin to be discriminated from one another. At this time there is most interest in identifying physical and chemical differences between comets with different dynamical characteristics, but since the orbits — and presumably the long-term orbital evolution — are generally chaotic we should expect to see evidence of comets from more than one source, even within an apparently clean dynamical group.

Following Bailey (1994), a straightforward interpretation of the variation of the rate of cometary accumulation versus heliocentric distance leads to the following picture. In the Jupiter-Saturn zone ($r \sim 5-10\,\mathrm{AU}$) the nebular density is sufficiently high that grain-grain collisions driven by turbulence in the gas produce bodies with sizes in the range 30–300 m on a timescale $\sim 1-10\,\mathrm{Myr}$, comparable with that expected for survival of a gaseous protoplanetary disc, while the largest particles may decouple from the turbulence and form a thin, dense dust disc close to the equatorial plane (Weidenschilling 1994). Such 'boulders' or Tunguska-sized cometary building blocks are hierarchical accumulations of interplanetary grains, themselves comprising micron-sized aggregates of the original submicron-sized, ice-covered interstellar grains which accreted during the presolar phases of nebular evolution.

Further out, in the Uranus-Neptune zone ($r \sim 20$–$30\,\mathrm{AU}$), grain growth in the presence of gas proceeds more slowly, leading to 'boulders' with sizes ranging only up to $\sim 10\,\mathrm{m}$. At still larger distances, in the protoplanetary disc beyond Neptune, grain-grain coagulation proceeds so slowly that it is difficult to see how particles much bigger than ordinary snowballs could be produced, i.e. aggregates in this region should have dimensions $\lesssim 1\,\mathrm{m}$ (cf. Kuiper 1951).

The details of how the nebular gas is removed are not well understood, but observational studies of young stellar systems suggest that in most cases this happens within $\sim 10\,\mathrm{Myr}$ (Strom et al. 1989, Beckwith et al. 1990, Gahm et al. 1994). Although Jupiter and Saturn must certainly have formed by this time, there is also evidence (Lissauer 1993) that even Uranus and Neptune had substantially formed on a similar timescale; how this happens so fast (cf. Greenberg et al. 1984, Greenberg 1985, 1989) remains an interesting difficulty for the standard planetesimal theory.

Leaving this problem aside, grain-grain coagulation in the absence of gas leads to damping of the particle velocities and to formation of a dense, dynamically cold dust disc. Such a structure is subject to local gravitationally instability, leading to 'planetesimals' with a size that, at least initially, is only weakly dependent on heliocentric distance r, varying roughly as $r^{2-\alpha}$ for a disc with surface density $\Sigma \propto r^{-\alpha}$. A typical model (e.g. Weidenschilling 1977, Bailey 1994) has $\alpha \sim 3/2$. As shown by Yamamoto & Kozasa (1988) and Yamamoto et al. (1994), the predicted radii of these bodies are in remarkably good agreement with those of the largest comets (e.g. Chiron) and the observed objects in the Kuiper belt (Jewitt & Luu 1995). The theory predicts the formation of bodies with diameters on the order of 200 km beyond Neptune, and since in this region the planetesimal-planetesimal collision timescale is greater than the age of the solar system, this should be close to their final size. We emphasize that in this region such bodies should be composed of relatively small ($\lesssim 1\,\mathrm{m}$) accumulations of micron-sized aggregates of protoplanetary solids, themselves comprising even smaller interstellar dusts and ices.

The evolution of planetesimals in the inner disc is more complex. First, the bodies produced by gravitational instability are smaller than those further out (assuming $\alpha \simeq 3/2$), for example ~ 50–$100\,\mathrm{km}$ in the Jupiter-Saturn zone and ~ 100–$200\,\mathrm{km}$ in the Uranus-Neptune zone, and composed of larger building blocks the smaller the heliocentric distance ($\sim 10^2$–$10^3\,\mathrm{m}$ in the Jupiter-Saturn zone, 10–$10^2\,\mathrm{m}$ in the Uranus-Neptune zone). Secondly, the

planetesimal-planetesimal collision timescale is relatively short, varying roughly as $r^{7/2}$ (Yamamoto & Kozasa 1988), so further coagulation should produce
much larger particles.

The evolution of planetesimals in the absence of gas thus leads to a further
level in the complexity of cometary accretion, and may even produce bodies
with sizes up to $\sim 10^3$ km — with obvious implications for physical and
thermal processing of the original building blocks. In fact, internal heating
may be significant for bodies as small as several tens of kilometres in radius
(Yabushita 1993a,b). Further evidence that the early solar system must have
contained Pluto-sized objects has been reviewed by Stern (1991). Despite the
final stages of their accumulation occurring in the absence of gas, such bodies
would evidently be 'cometary' in form when heated, and by the same token
such 'comets' could be scattered outwards to produce the observed Oort
cloud by gravitational interactions with the growing planets (e.g. Fernández
1985, Shoemaker & Wolfe 1984, Duncan et al. 1987, Bailey et al. 1990).

An alternative, possibly more consistent scenario, is that owing to the
prior formation of the outer planets local gravitational instability never
occurred. In this case the dust-and-ice particles left behind at the time of
nebular dissipation continued to coagulate, but now in a gas-free environment dominated by planetary perturbations (not unlike the present solar
system), until the process was terminated either by collisional fragmentation or dynamical ejection to the Oort cloud (Greenberg 1985). In either case
the protoplanetary disc ($r \sim 5$–30 AU) would produce cometary objects with
a broad size distribution ranging up to the largest bodies in the Kuiper belt,
possibly including bodies with sizes $\sim 10^3$ km or more. Such bodies would be
composed of bigger building blocks the closer to the sun they formed.

In summary, this picture for comet formation makes the following predictions. On the smallest scales, comets should comprise aggregates of ice-
covered submicron-sized interstellar grains, with sizes ranging from ~ 1–
10^2 μm and containing ices representative of the evolution and collapse of
the molecular cloud that led to the formation of the sun and protoplanetary disc. Leaving aside physical and thermal alteration, the comets which
accreted in the Jupiter-Saturn zone should contain building blocks with
sizes ranging from ~ 30–300 m, and a variety of interplanetary dusts and
ices themselves containing particles of fundamentally interstellar composition. Comets formed further out comprise smaller building blocks, which
may also be identified as having coagulated in a gaseous environment by
the intimate contact between interplanetary dust and ices. In the outer
solar system beyond ~ 50 AU these aggregates of interplanetary grains are
finally brought together by gravitational instability which leads to bodies
$\sim 10^2$ km in radius. This stage of accretion proceeds in the absence of gas
and so produces bodies with a low mean density owing to voids within the
loosely bound rubble pile. At intermediate heliocentric distances the pic-

ture depends on whether Uranus and Neptune are produced before, or as a result of, the formation of planetesimals. If the former, then coagulation of boulder-sized particles, driven by planetary perturbations, continues to its natural limit: ejection of comets to the Oort cloud and interstellar space, or fragmentation of particles by collisions. Alternatively, gravitational instability sets a new scale to the bodies produced, smaller at smaller heliocentric distances for $\alpha < 2$, and the evolution leading to observed planets proceeds by accumulation of planetesimals. Detailed predictions need to consider a variety of physical and thermal effects on the comets produced by these processes, for example the effects of comet-comet collisions and impacts by meteoroids, tidal disruption by close planetary encounters, and heating by both internal and external processes.

3. Dynamics

Following this picture, the comets formed in the region from 5–30 AU are scattered to the Oort cloud (both the outer cloud and inner core) and interstellar space. The fate of bodies formed beyond the planetary system is less certain, but most (at least those with initial orbits $\lesssim 500$ AU) should remain close to their initial sites of formation for at least the age of the solar system (Stern 1990). In general, substantial mixing of the orbits is the rule: for example, comets could be handed down to Jupiter-family orbits before eventual ejection or placement in the Oort cloud, or bodies formed near Jupiter could be perturbed into longer period orbits with perihelia close to Saturn or beyond, to be subsequently ejected to the Oort cloud together with particles directly formed in the Uranus-Neptune region. It seems likely, therefore, that both the inner core and outer Oort cloud should contain different classes of comet, bearing signatures of formation representing different parts of the protoplanetary disc. Similarly, the cometary mass function should include the largest planetesimals in the protoplanetary disc. Although the existence of 'Plutos' in the Oort cloud remains speculative, several arguments (e.g. Hughes 1990, Bailey et al. 1994a) suggest that such bodies may make a significant contribution to the total 'cometary' mass — and by the same token to the distribution of cometary objects in interstellar space. However, following the planetesimal theory the size of the bodies in the Kuiper belt should range only up to that of the first-formed planetesimals (Yamamoto et al. 1994).

It is not possible, in this brief review, to cover all the new results in cometary dynamics (cf. Bailey et al. 1994a, Bailey 1995), but it is worth emphasizing the sungrazing phenomenon. This has important implications for the interpretation of solar abundances, since the outer layers of the sun must have been contaminated by the accretion of a large mass of comets

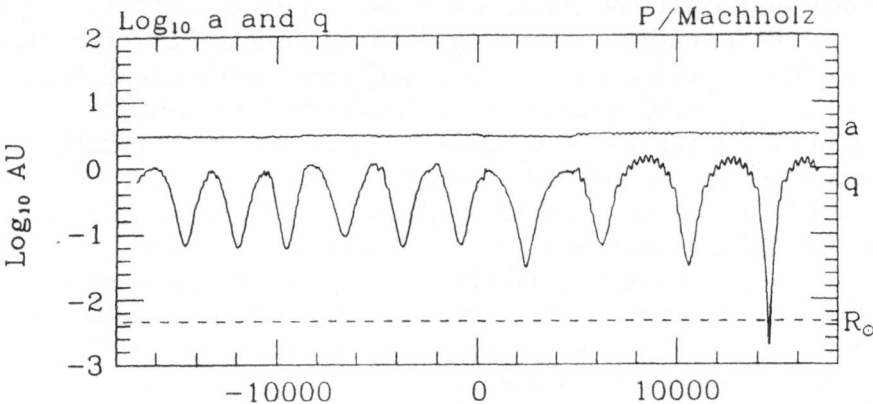

Fig. 1.   Semi-major axis and perihelion distance of P/Machholz, showing a collision with the sun occurring within ∼12,000 yr. After Bailey et al. (1992).

or planetesimals during the early evolution of the solar system, leading to systematic differences between the composition of the solar core and its convective envelope. As discussed by Joss (1974), and more recently by Levy & Ruzmaikina (1994), chemically inhomogeneous solar models provide a possible solution to the solar neutrino problem, while the accretion of comets and planetesimals (e.g. during and after the late heavy bombardment of the inner planets) due to the evolution of objects on to sun-colliding orbits provides a mechanism to contaminate the outer envelope.

Examples of the sungrazing phenomenon are found among virtually every type of cometary orbit: Halley-types (e.g. P/Machholz, P/Halley and Damocles; Bailey et al. 1992, Asher et al. 1994a, Bailey et al. 1994a), Jupiter-family (e.g. P/Encke; Levison & Duncan 1994, Bailey 1995), and asteroids on orbits close to those of the Taurid Complex or close to mean-motion resonances (Valsecchi et al. 1994, Farinella et al. 1994, Yoshikawa 1992). Figures 1 and 2 show examples of the general phenomenon, while a further instance (Figure 3), this time with the effects of the sun and Jupiter interchanged, is provided by the collision with Jupiter of P/Shoemaker-Levy 9. As discussed by Bailey et al. (1994b), the effects of secular perturbations lead to an enhancement in the rate of cometary collisions with Jupiter by about an order of magnitude compared with the prediction of a random collision model.

The surprisingly widespread nature of the sungrazing phenomenon has several implications in addition to that for the solar composition. The rate of cometary collisions with the sun is increased by a factor $\gtrsim 10$ (Bailey et al. 1992), but close solar passages may also break comets apart, especially the largest objects that dominate the cometary mass. Following the theory of the P/Shoemaker-Levy 9 trail (Scotti & Melosh 1994), cometary break-up

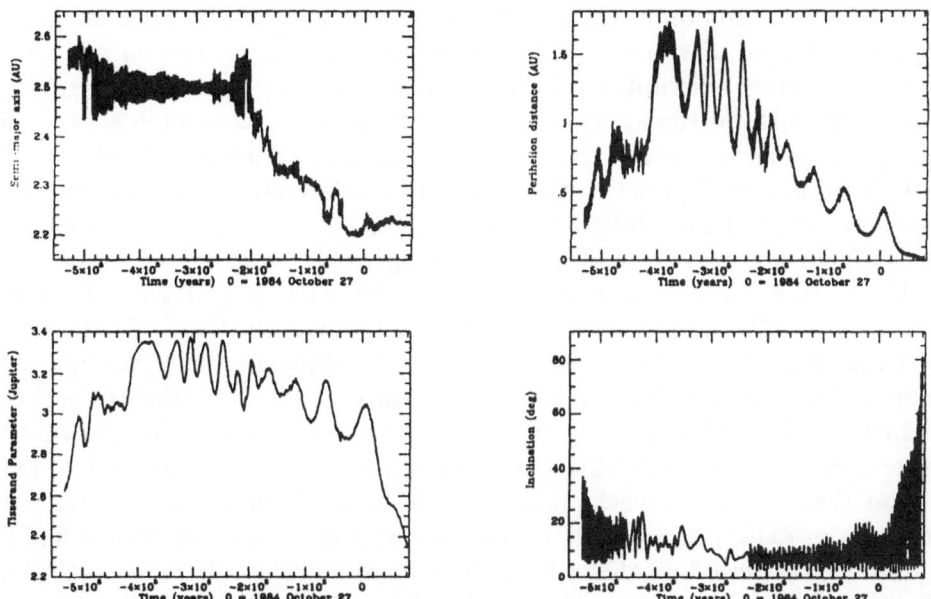

Fig. 2. Semi-major axis, perihelion distance, Tisserand parameter and inclination of a test particle with initial elements similar to those of P/Encke. The plot shows the evolution from -0.5 Myr to $+0.1$ Myr and illustrates large changes in perihelion distance driven by secular perturbations. This particle collides with the sun $\sim$80,000 yr in the future.

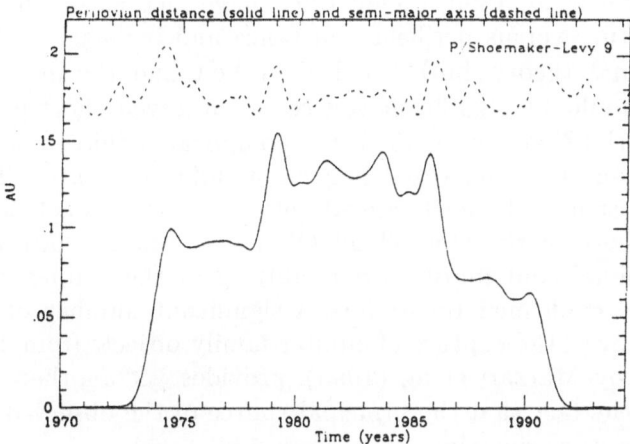

Fig. 3. Jovicentric semi-major axis and perijovian distance of P/Shoemaker-Levy 9. Secular perturbations similar to those affecting P/Machholz (though with the sun and Jupiter interchanged) cause this object periodically to come close to Jupiter, culminating in the collision of 1994. After Bailey et al. (1994b)

during tidal disruption will produce streams of debris moving in orbits of similar perihelion distance and orientation but different orbital periods, the range being a function of the size of the nucleus and the distance of closest approach during disruption. For an orbit of semi-major axis a_0 and perihelion distance q, disruption of a cometary nucleus of diameter d produces debris with orbital semi-major axes a in the range $a = a_0/(1 \pm a_0 d/q^2)$. If $a_0 d/q^2 \ll 1$, as is usually the case (e.g. ordinary short-period orbits such as P/Machholz or P/Encke), this produces a relatively short trail of debris similar to that of the observed P/Shoemaker-Levy 9 trail, with a length L at the following aphelion given by $L \simeq 4da_0^2/q^2$.

For a 10-km diameter nucleus in an orbit with $q \simeq 1.5R_\odot$ (for tidal disruption) and a semi-major axis $a \simeq 3\,\mathrm{AU}$ (similar to that of P/Machholz or P/Encke), the formula gives $L \simeq 0.05\,\mathrm{AU}$, implying the formation of a short trail at the aphelion immediately following tidal disruption, orientated almost perpendicular to the true orbits of its particles about the sun. On the other hand, if the tidally disrupted comet were large and the orbit of long period then the same mechanism provides an explanation for the range of semi-major axes of the comets in the Kreutz group. If the progenitor had an initial orbital period on the order of 800 yr and a nucleus diameter ~100 km, then the debris stream should contain orbits with periods ranging upwards from about 250 yr, roughly consistent with the observed spread of orbital periods in the Kreutz group.

An outstanding problem in cometary dynamics remains the origin of the Jupiter family. Previous discussions (e.g. Duncan et al. 1988, Stagg & Bailey 1989, Quinn et al. 1990, Bailey 1992) have focused on the efficiency of cometary capture versus perihelion distance and inclination. According to the planetesimal theory, both the Kuiper belt and the inner core of the Oort cloud should be significant sources of observed short-period comets, but recent work (Zheng et al. 1995) has complicated this picture by emphasizing that even the outer cloud may be a sufficient source. The observed number of short-period comets already places a strong constraint on models of the inner core of the Oort cloud (Bailey & Stagg 1990), strengthened by any additional contribution from splitting or the Kuiper belt, or if the outer cloud is confirmed to produce a significant number of short-period comets. Moreover, the capture of Jupiter-family objects from Trojan orbits, as suggested by Marzari et al. (1995), provides yet another complication, emphasizing the fact that the principal source of the observed Jupiter family remains an open question.

In summary, our understanding of cometary dynamics is undergoing rapid progress, with increasing attention paid to the issues of secular perturbations, cometary break-up and fragmentation, and further sources of short-period comets in addition to the observed near-parabolic flux. The existence of such sources emphasizes the connection between theories of comet formation and the origin and evolution of the solar system, while the disintegration of comets into meteoroid streams, and their break-up into tidal trails, strongly suggests the existence of streams of cometary material in the solar system in a form which is difficult to detect using present techniques. As described by Bailey et al. (1994a), the presence of such streams has potentially important implications for understanding the short-term variations, on timescales of decades to centuries, observed in the rate of accretion of cometary debris by the earth.

4. Implications

The hierarchical accretion of originally ice-covered interstellar dust grains provides a plausible picture for the formation of comets as a by-product of steps leading to planet formation. The theory predicts several distinct reservoirs for observed comets, including the outer Oort cloud, the inner Oort cloud, the trans-Neptunian cometary disc and Kuiper belt. Comets are predicted to contain (i) clusters of presolar, ice-covered interstellar grains ranging in size up to about $1\,\mu$m; (ii) aggregates of such clusters in the form of 'snowballs' or 'boulders' with sizes $\sim 10^{-2}$–10^2 m depending on heliocentric distance; (iii) larger accumulations of such bodies formed by gravitational instability in the dust disc, with sizes in the range ~ 10–100 km (again a function of heliocentric distance); and (iv) aggregates of these particles or 'planetesimals' ranging in size up to the largest protoplanetary objects. Beyond the observed planetary system, in the region of the Kuiper belt, the comets have sizes limited to those of the first-formed planetesimals.

The orbits of comets, and indeed all planet-crossing bodies, are strongly chaotic, so the comets in each of the predicted cometary reservoirs should be relatively well-mixed in terms of their initial formation sites.

Secular perturbations and secular resonances lead to large changes in the perihelion distances of short-period comets and asteroids, frequently allowing evolution into a sungrazing or sometimes Jupiter-grazing trajectory. Cometary disruption whilst in such orbits is a significant cometary end-state, producing streams of cometary debris in the inner solar system and raising the possibility that observed comets may themselves be the result of the break-up or disruption of larger progenitors. The Jupiter family, at least in part, may be enhanced by cometary splitting (Rickman 1990, Pittich & Rickman 1994) or by the break-up of an exceptionally large progenitor (Clube & Napier 1984, Clube 1992), while the Kreutz group provides a concrete example of the results of a tidal disruption.

In summary, comets and their disintegration products bear a complex interrelationship to the planets. They represent a significant time-dependent hazard to civilization as a result of the concentration of cometary debris in streams (e.g. Clube & Napier 1984, 1986; Asher & Clube 1993, Asher et al. 1994b, Bailey et al. 1994a), while on a theoretical level, the questions how to discriminate between different theories of cometary origin, and whether the different types of cometary dust might in principle constrain theories of the origin of the early solar system remain to be addressed.

Acknowledgements

I thank the organizers of the meeting 'Small bodies in the Solar System and their Interactions with the Planets' for the invitation to present this review, and colleagues particularly V. Emel'yanenko and J. Scotti for discussions.

References

Amari, S., Hoppe, P., Zinner, E., Lewis, R.S., 1993. *The isotopic compositions and stellar sources of meteoritic graphite grains.* Nature, **365**, 806–809.

Asher, D.J., Clube, S.V.M., 1993. *An extraterrestrial influence during the current glacial-interglacial.* Q. Jl. Roy. Astron. Soc., **34**, 481–511.

Asher, D.J., Bailey, M.E., Hahn, G., Steel, D.I., 1994a. *Asteroid 5335 Damocles and its implications for cometary dynamics.* Mon. Not. Roy. Astron. Soc., **267**, 26–42.

Asher, D.J., Clube, S.V.M., Napier, W.M., Steel, D.J., 1994b. *Coherent catastrophism.* Vistas in Astronomy, **38**, 1–27.

Bailey, M.E., 1987. *Giant grains around protostars.* Q. Jl. Roy. Astron. Soc., **28**, 242–247.

Bailey, M.E., 1991. *Comets and molecular clouds: the sink and the source.* In: Molecular Clouds, eds. James, R.A., Millar, T.J., 273–289. Cambridge University Press.

Bailey, M.E., 1992. *Origin of short-period comets.* Cel. Mech. Dyn. Astron., **54**, 49–61.

Bailey, M.E., 1994. *Formation of outer solar system bodies: comets and planetesimals.* In: Asteroids, Comets, Meteors 1993, eds. Milani, A., Di Martino, M., Cellino, A., IAU Symp. No. 160, 443–459. Kluwer, Dordrecht, The Netherlands.

Bailey, M.E., 1995. *Recent results in cometary astronomy: implications for the ancient sky.* Vistas in Astronomy, in press.

Bailey, M.E., Stagg, C.R., 1990. *The origin of short-period comets.* Icarus, **86**, 2–8.

Bailey, M.E., Clube, S.V.M., Napier, W.M., 1990. *The Origin of Comets.* Pergamon Press, Oxford.

Bailey, M.E., Chambers, J.E., Hahn, G., 1992. *Origin of sungrazers: a frequent cometary end-state.* Astron. Astrophys., **257**, 315–322.

Bailey, M.E., Clube, S.V.M., Hahn, G., Napier. W.M., Valsecchi, G.B., 1994a. *Hazards due to giant comets: climate and short-term catastrophism.* In: Hazards due to Comets and Asteroids, eds. Gehrels, T., Matthews, M.S., 479–533. University of Arizona Press, Tucson.

Bailey, M.E., Emel'yanenko, V.V., Scotti, J.V., 1994b. *Origin and dynamical evolution of Comet P/Shoemaker-Levy 9.* Unpublished work, submitted to Nature.

Beckwith, S.V.W., Sargent, A.I., 1993. *The occurrence and properties of disks around young stars.* In: Protostars and Planets III, eds. Levy, E.H., Lunine, J.I., 521–541. University of Arizona Press, Tucson.

Clayton, D.D., 1982. *Cosmic chemical memory: a new astronomy.* Q. Jl. Roy. Astron. Soc., **23**, 174–212.

Clayton, D.D., Scowen, P., Liffman, K., 1989. *Age structure of refractory interstellar dust and isotopic consequences.* Astrophys. J., **346**, 531–538.

Clube, S.V.M., 1992. *The fundamental role of giant comets in Earth history.* Cel. Mech. Dyn. Astron., **54**, 179–193.

Clube, S.V.M., Napier, W.M., 1984. *The microstructure of terrestrial catastrophism.* Mon. Not. Roy. Astron. Soc., **211**, 953–968.

Clube, S.V.M., Napier, W.M., 1986. *Mankind's future: an astronomical view. Comets, ice ages and catastrophes.* Interdisciplinary Sci. Rev., **11** (No. 3), 236–247.

Daniels, P.A., Hughes, D.W., 1981. *The accretion of comet dust — a computer simulation.* Mon. Not. Roy. Astron. Soc., **195**, 1001–1009.

Donn, B., 1991. *The accumulation and structure of comets.* In: Comets in the Post-Halley Era, eds. Newburn Jr., R.L., Neugebauer, M., Rahe, J., IAU Coll. No. 116, Vol. 1, 335–359. Kluwer, Dordrecht, The Netherlands.

Donn, B., Hughes, D.W., 1986. *A fractal model of a cometary nucleus formed by random accretion.* In: 20th ESLAB Symposium on the Exploration of Halley's Comet, eds. Battrick, B., Rolfe, E.J., Reinhard, R., ESA SP-250, Vol. III, 523–524. ESA Publications, ESTEC, Noordwijk, The Netherlands.

Duncan, M., Quinn, T., Tremaine, S.D., 1987. *The formation and extent of the solar system comet cloud.* Astron. J., **94**, 1330–1338.

Duncan, M., Quinn, T., Tremaine, S.D., 1988. *The origin of short-period comets.* Astrophys. J. Lett., **328**, L69–L73.

Farinella, P., Froeschlé, C., Froeschlé, C., Gonczi, R., Hahn, G., Morbidelli, A., Valsecchi, G.B., 1994. *Asteroids falling into the sun.* Nature, **371**, 314–317.

Fernández, J.A., 1985. *The formation and dynamical survival of the comet cloud.* In: Dynamics of Comets: Their Origin and Evolution, eds. Carusi, A., Valsecchi, G.B., IAU Coll. No. 83, 45–70. Reidel, Dordrecht, The Netherlands.

Gahm, G.F., Zinnecker, H., Pallavicini, R., Pasquini, L., 1994. *A search for cold dust around post-T Tauri candidates.* Astron. Astrophys., **282**, 123–126.

Greenberg, J.M., 1982. *What are comets made of? A model based on interstellar dust.* In: Comets, ed. Wilkening, L., IAU Coll. No. 61, 131–163. University of Arizona Press, Tucson.

Greenberg, J.M., 1986. *Fluffy comets.* In: Asteroids Comets Meteors II, eds. Lagerkvist, C.-I., Lindblad, B.A., Lundstedt, H., Rickman, H., 221–223. Reprocentralen HSC, Uppsala.

Greenberg, J.M., 1988. *The interstellar dust model of comets: post Halley.* In: Dust in the Universe, eds. Bailey, M.E., Williams, D.A., 121–143. Cambridge University Press.

Greenberg, J.M., Hage, J.I., 1990. *From interstellar dust to comets: a unification of observational constraints.* Astrophys. J., **361**, 260–274.

Greenberg, R., 1985. *The origin of comets among the accreting outer planets.* In: Dynamics of Comets: Their Origin and Evolution, eds. Carusi, A., Valsecchi, G.B., IAU Coll. No. 83, 3–10. Reidel, Dordrecht, The Netherlands.

Greenberg, R., 1989. *Planetary accretion.* In: Origin and Evolution of Planetary and Satellite Atmospheres, eds. Atreya, S.K., Pollack, J.B., Matthews, M.S., 137–164. University of Arizona Press, Tucson.

Greenberg, R., Weidenschilling, S.J., Chapman, C.R., Davis, D.R.; 1984. *From icy planetesimals to outer planets and comets.* Icarus, **59**, 87–113.

Hughes, D.W., 1990. *Cometary absolute magnitudes, their significance and distribution.* In: Asteroids Comets Meteors III, eds. Lagerkvist, C.-I., Rickman, H., Lindblad, B.A., Lindgren, M., 327–342. Reprocentralen HSC, Uppsala.

Jewitt, D.C., Luu, J.X., 1995. *The solar system beyond Neptune.* Astron. J., in press.

Joss, P.C., 1974. *Are stellar surface heavy-element abundances systematically enhanced?* Astrophys. J., **191**, 771–774.

Kuiper, G.P., 1951. *On the origin of the solar system.* In: Astrophysics, eds. Hynek, J.A., 357–424. McGraw-Hill, New York.

Levison, H.F., Duncan, M.J., 1994. *The long-term dynamical behavior of short-period comets.* Icarus, **108**, 18–36.

Levy, E.H., Ruzmaikina, T.V., 1994. *The possibility of forming an inhomogeneous sun and the solar neutrino deficit.* Astrophys. J., in press.

Lissauer, J.J., 1993. *Planet formation.* Annu. Rev. Astron. Astrophys., **31**, 129–174.

Marzari, F., Farinella, P., Vanzani, V., 1995. *Are Trojan collisional families a source for short-period comets?* Astron. Astrophys., in press.

Mathis, J., 1994. *Dusting off the calculations*. Nature, **372**, 225–226.

Ott, U., 1993. *Interstellar grains in meteorites*. Nature, **364**, 25–32.

Pittich, E.M., Rickman, H., 1994. *Cometary splitting — a source for the Jupiter family?* Astron. Astrophys., **281**, 579–587.

Quinn, T., Tremaine, S., Duncan, M., 1990. *Planetary perturbations and the origin of short-period comets*. Astrophys. J., **355**, 667–679.

Rickman, H., 1990. *Origin and evolution of the Jupiter family*. In: Nordic-Baltic Astronomy Meeting, eds. Lagerkvist, C.-I., Kiselman, D., Lindgren, M., 257–273. Reprocentralen HSC, Uppsala.

Scotti, J.V., Melosh, H.J., 1994. *Tidal breakup and disruption of P/Shoemaker-Levy 9: estimate of progenitor size*. Nature, **365**, 733–735.

Shoemaker, E.M., Wolfe, R.F., 1984. *Evolution of the Uranus-Neptune planetesimal swarm*. Lunar Planet. Sci. Conf., **XV**, 780–781.

Stagg, C.R., Bailey, M.E., 1989. *Stochastic capture of short-period comets*. Mon. Not. Roy. Astron. Soc., **241**, 507–541, and Microfiche MN 241/1.

Stern, S.A., 1990. *External perturbations on distant planetary orbits and objects in the Kuiper disk*. Cel. Mech. Dyn. Astron., **47**, 267–273.

Stern, S.A., 1991. *On the number of planets in the outer solar system: evidence of a substantial population of 1000-km bodies*. Icarus, **90**, 271–281.

Strom, K.M., Strom, S.E., Edwards, S., Cabrit, S., Skrutskie, M.F., 1989. *Circumstellar material associated with solar-type pre-main-sequence stars: a possible constraint on the timescale for planet building*. Astron. J., **97**, 1451–1470.

Valsecchi, G.B., Morbidelli, A., Gonczi, R., Farinella, P., Froeschlé, C., Froeschlé, C., 1994. *The dynamics of objects in orbits resembling that of P/Encke*. Icarus, in press.

Weidenschilling, S.J., 1977. *The distribution of mass in the planetary system and solar nebula*. Astrophys. Space Sci., **51**, 151–158.

Weidenschilling, S.J., 1994. *Origin of cometary nuclei as 'rubble piles'*. Nature, **368**, 721–723.

Yabushita, S., 1993a. *Thermal evolution of cometary nuclei by radioactive heating and possible formation of organic chemicals*. Mon. Not. Roy. Astron. Soc., **260**, 819–825.

Yabushita, S., 1993b. *The transport of super-volatiles through cometary nuclei by radioactive heating*. Memoirs of Faculty of Engineering, Kyoto Univ., **55**, 171–182.

Yamamoto, T., Kozasa, T., 1988. *The cometary nucleus as an aggregate of planetesimals*. Icarus, **75**, 540–551.

Yamamoto, T., Mizutani, H., Kadota, A., 1994. *Are 1992 QB1 and 1993 FW remnant planetesimals?* Publ. Astron. Soc. Japan, **46**, L5–L9.

Yoshikawa, M., 1992. *Numerical investigation of motions of resonant asteroids in the three-dimensional space*. Cel. Mech. Dyn. Astron., **54**, 287–290.

Zheng, J.Q., Valtonen, M.J., Mikkola, S., Korpi, M., Rickman, H., 1995. *Orbits of short-period comets captured from the Oort cloud*. This volume.

ON THE AGING OF COMETS

Harvard-Smithsonian Center for Astrophysics

Abstract. This study is based primarily on the calculations of comet orbits over $\sim 10^6$ years for 160 short-period comets by Harold F. Levison and Martin J. Duncan from which there are calculated "ablation AGES". There are positive statistical correlations (having many deviations) with radial nongravitational forces, comet activity measures, and dust-to-gas ratios in the spectra, in the sense that comets of greater "AGES" tend to be less active and to show less dust in their spectra than comets of lesser "AGES".

1. Introduction

Since Oort and Schmidt (1951) began searching for physical differences between "new" and "old" comets, the search has continued with negligible success. Probably new comets with $1/a(\text{original}) < 10^{-4}$ $(\text{AU})^{-1}$ tend to brighten at larger solar distances than other comets when they approach the Sun because of surface activity induced by cosmic rays (Donn, 1976; Whipple, 1977). Also the extreme lack of dust about P/Encke, undoubtedly one of the oldest periodic comets, has aroused interest.

Comparing the continuum to molecular emissions (the Dust/Gas ratio) among some 85 comets of various "ages" as measured by their values of $1/a$, Donn (1977) found "no significant difference among the patterns of relative intensity distribution among four age groups...."

Intensive search for spectroscopic differences among various comet groups led A'Hearn and Millis (1980) to conclude that they could find no correlation of CN, C_2, C_3, OH and gas-to-dust ratio with the dynamical age of a comet. In like manner, from IUE observations of faint comets, Weaver et al. (1981) reached the same conclusion: "All of the cometary spectra are remarkably similar which suggests that these comets have a common composition and origin."

From a study of the spectra of 17 comets, Newburn and Spinrad (1984, 1985) could not find that the mass loading of Dust/Gas varied from new to evolved comets. Similarly Cochran et al. (1992) concluded: "When speaking of the gas in the coma of a comet, it appears that comets must have been formed under remarkably uniform conditions and they must have evolved and formed their comae in a similar fashion...and that the interior of a comet must have been reasonably uniform."

Another characteristic of the Halley family has been found by Hughes (1987) and by Kresák and Kresáková (1987). The former showed that the cumulative number of comets with absolute magnitudes, $\overline{H}_{10}$, fainter than

Earth, Moon, and Planets **72**: 69-78, 1996.

successive values of $\overline{H}_{10}$, increased more rapidly for both short-period (SP, $P < 15$ yr) comets and long-period (LP with $P > 200$ yr) comets than for comets in the range $15 < P < 200$ yr. The Kresáks found similar results for the comets with periods in the range 20 to 200 yr compared with the SP comets of $P < 20$ yr (the former being called Halley-type comets).

These relationships led me to embark on this current research. Harold Levison and Martin Duncan (1994a) very generously forwarded to me copies of their enormous computational results for 160 comets with periods less than 200 yr. Each orbit was integrated forwards and backwards with 4 starting orbits, continuing until ejection from the solar system to > 50 AU. A record was kept of the last planet encountered and sun grazing within $q < 0.01$ AU. The graphs include the semimajor axis a (in AU), eccentricity e, and inclination i (deg). Perturbations were calculated for the seven planets, Venus to Neptune and neglected correction for general relativity and the quadrapole moment of the Earth-Moon System. This massive effort is thus the foundation for the remainder of this paper, a search for evidence of aging among comets.

2. Methodology

From the graphs of the calculated backward elements, starting with the actual elements of the comets, I read the time elapsed, $t(q_i)$ and $a(q_i)$ at q_i. This was followed by a calculation of a mass flux $\Delta M(q)$ per period derived from Table I below, which was evaluated by the equation

$$\Delta M = \int_0^P f(t)\, dt = \frac{1}{9\sqrt{p}} \int_0^{2\pi} \frac{r^2 d\nu}{x^{1.9}(1 + x^5)^{4.6}} \tag{1}$$

where $x = r(\text{AU})/2.808$, $\nu = $ true anomaly, $r = $ solar distance, $p = a(1-e^2)$, and the flux in the integral is a slightly simplified version of the flux used by Marsden and Williams (1994, as calibrated by Delsemme and Delsemme). The flux was corrected by $r^{1/4}$ to allow for the velocity of gas ejection.

In practice it was found that ΔM was almost independent of e so that it could be taken as $\Delta M(q) = \Delta M$, far within the accuracy of the measures.

An "AGE" for each comet could then be derived from the measured $t(q_i)$ and $a(q_i)$ with $P(q_i) = a(q_i)^{3/2}$ and $\Delta M(q_i)$ taken from Table I by the equation

$$\text{"AGE"} = \sum_i \frac{t(qi)\Delta M(qi)}{P(qi)} \times \frac{1}{20} \tag{2}$$

TABLE I

q (AU)	$\Delta M(q)/2$	q (AU)	$\Delta M(q)/2$
0.01	11.08	0.2	3.38
0.02	8.40	0.4	2.59
0.03	7.15	0.6	2.23
0.04	6.38	0.8	2.01
0.05	5.84	1.0	1.83
0.06	5.43	1.5	1.36
0.07	5.11	2.0	0.71
0.08	4.84	2.5	0.19
0.09	4.62	3.0	0.03
0.10	4.44	3.5	0.00

TABLE II

Distribution of "AGES"

	Among 160 ($P < 200$ yr)	Among Halley Type (TC* < 2.0)
Small (≤ 100)	54%	1 (5%)
100–1000	24%	8 (38%)
>1000	22%	12 (57%)
P/Encke 65,000		P/Crommelin 3,000
P/Halley 1,500		P/Grigg-Skjellerup 130

*Where TC $\equiv$ Tisserand Criterion

3. Results and Conclusions

The "AGES" have an arbitrary correction factor of 1/20, possibly corresponding to 5 percent efficiency in the use of solar energy received on their surfaces. They, indeed, have at best, relative values. Their possible intrinsic values (and units) can only be found by comparison with observations.

The observations of possible interest are those that measure cometary activity such as non-gravitational forces, magnitude variations, activity indices, and dust-to-gas ratios in their spectra. The 70 short-period comets for which such measures were available appear in Table III, which lists in successive columns:

TABLE III

Comet	q(AU)	P_{yr}	i	T-C	FLW "AGE"	A_1	Donn D/G	ψ	ΔAI m	RFKF Age
Arend	1.85	8.0	19°9	2.7	260	+0.14	-	-	-	-
ArRiq	1.45	6.8	17.8	2.7	1150	-	-	-	-1.5	-
AshJA	2.30	7.5	12.2	2.9	Y	+0.06	-	-	+0.9	-
Borre	1.36	6.9	30.3	2.6	130	+0.14	-	0.29	-	9
BroMe	0.48	70.5	19.3	1.1	750	+0.02	-	-	-	-
Brook2	1.84	6.9	5.5	2.9	Y	+1.46	-	0.26	-	-
ChuGe	1.30	6.6	7.1	2.7	Y	+0.07	-	-	-	2
Clark	1.56	5.5	9.5	2.9	9	+0.99	-	-	-	-
ComSo	1.83	8.8	13.0	2.7	Y	+0.58	-	-	-	4
Cromm	0.74	27.4	29.1	1.5	3000	-0.08	-	-	-	104
Danie	1.65	7.1	20.1	2.7	Y	+0.72	-	-	-	-
dArr	1.29	6.4	19.4	2.7	Y	+0.11	M	-	-	1
DenFu	0.78	9.0	8.6	2.2	2900	-0.01	-	-	-	-
dViSw	1.62	6.3	3.6	2.9	Y	-	H	-	-	-
dToHa	1.20	5.2	2.9	2.9	Y	+0.57	-	-	-	-
dTND	1.72	6.4	2.8	2.9	Y	+0.95	-	-	-	-
Encke	0.33	3.3	11.9	3.0	65000	-0.08	L	0.06	-1.0	144
Faye	1.59	7.3	9.1	2.8	Y	+0.29	H		-1.0	12
Finla	1.09	7.0	3.7	2.6	Y	+0.22	-	-	-	-
Forbe	1.48	6.3	7.2	2.9	Y	+0.44	-	0.18	-	10
Gehre	2.36	8.0	6.7	2.9	50	-	-	0.16	-	-
GiaZi	1.03	6.6	31.8	2.5	Y	+0.22	H	-	-0.2	12
GriSk	1.00	5.1	21.1	2.8	130	-0.00	-	0.10	-	-
Gunn	2.47	6.8	5.8	2.3	Y	+2.23	-	-	-	-
Halle	0.59	76.0	162.2	-0.6	1500	+0.10	M	-	-	128
Harri	1.60	6.8	8.7	2.8	350	+0.32	-	-	-	-
HarAb	1.77	7.6	10.2	2.8	Y	+0.34	-	-	-	-
Holme	2.17	7.1	19.2	2.9	56	+0.09	-	-	-	-
HoMrP	0.54	5.3	4.2	2.6	900	-0.05	-	-	-2.5	16
JacNe	1.44	8.5	14.1	2.6	90	+0.42	-	-	-	-
Johns	2.31	7.0	13.7	2.9	Y	+0.64	-	-	-	-
KeaKw	2.22	9.0	9.0	2.8	Y	+3.18	H	0.14	+0.1	-
Kohou	1.78	6.6	10.5	2.9	Y	+3.96	-	0.07	-	-
Kopff	1.58	6.5	4.7	2.9	Y	+0.49	L	-	-	5
Kowa2	1.50	6.4	15.8	2.8	Y	-2.37	-	-	-	-
Longm	2.41	7.0	24.4	2.9	190	-0.08	-	-	-	-
Melli	0.19	145.	32.7	0.6	830	-	L	-	-	-
MetBr	1.59	7.8	13.0	2.7	610	+0.58	-	-	-	-

continued on next page

continued from previous page

Comet	q(AU)	P$_{yr}$	i	T-C	FLW "AGE"	A$_1$	Donn D/G	ψ	ΔAI m	RFKF Age
Neuj1	1.55	18.2	14°.2	2.2	Y	-	L	-	-	-
Neuj3	1.98	10.6	3.9	2.6	1440	+1.61	-	-	-	-
Olber	1.18	69.6	44.6	1.3	250	+0.48	-	-	+1.1	-
PerMr	1.27	6.7	5.9	2.7	Y	-0.08	-	-	-	-
PonBr	0.77	70.9	74.2	1.1	910	-0.09	-	-	+0.5	-
PonWi	1.26	6.4	22.3	2.7	Y	+0.05	H	-	-1.9	-
Rein1	1.87	7.3	8.1	2.8	230	+0.18	-	-	-	-
Rein2	1.94	6.7	7.0	2.9	140	+0.10	-	-	-1.8	-
Schau	1.21	8.3	11.8	2.5	10	+0.55	M	-	-	-
ShaSc	2.33	7.5	6.1	2.9	Y	+3.24	-	-	+1.7	-
ScWa1	5.77	14.9	9.4	3.0	250	-	H	-	-	-
ScWa2	2.01	6.4	3.8	3.0	440	+1.28	-	-	-0.1	-
ScWa3	0.94	5.4	11.4	2.8	90	+0.58	-	-	-	-
Schau	1.21	8.3	11.8	2.5	10	+0.55	-	-	+0.8	103
SteOt	1.25	61.9	18.0	1.9	14	+0.17	-	0.41	-	102
SwiGe	1.36	9.2	9.3	2.5	5300	+0.56	-	0.09	-	-
Temp2	1.38	5.3	10.6	3.0	140	+0.01	H	-	-	21
TemSw	1.15	5.7	5.4	2.8	27	+0.04	-	-	-	-
TemTu	0.98	32.9	162.7	-0.6	7060	0.00	-	-	-	-
Tsuc1	1.50	6.6	7.0	2.8	100	+0.46	-	-	-	-
Tsuc2	1.78	6.8	10.5	2.9	140	-0.86	-	-	-	-
Tutt1	1.02	13.7	54.7	1.6	3300	+0.14	-	0.08	-	111
TuGiK	1.07	5.5	9.2	2.5	550	+0.90	M	-	-0.6	-
Väis1	1.80	10.9	11.6	2.5	2860	-0.04	-	-	-	-
Westp	1.25	61.9	40.9	1.4	37000	-	L	-	-	-
WeKoI	1.57	6.4	30.6	2.7	10	-0.05	-	-	-	-
Whipp	3.08	8.5	9.9	3.0	Y	+0.52	-	-	-	-
Wild2	1.58	6.4	3.2	2.9	82	-0.01	-	-	-	-
Wild3	2.29	6.9	15.5	2.9	Y	-	-	0.14	-	-
Wilk	0.62	187.	26.0	1.0	420	-	L	-	-	-
Wirta	1.08	5.5	11.7	2.8	Y	+0.79	-	-	-	-
Wolf	2.42	8.2	27.5	2.7	Y	+0.42	-	-	-	-
WolHa	1.61	6.5	18.5	2.8	Y	+0.28	-	-	-	6

1) The comet's name in 5 letters and numbers; 2) q; 3) P; 4) i; 5) TC; 6) "AGES"; 7) A_1, the radial nongravitational acceleration by Marsden and Williams (1994); 8) the Dust/Gas ratio by Donn (1977); 9) the quantity, ψ, defined by Newburn and Spinrad (1985) as the dust to gas ratio by mass; 10) the activity index of Whipple (1992a) corrected for magnitude; 11) the age determined by Rickman et al. (1991) from perturbations of the perihelion distance.

The table is limited to reduce journal space. The direct values of radial nongravitational forces, A_1, are compared with the "AGES" in Table IV. The values of A_1 were derived from the 1994 catalogue by Marsden and Williams. As Marsden pointed out to me, the comets with large q tend to show unusually large values of A_1 because of computational problems. Hence the straight means of A_1 are not very representative. Hence means were made by weighting according to the number of determinations and also by using only the "best" values defined as two or more consistent determinations. A further indicator is the number of negative values, because small errors in determinations affect small values of A_1 more seriously to produce negative values

Rickman et al. (1991) have derived comet ages from perturbations of the perihelion distance for comparison with nongravitaitonal forces. Their Table 3 contains ages from 17 comets, the values of which are listed in column 11 of Table III above. Their determinations of age are bunched in two groups; the first of eleven, 1 to 21, which I shall call "young" and the second, 102 to 158, which I shall call "old" in Table V, for comparison with the "AGES" of this paper.

TABLE IV

Mean Values of A_1

"AGES"	No.	Means	Weighted Means
≤ 100	37	$+0.59 \pm 0.18$	$+0.59 \pm 0.16$
$10^2 - 10^3$	14	$+0.19 \pm 0.12$	$+0.24 \pm 0.12$
$> 10^3$	14	$+0.24 \pm 0.18$	$+0.16 \pm 0.16$
		Best Values	Negative Values
≤ 100	9	$+0.56 \pm 0.22$	$4/37 = 11\%$
$10^2 - 10^3$	6	$+0.26 \pm 0.21$	$4/13 = 31\%$
$> 10^3$	3	$+0.06 \pm 0.07$	$4/9 = 44\%$

TABLE V

"AGES" compared with those of Rickman et al.
(1991)

"young"	900,Y,Y,Y,Y,Y,Y,130,Y,Y,140
"old"	6500,10,3300,3000,14,1500

TABLE VI

Activity Index (FLW 1992a)

"AGE"	N. Comets	ΔAI
≤ 100	7	$+0.06 \pm 0.46$
> 100	8(all)	-0.80 ± 0.42
> 100	6*	-1.33 ± 0.30

*without Olbers (+1.1) and Pons-Brooks (+0.5).

In Table V there are only two or three major discrepancies out of 17 cases, a surprisingly good agreement considering the diverse nature of the methods.

Because greater activity seems to be a characteristic of younger comets, the relative activity indices (ΔAI, Whipple 1992a) of 15 comets are compared with the "Ages" of Table I in Table VI. Here the comets of greater "AGES" are systematically less active than those of lesser "Ages". The two Halley-type comets appear to be younger in terms of activity.

I find no correlation between the "AGES" and the volatility index (Whipple 1992b).

Another activity measure for comets is the difference between the nuclear absolute magnitude and total absolute magnitude at the time of major activity. Such measures are tabulated by Nakano and Green in their *International Comet Quarterly Handbook*. A summary of the measures was kindly provided by D. Green (private communication) and averages are shown in Table VII, although the data are not given in Table III.

The eleven Halley-type (TC < 2.0) comets in Table III have an average "Age" of 5000±3300, which is distorted by P/Westphal (37,000). Without it the average is 1800±700 placing them at a respectable middle age. Eight with values of A_1 average $\langle A_1 \rangle = 0.09 \pm 0.06$, on the older side of Table IV.

The more "aged" comets brighten less than the less "aged" comets by a significant amount of over a magnitude.

TABLE VII

Magnitude (Nucleus-Coma) Differences
from Catalogues by Nakano and Green

"Age"	No.	$\overline{\Delta \text{Mag}}$.
<1000	110	5.57±0.17
>1000	28	4.47±0.28

TABLE VIII

Dust/Gas vs. "AGES"

B. Donn (1977)		
Dust/Gas	"AGES"	Mean
High	250, Y, Y, Y, Y, 140	63
Medium	1500, 10, Y, 550	518
Low	830, 37000, Y, 65000, Y, 420	17,200

Newburn and Spinrad (1985), Mass Loading Dust/Gas = ψ

Dust/Gas(ψ)	"AGES"	Mean
>0.20	14, 130, Y	51
0.10-0.20	130, Y, Y, Y, 500	132
<0.10	65000, 3300, 5300, Y	18,400

(Y = young, taken as 10 in averages)

Although the Dust/Gas ratio has shown no correlation with "New" vs "Old" comets, the "AGES" from orbital data show a strong correlation in the expected sense, *viz.*, very old comets such a P/Encke, seem to develop a relatively strong crust that is not easily fragmented by outflowing gas. The data are given in Table VIII for the S-P comets treated by Donn (1977) and by Newburn and Spinrad (1985).

I could find no convincing evidence for correlations of "AGES" with spectroscopic composition of comets nor with the frequency distribution of absolute magnitudes except for the peculiar properties of the Halley-type comets, mentioned in the *Introduction*.

In conclusion, the arbitally determined "AGES" correlate statistically (having many blatant discrepancies) with radial nongravitational forces, brightness activities and dust-to-gas ratios of S-P comets in the sense that older comets are less active and show less dust in their spectra. The discrepancies may arise from chaotic factors in the orbital motions, from my inexact reading of the graphs by Levison and Duncan and/or by a likely characteristic of comets at any given time *not to act their age.*

Acknowledgements

I am extremely indebted to Harold F. Levison and Martin J. Duncan for the early use of their extraordinary calculations. I wish to thank Daniel W.E. Green for providing the magnitude summary and also thank Brian Marsden and John Chambers for computing and other advice. This effort has been supported by a grant from the Planetary Geology and Geophysics Program of the National Aeronautics and Space Administration.

References

A'Hearn, M.F. and Millis, R.L.: 1980, 'Abundance Correlations Among Comets', *Astron. J.* **85**, 1528

Cochran, A.L.: 1987, 'Another Look at Abundance Correlations Among Comets', *Astron. J.* **92**, 231

Cochran, A.L., Green, J.R., and Barker, E.S.: 1987, 'The Relationship Between Low Activity Comets and More Active Comets', in E.J. Rolfe and B. Battrick, eds., *Symposium on the Diversity and Similarity of Comets, ESA SP-278*, ESA: Brussels, p. 151

Cochran, A.L., Barker, E.S., Ranseyer, Tod, F., and Storrs, A.D.: 1992, 'The McDonald Observatory Faint Comet Survey. Gas Production in 17 Comets', *Icarus* **98**, 151

Donn, B.: 1976, 'Panel Discussion', in B. Donn et al., eds., *The Study of Comets, NASA SP-393*, NASA: Greenbelt, p. 611-617

Donn, B.: 1977, 'A Comparison of the Compositions of New and Evolved Comets', in A.H. Delsemme, ed., *Comets, Asteroids, Meteorites*, Univ. Toledo Press: Toledo, p. 15

Hughes, D.W.: 1987, 'Cometary Magnitude Distribution: The Tabulated Data', in E.J. Rolfe and B. Battrick, eds., *Symposium on the Diversity and Similarity of Comets, ESA SP-278*, ESA: Brussels, p. 43-48

Kresák, L. and Kresáková, M.: 1987, 'The Absolute Total Magnitude of Periodic Comets and Their Variations', in E.J. Rolfe and B. Battrick, eds. *Symposium on the Diversity and Similarity of Comets, ESA SP-278*, ESA: Brussels, p. 37

Levison, H.F. and Duncan, M.J.: 1994a, 'The SWRI Catalogue of Short-Period Comet Orbit Integrations', Preprint, Planetary and Astrophysical Sciences, Southwest Research Institute, San Antonio.

Levison, H.F. and Duncan, M.J.: 1994b, 'The Long-Term Behavior of Short-Period Comets', Preprint, Planetary and Astrophsyical Sciences, Southwest Research Institute, San Antonio.

Marsden, B.G., Sekanina, Z., and Yoemans, D.K.: 1973, 'Comets and Nongravitational Forces. V.', *Astron. J.* **78**, 211

Marsden, B.G. and Williams, G.V.: 1994, *Catalogue of Cometary Orbits*, Minor Planet Center, Smithsonian Astrophysical Observatory: Cambridge

Newburn, Jr. R.L. and Spinrad, H.: 1984, 'Spectrophotometry of 17 Comets. I. The Emission Features', *Astron. J.* **89**, 289-309

Newburn, Jr. R.L. and Spinrad, H.: 1985, 'Spectrophotometry of 17 Comets. II. The Continuum', *Astron. J.* **90**, 2591-2608

Oort, J.H. and Schmidt, M.: 1951, 'Differences between New and Old Comets', *Bull. Ast. Inst. Neth.* **11**, 259

Rickman, H., Froeschlé, C., Kamél, L., and Festou, M.C.: 1991, 'Nongravitational Effects and the Aging of Periodic Comets', *Astron. J.* **104**, 1446

Storrs, A.D., Cochran, A.L., and Barker, E.S.: 1992, 'Spectrophotometry of the Continuum in 18 Comets', *Icarus* **98**, 163

Weaver, H.A., Feldman, P.D., Festou, M.C., A'Hearn, M.F., and Keller, H.U.: 1981, 'IUE Observations of Faint Comets', *Icarus* **47**, 449

Whipple, F.L. and Stefanik, R.P.: 1966, 'On the Physics and Splitting of cometary Nuclei', *Nature and Origin of Comets*, Univ. de Liège, **37**, 33.

Whipple, F.L.: 1977, 'The Constitution of Cometary Nuclei', in A.H. Delsemme, ed., *Comets, Asteroids, Meteorites*, Univ. Toledo Press: Toledo, p. 25

Whipple, F.L.: 1992a, 'A New Activity Index for Comets', in A.W. Harris and E. Bowell, eds., *Asteroids, Comets, Meteors, 1991*, Lunar and Planetary Institute: Houston, p. 633

Whipple, F.L.: 1992b, 'A Volatility Index for Comets', *Icarus* **98**, 108

ON THE IMPORTANCE OF DUST IN COMETARY NUCLEI

E.K. KÜHRT

Deutsche Forschungsanstalt für Luft- und Raumfahrt,
Rudower Chaussee 5, 12489 Berlin, Germany

and

H.U. KELLER

Max-Planck-Institut für Aeronomie, 37191 Katlenburg-Lindau, Germany

Abstract. The icy conglomerate model introduced by Whipple more than 40 years ago has been widely accepted in cometary science because it is able to describe numerous cometary phenomena. In this model comets are described as a conglomerate of ices and dust where the ices represent the major component. However, some recent observations seem to favour dust rich comets. The purpose of this paper is to summarize the observational facts supporting the dominance of refractories in comets and to discuss the consequences of a dust dominated nucleus for cometary physics.

1. Introduction

The space missions to comet P/Halley in 1986 and refined astronomical observations during the last years provided interesting new details of cometary phenomena. However, comets still seem to be the most mysterious members of the Solar System. Numerous observational facts like erratic activity, outbursts, splittings, sudden disappearances of comets, or their diversity in appearance are not well understood. A reason for this unsatisfactory situation is that the formation, structure, and composition of comets – their physical nature – are only roughly understood. Some information on the formation and evolution of comets can be derived from their orbits and their composition. Whereas dynamical studies to the accretion times for the outer planets favour the Uranus-Neptune region as place of birth of comets (Safronov, 1969), cosmochemists (Yamamoto, 1991) explain the existence of strongly volatile molecules like S_2 and CO with cometary formation in the Kuiper belt beyond 50 AU. The composition of comets is deduced from spectrometric measurements in the cometary comae. The problem is that one observes with a few exceptions only the daughter molecules (dissociation products) and therefore has to speculate about the parent molecules in the nucleus. A major hint for the cometary structure comes from density estimations. They have, however, large error bars. In the case of comet Halley's density, estimations vary from 300 to 1500 kg m^{-3}. The abundance and composition of cometary dust has been the subject of ground based infrared-measurements (Zarnecki, 1990) and of in situ observations by means of infrared (Encrenaz and Knacke, 1991) and mass (Brownlee and

Earth, Moon, and Planets **72**: 79-89, 1996.
© 1996 *Kluwer Academic Publishers.*

Kissel, 1990). The physical and chemical properties of cometary dust were reviewed by McDonnell et al. (1991) and Jessberger and Kissel (1991). The ratio of dust to volatiles in comets has been controversially discussed. This basic quantity allows one to discriminate between various types of models.

The most widely accepted model for comets is the icy conglomerate model proposed by Whipple (1950) more than 40 years ago. It assumes that volatiles are the major components in comets and has been successfully applied in describing the basic features of comets as activity, non-gravitational forces, the nature of dust and gas tails. However, it fails to explain some recent observations. The result of the Giotto mission that activity is only evident on about 20% of the illuminated surface of comet P/Halley (Keller et al., 1987) and the discovery of dust trails on orbits of several comets by the IRAS satellite (Campins et al., 1990) can be hardly understood within the icy conglomerate model. More recent models (Keller, 1989; Sykes and Walker, 1992) start from the dominance of dust in cometary nuclei.

The objective of this paper is to review the observations and measurements that provide arguments for a modification of the icy conglomerate model. The physical properties of dust dominated nuclei are discussed.

2. Observations

Refined ground based measurements in addition to the VEGA- and GIOTTO-missions to comet P/Halley and the extended GIOTTO-mission to comet P/Grigg-Skjellerup (McDonnell et al., 1993) have related more and more sophisticated details of comets. In the following some recent observations are discussed.

2.1. OBSERVED VS. REAL DUST/GAS RATIO

The dust/gas mass ratio is an important parameter for the characterization of comets. The dust mass in the coma is commonly derived from ground based measurements in the optical wavelength range (Jewitt, 1991). Newburn Jr. and Spinrad (1989), Storrs et al. (1992), Singh et al. (1992), and Sekanina (1991a) interpreted measurements of comets (partly coherent, partly heterogenous samples) and found mass ratios between 0.1 and 1. These results seem to support the icy conglomerate model. However, several observational circumstances lead to the tendency that optical measurements generally underestimate the amount of dust in comets.

A basic problem for the determination of the emitted amount of dust by scattered light is that only particles comparable in size to the wavelength of light are detectable. Grains much larger than 1 μm or smaller than 0.1 μm are not efficient at scattering optical wavelengths. Commonly the mass release q of dust with the density ρ and the size distribution function $f(a)$ is given by

$$q = \int_{a_{min}}^{a_{max}} \frac{4}{3} \pi a^3 \rho(a) f(a) \, da , \qquad (1)$$

where a_{min} and a_{max} are the minimum and maximum sizes of the grains, respectively (Singh et al., 1992). $f(a)$ is often given by (Brin and Mendis, 1979)

$$f(a) = C a^{-\gamma} \qquad (2)$$

where γ has been derived by the same authors to be 3.5. The normalization constant C can be directly related to observations. For such a size distribution the main mass is represented by the small grains. Further, neither a_{min} nor a_{max} can be determined from optical measurements. From in situ dust measurements during the Halley missions (McDonnell et al., 1991), from the analysis of the dust jets (Knollenberg, 1994), from the discovery of cometary dust trails by the IRAS satellite (Campins et al., 1990), and from theoretical investigations (Coradini and Magni, 1977) we have learned that the dust mass represented by large grains in the millimetre to decimetre range has been strongly underestimated by extrapolation of results from scattered light. Knollenberg found that an exponent $\gamma = 2.5$ in Eq. (2) fits the observed dust jets well resulting in a size distribution where the major mass is represented by large grains. The observations of fireballs, radar measurements (Campbell et al., 1989), and meter-sized particles found in dust trails (Campins et al., 1990) demonstrate that maximum sizes of grains far beyond the optical wavelength are emitted from comets. Sykes and Walker (1992) derive a dust/ice ratio of about 3 including the large particles found in the cometary trails.

A further problem of estimating the dust/ice ratio in comets from the relations in the coma results from inhomogeneities of the nucleus. Dust should be released preferably from active regions that probably differ from inactive areas in their dust to ice ratio (Kührt and Keller, 1994). Therefore, the ratio in jets cannot be expected to be representative of the whole nucleus.

2.2. ACTIVITY OF COMETS

One of the most surprising results of the space mission to comet P/Halley was that the activity was only evident on about 20% of the illuminated surface (Keller et al., 1986). The dominance of inactive regions is also supported by IR-measurements where surface temperatures as high as 400 K were found (Combes et al., 1988). This value is much higher than the sublimation temperature of water ice (about 200 K). Ground based observations of many different comets recently surveyed by Sekanina (1991b) has shown that the restriction of activity to a minor fraction of the surface is a general feature even near perihelion.

This phenomenon is not clearly understood up to now. Models based on the icy conglomerate concept [e.g., Brin (1980), Fanale and Salvail (1984), Rickman et al. (1990), Orosei et al. (1995)] cannot explain the existence of stable inactive areas over wide parts of the cometary surface. Permanent mantles can only be generated by this kind of models for comets on Halley-like or Encke-like orbits if special geometrical orientations of the spin axis, extreme thermo-physical parameters, or doubtful model approximations are assumed (Kührt and Keller, 1994).

2.3. SPLITTING OF COMETARY NUCLEI

Splitting of cometary nuclei has been observed in many cases. It seems to occur anywhere on the cometary orbit (Sekanina, 1982). No correlation with physical parameters of the comets or their orbits has been established. Generally no particular cause is apparent with the exception of tidal disruptions of comets closely approaching Jupiter or the Sun. The most recent spectacular event of this kind was the split of comet P/Shoemaker-Levy 9 caused by tidal forces of Jupiter. Separation of small pieces seems to be frequent (Chen and Jewitt, 1994). Sometimes the break-up of a comet leads directly to its fading and loss. Splitting of cometary nuclei is probably their dominant loss mechanism. The occurrences, dynamics, and activity of the broken off nuclei were analyzed in detail in a series of papers by Sekanina (1977), Sekanina (1978) and Sekanina (1979). The frequent occurrences even at large heliocentric distances ($r_h > 9$ AU!) confirm that cometary nuclei are fragile. In some cases increased activity could be observed before or during the splitting (flare up). Some fragments have long lifetimes, some short ones. In most cases the medium and long term activity of the multitude of nuclei is hardly enhanced if compared to that before the splitting. This indicates that the fraction of active areas on the new surfaces stemming from the interior of the nucleus is similar to that of the surface of the original nucleus. Hence, the nuclei are heterogenous in the dust to ice ratio (see Sect. 3) and inert volumes predominate.

3. The icy dirt ball model and its consequences

Stimulated by the GIOTTO images Keller (1989) suggested the concept of an icy dirt ball. The microstructure of a cometary nucleus is here characterized by refractory material rather than by ice*. From analyzing IRAS mea-

---

* Whipple used the descriptive expression *Tundra* model in his summary of the meeting.

surements of cometary dust trails Sykes and Walker (1992) came to a similar
conclusion. Kührt and Keller (1994) investigated the physical behaviour of
dust dominated nuclei.

3.1. ACTION OF COHESIVE FORCES

Cohesive forces act between dust grains because they touch each other. In
contrast to cohesive bonding between ice grains a dust conglomerate cannot
be eroded by thermal energy. This is an important consequence of the icy
dirt ball model.

The importance of binding forces for cometary modelling has often been
mentioned but corresponding effects have rarely been included in the models.
The significance of cohesive forces becomes readily apparent if one compares
their strengths to those of cometary gravity and vapour pressure forces.
Vapour pressure does not exceed 100 Pa even for the very high sublimation
temperature of 250 K (Kührt and Keller, 1994). Chokshi et al. (1993) analyzed van der Waals forces between grains in the primordial nebula. Strength
values ranging from 10^2 to 10^5 Pa can be derived for conglomerates of mm-
sized to μm-sized grains in agreement with measurements. Whipple (1982)
derived an upper limit for the tensile strength of comets of 10^4 Pa based on
the analysis of cometary spins and size statistics of the nuclei. A strength
of 10^2 to 10^3 Pa was found for lunar regolith from investigations during the
Apollo program (Mitchell et al., 1973). Saunders et al. (1986) and Storrs
et al. (1988) found a tensile strengths of filamentary sublimate residues of
about 10^4 Pa from laboratory investigations. Fireballs that probably orig-
inate from comets have a mechanical strength of 10^3 to 10^6 Pa (Wetherill
and ReVelle, 1982).

Therefore, the cohesive strength within a matrix structure of refractory
material generally exceeds the vapour pressure of water ice in comets.

3.2. COMETARY SURFACE CRUSTS

A consequence of the cohesiveness of nuclei is the depletion of the outermost
surface layers from volatiles. A stable crust is formed. In contrast to loose
mantles that are described by numerous models [for a review see Kührt and
Keller (1994)] cohesive crusts are stable against the vapour pressure even
at heliocentric distances smaller than 1 AU. It should be emphasized that
the cohesiveness within a dust layer accumulated at the surface but without
bonds to the interior does not stabilize the surface layer.

Kührt and Keller (1994) developed a thermal model for stable cometary
crusts. They found that:

- The thickness of surface crusts is between 10 cm and 10 m. It depends
  mainly on the value of the heat conductivity. Porosity, pore size, and
  orbit parameters are of minor importance.

- The vapour pressure exceeds gravitational pressure for all reasonable parameters. This means that loose dust mantles are blown off. Cohesive
  forces may withstand the gas pressure, large parts of the cometary surface are covered by a stable cohesive crust.

3.3. ACTIVITY AND INHOMOGENEITY OF COMETARY NUCLEI

A crust on the surface drastically reduces cometary activity because it
shields the volatiles below from insolation. Figure 1 shows the development
of the (maximum) diurnal gas flux through a well conducting and a badly
conducting crust for a comet Halley-like orbit. Activity is strongly quenched
to less than 1% of the free sublimation level depending on the heat conductivity in the refractory crust. Therefore, crusts can explain the observed
stable inactive regions on comets (see Sect. 2) in a natural way. It can be further seen that the thermal inertia of the crust causes a hysteresis behaviour
of the activity.

On the night side of a comet free sublimation stops whereas the low
activity through the crust is hardly depressed because the crust stores the
heat (Fig. 2). A low night side activity level of 0.1% to 1% is consistent with
the images taken during the Giotto fly-by (Knollenberg, 1994).

The icy dirt ball model and the formation of a stable crust do not explain
the strongly localized jet-like cometary activity (Sekanina, 1993). A plausible explanation for the variable behaviour of different surface areas is the
assumption of a structurally inhomogeneous nucleus. This picture is compatible with a comet consisting of several cometesimals formed under different
conditions in the solar nebula resulting in a varying dust/ice ratio within the nucleus. The dust/ice ratio governs the formation of stable crusts
and spots of relatively stable activity, respectively. According to this picture activity originates from regions where the volatile component (ice) is
so abundant that stable insulating crust cannot build up. Observations indicate that cometary activity sources are typically stable over many orbits
(Sekanina, 1993). An active region embedded in an inactive area produces
the observed jet-like activity (Knollenberg, 1994).

Cohesive, dust dominated clusters smaller than the critical size beyond
which they become bound by gravity (typically decimetre to metre size) can
leave the nucleus after the volatiles in their pores have been sublimated.
Then the heat wave can erode the dust-ice bridges below the dust cluster.
The escaping fluffy dust agglomerates form the dust trails or reach Earth

MAX. H2O-SUBLIMATION

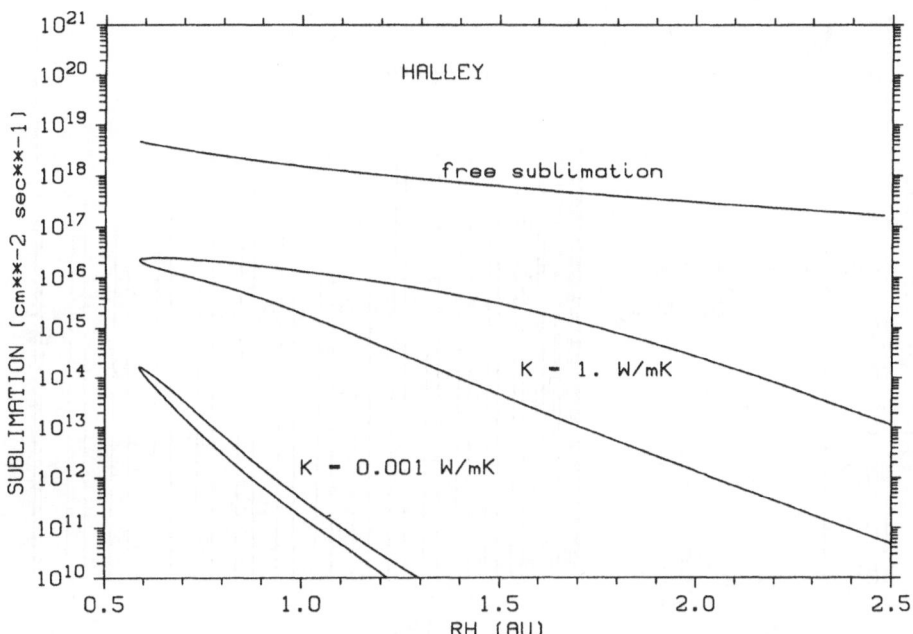

Fig. 1.   Modelled maximum diurnal sublimation rates from the equator of a rotating cometary nucleus are shown. The curves depict cases of free sublimation and of an area with a crust. Two extreme values are taken for the heat conductivity corresponding to different thicknesses of the crust. An orbit similar to that of comet Halley, an obliquity of 0, and a rotation period of 50 h have been assumed.

as fireballs. Larger boulders can be separated from the nucleus by splitting processes that seem to be common (Sekanina, 1982; Chen and Jewitt, 1994).

Consequently, the introduced model of an inhomogeneous nucleus structure can explain several features of cometary activity.

3.4. AGING OF COMETS

From the analysis of non-gravitational forces Rickman et al. (1991) derived that the relative part of the inactive surface area becomes larger the older the comets are and the lower their perihelion distance is. Kresák (1991) and Rickman et al. (1991) found that comets show the tendency of fading with age. This is consistent with the scenario described above. Crusted areas remain inactive and active regions become less important because they disappear after the volatiles are consumed. The last stage of such a development

H2O-SUBLIMATION

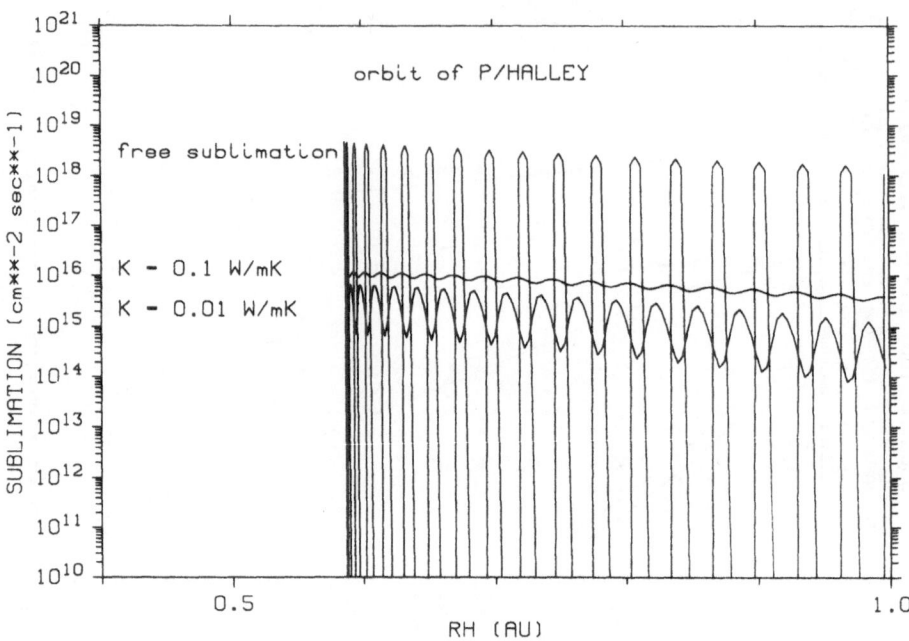

Fig. 2. Diurnal variation of the activity at perihelion; same conditions as in Fig. 1.

is a dormant body with an asteroid-like behaviour. It should still contain ice but it has no or rather low activity because it is completely covered by a depleted crust. Several candidates for such objects have been identified (Yeomans, 1991).

4. Conclusions

Despite the progress in the observational techniques the nature of comets is not yet well understood. Speculations about their origin and structure will prevail at least until the in situ measurements of the planned ROSETTA mission. Without doubt the physical consistence of the microstructure of cometary nuclei is a key point for the understanding of origin, activity, and death of comets.

The introduced model postulates an inhomogeneous, dust rich comet and can explain some observational facts such as the restricted activity and some aspects of cometary aging. General consequences of the presented investigations are:

- The cohesive strength of cometary nuclei is orders of magnitude stronger than gravitational pressure and higher than the vapour pressure ($P_{coh} > P_{vap} > P_{grav}$) and must be incorporated in cometary models.

- Comet models based on heterogeneous (varying dust to ice ratio) cohesive nuclei can explain the observations. Structural inhomogeneities and a varying dust/ice ratio can be the key to the understanding of local activity and large scale inactivity of comets. Dust and ice clusters may be substructures of cometesimals or cometesimals from different origins. Ice rich clusters on the surface yield the active areas. Depleted dust clusters too big to be removed by gasdynamic forces form the stable crust. Small dust clusters can be ejected after they have been depleted of volatiles and conduct the heat to the underlying ice.

- Dust/gas ratios derived from sampling the coma underestimate the mean ratio in the nucleus.

References

Brin, G.D.: 1980, Three models of dust layers on cometary nuclei. *Astrophys. J.* **237**, 265–279.

Brin, G.D. and D.A. Mendis: 1979, Dust release and mantle development in comets. *Astrophys. J.* **229**, 402–408.

Brownlee, D.E. and J. Kissel: 1990, The composition of dust particles in the environment of Comet Halley. In *Comet Halley, Investigations, Results, Interpretations Volume 2: Dust, Nucleus, Evolution.* (J. Mason and P. Moore, Eds.), pp. 89–98. Ellis Horwood Ltd., London.

Campbell, D.B., J.K. Harmon, and I.I. Shapiro: 1989, Radar Observations of Comet Halley. *Astrophys. J.* **338**, 1094–1105.

Campins, H., R.G. Walker, and D.J. Lien: 1990, IRAS Images of Comet Temple 2. *Icarus* **86**, 228–235.

Chen, J. and D. Jewitt: 1994, On The Rate at Which Comets Split. *Icarus* **108**, 265–271.

Chokshi, A., A.G.G.M. Tielens, and D. Hollenbach: 1993, Dust coagulation. *Astrophys. J.* **407**, 806–819.

Combes, M., V.I. Moroz, J. Crovisier, T. Encrenaz, J.P. Bibring, A.V. Grigoriev, N.F. Sanko, N. Coron, J.F. Crifo, R. Gispert, D. Bockelée-Morvan, Yu.V. Nikolsky, V.A. Krasnopolsky, T. Owen, C. Emerich, J.M. Lamarre, and F. Rocard: 1988, The 2.5 to 12 m spectrum of comet Halley from the IKS-Vega Experiment. *Icarus* **76**, 404–436.

Coradini, A. and G. Magni: 1977, Grains Accretion Processes in a Proto-Planetary Nebula. II. Accretion Time and Mass Limit. *Astrophys. Space Sci.* **48**, 79–87.

Encrenaz, T. and R. Knacke: 1991, Carbonaceuous compounds in comets: infrared observations. In *Comets in the Post-Halley Era.* (R. Newburn and J. Rahe, Eds.), pp. 107–137. Kluwer Academic Publishers, Dordrecht, The Netherlands.

Fanale, F.P. and J.R. Salvail: 1984, An idealized short period comet model: surface insolation, H2O flux, dust flux and mantle development. *Icarus* **60**, 476–511.

Jessberger, E.K. and J. Kissel: 1991, Chemical properties of cometary dust and a note on carbon isotopes. In *Comets in the Post-Halley Era.* (R. Newburn and J. Rahe, Eds.), 2, pp. 1075–1092. Kluwer Academic Publishers, Dordrecht, The Netherlands.

Jewitt, D.: 1991, Cometary Photometry. In *Comets in the Post-Halley Era*. (R. Newburn and J. Rahe, Eds.), 1, pp. 19–65. Kluwer Academic Publishers, Dordrecht, The Netherlands.

Keller, H.U.: 1989, Comets – Dirty Snowballs or Icy Dirtballs?. In *Proceedings of an International Workshop on Physics and Mechanics of Cometary Materials*. (J. Hunt and T.D. Guyeme, Eds.), ESA SP–302, pp. 39–45. ESA Publications Division, ESTEC, Noordwijk.

Keller, H.U., W.A. Delamere, W.F. Huebner, H.J. Reitsema, H.U. Schmidt, W.K.H. Schmidt, F.L. Whipple, and K. Wilhelm 1986. Dust Activity of Comet Halley's Nucleus. In *20th ESLAB Symposium on the Exploration of Halley's Comet*. (B. Battrick, E.J. Rolfe, and R. Reinhard, Eds.), ESA-SP 250 II, pp. 359–362. ESA, Publications Division, ESTEC, Noordwijk.

Keller, H.U., W.A. Delamere, W.F. Huebner, H.J. Reitsema, H.U. Schmidt, F.L. Whipple, K. Wilhelm, W. Curdt, J.R. Kramm, N. Thomas, C. Arpigny, C. Barbieri, R.M. Bonnet, S. Cazes, M. Coradini, C.B. Cosmovici, D.W. Hughes, C. Jamar, D. Malaise, K. Schmidt, W.K.H. Schmidt, and P. Seige: 1987, Comet P/Halley's nucleus and its activity. *Astron. Astrophys.* **187**, 807–823.

Knollenberg, J.: 1994, *Modellrechnungen zur Staubverteilung in der inneren Koma von Kometen unter spezieller Berücksichtigung der HMC-Daten der GIOTTO-Mission*. Ph.D. thesis. Georg-August-Universität, Göttingen, Germany.

Kresák, L.: 1991, The ageing of Comet Halley and other periodic comets. In *Comet Halley, Investigations, Results, Interpretations Volume 2: Dust, Nucleus, Evolution*. (J. Mason and P. Moore, Eds.), pp. 259–270. Ellis Horwood Ltd., London.

Kührt, E. and H.U. Keller: 1994, The Formation of Cometary Surface Crusts. *Icarus* **109**, 121–132.

McDonnell, J.A.M., P.L. Lamy, and G.S. Pankiewicz: 1991, Physical properties of cometary dust. In *Comets in the Post-Halley Era*. (R. Newburn and J. Rahe, Eds.), 2, pp. 1043–1073. Kluwer Academic Publishers, Dordrecht, The Netherlands.

McDonnell, J.A.M., N. McBride, R. Beard, E. Bussoletti, L. Colangeli, P. Eberhardt, J.G. Firth, R. Grard, S.F. Green, J.M. Greenberg, E. Grün, D.W. Hughes, H.U. Keller, J. Kissel, B.A. Lindblad, J.-C. Mandeville, C.H. Perry, K. Rembor, H. Rickman, G.H. Schwehm, R.F. Turner, M.K. Wallis, and J.C. Zarnecki: 1993, Dust particle impacts during the Giotto encounter with Comet Grigg-Skjellerup, *Nature* **362**, 732–734.

Mitchell, J.K., W.D. Carrier, N.C. Costes, W.N. Houston, R.F. Scott, and H.J. Hovland: 1973, *Soil Mechanics*. NASA Conference Publication, NASA SP–330.

Newburn Jr., R.L. and H. Spinrad: 1989, Spectrophotometry of 25 comets: Post-Halley updates for 17 comets plus new observations for eight additional comets. Plus Eight Additional Comets. *Astron. J.* **97**, 552–569.

Orosei, R., F. Capaccioni, M.T. Capria, A. Coradini, S. Espinasse, C. Federico, M. Salomone, and G.H. Schwehm: 1995, Gas and Dust Emission from a Dusty Porous Comet. Preprint.

Rickman, H., J.A. Fernández, and B.Å.S. Gustafson: 1990, Formation of stable dust mantles on short-period comet nuclei, *Astron. Astrophys.* **237**, 524–535.

Rickman, H., C. Froeschlé, L. Kamél, and M.C. Festou: 1991, Nongravitational Effects and the Aging of Periodic Comets. *Astron. J.* **102**, 1446–1463.

Safronov, V.S.: 1969, *Evolution of the Protoplanetary Cloud and Formation of the Earth and Planets*. NASA Tech. Memorandum. TT F-677.

Saunders, R.S., F.P. Fanale, T.J. Parker, I.B. Stephens, and S. Sutton: 1986, Properties of filamentary sublimation residues from dispersions of clay in ice, *Icarus* **66**, 94–104.

Sekanina, Z.: 1977, Relative motions of fragments of the split comets. I. A new approach, *Icarus* **30**, 574–594.

Sekanina, Z.: 1978, Relative motions of fragments of the split comets. II. Separation velocities and differential decelerations for extensively observed comets, *Icarus* **33**, 173–185.

Sekanina, Z.: 1979, Relative motions of fragments of the split comets. III. A test of splitting and comets with suspected multiple nuclei, *Icarus* **38**, 300–316.

Sekanina, Z.: 1982, The Problem of Split Comets in Review. In *Comets: Gases, Ices, Grains and Plasma.* (L.L. Wilkening, Ed.), pp. 251–287. University of Arizona Press, Tucson, Arizona.

Sekanina, Z.: 1991a, Comprehensive Model for the Nucleus of Periodic Comet Tempel 2 and Its Activity. *Astron. J.* **102**, 350–388.

Sekanina, Z.: 1991b, Cometary activity, discrete outgassing areas, and dust-jet formation. In *Comets in the Post-Halley Era.* (R. Newburn and J. Rahe, Eds.), 2, pp. 769–823. Kluwer Academic Publishers, Dordrecht, The Netherlands.

Sekanina, Z.: 1993, Orbital anomalies of periodic comets Brorsen, Finlay, and Schwassmann-Wachmann 2, *Astron. Astrophys.* **271**, 630–644.

Singh, P.D., A.A. de Almeida, and W.F. Huebner: 1992, Dust release rates and dust-to-gas mass ratios of eight comets, *Astron. J.* **104**, 848–858.

Storrs, A.D., F.P. Fanale, R.S. Saunders, and J.B. Stephens: 1988, The formation of filamentary sublimate residues (FSR) from mineral grains, *Icarus* **76**, 493–512.

Storrs, A.D., A.L. Cochran, and E.S. Barker: 1992, Spectrophotometry of the Continuum in 18 Comets, *Icarus* **98**, 163–178.

Sykes, M.V. and R.G. Walker: 1992, Cometary Dust Trails, *Icarus* **95**, 180–210.

Wetherill, G.W. and D.O. ReVelle: 1982, Relationships between comets, large meteors, and meteorites. In *Comets: Gases, Ices, Grains and Plasma.* (L.L. Wilkening, Ed.), pp. 297–319. Univ. of Arizona Press, Tucson, Arizona.

Whipple, F.L.: 1950, A Comet Model I. The Acceleration of Comet Encke, *Astrophys. J.* **111**, 375–394.

Whipple, F.L.: 1982, The Rotation of Comet Nuclei. In *Comets: Gases, Ices, Grains and Plasma.* (L.L. Wilkening, Ed.), pp. 227–250. University of Arizona Press, Tucson, Arizona.

Yamamoto, T.: 1991, Chemical theories on the origin of comets. In *Comets in the Post-Halley Era.* (R. Newburn and J. Rahe, Eds.), pp. 361–376. Kluwer Academic Publishers, Dordrecht, The Netherlands.

Yeomans, D.K.: 1991, A Comet Among the Near-Earth Asteroids?, *Astron. J.* **101**, 1920–1927.

Zarnecki, J.C.: 1990, Infrared studies of Comet Halley during the 1985–1986 apparition, In *Comet Halley, Investigations, Results, Interpretations Volume 2: Dust, Nucleus, Evolution.* (J. Mason and P. Moore, Eds.), pp. 51–63. Ellis Horwood Ltd., London.

PROCESSING OF COMETARY GRAINS AT THE NUCLEUS SURFACE

MAX K. WALLIS and SIRWAN AL-MUFTI

School of Mathematics, University of Wales/Cardiff, Wales UK

Abstract. Cometary material inevitably undergoes chemical changes before and on leaving the nucleus. In seeking to explain comets as the origin of many IDPs (interplanetary dust particles), an understanding of potential surface chemistry is vital. Grains are formed and transformed at the nucleus surface; much of the cometary volatiles may arise from the organic material. In cometary near-surface permafrost, one expects cryogenic chemistry with crystal growth and isotope. This could be the hydrous environment where IDPs form. Seasonal and geographic variations imply a range of environmental conditions and surface evolution. Interplanetary dust impacts and electrostatic forces also have roles in generating cometary dust. The absence of predicted cometary dust 'envelopes' is compatible with the wide range of particle structures and compositions. Study of IDPs would distinguish between this model and alternatives that see comets as aggregates of core-mantle grains built in interstellar clouds.

1. Introduction

The postulate that cometary grains are chemically transformed in the coma, emitting gases under solar heating (Wallis et al. 1986), is now a validated concept. Grains smaller than infra-red wavelengths become superheated to 4-600 K, causing some chemical pyrolysis of the organic material. The inactive dark comet surface does not get so hot, with temperatures $\sim$ 340 K (at 1 AU; 370 K subsolar on Halley at Vega's flyby at 0.8 AU), while the active gassing surface may be closer to the $\sim$ 200 K of sublimating ice. Organics emit some gas at the hotter levels, and surface-molecule reactions proceed throughout the temperature range; indeed the latter occur in much colder situations, as recognised for forming the complex molecules of molecular clouds.

Hydrated minerals could form at $\geq$ 300 K via reactions with monolayers of evaporating H_2O (Nelson et al. 1987). But such chemical changes occur even at cryogenic temperatures. The hydrated mineral phases apparently common to cometary and interplanetary grains suggest low-T diagenesis. Drawing on studies of permafrost, Rietmeijer (1985) has argued for hydro-cryogenic processing via interfacial molecular H_2O layers of order 1nm thick (above 193 K), that provide a medium for atom migration and slow chemical changes. Mineral particles exposed in sublimating ice environments have similar H_2O layers, though how quickly chemical processes proceed is unknown. Rietmeijer inferred such cryogenic alteration from features of interplanetary dust grains (IDPs) – silicate whiskers, high D:H fractions, enhanced ^{13}C

Earth, Moon, and Planets **72**: 91-97, 1996.

in carbonates, and hydrous pyrolysis of hydrocarbons (reviewed by Gibson 1992). The time scale is $> 10^4$ yr under antarctic conditions at 233-268 K; but in a comet, the short bursts of perihelion warming could accelerate changes, with a time-scale of weeks at 300 K (Nelson et al. 1987). Thus, IDPs might come from comets, but proof of the link is lacking.

The Vega and Giotto dust spectrometers showed (Dikov 1989, Brownlee and Kissel 1990) three populations of grains – a fraction of mineral grains, some purely organic ('CHON') and some mixed. The mixed composition is said to show evidence for core-mantle grains, supposed to form in interstellar clouds, though this is unconvincing in the absence of quantitative argument. Moreover, chondritic mineral components of IDPs appear related to asteroidal fragments (Gibson 1992). The question arises – might mixed grains form in situ in the comet surface layer as a natural consequence of sublimation? And could the relative populations be explained as simply following from water-rich kerogenous (complex organic) material with an admixture of mineral dust?

The inference of 'POM' – polyoxymethylene, or more generally, polysaccharides – from the gas mass spectrum in Halley's coma, has been used to support the idea that organics are processed and polymerised on the surfaces of interstellar grains (Huebner et al. 1987). But polymerisation could also occur in the cometary surface material. Simple organics can polymerise on suitable mineral substrates, with subsequent release as linear POM molecules, whereas pyrolysis of organics generally produces complex crosslinkages and so aromatic (ring) structures. We might expect both transformations on the comet surface; detailed studies could establish their relative importance.

2. Experimental Studies of Surface Reactions

Several laboratory studies have shown that sublimating mixtures do not simply evaporate, but grow complex structures (Kaimakov et al. 1981, Ibadinov 1989). Less volatile material tends to condense into highly porous matrixes, that fracture and are expelled erratically. Molecules that find energetically-favourable sites on surfaces tend to stick there; so crystalline structures grow in suitable materials. These characteristics are true of both salts and organics. Nucleic acids are found to grow spirally, forming filaments with one end attached to the subliming surface. Attachment of such fragile structures depends on low gravity and gas forces. Such favourable conditions would exist on comet surfaces too. The KOSI experiments (Grün et al. 1990) on simple gas mixtures confirmed fractionation of the material. Indeed, gases diffuse into the colder interior of a porous matrix and enrich it with volatiles.

If gases depositing on mineral grains in the surface layer bind to the substrate or polymerise, mixed 'core mantle' grains result. Gases that bind lightly would sublimate in the inner coma (lower vapour pressure; super-heating). In principle, such grains could constitute the observed H_2CO-source, but a practical abundant substrate needs to be demonstrated. The association of organic matter with layer silicates in the Halley-PUMA data (Dikov et al. 1989) is indicative. Another indication is the Si-O feature — much broader than from crystalline silicates and relatively uncommon; the conventional explanation is amorphous silicate smoke, but hydrous changes are likely on the surface and strong contamination is inevitable in the organic-rich coma. Siliceous organic compounds, often hidden under organic coatings, provide a natural explanation.

3. Weathering and Growth of a Crusted Surface

Periodic sunlight and atmospherics combine to transform the surface material, though in different ways from on Earth. Cometary rotation periods from ~ 10 hours to days are indicated. In the active gassing regions, less volatile grains that are too large (mm-sized) to be blown away by gases in the evening and early morning fall back to consolidate into a thin crust, before being blown off again around noon. Evidence from Halley for such diurnally periodic emissions is clear, though deviations are strong. The existence of areas such as ravines and shadowed hillsides where gases from neighbouring regions condense to initiate fragile crusts, cause erratic changes superposed on the diurnal cycle. One mechanism for accumulating less volatile grains arises because the gas flow is non-uniform. Grains of mm to cm sizes, that are marginally blown off from active regions, tend to fall back as 'hail'. This arises because the gas drag $\sim \rho u^2$ first increases with altitude as the flow goes more supersonic ($M \approx 1$ at the surface), but ultimately decreases as the gas flux ρu spreads from initial emission 'jets' (perhaps 20° wide). The direct evidence of wind-blown dust on Halley's nightside (Thomas & Keller 1991) indicates that the gas flow is far from the simple radial model on which all grains lifted from the surface are ejected. However, the hail falls globally rather than close to active regions, so helps build and consolidate the 'inactive' crust.

Seasonal changes around the comet orbit seem therefore to be more important (Wallis 1986). Crust forms in active regions during the 'winter' season when gas emission is low, being consolidated by hail and gas condensation in the more shaded areas. But such crust is fragile and perhaps only mm thick, so often blows off during the summer season. It sometimes becomes stronger through being 'toasted' in sunlight (shedding O and N, in favour of stronger carbon bonding – giving poorly graphitised hydrocarbons as in IDPs); it may persist through a whole summer and suppress the active region. When only part of an active region is covered, the edges of the crust are subject to wind erosion and tend to crumble and retreat during the summer. The apparent directionality of the dust jets implies the km-scale active regions are hollows several 100 m deep (Keller et al. 1994), so presumably remain active over tens of apparitions. In this case, crust build-up on the shadier parts of the crater sides would be common.

4. Types of Cometary Grains

The very low albedo crust that covers some 80% of Halley is explicable as organic material porous down to the micro-scale, with many cross-linkages formed via pyrolysis at $\succeq$ 350 K daytime, around perihelion. It would be durable against cometary snow; falling snowflakes fragment on impact, there being no gas pressure to brake or lift them – then sublimate when exposed to brighter sunlight. So except for occasional impacts of interplanetary dust, the crust is unlikely to be a source of cometary dust.

Stern (1988) has considered dust impacts; interstellar dust particles are rare and impact at high speed, burrowing into the porous material and ejecting $\succeq$ 100 times their own mass (Wallis & Wickramasinghe 1991). Interplanetary 10 μm dust out beyond Jupiter with flux $10^{-5} - 10^{-6}/(m^2 s)$ impacts at $\precsim$ 1 km/s and fragments the surface to $\sim$ 100 μm over 10 yr. With escape speeds of $\sim$ 10 m/s, some is ejected but other remains as loose microdust. This may stick, aided by cold-welding, but tends to erode via electrostatic forces. Beyond a few AU where gas density becomes low, the solar wind penetrates to the nucleus. The surface is charged to about 7 V and a single charge on grains $\precsim$ 0.3 μm suffices to levitate them, whereupon subsequent charging gives sufficient repulsion to escape (Mendis et al. 1981).

In the expanding gases, nucleation theory implies that some condensation occurs (Crifo 1987). H_2O-ice condensates re-sublimate, but some stable smoke-like particles could result from the complex mixtures, maybe forming aggregates as organics stick readily together. Such processes facilitate fractionation and are heterogeneous, being sensitive to density. We thus envisage that cometary grains could comprise a) fragments of temporary crust, from shadowed active regions (a matrix of smaller grains), b) crystals formed at the surface of the sublimating mixture, c) mineral dust grains with or without coatings of organics, and d) condensates in the near-surface coma.

5. Cometary Dust Envelopes

The concept of a dust particle envelope, marking the maximum sunward distance attained by all but the smallest grains, was not validated by Halley (though Vaisberg (1991) has tried to identify such structures). Hydrodynamic theory gives the terminal speeds of grains mass M blown out by the gas of momentum flux at the surface P_s as

$$v_t \sim \sqrt{(P_s < A > /M)} \qquad \text{if the gas } P_s = \rho u^2 \sim r^{-2}. \qquad (1)$$

A is the surface area offered to the gas flow, so an average is taken appropriate to a tumbling irregular grain. The radiative acceleration away from the sun is

$$g \sim Q < A > /M, \qquad \text{with } Q = Q_{abs} + < (1 - \cos\theta)Q_{scatt}(\theta) > \qquad (2)$$

The radiative efficiency Q is hardly dependent on size for particles larger than $1\,\mu m$ (average over solar wavelengths). The shape average $< A >$ is the same in both cases if the grains are tumbling randomly relative to both streams of photons and of gas, so then the envelope scale is

$$L = v_t^2/g \propto P_s/Q. \qquad (3)$$

This scale appears therefore independent of size, shape and density of the grains. However, Q does vary somewhat with composition (Burns et al. 1979). P_s varies with insolation, so could be taken to vary with solar zenith angle; for a spherical gassing nucleus, this would still give an envelope, though deviating from the classical parabola. However, on our arguments above, P_s is highly variable on the local scale and in time, as active gassing regions come into sunlight and temporary crust breaks up. This would be the major reason for the non-uniqueness of L and therefore the non-appearance of dust envelopes.

6. Discussion

Attempts to infer pristine interstellar material from cometary dust data appear premature. The absence of Greenberg's (1992) core-mantle and birds-nest structures in IDPs but variety of grains are evidence of transformation in cometary surface crusts. Grains are formed and transformed at the nucleus surface while much of cometary gases arise from transforming organic material at the surface and in the coma. Experiments show that sublimating mixtures undergo chemical fractionation. Residues of more refractory components crystallise at the exposed surface, depending on sublimation conditions. Organic molecules may stick to mineral substrates and build refractory polymer coatings. Just under the surface crust is a permafrost layer, changing slowly over years to millennia, within which molecular layers of H_2O facilitate cryogenic alteration and crystal growth with enrichment of ^{13}C and D. This is the hydrous environment in which complex aggregate IDPs form. Most of the grains analysed by the Halley probes come from 'active' regions of rapid evaporation, so do not have time to hydrate and become precursors of the minor class of IDPs of highly porous, anhydrous minerals (Brownlee and Kissel 1990).

On this hypothesis, cometary grains blown into space are a combination of surface condensates and interstellar grains, the latter originating from submicron condensates of stellar envelopes. Because of the diverse geography and insolation on a comet, the retention of condensates on the surface may be for minutes, days or months; diurnal heating and cooling results in some larger and more refractory grains. Adhesion of grains leads to a fragile crust, which is consolidated or destroyed under orbital and seasonal changes. The mechanical forces tending to fragment the crust and eject these grains – breezes and micro-hail – are very weak, and electrostatic forces can play a role. Interplanetary dust impacts also erode the surface to mm-depths. The size distribution of cometary coma grains is inferred to vary with orbit character and cometary activity, and to depart from the commonly assumed power law. Non-uniformities in composition and aging of the condensate grains, as well as in gassing rates, explain the absence of bounding 'mass envelopes' in the dust comas.

Significant contributions from such varied sources are unlikely to be compatible with a simple size spectrum. In the Halley data (McDonnell et al. 1987), two apparent changes in slope of that spectrum (at $0.3\,\mu$m and $100\,\mu$m) could be considered as evidence for populations from different sources. In conclusion, interpretations of in-situ and astronomical data on cometary grains and IDPs require us to transcend the old 'pristine material' paradigm and look to surface chemistry in relevant environments for explanations of the variety of structures and compositions.

References

Brownlee D E, Kissel J: 1990, in *Comet Halley Investigations, Results, Interpretations 2*, ed. J Mason, Ellis Horwood, England, pp. 89-98
Burns J A, Lamy P L, Soter S: 1979, *Icarus* **40**, 1-48
Crifo J F: 1987, *Astron. Astrophys.* **187**, 438-450
Dikov Yu P + 7 co-authors: 1989, *Adv. Space Res.* **9**(3), 253-258
Gibson E K: 1992, *J. Geophys. Res.* **97**, 3865-3875
Greenberg J M: 1990, in *Comet Halley Investigations, Results, Interpretations 2*, ed. J Mason, Ellis Horwood, England, pp. 99-120
Grün E + 25 co-authors: 1993, *J. Geophys. Res.* **98**, 15091-15104
Huebner W F + 7 co-authors: 1987, in *Diversity & Similarity of Comets*, ESA SP-278, pp. 163-167
Ibadinov Kh I: 1989, *Adv. Space Res.* **9**(3), 97-112
Kaimakov E A, Lizunkova I S, Dranevich V A: 1981, *Sov. Astron. Lett.* **7**, 63-65
Keller H U, Knollenberg J, Markiewicz W J: 1994, *Planet. Space Sci.* **42**, 367-382
McDonnell J A M + 27 co-authors: 1987, *Astron. Astrophys.* **187**, 719-741

Mendis D A, Hill J R, Houpis H L F, Whipple E C: 1981, *Astrophys. J.* **249**, 787

Rietmeijer F J M: 1985, *Nature* **313**, 293-294

Rietmeijer F J M, Mackinnon I D R: 1987, in *Diversity & Similarity of Comets*, ESA SP-278, pp. 363-368

Stern S A: 1988, *Icarus* **73**, 499-507

Thomas N, Keller H U: 1991, *Astron. Astrophys* **249**, 258

Vaisberg O L: 1990, in *Comet Halley Investigations, Results, Interpretations 2*, ed. J Mason, Ellis Horwood, England, pp. 33-44

Wallis M K: 1986, in *Comet Nucleus Sample Return*, ESA SP-249, pp. 63-67

Wallis M K, Wickramasinghe N C: 1991, *Space Sci. Rev.* **56**, 93-97

Wallis M K, Rabilizirov R, Wickramasinghe N C: 1987, *Astron. Astrophys.* **187**, 801-806

REFERENCES

DUST OUTFLOW VELOCITY AND THE GAS-TO-DUST RATIO IN COMETS

ALEX D. STORRS

Space Telescope Science Institute

Abstract. The observational determination of coma outflow velocity for gaseous species is fairly straightforward using high-resolution spectroscopy. The determination of the outflow speed of the dust is much more difficult. Most sources cite Bobrovnikoff (1954). This brief report is not a strictly refereed publication, however, and mixes data from different comets.

We present here a simple analysis of some data from the International Halley Watch (IHW) archive. Differences between continuum images from successive nights show dust jets and shells clearly. Their motion is apparent to first order from the edges of the features. The component of the dust outflow velocity perpendicular to the observer's line of sight may thus be determined. This is of course a lower limit on the dust outflow velocity. Many measurements, at different heliocentric distances (R), allow determination of the heliocentric dependence of the dust outflow velocity.

We find that the dust outflow velocity in comet P/Halley varied as $R^{-0.41}$. If data from an outburst at 14 AU (Sekanina et al. 1992) is included in the fit, this dependence becomes $R^{-0.55}$. This confirms the canonical (e.g. Delsemme 1982) inverse-square-root law, and supports the conclusion of Storrs et al. (1992) on the variability of cometary gas-to-dust ratios.

1. Introduction

It is commonly assumed that the dust outflow velocity (v) in cometary comae varies as the inverse square root of the heliocentric distance (R) (Delsemme 1982). This result in turn was based on an analysis of 57 observations of the motions of jets and expanding haloes by Bobrovnikoff (1954), and agrees with a simple thermal expansion law. Bobrovnikoff's publication, however, is in the report of the Perkins Observatory, which consists of relatively informal summaries of the work of the staff. The report mentions that the results are based on "...the motion of jets and the expansion of halos..." presumably from the inspection of photographs of many different comets (the report mentions 57 "cases", but leaves unclear whether these are comets or observations). The fit to this data is excellent and agrees with a simple model for dust outflow, and so has never been followed up.

In addition, a hydrogen model of the dynamics of a cometary coma indicates that dust velocity should vary more nearly as the inverse of heliocentric distance ($v \propto R^{-1}$). The model is presented in some detail in Wallis (1982). We present here observational data to resolve the discrepancy between the models.

Earth, Moon, and Planets **72**: 99-102, 1996.
© 1996 *Kluwer Academic Publishers.*

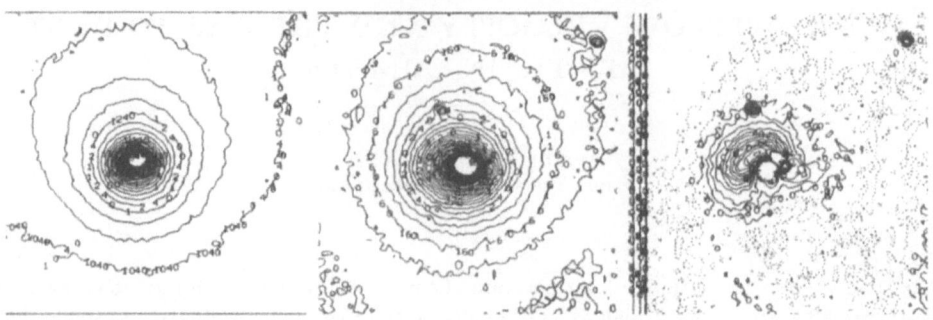

Fig. 1. left, IHW image nnsn0640, taken 20 Dec. 1985 with an RG665 (continuum) filter.
center, IHW image nnsn0665, taken 24 hours later. right, Difference between nnsn0665
and nnsn0640, after sky subtraction and brightness matching. Note prominent jet to left
of nucleus. North is up, east to the left, image sections are 3.2 arcmin across.

2. Dust Outflow Velocity

Bobrovnikoff's results apparently averaged the dust outflow velocity of all
comets observed at a given R. This dependence may vary among comets. To
minimize the effects of this variation, the dependence of dust outflow velocity
with R is here determined for one well observed comet, P/Halley.

The International Halley Watch (IHW) near-nucleus studies network
(nnsn) data base was searched for images that showed signs of expanding
jets or haloes. The technique used was to take two continuum images made
about 24 hours apart, subtract the background, scale the peak brightnesses,
align the images, and subtract.

This process is demonstrated in Figure 1. The first panel is a contour
plot of IHW archive image nnsn0640, made on 20 Dec. 1985 at 0325 UT
with an RG665 filter. The second panel shows image nnsn0665, made on
21 Dec. 1985 at 0351 UT. The third panel shows the result of scaling and
subtracting the two images.

The prominent near nucleus brightness in the first image leads to a generally negative area in the difference image near the peak (which of course
has a value of zero). This has resolved into a broad jet, which is positive in
the difference image. The outer edge of this jet has expanded a maximum of
35 pixels which leads to a projected outflow velocity (v) of 0.231 km/sec.

A non-exhaustive search of the IHW nnsn database turned up six unam-
biguous jets. Many more exist, of course, but the temporal difference method
used here is sensitive only to large changes. In this weakness it mimics
Bobrovnikoff's measurements as well. These outflow velocity points are plotted against the heliocentric distance at which they were observed, in Figure
2. A power law fit to this data gives $v \propto R^{-0.41}$. If data from an outburst
at 14 AU is included ($v=0.045$ km/sec, Sekanina et al. 1992) this exponent
becomes -0.55. This point is included in Figure 2.

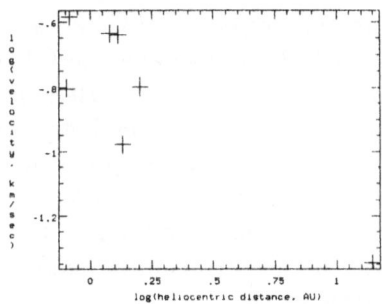

Fig. 2.   Dust outflow velocity as a function of heliocentric distance (R). Note that values are lower limits due to projection effects. Point at $\log(R) = 1.15$ (R=14AU) is from Sekanina et al. (1992).

3. Discussion

The outflow velocities presented here are lower limits due to projection effects. A more exhaustive search, perhaps using a more sensitive algorithm than the temporal difference method, would probably turn up more data points and better define the upper envelope of the v, R field. If we attempt to limit the effects of projection by fitting only the upper four points in Fig. 2, we obtain $v \propto R^{-0.66}$. However, the paucity of data casts some doubt on this last measurement.

It is also possible that the large outbursts such as are detected here are unusual in the dust terminal velocity achieved. It may be that the steady state dust outflow follows a different process and therefore a different dependence on heliocentric distance. This more complex hypothesis is untestable at present. Strictly speaking, we have only determined the heliocentric dependence of the dust outflow velocity in outbursts.

4. Gas-To-Dust Ratios

Recently Schleicher (1993) has detected spectrographic differences among comets. Schleicher shows that there appear to be two types of comets, the "normal" comets with approximately equal production rates of C_2 and OH, and "depleted" comets, with much less C_2 produced than OH (similar variation with respect to CN is seen). Members of this latter category include P/Giacobini-Zinner, P/Borelly, P/Swift-Gehrels, and (the extreme case) P/Wolf-Harrington.

To further parameterize possible differences among comets, Storrs et al. (1992) determined the gas-to-dust ratio (GDR) for well-observed comets in the McDonald Observatory Faint Comet Survey (FCS) (Cochran et al.

1992). These comets were observed several times, at varying heliocentric distances. Storrs et al. found that on average the GDR appeared to decrease with increasing heliocentric distance, with GDR $\propto R^{-0.59}$. Storrs et al. interpret this as observational evidence of outgassing primarily through a porous mantle: as more heat is added to the nucleus, more gas is produced but the dust that would accompany the gas in the case of free sublimation is largely trapped by the mantle. Thus the GDR determined by "usual" methods should not necessarily be used to compare different comets, observed at different heliocentric distances.

This conclusion assumes $v \propto R^{-0.5}$. If $v \propto R^{-1}$, the observed variation in GDR with R disappears and measurements of a given comet's GDR at various heliocentric distances can be averaged to give an indication of the volatile-to-refractory ratio of the nucleus. If this is done then the only comet in the "depleted" group of Schleicher (1993) that had its GDR determined by Storrs et al. (P/Giacobini-Zinner) has a GDR near the average of the sample. The observational data presented here (as well as that of Bobrovnikoff (1954)) indicate, however, that outflow velocity $v \propto R^{-0.5}$ and so the original conclusions of Storrs et al. (1992) (e.g. GDR $\propto R^{-0.59}$) are valid.

5. Conclusions

The dependence of v on R for comet P/Halley appears to follow the inverse-square-root law mentioned by Delsemme (1982) and others. This is also in agreement with Combi (1989) who combined data from various sources with his Monte-Carlo coma model and found a general agreement with an inverse-square-root law. Combi and others have also pointed out that this result is in agreement with energy equilibrium calculations. This observational confirmation of the inverse-square-root dependence of dust outflow velocity on heliocentric distance confirms the results of Storrs et al. (1992) that the canonically measured gas-to-dust ratio in comets varies with heliocentric distance in approximately the same manner.

References

Bobrovnikoff: (1954), *A. J.* **59 no. 1221**, 356–358
Cochran, A.L., Barker, E.S., Ramseyer, T.F., and Storrs, A.D.: (1992), *Icarus* **98**, 151–162
Combi, M.: (1989). *Icarus* **81**, 41–50
Delsemme, A.: (1982), *Comets*, ed. Wilkening, pp. 85–130
Schleicher, D.G.: (1993), *IAU Symposium 160: "Asteroids, Comets, and Meteors 1993"*, ed. A. Milani, M. di Martino, and A. Cellino, pp. 415–428
Sekanina, Z., Larson, S.M., Hainaut, O., Smette, A., and West, R.M.: (1992), *Astron. & Astrophys.* **263**, 367–386
Storrs, A.D., Cochran, A.L., and Barker, E.S.: (1992), *Icarus* **98**, 163–178
Wallis, M.: (1982), *Comets*, ed. L.L. Wilkening, pp. 357–369

INTERPRETATION OF HMC IMAGES BY
A COMBINED THERMAL AND GASDYNAMIC MODEL

J. KNOLLENBERG and E. KÜHRT
Deutsche Forschungsanstalt für Luft- und Raumfahrt,
Institut für Weltraumsensorik,
Rudower Chaussee 5, 12489 Berlin, Germany

and

H.U. KELLER
Max-Planck-Institut für Aeronomie, 37191 Katlenburg-Lindau, Germany

Abstract. Images of comet Halley's nucleus taken by the HMC camera during the GIOT-TO encounter in 1986 show that a major part of the total dust production is localized in a few active areas which are the sources of gas-dust jets. The global dust distribution in the inner coma is dominated by two main jets roughly directed to the sun. A combination of a 1D thermal nucleus model with an axisymmetric continuum model of the jet outflow was used to investigate the properties of the inner coma. Detailed investigations show that the characteristics of the observed jets can be reproduced by outgassing from free sublimating active areas of a few km in diameter, a dust to gas ratio of 1 – 2.5 and a size distribution dominated by the larger grains. It is further shown that most of the observational constraints provided by the HMC data can be met simultaneously by a model of three jets superimposed on a weak background.

1. Introduction

The spacecraft fly-bys of Halley's comet in 1986 have demonstrated that the dynamics and optical appearance of the inner coma is dominated by jets of gas and dust (Keller *et al.*, 1986; Sagdeev *et al.*, 1987). Images taken by the HMC camera showed that most of the dust production is localized in three active areas which did not cover more than about 20% of the illuminated surface (Keller *et al.*, 1987). These areas are the source regions of the observed gas-dust jets. Reitsema *et al.* (1989) showed that the angular dependence of the intensity can be well described by the sum of three Gaussians and a constant. The halfwidths of the two main jets visible in the HMC images were determined to 31° and 37° (Keller *et al.*, 1994). In spite of the observation that the jets are roughly directed to the sun and that no clear evidence for nightside activity could be detected in the images, a surprisingly high level of intensity was measured on the antisunward side of the images. The ratio of the total intensities integrated over the sunward and antisunward hemispheres was determined to $D/N = 3.2$ (Keller and Thomas, 1989). On the other hand, the jump of the intensity across the nightside terminator was only about a factor of 2–2.5, thus providing evidence against the hypothesis that a jet directed away from HMC could be responsible for the low D/N-value observed.

Earth, Moon, and Planets **72**: 103-112, 1996.
© 1996 *Kluwer Academic Publishers.*

Because existing jet models suffer from unrealistic assumptions, like a dust concentration in μm-sized grains [e.g. Kitamura (1987); Körösmezey and Gombosi (1990)] or a large nightside production (Kitamura, 1986), they are not able to explain the observations. Therefore, the objective of this paper is to present a new cometary jet model and to give a quantitative interpretation of the HMC images. Because reliable information about the position and extent of the active regions and of the three-dimensional dust distribution could not be deduced from the measurements, it was decided to develop a new axisymmetric model.

2. Model

The physical model of the inner coma can be described by the expansion of a mixture of gas and dust emanating from active regions on the nucleus surface into a vacuum or a radial background flow. The gas phase is characterized by a constant adiabatic exponent $\gamma_{ad} = 4/3$ (corresponding to water vapour) and is treated as an inviscid, compressible continuum.

At the nucleus surface a gas production rate was prescribed as a function of the polar angle Θ with respect to the jet axis:

$$Z(\Theta) = Z_j f(\Theta) + Z_b. \tag{1}$$

Two extreme cases for the spatial distribution of the production rate inside the active region were considered, a constant [$f(\Theta) = 1$ for $\Theta < \Theta_j$, $f(\Theta) = 0$ otherwise] and a Gaussian distribution of production rate [$f(\Theta) = \exp(-(\Theta/\Theta_j)^2)$]. The actual values of the outgassing rates were calculated using a sophisticated thermal model described in detail by Kührt and Keller (1994). The thermal calculations were carried out for the orbit of P/Halley, a heliocentric distance of 0.89 AU, a rotation period of 55 h and different values of the thermal conductivity of the dust crust in the range $K_c = 0 - 1$ W K^{-1} m^{-1}. This model provides the gas flux as a function of latitude and local time but for values of $K_c > 0.1$ W K^{-1} m^{-1} the background production Z_b could be reasonably approximated by a constant. The deviations from isotropic outgassing were less than 15% in these cases. Experiments with different rotation periods showed that Z_b is not critically dependent on the rotational state of the nucleus. The gas production rate on the jet axis was determined by the energy balance of a freely sublimating icy surface to $Z_j = 2\,10^{22}$ mol. m^{-2} s^{-1}. The symmetry conditions were used on the axes, for $\Theta = 0°$ and $\Theta = 180°$ and the free outflow conditions were applied at the outer boundary.

The results presented below were calculated with a grain size distribution which was approximated by 11 logarithmically spaced discrete grain sizes between 1 μm and 1 mm. The grains are considered as spheres with a density of $1 \, \mathrm{g \, cm^{-3}}$ and their interaction with the surrounding gas stream is described by the theory of free molecular flow (Hayes and Probstein, 1959). The distribution of mass between the different size classes is characterized by an exponent γ. The contribution of the individual size classes to the total mass loading κ is then given by :

$$\kappa_i = \kappa_m \left(\frac{a_i}{a_m} \right)^{3-\gamma} \tag{2}$$

where a_m and κ_m are the radius and the dust to gas ratio of the largest grains.

The numerical solution of the coupled hyperbolic system of conservation laws of gas and dust flow is achieved by an explicit second order Godunov-type scheme. The continuum equations were integrated in time until a steady state was reached. Calculations were performed on a fixed grid of 180×51 points with an angular spacing of $\Delta\Theta = 1°$ and an increasing radial mesh size with distance from the nucleus. The nucleus is assumed to be spherical with a radius of $R_N = 6$ km. The width of the first computational cell above the nucleus was 0.1 km and the outer boundary was set at a cometocentric distance of 180 km. A more detailed description of the numerical procedure is given by Knollenberg (1994).

For a direct comparison of the model results with the HMC images the calculated density values were converted to relative intensities by integrating along the line of sight and weighting appropriately with the particle cross sections. Under the assumptions of an optically thin coma (e.g. Chick and Gombosi, 1992) and identical scattering phase functions for all grain size classes, this quantity is proportional to the intensity of the light scattered by the dust grains in the coma.

3. Results

Several model runs were performed to study the influence of the dust to gas ratio, the grain size distribution, the extent of the active region and the thermal conductivity of the dust crust on the formation of dust jets. Furthermore, the influence of the viewing geometry on the optical appearance of the inner dust coma was investigated.

Fig. 1 shows contours of the calculated intensity of two models, both with $\Theta_j = 15°$, a dust to gas ratio $\kappa = 1$ and a size distribution exponent $\gamma = 2.5$ for three different aspect angles (angle jet axis-observer) $\phi = 90°, 107°$ and $135°$. The spherical nucleus is located in the centre and the jet axis is the negative x-direction. The contours on the left (Fig. 1a, c and e) were calculated with a thermal conductivity $K_c = 0$ implying a diminishing background production. In this case a smooth decrease from the maximum in the centre to the wings of the jet is obvious (Fig. 1a). This is caused by the strong lateral expansion of the gas jet into the vacuum until the maximum expansion angle is reached (see Koppenwallner *et al.*, 1986). But because the dust particles decouple from the gas flow a few km above the nucleus the dust jet is more confined and no significant flow of grains across the terminator is generated, thus giving a value of $D/N = \infty$. A close inspection reveals that the angular dependence of the intensity can be approximated by a Gauss-function (Knollenberg, 1994).

The results in the right column (Fig. 1b, d and f) were calculated with a thermal conductivity of the crust of $K_c = 0.2 \, \mathrm{W \, m^{-1} \, K^{-1}}$, resulting in a background production of $Z_b = 0.2\%$ of Z_j. In this case the lateral expansion of the gas is stopped by the ambient pressure in a contact discontinuity. The angular position of the contact surface is determined by the ratio of the gas production rate inside the active region to that of the background. In our example it forms an angle of 80° with the axis. Furthermore, dust is accumulated here, thus forming the narrow cone visible in Fig. 1b. By tracing test particles in the calculated gas flow field, it can be shown that the majority of the dust which is concentrated in this cone stems from an annulus immediately surrounding the active region. On the nightside of the nucleus the undisturbed radial background flow is evident. The low gas production rate causes considerably lower dust particle velocities compared with the velocities assumed on the jet axis (by a factor of 6–15, depending on the grain size). Thus, a surprisingly high intensity is apparent in the antisunward hemisphere and a ratio of $D/N = 2.3$ results.

In Figure 1 c–f, the effect of a changing aspect angle on the optical appearance of the inner dust coma is demonstrated. For the isolated jet with no background production the shape of the contours remains, but they are slightly broadened and some dust appears in projection on the antisunward side of the nucleus. For $\phi = 135°$ a value of $D/N = 25$ is reached and the jump in intensity by crossing the nightside limb is $I_\infty/I_c = 10^4$ ($I_c =$ intensity just above the nucleus surface). In the case of the surrounded jet a deviation of the aspect angle from 90° causes the narrow dust cone to develop into a broad maximum (Fig. 1d). A further increase then smears out the dust cone completely. The sunward to antisunward ratio remains fairly constant in this range of aspect angles, varying only between $D/N = 2.2 - 2.5$. The intensity jump can be evaluated to be $I_\infty/I_c = 2.7$ for $\phi = 107°$ (corresponding to an exactly sunward jet in the HMC images). Then it is more rapidly increasing with increasing aspect angle, reaching a value of $I_\infty/I_c = 6$ for $\phi = 135°$.

Because the strength of the gas-dust interaction is mainly determined by the total cross section of the dust particles, a variation of the dust/gas ratio κ and of the size distribution exponent γ has consequences for the resulting halfwidths of the jets. For the same model as discussed above the influence

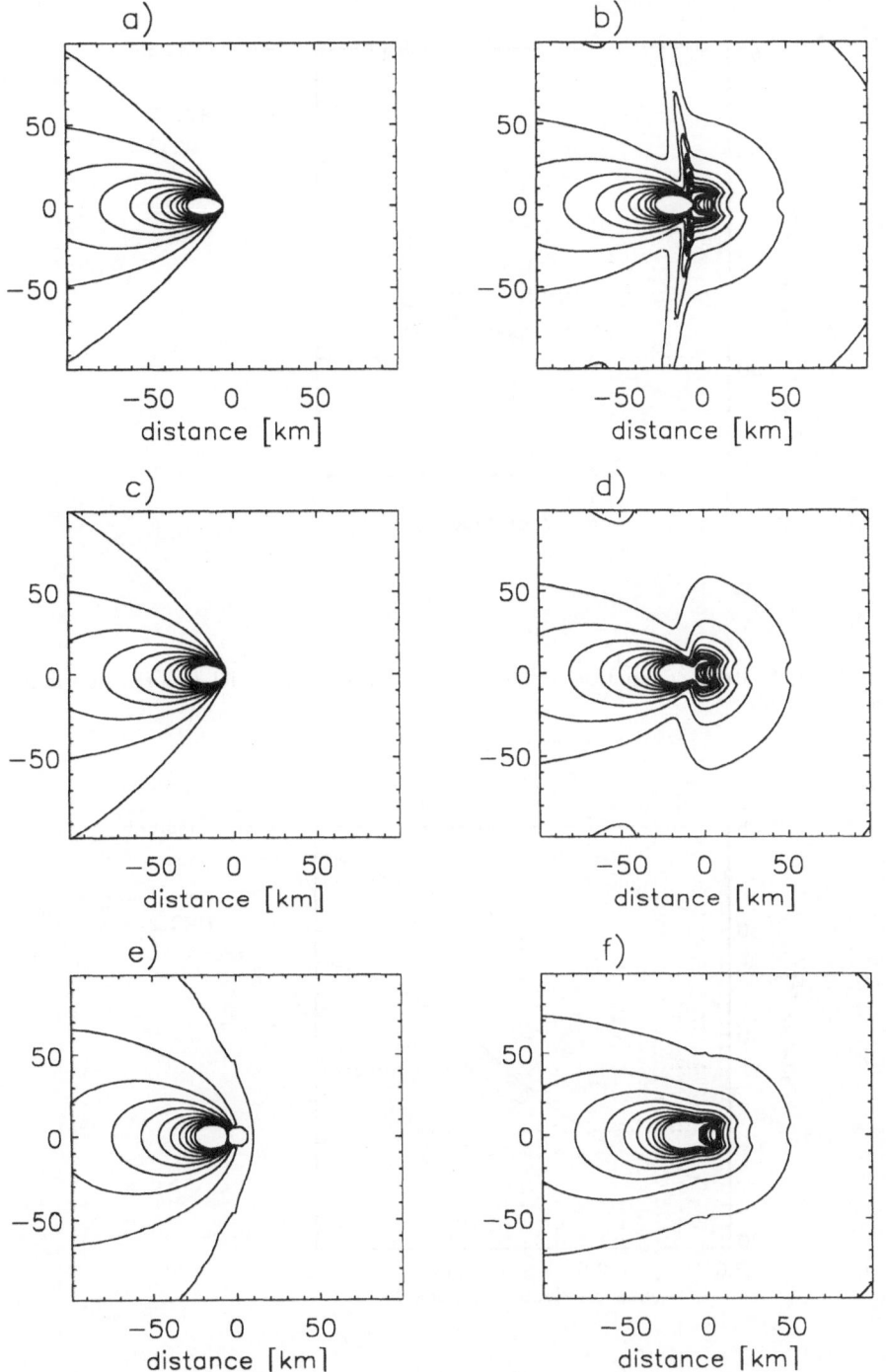

Fig. 1. Contours of intensity for different aspect angles: a), c) and e) $K_c = 0$;   b), d) and f) $K_c = 0.2 \, \mathrm{W \, K^{-1} \, m^{-1}}$;   a), b) $\phi = 90°$;   c), d) $\phi = 107°$;   e), f) $\phi = 135°$

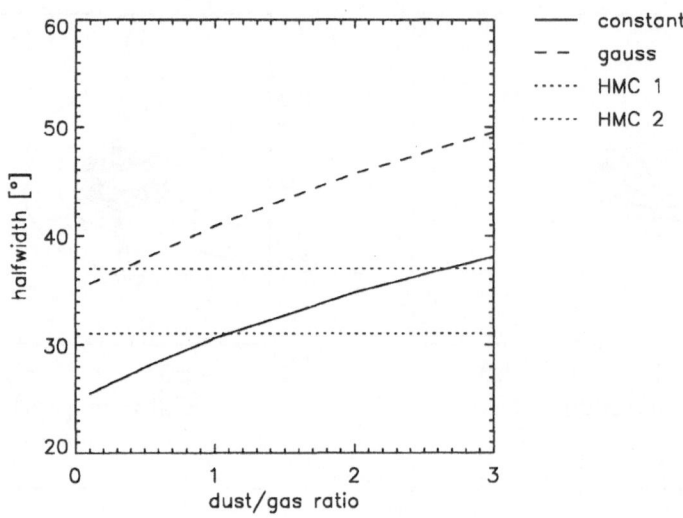

Fig. 2. Jet halfwidth as a function of the dust to gas ratio for a model with size distribution exponent $\gamma = 2.5$

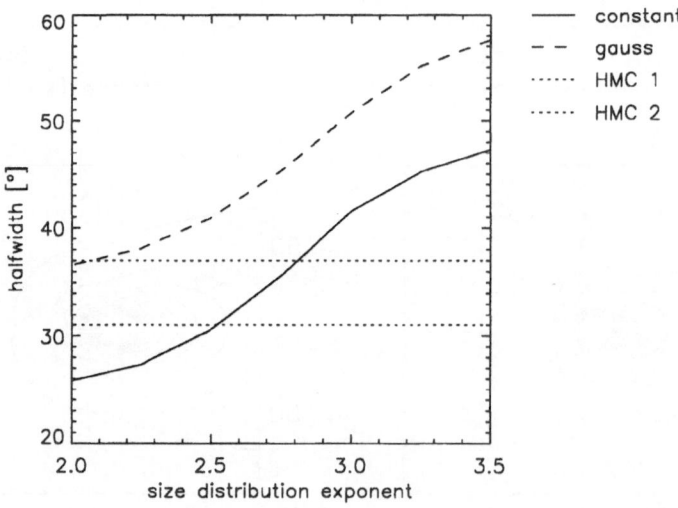

Fig. 3. Jet halfwidth as a function of the size distribution exponent γ for a dust to gas ratio $\kappa = 1$

of these parameters is shown in Figures 2 and 3 ($\phi = 90°$). In addition, the results of another series of model runs are presented (dashed lines) where the gas and dust production inside the active area was assumed to be a Gauss function of the polar angle [$f(\Theta) = \exp(-(\Theta/\Theta_j)^2)$]. For comparison the values determined for the two strong jets visible in the HMC images are included in the plots. Because of the larger lateral pressure gradient just above the active region the Gaussian jets are in all cases about 10 – 12° (or 30–40%) broader than the dust jets originating from a uniformly active source. It can be seen from these figures that the most important parameter which determines the halfwidth of the jets is the distribution of grain sizes.

To conclude, Figure 4 shows a comparison of HMC data (4a) with a superposition of three axisymmetric model jets (4b) and an additional weak background (4c). The sun is to the left and 17° behind the image plane. The jets are rotated in the image plane by −43°, 15°, 85° with respect to the sun direction (counted clockwise). In the case of the southern (−43°) jet a larger active area was assumed ($\Theta_j = 20°$); therefore its total dust production is 1.8 times higher than that of the left going jet, which is again three times stronger than the weak upgoing one ($\kappa = 0.3$). The axes of the two strong jets are lying in the image plane whereas the weak northern jet is tilted by 45° out of the image plane (to the observer). It can be seen that the shape of the contours on the dayside is quite well approximated by the 3-jet model. Deviations are only significant in the vicinity of the nucleus where the irregular topography and processes like fragmentation of grains are important. But because only some dust of the weak northern jet is projected onto the nightside, the D/N ratio is much higher than observed ($D/N = 15$). This discrepancy to the data can be removed by introducing a small background activity of the order of 0.2% of Z_j, now giving a value $D/N = 3.9$ (Fig. 4c). The slight difference in the shape of the contours on the nightside is caused by the assumed spherical shape of the model nucleus and could be removed by a more realistic approximation of the nucleus.

4. Discussion and Conclusions

The model calculations have shown that dust jets originating from source areas with strongly enhanced activity compared to the background develop a Gaussian shape, independent of the specific distribution of production rate inside the source region. This is in accordance with the observation of the global dust distribution in the inner coma (e.g. Keller *et al.*, 1994). Figures 2 and 3 show that the observed halfwidths of the two main jets can be explained by a model with a uniformly active region, a size distribution exponent $\gamma = 2.5 - 2.8$ and a dust to gas ratio in the range of

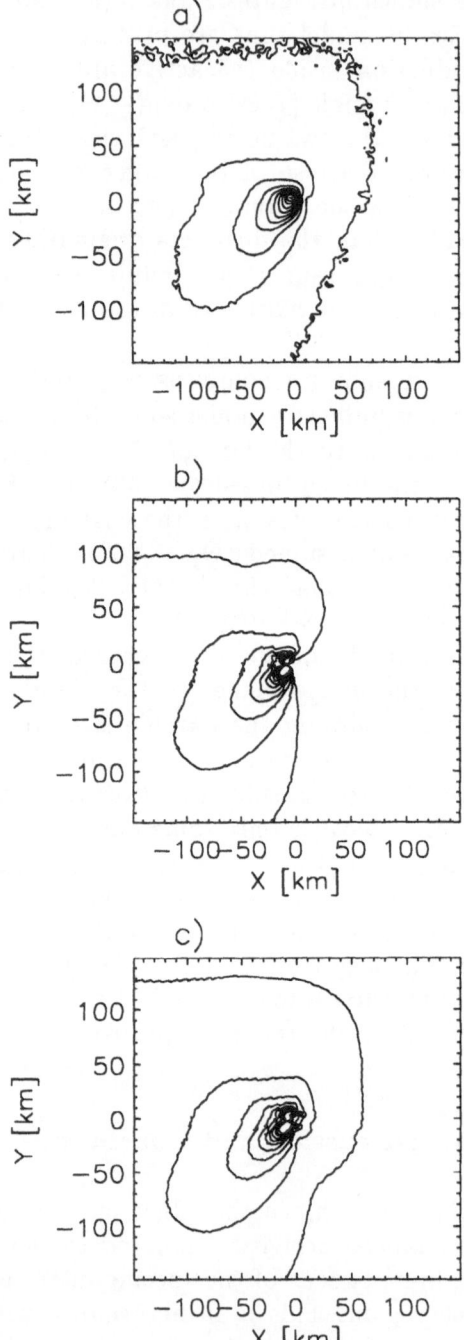

Fig. 4.   Comparison of HMC data with model results: a) HMC image 3128;   b)
Model of three jets without background;   c) Model of three jets with weak background
production

$\kappa = 1 - 2.5$. This implies a grain size distribution where the mass is dominated by the larger grains, consistent with the in-situ measurements in the coma of P/Halley (McDonnell et al., 1989). A size distribution with a significantly higher exponent, as deduced from optical ground based measurements [e.g. $\gamma = 3.5$ (Brin and Mendis, 1979)] would lead to much broader dust jets ($> 50°$) and, thus, cannot be representative for the coma of P/Halley at the GIOTTO encounter. The calculations presented above suggest to explain the low D/N-ratio observed by an approximately isotropic weak background production of gas and dust from the "inactive" areas. The background production needed is of the order of 0.1–0.3% of the free sublimation rate, a value provided by thermal models with reasonable thermal conductivities of the dust crust. In addition, such a low production rate is within the observational constraint given by the ratio of the lowest intensity measured above the nightside of the nucleus I_c to the maximum intensity I_{max} inside an active region of $I_c/I_{max} = 0.01$. Furthermore, it seems to be unlikely that the projection of a jet directed to or away from HMC could contribute much to the observed intensity on the nightside. First, the halfwidth of the jets increase with the deviation from $\phi = 90°$ giving values within the range of the observations only for $\phi = 90° \pm 35°$ and, second, the jump in intensity across the nightside limb would not be compatible with the observed $I_\infty/I_c = 2 - 2.5$ for $\phi > 110°$. In addition, the shape of the contours changes significantly, if the axes of the two main jets are more than 30° out of the image plane. In contrast, the dust emission of the weak and broad upgoing jet seems to be tilted with respect to the image plane. This could provide the observed halfwidth without the need of an increased dust loading and could serve as an explanation of the observed intensity gradient above the nucleus in the north-south direction (Keller and Thomas, 1989).

This analysis of the global distribution of dust in the inner coma of P/Halley therefore supports an inhomogeneous comet nucleus model where a few ice dominated active areas are surrounded by regions covered by a porous dust crust which chokes the activity to quite low values.

References

Brin, G.D. and Mendis, D.A.: 1979, Dust release and mantle development in comets, Astrophys. J. 229, 402–408.

Chick, M.K. and Gombosi, T.I.: 1992, Multiple Scattering of Light in a Spherical Shell Cometary Atmosphere with an Axisymmetric Jet, Icarus 98, 179–194

Hayes, W.D. and Probstein, R.F.: 1959, Hypersonic Flow Theory, Academic Press, New York, USA.

Keller, H.U. and Thomas, N.: 1989, Evidence for near-surface breezes on comet P/Halley. Astron. Astrophys. 226, L9–L12.

Keller, H.U., Arpigny, C., Barbieri, C., Bonnet, R.M., Cazes, S., Coradini, M., Cosmovici, C.B., Delamere, W.A., Huebner, W.F., Hughes, D.W., Jamar, C., Malaise, D., Reitsema, H.J., Schmidt, H.U., Schmidt, W.K.H., Seige, P., Whipple, F.L, and Wilhelm,

K.: 1986, First Halley Multicolour Camera imaging results from Giotto. *Nature* **321**, 320–326.

Keller, H.U., Delamere, W.A., Huebner, W.F., Reitsema, H.J., Schmidt, H.U., Whipple, F.L., Wilhelm, K., Curdt, W., Kramm, J.R., Thomas, N., Arpigny, C., Barbieri, C., Bonnet, R.M., Cazes, S., Coradini, M., Cosmovici, C.B., Hughes, D.W., Jamar, C., Malaise, D., Schmidt, K., Schmidt, W.K.H., and Seige, P.: 1987, Comet P/Halley's nucleus and its activity. *Astron. Astrophys.* **187**, 807–823.

Keller, H.U., Knollenberg, J., and Markiewicz, W.J.: 1994, Collimation of Cometary Dust Jets and Filaments, *Plan. Sp. Sc.* **42**, 367–382.

Kitamura, Y.: 1986, Axisymmetric Dusty Gas Jet in the Inner Coma of a Comet. *Icarus* **66**, 241–257.

Kitamura, Y.: 1987, Axisymmetric Dusty Gas Jet in the Inner Coma of a Comet. II. The Case of Isolated Jets. *Icarus* **72**, 555–567.

Knollenberg, J.: 1994, Modellrechnungen zur Staubverteilung in der inneren Koma von Kometen unter spezieller Berücksichtigung der HMC-Daten der GIOTTO-Mission. Ph.D. Thesis, Georg-August Universität zu Göttingen, Göttingen, Germany.

Koppenwallner, G., Boettcher, R.D., Detleff, G., and Legge, H.: 1986, Rocket exhaust plume flow into space. ESA/ESTEC ESA-265.

Körösmezey, A. and Gombosi, T.I.: 1990, A Time-Dependent Dusty-Gas Dynamic Model of Axisymmetric Cometary Jets. *Icarus* **84**, 118–153.

Kührt, E. and H.U. Keller: 1994, The Formation of Cometary Surface Crusts. *Icarus* **109**, 121–132.

McDonnell, J.A.M., Green, S.F., Grün, E., Kissel, J., Nappo, S., Pankiewicz, G., and Perry, C.H.: 1989, In-situ exploration of the dusty coma of comet P/Halley at Giotto's encounter: flux rates and time profiles from 10^{-19} kg to 10^{-5} kg. *Adv. Space Res.* **9**, (3)277–(3)280.

Reitsema, H.J., Delamere, W.A., Williams, A.R., Boice, D.C., Huebner, W.F., and Whipple, F.L.: 1989, Dust Distribution in the Inner Coma of Comet Halley: Comparison with Models. *Icarus* **81**, 31–40.

Sagdeev, R.Z., Smith, B., Szegö, K., Larson, S., Tóth, I., Merényi, E., Avanesov, G.A., Krasikov, V.A., Shamis, V.A., and Tarnapolski, V.I.: 1987, The spatial distribution of dust jets seen during the Vega 2 flyby. *Astron. Astrophys.* **187**, 835-838.

EFFECTS OF SHAPE AND SPIN ON THE TIDAL DISRUPTION OF P/SHOEMAKER-LEVY 9

A.W. HARRIS

Jet Propulsion Laboratory, California Institute of Technology,
MS 183-501, Pasadena, CA 91109 U.S.A.

Abstract. I derive an approximate criterion for the tidal disruption of a "rubble pile" body as it passes close to a planet (or the sun):

$$\rho_c \approx \left[2\rho_p \left(\frac{R_p}{r} \right)^3 + \left(\frac{\omega}{\omega_0} \right)^2 \right] \left(\frac{a}{b} \right),$$

where ρ_c is the critical density below which the body will be disrupted, ρ_p is the density of the planet (or sun), R_p is the radius of the planet, r is the periapse distance, ω is the rotation frequency of the body, ω_0 is the surface orbit frequency about a body of unit density, and a/b is the axis ratio of the body, considered as a prolate ellipsoid. For P/Shoemaker Levy 9, in its passage close to Jupiter in 1992, this expression suggests that the critical density is ~1.2 for a spherical, non-spinning nucleus, but could be >2.5 for a 2:1 elongate body with a typical rotation period of ~10 hours.

1. Introduction: a generalized Roche Limit

The *classical* Roche limit is defined as the lower limit of the density of a homogeneous, hydrostatic (that is, synchronously co-rotating) fluid body in a circular orbit about another body at a given orbital distance, or analogously, it is the minimum orbital distance for a given density body for which there is a closed hydrostatic equilibrium figure. The solution to this idealized problem is itself quite complex (see, for example, Chandrasekhar 1969). To further complicate matters, practical applications, such as the question of how close to a planet can satellites exist, or what is the minimum density of a small body (comet nucleus) passing near a planet or the sun which can resist disruption, depend on additional parameters, such as the shape, direction and rate of spin, and its material properties.

It is sometimes useful to state a *generalized* Roche Limit, as follows:

$$\frac{r}{R_p} = f \left(\frac{\rho_p}{\rho_c} \right)^{1/3}, \tag{1}$$

or equivalently,

$$\rho_c = f^3 \rho_p \left(\frac{R_p}{r} \right)^3, \tag{2}$$

Earth, Moon, and Planets **72**: 113-117, 1996.

where ρ_p and R_p are the density and radius of the planet, respectively; r is the orbital radius, or periapse of an elliptical or hyperbolic orbit, and ρ_c is the density of the secondary body (the comet nucleus in this case). The constant f allows a comparison of disruption criteria derived under a variety of assumptions. Table I is a summary several such models.

TABLE I

No.	Model	Rotation state	f	f^3
1	Hydrostatic fluid (Classical Roche)	synchr. rotating	2.46	14.8
2	◯◯	synchr. rotating	2.88	24
3	◯◯	non-rotating	2.52	16
4	◯•	synchr. rotating	1.44	3
5	◯•	non-rotating	1.26	2
6	Boss *et al.* (1991)	non-rotating	1.31-1.47	2.3-3.2
7	Sridher & Tremaine (1992)	non-rotating	1.69	4.8
8	Ziglina (1978)	synchr. rotating?	1.4	3

The first line in the table is the classical limit for fluid bodies, where $f = 2.46$. A commonly employed model in elementary textbooks is to derive the limit where two spheres of equal size and density will be pulled apart by tidal forces. For the case of spheres radially aligned and synchronously rotating in circular orbit, $f = 2.88$; for non-rotating bodies, radially aligned as they pass periapse (perhaps of a hyperbolic orbit), $f = 2.52$. One can see the appeal of this simplified model: it yields close to the "right" answer, compared to the classical limit. Another simple model is to consider when a small test particle, radially aligned on a larger sphere, would be separated off of the larger body by tidal force. This model yields a strikingly lower value of f, 1.44 or 1.26, for rotating and non-rotating cases, respectively. Finally, we can examine the results of several numerical or analytical calculations. Boss *et al.* (1991) numerically calculated disruptions of inviscid planetesimals passing near planets, to obtain a range of 1.31 to 1.47 for f; Sridhar and Tremaine (1992) obtain an estimate of $f = 1.69$ for viscous bodies, and Ziglina (1978) analytically obtained a value of $f = 1.4$ as a criterion for whether the pieces of a fractured sphere would disperse or remain as an undispersed "rubble pile".

2. The Roche Limit of a rotating spherical "rubble pile"

It is noteworthy that the various limits derived for solid materials (models 6, 7, 8) yield values of f which are generally similar to the values for the simple "test particle on a sphere" models (4, 5), but markedly different from the fluid or "two equal sphere" models (1, 2, 3). One way to understand this is that solid materials, even a "rubble pile" in which the body as a whole has only compressive strength, tends to remain relatively undeformed until it yields catastrophically. This yielding generally occurs when the local acceleration due to gravity plus tides plus spin reverses so that loose material levitates off the surface, at least somewhere on the body. Dobrovolskis (1982) showed that for not too irregular bodies, the internal stress pattern is such that the "angle of repose" of typical regolith materials ($\sim$45°) is not exceeded until the stress becomes tensile. That is, slumping does not occur until the body starts to be pulled apart. Thus we expect that the disruption criterion for a spherical "rubble pile" should be that given by models 4 or 5 in Table I. Model 4 appears to be the same result as Ziglina (1978). To derive this criterion, we set the sum of the three accelerations, gravitational, tidal, and centrifugal, equal to zero:

$$-\omega_0^2 \rho_c a + 2\omega_0^2 \rho_p \left(\frac{R_p}{r}\right)^3 a + \omega^2 a = 0, \tag{3}$$

where a is the radius of the body, ω is its spin frequency, and ω_0 is the surface orbit frequency about a sphere of unit density. Thus we obtain the Roche limit in terms of density:

$$\rho_c = 2\rho_p \left(\frac{R_p}{r}\right)^3 + \left(\frac{\omega}{\omega_0}\right)^2. \tag{4}$$

For a synchronously rotating body in circular orbit, $\left(\frac{\omega}{\omega_0}\right)^2 = \rho_p \left(\frac{R_p}{r}\right)^3$, so we obtain $f^3 = 2$ and 3 for the non-rotating and synchronously rotating cases, respectively.

3. The Roche Limit for a spinning prolate ellipsoidal "rubble pile"

For an elongate body, we can apply the same balance of accelerations as for the sphere. The least stable point on a prolate body will be the end of the long axis. The gravitational acceleration at the tip of the long axis can be obtained by integrating the acceleration over the volume of the ellipsoid. The result involves an elliptic integral, but turns out to be very close to being equal to the gravitational acceleration at the surface of a sphere of the same density and radius equal to the long axis, a, of the prolate ellipsoid, times the axis ratio, b/a. Thus for a 2:1 elongate body, the acceleration of gravity at the tip is ~1/2 that of a sphere of the same density and radius a. Equivalently, the acceleration of gravity would be about the same as the sphere for a prolate spheroid if the density were increased by a factor a/b. Thus for a prolate spheroid, at a time when the long axis of the body is aligned in the direction of the planet, the disruption criterion equivalent to (4) becomes:

$$\rho_c \approx \left[2\rho_p \left(\frac{R_p}{r} \right)^3 + \left(\frac{\omega}{\omega_0} \right)^2 \right] \left(\frac{a}{b} \right). \tag{5}$$

For P/Shoemaker Levy 9, passing $r \approx 1.3\ R_p$ from Jupiter:

$$\rho_c \approx \left[1.22 + \left(\frac{3.3^h}{P_{rot}} \right)^2 \right] \left(\frac{a}{b} \right). \tag{6}$$

Thus for a non-rotating sphere, $\rho_c \approx 1.2$, but for a 2:1 elongate nucleus, $\rho_c \approx 2.4$ for a non-rotating body, and even more for a rotating one. It is of course possible for a less dense body to not be disrupted, depending on the orientation as it passes periapse, so the expression (6) provides an upper limit of density for *possible* disruption. One can compare these density limits to the density, ~2.0, of the Martian satellite Phobos, which is generally considered to be a rocky, although likely porous, body. It therefore appears that for bodies of plausible shapes and spins, one cannot reliably infer that the original nucleus of P/Shoemaker Levy must have had an ice-like density rather than a rock-like one.

Acknowledgment

This research at the Jet Propulsion Laboratory, California Institute of Technology, was carried out under contract with NASA.

References

Boss, A. P., Cameron, A. G. W., and Benz, W.: 1991, 'Tidal Disruption of Inviscid Planetesimals', *Icarus* **92**, 165-178.

Chandrasekhar, S.: 1969, *Ellipsoidal Figures of Equilibrium*, Yale Univ. Press, New Haven, CT.

Dobrovolskis, A. R.: 1982, 'Internal Stresses in Phobos and Other Triaxial Bodies', *Icarus* **52**, 136-148.

Sridhar, S., and Tremaine, S.: 1992, 'Tidal Disruption of Viscous Bodies', *Icarus* **95**, 86-99.

Ziglina, I. N.: 1978, 'Tidal Destruction of Bodies Near Planets', *Phys. Solid Earth* **14**, 467-471.

UNUSUAL COMETS (?) AS OBSERVED FROM THE HUBBLE SPACE TELESCOPE

KAREN J. MEECH
Institute for Astronomy
2680 Woodlawn Drive, Honolulu, HI 96822

and

HAROLD A. WEAVER
Space Telescope Science Institute
3700 San Martin Drive, Baltimore, MD 21218

Abstract. In separate projects, the Hubble Space Telescope has been used to assess the nature of 3 unusual objects: Chiron, Pholus and P/Shoemaker-Levy 9. This paper will compare these objects and discuss how the unique capabilities of the HST may be used to address the issue of cometary activity in each. Chiron, which has exhibited obvious cometary characteristics for several years, might have a bound dust coma that is unresolvable from the ground. In an attempt to directly observe this bound coma, we have obtained a series of images of Chiron with the HST Planetary Camera. Inner coma structure out to 0.″2 has been detected. From these observations we infer a low bulk nucleus density for Chiron. Both HST and ground-based images of 5145 Pholus have been obtained to search for evidence of activity. The ground-based data give the most sensitive limits; however, it is shown that the WFPC-2 on HST can give limits 2–3 orders of magnitude more sensitive than conventional ground–based limits. Finally, as part of a collaborative effort, we have been obtaining HST observations of SL9 in order to determine the fragment sizes and to assess their nature (*i.e.*, cometary vs. asteroid). Both ground–based observations from the UH 2.2m telescope on Mauna Kea and HST observations show that the near–nucleus dust is redder than the sun. While FOS spectra did not detect OH emission, the WFPC-2 HST data show that the inner coma remained very circular from July 1993 up until 2 weeks prior to impact, implying continued production of dust.

Key words: Nucleus – Hubble Space Telescope – Comet Activity

1. Introduction

Although comet nuclei are generally not resolvable objects, the Hubble Space Telescope (HST), even prior to the repair of the optics, was a valuable tool for ascertaining information about the cometary nature of small Solar System bodies. In particular, the HST affords access to spectral regions diagnostic of cometary activity which are not easily accessible from the ground, specifically in the near UV. Secondly, the resolution afforded by HST is instrumental for understanding processes in the dust coma near the nucleus, and is ideally suited to very sensitive searches for activity near the nucleus. In this paper, we discuss how the HST has been used to look at cometary processes for three unusual cometary candidates: 2060 Chiron, 5145 Pholus and P/Shoemaker–Levy 9 (P/SL9).

Earth, Moon, and Planets **72**: 119-132, 1996.
© 1996 *Kluwer Academic Publishers.*

2060 Chiron is classified as a Centaur, one of three outer Solar System small bodies with an orbit in the vicinity of Saturn. Originally classified as a distant asteroid, the unusual orbit lead to speculation about the possibility that Chiron was a comet shortly after its discovery (Kowal *et al.*, 1979). Chiron's cometary nature was first inferred from its excess heliocentric light curve brightness (Hartmann *et al.*, 1990) which suggested activity beyond 12 AU, and was confirmed by the first detection of the coma by Meech and Belton (1989). As more observations became available, it became apparent that the long–term light curve exhibited an unusual slow outburst behavior, super-imposed upon which were irregular short–term brightness fluctuations. In addition to the activity at large heliocentric distances, r, Chiron was known to have an unusually large nucleus (radius $\lesssim$ 150-200 km) from ground–based, IRAS thermal infrared measurements and submillimeter observations (Lebofsky *et al.*, 1984; Spencer, *et al.*, 1989; Sykes and Walker, 1991; and Jewitt and Luu, 1992). More recent thermal IR detections and occultation measurements have suggested a radius near 90 km, still large for a comet nucleus (Buie *et al.*, 1993; Campins *et al.*, 1994). Based both on the large nucleus and the unusual light curve behavior, Meech and Belton (1990) developed a model for the coma which suggested the presence of a bound inner coma which was populated by grains on radiation–pressure perturbed ballistic trajectories which had residence times of months. In this scenario, the observed coma on Chiron and the short–term brightness fluctuations were the result of small escaping dust grains, whereas the long–term brightness variations were the result of the gravitational trapping of dust in an inner coma as populated by discrete sources on the nucleus. The model represented the observed light curve well, however, critics argued that the unobservable bound coma, was not photometrically necessary in order to explain the profiles (Luu and Jewitt, 1990), and that numerical integrations suggested that the grain lifetimes would be short and could not maintain the coma (Stern *et al.*, 1992). The unique ability of the HST to resolve the inner few tenths of an arcsec of the coma of Chiron was required to clarify this issue.

Another small body for which the HST can play a unique role in understanding its cometary nature is 5145 Pholus, a Centaur in an orbit which is very similar to Chiron. Discovered with the Spacewatch camera (Scotti, 1992), it was immediately realized that this outer Solar System object was redder than any other minor planet or comet (Mueller and Tholen, 1992), and that like Chiron, it was quite large, with a radius near 90 km (Howell, *et al.*, 1992; Davies *et al.*, 1993a). Shortly after discovery, Hainaut and Smette (1992) obtained deep images of Pholus to search for coma, and spectra were obtained by several groups to investigate the nature of the unusually red surface material. With an orbit similar to that of Chiron, it was hoped that evidence of cometary activity might be found; however, no coma was detected. Four groups have suggested that the spectrum might be matched by mixtures of organic tholins (Fink, *et al.*, 1992; Mueller *et al.*, 1992; Davies, *et al.*, 1993b; Hoffman, *et al.*, 1993), which are formed from irradiated CH_4-bearing ices. Wilson *et al.* (1994) has used Hapke scattering theory to match the Pholus spectrum with a mixture of tholins and water ice. They address several methods of its production on Pholus, and suggest that the spectrum is compatible with subsurface irradiation, implying that if Pholus does contain ices, it has not yet been active. Interestingly, the recently discovered Centaur, 1993 HA_2 (Rabinowitz, 1993) has an extremely red color similar to that of Pholus (Davies, 1994). This may have interesting evolutionary implications for comets which have resided in the Oort cloud or Kuiper Belt versus Chiron–like objects, which have spent some time in the inner Solar System, for which activity has removed the primitive irradiation mantle. There was thus great interest in placing very stringent limits on the amount of activity which could be present, and the resolution of HST enables the strongest limits to be placed on the existence of a dust coma.

Finally, it was hoped that the HST would be able to resolve the issue of whether or not P/Shoemaker–Levy 9 was in fact a split comet or asteroid. P/SL9 was discovered in March 1993 by Shoemaker *et al.* (1993), and by late May of the same year the orbital accuracy was high enough to infer that this object would impact Jupiter around 21 July 1994 (Marsden, 1993). Because of this, an intense observing effort began in order to determine the sizes of the nuclei, their composition and density so that the effect of the impacts with Jupiter could be adequately modelled. As discussed by Weaver *et al.* (1994a), there was initial circumstantial evidence that this object was a comet, based both on its low inferred nucleus density (Sekanina, 1993) and the circular appearance of the comae which is typical of cometary activity but difficult to create from tidal splitting. Both imaging and spectroscopic HST observations were made by Weaver *et al.* on 1993 July 1. While the radial surface brightness profiles of the comae were flatter than the canonical p^{-1} coma expected from isotropic sublimation, this did not rule out cometary activity, instead suggested possible non-steady state dust production. The non-detection of OH fluorescence placed an upper limit to the water production at 2×10^{27} molec sec^{-1}, which was similar to the observed value for P/Halley at $r = 4.7$ AU preperihelion, and thus did not exclude a volatile component.

2. Observations

2.1. CHIRON

HST observations of Chiron with a total integration time of 5,280 sec were made with the Planetary Camera (PC) on 1993 February 22, and 23, and

on March 8 near its minimum geocentric distance. Additional observations
of a PSF star were made on each date for deconvolution of these aberrated
images. All of the observations were obtained with the F555W filter, which
is similar to the V bandpass. At the time of the observations, Chiron was
at $r = 9.31$ and 9.29 AU and geocentric distances, $\Delta = 8.35$ and 8.41 AU,
respectively. Each PC pixel is $0.''044$ on a side, and this projected to a distance of approximately 260 km at Chiron. The data were processed through
the standard STScI pipeline processing programs combined with the best
known calibrations files as of 1994 February. Cosmic rays were removed
from the images, and then all images were registered on the peak of Chiron
after background subtraction. The CLEAN routine, using the modifications
by Keel (1991) of the 2–dimensional radar aperture synthesis image reconstruction techniques developed by Högbom (1974), was used for the image
deconvolution. The details of the observations, reductions and deconvolution
process are discussed in Meech et al. (1994a).

2.2. PHOLUS

A total of nine HST images of Pholus were obtained on 1992 May 2 using
the PC. Images were obtained in both the F785LP (essentially the I-band)
and the F555W filter. At the time of the HST observations, Pholus was
at $r = 8.74$ AU and $\Delta = 8.93$ AU, which gave a corresponding PC pixel
scale of 278 km pix^{-1} at Pholus. As in the case of Chiron, the data were
processed through the standard STScI pipeline system. Additional ground-
based observations were made of Pholus on 1992 Jan 31 using the UH 2.2m
telescope and on 1992 March 7 using the Cerro Tololo Interamerican Observatory 4m telescope. Pholus was imaged through the Mould R bandpass in
conditions of moderately poor seeing ($1.''2$–$1.''7$), with plate scales of $0.''351$
pix^{-1} and $0.''467$ pix^{-1}, respectively. During the January run, Pholus was at
a distance of $r = 8.705$ AU, $\Delta = 7.738$ AU, and in March the distances were
$r = 8.712$ AU, $\Delta = 8.041$ AU. The specifics of the data acquisition, standard
processing, and calibrations are discussed in Meech et al. (1994b).

2.3. P/SHOEMAKER–LEVY 9

The HST observations which began in 1993 (see Weaver et al. 1994a) continued (after the successful servicing of HST) throughout the spring of 1994 in a
campaign to follow the evolution of the dust coma as the comet approached
the impact with Jupiter (see Weaver et al., 1994b for a complete discussion
of this program). Observations were made during 8 periods from 1994 January 27 through July 20 using both the Wide–Field Planetary Camera 2
(WFPC–2), which had plate scales of $0.''0995$ pix^{-1} and the Faint Object
Spectrograph (FOS). The observations were made primarily through the

F702W filter (R bandpass) and in addition a few images were made in the F555W filter. Spectra were obtained using the G270H grating which covers the spectral range from 2223–3278 Å. During the period of observation, the geocentric distance varied from 4.4 to 5.5 AU. Additional ground–based observations were made using the UH 2.2m telescope with the TEK 2048 CCD on the nights of 1994 January 15, 17 and 18 UT, when the comet was at $r = 5.39$ AU and $\Delta = 5.55$ AU. At Cassegrain focus the plate scale was $0.''219$ pix^{-1}. The data were obtained under photometric conditions with sub–arcsec seeing ($\approx 0.''8$, FWHM). The reductions were done in a standard manner, using flats obtained on the twilight sky, which enabled flattening to better than 0.5% across the CCD. The transformation to the standard system used the standards from Landolt (1992) from which nightly extinction and transformation coefficients were determined. Photometry was performed using apertures of $0.''8$ and $2.''0$ on each of the nuclei. The $0.''8$ aperture was necessary to separate nuclei P$_1$ and P$_2$. The photometric measurements were done using the program "basphotc", written by M. Buie (see Buie and Bus, 1992), which allows the sky to be determined in a location which is far from the comet to avoid coma contamination. The sky subtraction was done on clear sky near the comet train, just to the southeast of the comet.

3. Discussion

3.1. CHIRON

The radial extent of the bound coma, or exopause boundary for Chiron, depends upon the mass of Chiron, as well as the grain sizes and scattering properties as defined by β. For a nucleus the density of water–ice (see Eqs. 1 and 2), this was expected to lie between 2,000 and 5,000 km from Chiron during the 1993 close approach, corresponding to $0.''3$–$0.''8$ from the nucleus. The exopause distance is given by:

$$r_{exp} = r[M_N/(M_\odot \beta)]^{0.5} \tag{1}$$

where $M_N = 4\,\pi\,R_N^3\,\rho_N\,/\,3$ [kg] is the nucleus mass for a nuclear density of ρ_N [kg m^{-3}] and radius R_N [m] and $M_\odot$ [kg] is the solar mass. Here β is the ratio of the solar radiation pressure acceleration on the grain to the solar gravity and is given by:

$$\beta = 5.7398 \times 10^4 Q_{rp}/(\rho_{gr}a_{gr}) \tag{2}$$

where Q_{rp} is the radiation pressure scattering efficiency, ρ_{gr} [kg m^{-3}] is the grain density and a_{gr} is the grain radius.

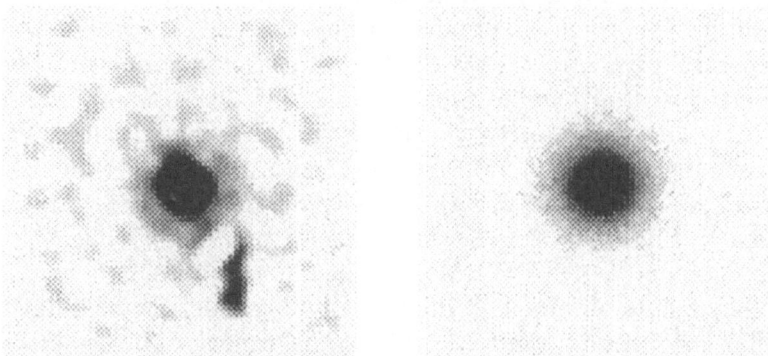

Fig. 1.   Deconvolved composite HST image of Chiron from 1993 Feb 23 showing the exopause boundary extending 0."2 (1,200 km) from the center (left). The dark blob in the lower right of the image is a CCD defect. On the right is a simulated image of Chiron and its bound coma.

The deconvolved HST composite image of Chiron, shown in Figure 1, clearly shows material close to the nucleus, extending out to about 1,200 km. Extensive testing of the deconvolution procedure was undertaken with the use of simulated Chiron images in order to assure the reality of the observed feature. This is discussed in some detail in Meech *et al.* (1994a). Figure 2 is the gradient of the Chiron surface brightness profile which shows the change in slope that is the expected exopause signature. At 0."2, the angular extent of the exopause was much closer to the nucleus than expected, and given the new better constraints on the nucleus size from occultation measurements (Buie, 1993) and from IR observations (Campins *et al.*, 1994), unless the grains are unusually small, it is probable that Chiron's density is quite low. Combining the above equations, substituting in the best estimates for R_N and r_{exp}, and assuming $Q_{rp} \approx 1$ gives

$$\rho_N = 0.27/(\rho_{gr} a_{gr}) \tag{3}$$

The trapped grain sizes are constrained not to be predominantly sub–micron because of the observed neutral color of the coma which represents the escaped particles; hence the mean coma particle size is at least microns in size. This implies that the bound particles, those which did not achieve escape velocity, must be larger. For comparison, stratospheric micromete-orites (Brownlee particles) have bulk densities in the range from 300 to 6,200 kg m^{-3}, with a mean value near 2,000 kg m^{-3} (Love *et al.*, 1994). This implies a nucleus density between 40–900 kg m^{-3}, with values biassed toward the low end for typical Brownlee particle dust densities. The impli-cations of the low density of Chiron in terms of its formation condition, and the possibility that the bound coma might represent sublimating grains are discussed fully in Meech *et al.* (1994a).

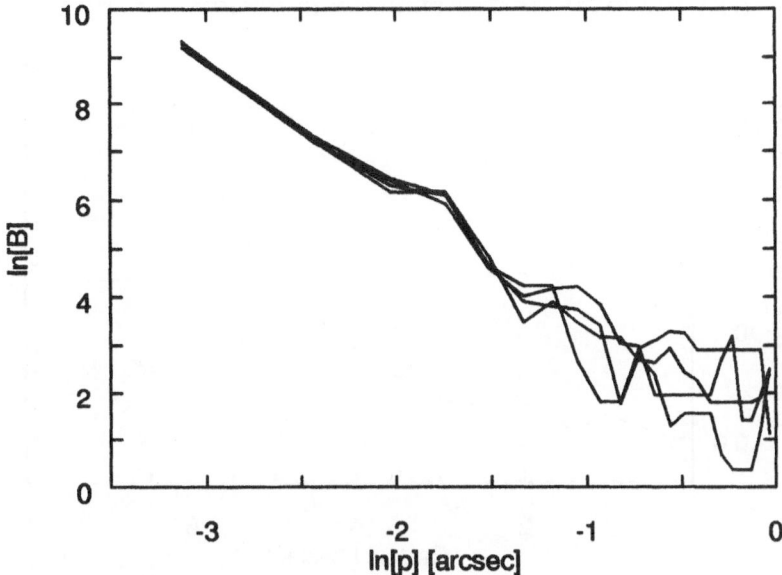

Fig. 2.   Relative surface brightness profile gradients of Chiron's inner coma from HST on 1993 Feb 22, 23, March 8, and the mean of all three.

The ground–based observations of Luu and Jewitt, from which they inferred that the bound coma was not photometrically necessary, are in fact consistent with the HST observations. Their analysis was unable to take account of the changing surface brightness profile in the innermost coma of Chiron, and they therefore came to an incorrect conclusion about its existence. In this case, the HST was instrumental in understanding a unique form of cometary activity on Chiron, and from these observations we will eventually be able to place exciting constraints on a comet nucleus density. Cycle 5 HST time has been awarded to continue this investigation.

3.2. PHOLUS

Luu and Jewitt (1992) have proposed a method of searching for low–level activity near asteroids, by comparing observed asteroid profiles with seeing–convolved models of nuclei plus varying amounts of coma. In their formulation, the most sensitive constraints on the amount of scattered light from the coma dust is found in the profile wings, far from the core of the image. However, it is far from the core where the precise determination and removal of the night sky brightness is the most critical. With this technique they were sensitive to mass loss rates > 0.1 kg sec^{-1}. With a different approach, however, and with the use of the HST to probe closer into the region close

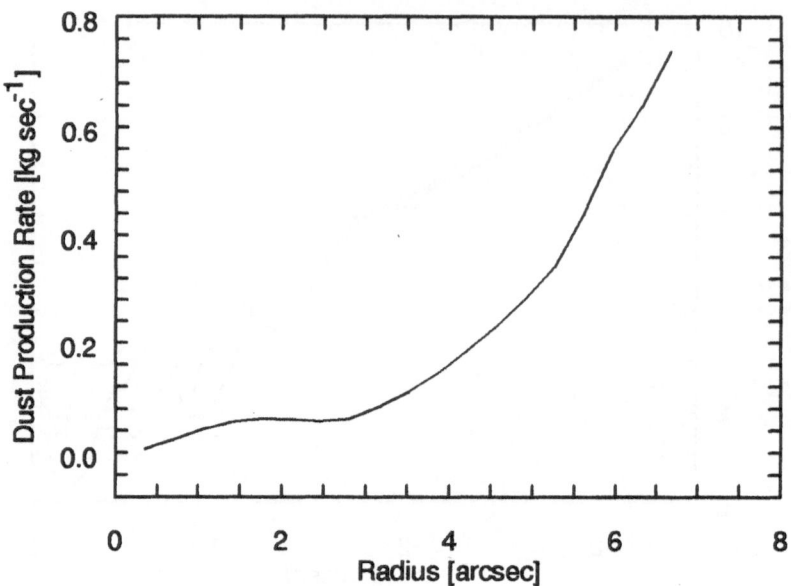

Fig. 3.   Ground–based dust production limits for Pholus

to the core of the stellar image, it is possible to improve upon this mass loss limit by more than 2 orders of magnitude. For the ground–based data, the azimuthally averaged surface brightness profiles were computed for both Pholus and for reference field stars (see Meech *et al.*, 1994b, for details on this procedure). For an untrailed asteroidal object with no coma, the subtraction of the normalized stellar profile fluxes from those of the asteroid should yield a value of zero with an associated error. One can use 3σ of this error as the limiting possible maximum flux contributed from scattered coma light. This flux will be given by:

$$F = S_\odot \pi a_{gr}^2 p_v Q \phi / 2 r^2 \Delta^2 v_{gr} \tag{4}$$

where $S_\odot$ is the solar flux through the bandpass [W m^{-2}], a_{gr} [m] the grain radius, p_v the grain albedo, Q [kg s^{-1}] the dust production rate, ϕ the projected size of the aperture [m], v_{gr} [m s^{-1}] the grain velocity, and r is in AU and Δ in m. It should be noted, that if one assumes a Bobrovnikoff relation for the terminal grain velocities, $v_{gr} = v_{bob} = 600 r^{-0.5}$, and recalls that $\phi = \Delta \phi' / 206265$ where ϕ' is the angular size of the aperture, that for a given observed flux, the dust production will vary as

$$Q \propto r^{1.5} \Delta^1 \tag{5}$$

such that the most sensitive limits will be made when the object is observed at the smallest heliocentric distance. In Figure 3, the dust production limit is plotted versus distance from the image core. It is apparent that the most sensitive limits are made close to the nucleus (the small bump in the profile is due to telescope wind shake–induced tracking errors). With seeing near 1.''7 for both the January and March ground–based observations, the most sensitive upper limits for dust mass loss for Pholus must be obtained outside the seeing disk, and they lie between $2''$–$3''$ at Q = 0.074 kg sec^{-1} (at $r =$ 8.7 AU in this case) as shown in Table I. Because the HST images were obtained with the WFPC-1 where the radius of 90% encircled energy was near $2''$, the technique does not give particularly sensitive limits. No field star for PSF deconvolution was observed and a Lucy deconvolution of the data with either synthetic PSF stars, or archival star images left unacceptable residuals in the surface brightness profile because of an inexact match to the aberrated image. The only technique available to this data set, then, was the one of Luu and Jewitt where dust models of the nucleus plus coma were created and then convolved with an HST PSF star to compare with profiles made from the HST data. Because of the lack of sensitivity in the core of the image where the sky subtraction introduces the least error, we were not able to place as sensitive limit with the HST as with the ground–based observations. However, because the WFPC-2 images with HST have the radius of 90% encircled energy at 0.''3, it is clear from Table I, that the sensitivity can be improved by another order of magnitude over the ground–based observations. The use of the HST is therefore the most sensitive means for detecting evidence of cometary activity in the form of a dust coma.

3.3. P/SHOEMAKER–LEVY 9

The HST observations did not unambiguously resolve the question of whether P/Shoemaker–Levy 9 was a comet or an asteroid, however, all the evidence is consistent with and favors the cometary origin, and this conclusion could not have been obtained from resolution–limited ground–based observations. The colors of the individual fragments as obtained from ground–based observations are shown in Table II. Although there is less coverage with the HST data, the average V–R color of the fragments from HST (V–R = 0.40±0.15) is consistent with the ground–based data taken at nearly the same time in mid–January 1994 (see Table III). All the colors are slightly redder than solar, and although slightly redder than the average cometary value, fall well within the expected range of cometary colors. The coma feature most suggestive of activity was the region within 2,000 km of the nucleus, which remained remarkably circular until 2 weeks prior to the impact with Jupiter, at which time gravitational distortions from Jupiter affected the shape. Throughout the period of HST observation, the separation of the nuclei

TABLE I

Pholus Dust Production Limits - 1992 January Ground-Based Data

Radius (arcsec)	Q (grain/sec)	Q (kg/s)	$Af\rho$ (m)
0.0000	0.5384D+12	0.0023	0.0003
0.3510	0.1881D+13	0.0079	0.0011
0.7020	0.6628D+13	0.0278	0.0041
1.0530	0.1226D+14	0.0513	0.0076
1.4040	0.1662D+14	0.0696	0.0103
1.7550	0.1835D+14	0.0769	0.0113
2.1060	0.1796D+14	0.0752	0.0111
2.4570	0.1671D+14	0.0700	0.0103
2.8080	0.1828D+14	0.0766	0.0113
3.1590	0.2465D+14	0.1033	0.0152
3.5100	0.3241D+14	0.1357	0.0200
3.8610	0.4265D+14	0.1787	0.0264
4.2120	0.5484D+14	0.2297	0.0339
4.5630	0.6834D+14	0.2863	0.0422
4.9140	0.8421D+14	0.3528	0.0520
5.2650	0.1022D+15	0.4280	0.0631
5.6160	0.1310D+15	0.5488	0.0810
5.9670	0.1661D+15	0.6957	0.1026
6.3180	0.1906D+15	0.7985	0.1178

along the "train" direction increased by an order of magnitude, but the inner coma did not exhibit a similar stretching. Without assuming unusual circumstances for the coma dust, continued production of dust throughout the observations is required to maintain the inner coma profiles.

4. Conclusions

From these three examples, it is clear that although the HST will not in general be able to resolve the nuclei of comets, owing to their small sizes, the resolution is instrumental for understanding the nature of the physical processes of cometary activity, and in many cases is uniquely suited to discriminating between objects which exhibit cometary outgassing and those which do not. As the outer Solar System is now being opened up to the study of small bodies, the HST will play a strong role in helping to understand the physics of the outer Solar System.

TABLE II

P/Shoemaker-Levy 9 Colors from Mauna Kea – 0.8" Aperture

Nucleus	V-R	R-I
1 = W	0.501 ± 0.061	0.446 ± 0.059
2 = V	0.476 ± 0.140	0.419 ± 0.144
3 = U	0.351 ± 0.477	
4 = T	0.551 ± 0.093	0.387 ± 0.101
5 = S	0.421 ± 0.040	0.395 ± 0.047
6 = R	0.458 ± 0.068	0.408 ± 0.069
8a = P_1	0.425 ± 0.102	0.278 ± 0.147
8b = P_2	0.336 ± 0.117	0.457 ± 0.166
9 = N	0.436 ± 0.128	
11 = L	0.438 ± 0.035	0.395 ± 0.029
12 = K	0.403 ± 0.020	0.458 ± 0.027
14 = H	0.458 ± 0.034	0.486 ± 0.033
15 = G	0.445 ± 0.029	0.393 ± 0.029
16 = F	0.419 ± 0.072	0.530 ± 0.073
17 = E	0.392 ± 0.062	0.527 ± 0.063
19 = C	0.473 ± 0.111	0.369 ± 0.121
20 = B	0.438 ± 0.094	0.402 ± 0.103
21 = A	0.536 ± 0.150	0.421 ± 0.145

TABLE III

Average Colors – Ground–Based

Color	8" aperture	2.0" aperture	Comets	Solar
V-R	0.431 ± 0.011	0.426 ± 0.021	0.392 ± 0.002	0.36
R-I	0.432 ± 0.012	0.368 ± 0.249	0.398 ± 0.002	0.28

5. Acknowledgements

The image processing in this paper has been performed using the IRAF program. IRAF is distributed by the National Optical Astronomy Observatories, which is operated by the Association of Universities for Research in Astronomy, Inc. (AURA) under cooperative agreement with the National Science Foundation. This work has been supported by NASA grant No.

NGL 12-001-057 and a grant from the Space Telescope Science Institute, No. GO-3769.01-91A.

References

Buie, M. W. (1993). "(2060) Chiron", *IAU Circ. No. 5898.*

Buie, M. W. and S. J. Bus (1992). "Physical observations of (5145) Pholus", *Icarus* **100**, 288-294.

Campins, H., C. Telesco, D. Osip, G. Rieke and M. Rieke (1994). "The Color Temperature of (2060) Chiron: a Warm and Small Nucleus", *Astron. J.*, submitted.

Davies, J. (1994). "1993 HA$_2$", *IAU Circ. No. 5997.*

Davies, J. J. Spencer, M. Sykes, D. Tholen and S. Green (1993). "1992 AD", *IAU Circ. No. 5698.*

Davies, J. K, M. V. Sykes, and D. P. Cruikshank (1993). "Near–Infrared Photometry and Spectroscopy of the Unusual Minor Planet 5145 Pholus (1992AD)", *Icarus* **102**, 166-169.

Fink, U., M. Hoffman, W. Grundy, M. Hicks and W. Sears (1992). "The Steep Red Spectrum of 1992 AD: An Asteroid Covered With Organic Material?", *Icarus* **97**, 145-149.

Hainaut, O. and A. Smette (1992). "1992 AD", *IAU Circ. No. 5450.*

Hartman, W. K., D. J. Tholen, K. J. Meech and D. P. Cruikshank (1990). "2060 Chiron, Colorimetry and Possible Cometary Behavior", *Icarus* **83**, 1-15.

Hoffman, M, U. Fink, W. Grundy and M. Hicks (1993). "Photometric and Spectroscopic Observations of 5145 Pholus", *JGR* **98**, 7403-7407.

Högbom, J. (1974). "Aperture Synthesis with a Non-Regular Distribution of Interferometer Baselines", *Astron. Astrophys. Supp.* **15**, 417-426.

Howell, E. R. Marciales, R. Cutri, M. Nolan, L. Leborsky and M. Sykes (1992). "1992 AD", *IAU Circ. No. 5449.*

Jewitt, D. and J. Luu (1992). "Submillimeter Continuum Observations of 2060 Chiron", *Astron. J.* **104**, 398-404.

Keel, W. C. (1991). "A Simple, Photometrically Accurate Algorithm for Deconvolution of Optical Images", *Pub. Astron. Soc. Pac.* **103**, 723-729.

Kowal, C. T., W. Liller and B. G. Marsden (1979). "The Discovery and Orbit of (2060) Chiron", in *Dynamics of the Solar System, IAU Symp 81,* ed. R. L. Duncombe, Dordrecht, D. Reidel, 245-250.

Landolt, A. (1992). "UBVRI Photoelectric Standard Stars in the Magnitude Range 11.5 < V < 16.0 Around the Celestial Equator", *Astron. J.* **104**, 340-371.

Lebofsky, L. A., D. J. Tholen, G. H. Rieke and M. J. Lebofsky (1984). "2060 Chiron: Visual and Thermal Infrared Observations", *Icarus* **60**, 532-537.

Love, S. G., D. J. Joswiak and D. E. Brownlee (1994). "Densities of Stratospheric Micrometeorites", *Icarus* **111**, 227-236.

Luu, J. X. and D. C. Jewitt (1990). "Cometary Activity in 2060 Chiron", *Astron. J.* **100**, 913-922.

Luu, J. X. and D. C. Jewitt (1992). "High Resolution Surface Brightness Profiles of Near-Earth Asteroids", *Icarus* **97**, 276-287.

Marsden, B. G. (1993). "Periodic Comet Shoemaker–Levy 9", *IAU Circ. No. 5801.*

Meech, K. J. and M. J. S. Belton (1990). "The Atmosphere of 2060 Chiron", *Astron. J.* **100**, 1323-1338.

Meech, K. J., M. W. Buie, N. Samarasinha, B. E. A. Mueller and M. J. S. Belton (1994a). "Planetary Camera Observations of Structures in the Inner Coma of Chiron", in preparation.

Meech, K. J., H. A. Weaver, K. Noll and B. Zellner (1994b). "Ground–based and HST Search for Coma Around 5145 Pholus", in preparation.

Mueller, B. E. A. and D. J. Tholen (1992). "1992 AD", *IAU Circ. No. 5434.*

Mueller, B. E. A., D. J. Tholen, W. K. Hartmann, D. P. Cruikshank (1992). "Extraordinary Colors of Asteroidal Object (5145) 1992 AD", *Icarus* **97**, 150-154.

Rabinowitz, D. L. (1993). "1993 HA$_2$", *IAU Circ. No. 5789*.

Sekanina, Z. (1993). "Disintegration Phenomena Expected During Collision of Comet Shoemaker–Levy 9 with Jupiter", *Science* **262**, 382-387.

Scotti, J. V. (1992). "1992 AD", *IAU Circ. No. 5434*.

Shoemaker, C. S., E. M. Shoemaker and D. H. Levy (1993). "Comet Shoemaker–Levy (1993e)", *IAU Circ. No. 5725*.

Spencer, J. R., L. A. Lebofsky and M. V. Sykes (1989). "Systematic Biases in Radiometric Diameter Determinations", *Icarus* **78**, 337-354.

Stern, S. A., A. A. Jackson, and D. C. Boice (1992). "Discrete Coma Particle Trajectories Around 2060 Chiron", in the *Workshop on the Activity of Distant Comets*, Ed. W. F. Huebner, H. U. Keller, D. Jewitt, J. Klinger and R. West, Lenggries, Germany, 140-152.

Sykes, M. V. and R. G. Walker (1991). "Constraints on the Diameter and Albedo of 2060 Chiron", *Science* **251**, 777-780.

Weaver, H. A. P. D. Feldman, M. F. A'Hearn, C. Arpigny, R. A. Brown, E. F. Helin, D. H. Levy, B. G. Marsden, K. J. Meech, S. M. Larson, K. S. Noll, J. V. Scotti, Z. Sekanina, C. S. Shoemaker, E. M. Shoemaker, T. E. Smith, A. D. Storrs, D. K. Yeomans, and B. Zellner (1994a). "Hubble Space Telescope Observations of Comet P/Shoemaker–Levy 9 (1993e)", *Science* **263**, 787-791.

Weaver, H. A., M. F. A'Hearn, C. Arpigny, D. C. Boice, P. D. Feldman, S. M. Larson, Ph. Lamy, D. H. Levy, B. G. Marsden K. J. Meech, K. S. Noll, J. V. Scotti, Z. Sekanina, C. S. Shoemaker, E. M. Shoemaker, T. E. Smith, S. A. Stern, A. D. Storrs, J. T. Trauger, D. K. Yeomans, B. Zellner (1994b). "The Hubble Space Telescope Observing Campaign on Comet P/Shoemaker–Levy 9", *Science*, submitted.

Wilson, P. D., C. Sagan and W. R. Thompson (1994). "The Organic Surface of 5145 Pholus: Constraints Set by Scattering Theory", *Icarus* **107**, 288-303.

THE MAIN BELT AS A SOURCE OF NEAR–EARTH ASTEROIDS

MARIO MENICHELLA, PAOLO PAOLICCHI

Dipartimento di Fisica, Università di Pisa, Piazza Torricelli 2, 56126 Pisa, Italy

and

PAOLO FARINELLA

Gruppo di Meccanica Spaziale, Dipartimento di Matematica, Università di Pisa, Via Buonarroti 2, 56127 Pisa, Italy

Abstract. We investigate the flux of main–belt asteroid fragments into resonant orbits converting them into near–Earth asteroids (NEAs), and the variability of this flux due to chance interasteroidal collisions. A numerical model is used, based on collisional physics consistent with the results of laboratory impact experiments. The assumed main–belt asteroid size distribution is derived from that of known asteroids extrapolated down to sizes of ≈ 40 cm, modified in such a way to yield a quasi–stationary fragment production rate over times ≈ 100 Myr. The results show that the asteroid belt can supply a few hundred km–sized NEAs per year, well enough to sustain the current population of such bodies. On the other hand, if our collisional physics is correct, the number of existing 10–km objects implies that these objects either have very long–lived orbits, or must come from a different source (i.e., comets). Our model predicts that the fragments supplied from the asteroid belt have initially a power–law size distribution somewhat steeper than the observed one, suggesting preferential removal of small objects. The component of the NEA population with dynamical lifetimes shorter than or of the order of 1 Myr can vary by a factor reaching up to a few tens, due to single large–scale collisions in the main belt; these fluctuations are enhanced for smaller bodies and faster evolutionary time scales. As a consequence, the Earth's cratering rate can also change by about an order of magnitude over the 0.1 to 1 Myr time scales. Despite these sporadic spikes, when averaged over times of 10 Myr or longer the fluctuations are unlikely to exceed a factor two.

Key words: Asteroids, Near–Earth asteroids, Cratering record

1. Introduction

It is well known that the population of interplanetary bodies which can collide with the Earth — near–Earth asteroids (NEAs), comets, meteoroids — is characterized by dynamical and collisional lifetimes much shorter than the age of the solar system, i.e. ranging from $\approx 10^5$ to 10^8 yr. Thus, in order to maintain a quasi–stationary abundance of bodies on such orbits, sources are needed to balance the loss rate due to hyperbolic ejections out of the solar system and to disruptive impacts with the planets, the Sun and other interplanetary objects. As a matter of fact, a significant, probably major fraction of the near–Earth population appears to be supplied from the "storage region" in the main asteroid belt, where fragments are continuously produced by interasteroidal collisions and sometimes injected into resonant, chaotic orbits undergoing large variations of eccentricity and

Earth, Moon, and Planets **72**: 133-149, 1996.

therefore amenable to planetary encounters. Although this transport mechanism is known since a long time as a matter of principle, only recently quantitative models of it have become available, thanks to a better understanding of both asteroidal collisions and resonant dynamics (Farinella et al. 1993a,b, 1994a,b; Morbidelli et al. 1994; Michel et al. 1994).

In particular, Farinella et al. (1993a) have estimated that the fraction of main–belt asteroid fragments ending up into either the 3/1 mean motion resonance with Jupiter or the ν_6 secular resonance (the two most effective dynamical *routes* from the main belt to the inner planet zone) range from about 1% to 4%, depending on the detailed assumptions adopted on the collisional physics — in particular, the ejection velocity distribution of fragments from hypervelocity impacts. With standard assumptions about the collisional lifetimes of main–belt asteroids — which approach the age of the solar system for bodies 100 km in diameter, and for smaller asteroids are roughly proportional to the square root of size (see Farinella et al. 1992a) — this model predicts a yield of the order of 100 fragments larger than 1 km in diameter per Myr into the resonances. This appears to be of the same order as the loss rate quoted above, even taking into account the recent findings about the frequency of NEAs reaching extreme orbital eccentricities and thus hitting the Sun (Farinella et al. 1994b).

The purpose of this paper is to refine this order–of–magnitude estimate of the NEA flux from the main belt by performing *ad hoc* simulations of the collisional evolution of main–belt asteroids through a suitable numerical model (Campo Bagatin et al. 1993, 1994a,b). The corresponding algorithm takes into account a variety of possible collisional outcomes (cratering, break–up, partial reaccumulation of ejecta), in agreement with the available experimental evidence on hypervelocity impacts (e.g., Giblin et al. 1994), and has been modified to give as an output the number of new fragments of any given size generated by chance asteroidal collisions over time. These simulations can be used both to estimate the average flux of new fragments injected into the resonant *routes* — by applying the injection efficiency factor of 0.01 to 0.04, derived as explained in Farinella et al. (1993a) — and to provide the first quantitative estimate of the time variability of this flux on different time scales. These estimates are important for many applications, e.g. studies of the equilibrium population of NEAs of different sizes, of the cratering rates on the terrestrial planets, of meteorite fall rates and exposure ages, and of possible "transient" processes related to a temporarily enhanced impact flux against the Earth.

The remainder of this paper is organized as follows. In Sec. 2 we describe our numerical collisional evolution code and the corresponding assumptions we have made to address the problems mentioned above. In Sec. 3 we will present and discuss some quantitative results of the numerical simulations carried out with the code. Sec. 4 will be devoted to a summary of some general conclusions on the significance of these results for the "demography" of NEAs and their impact rates against the inner planets.

2.   The collisional evolution model

An improved version of the numerical model developed by Campo Bagatin (1993) and described in details in Campo Bagatin et al. (1994a,b), has been used to estimate the creation rate of new fragments in the main asteroid belt and its variability. The original purpose of this model was that of studying the evolution of a population of colliding bodies, such as the asteroids, taking into account both cratering and catastrophic disruption events. Like in previous codes of this type (Davis et al. 1985, 1989; Farinella et al. 1992b), the bodies making up the overall evolving population are divided into a number of discrete size bins, which at every time step interact due to mutual collisions; as a consequence, the number of objects residing in all the bins is suitably updated. Collision rates are estimated in agreement with the average intrinsic collision probability for the real asteroid population calculated from Wetherill's (1967) formulae by Farinella and Davis (1992).

The novel feature of Campo Bagatin's code is the *a priori* derivation of a "collision matrix" which includes all the assumptions on the collisional physics, and whose elements C_{ijk} give the number of bodies in the k–th size bin generated (or lost) after a typical collision involving a projectile in the i–th bin and a target in the j–th bin. These numbers are derived from a semiempirical collisional model, as described in Davis et al. (1989) and Petit and Farinella (1993). By multiplying the collision matrix times the number of events involving, during every time step, objects belonging to the corresponding pair of bins, the variations of the bin populations over time are readily obtained.

We have now modified this program by devoting a specific attention to the problem of "rare" impact events: they are defined as those events whose probability is low enough that they are expected to happen a small number of times (< 1, or a few) *within one time step*, and for which using just average collision rates cannot account for the intrinsic random fluctuations of the collisional process. For instance, in the previous version of the code, when 0.1 events involving a given pair of target/projectile bins were expected, the number of fragments predicted for one such event by the collision matrix was just divided by 10 — neglecting the fact that collisions are in fact discrete events, which in a given interval either do not occur at all or occur an integer number of times. Note that these "rare" events must be distinguished from the "improbable" ones, those which might not take place even in the whole lifetime of the Solar System, and which typically involve pairs of sizeable asteroids both in the target and in the projectile role: these latter events represent of course an unavoidable source of uncertainty in our reconstructions of the past evolution of the asteroid belt. On the other hand, since in the current context we are interested just in relatively recent times (say, the last 10^8 yr), these "improbable" events have a very small probability of having taken place, and therefore can be safely neglected.

The "rare" events are not so critical for the overall history of the asteroid belt (however, we plan to devote a future study to analyzing their possible systematic effects on the results of the numerical simulations). On the other hand, they are extremely important for assessing what happens on a shorter time scale, represented in the code as a single time step or a small number of them. In particular, for what concerns the production of small and intermediate–sized fragments, which could be transformed into NEAs or meteoroids, the rare events can cause strong fluctuations in time. As previously done by Farinella et al. (1992b), the problem of rare events has been dealt with in this way: when, in a time step, a collision involving two given (i–th and j–th) bins is expected to happen a (real) number $x_{ij} < 5$ of times, a random number generator is called, which gives back an integer number n_{ij} (≥ 0), chosen according to a Poisson probability distribution having x_{ij} as the mean value. In this way we have always an integer number of "rare" collisions in a discrete time step (of course, when the expected number x_{ij} is > 5, the difference between x_{ij} and n_{ij} is not important). In the following we will make some comparisons between this procedure and the simpler, deterministic one using always real numbers of events and thus deriving a kind of "mean evolution".

In addition to introducing this probabilistic feature in the evolution code, in this work we adopted the following assumptions:

(i) The evolution is started basically at the present time, 4.5 Byr after the origin of the solar system (and of the asteroid belt). Consistent with this, the initial population is an approximation to the current one, derived from observations for large asteroids and extrapolated to smaller ones with a constant power–law exponent, as explained by Davis et al. (1994) and Campo Bagatin et al. (1994a,b). As pointed out by Cellino et al. (1991) and Farinella and Davis (1994), the observed population can be considered complete only for diameters larger than ≈ 40 km, and an increasing uncertainty affects the extrapolated size distribution at smaller sizes. At diameters of a few km, the smallest ones relevant for the results in the present context, the uncertainty in the size distribution can be estimated to be of about plus or minus a factor two.

(ii) The collisional physics parameters used to generate the matrix C_{ijk} are those corresponding to the "standard case" defined by Campo Bagatin et al. (1994a,b). Significant uncertainties affect many of these parameters, in particular because they have to be scaled from sizes typical of laboratory experiments up to asteroidal ones. However, on the basis of a number of tests, we believe that these uncertainties cannot change the derived fragment supply by orders of magnitude, and probably they affect even less its inferred time variability (which basically comes from the random character of the collisional process).

(iii) The numerical calculations are made by using 32 logarithmic size bins, spanning each a factor 4 in mass (1.587 in size), with central values ranging from 53 cm to 890 km in diameter. The largest bin contains only (1) Ceres, whereas the smallest one extends down to about 42 cm. The effects of this small–size cutoff on the shape of the evolving size distribution are discussed in details by Campo Bagatin et al. (1994b). Again, the corresponding uncertainty cannot change our results by more than a factor of a few.

We analyzed the collisional evolution predicted by the code for our assumed current population over some time, monitoring in particular the production of small and intermediate–sized fragments. We soon discovered that the assumed initial population is rather far from an equilibrium one in which at every size the collisional losses are approximately balanced by the input of new fragments and the shape of the size distribution keeps unchanged with time (see Dohnanyi 1969; Campo Bagatin et al. 1994b; Paolicchi 1994). Instead, we had a transient phase in which the production of fragments varied in a significant way in the different bins before stabilizing at quasi-stationary values, after a time of the order of several tens of Myr. Comparing the transient phase to the quasi–stationary regime, we found that in the former phase the fragment production rate was larger by factors of about 4 and 2 for fragments 100 m and 1 km in diameter, respectively, while it was smaller by about 8% at 10 km. Since the real asteroid size distribution has already had 4.5 Byr to reach a quasi–stationary regime, this is clearly an artifact of our collisional model and/or initial conditions, both of which are only rough approximations to the reality.

This problem compounds with the uncertainties described earlier (although it is in fact a consequence of them). We have chosen to estimate the fragment production rate after the initial population has relaxed to the quasi-stationary regime, so we evolved it for 100 Myr and then used the final population as a basis for our computations (it is interesting to note that this final population differs little from the initial one for sizes larger than a few tens of km, where the real asteroid size distribution is known in a reliable way). Since, as mentioned earlier, up to about 10 km in diameter the transient fragment production rates are higher than the quasi–stationary ones, this choice is likely to provide a lower bound to the actual fragment yield. At the current state of knowledge, however, we think that an overall uncertainty of a factor 2 to 4 cannot be avoided.

TABLE I

Results from the "mean evolution" simulations described in the text: fragment production rates, resonance yields, and equilibrium abundances in the "fast–track" and "slow–track" NEA populations.

Size bin (km)	Production rate (Myr^{-1})	Resonance yield (Myr^{-1})	Fast-track eq. pop.	Slow-track eq. pop.
0.07-0.11	1.25×10^7	2.5×10^5	2.1×10^5	1.1×10^6
0.69-1.10	1.85×10^4	370	300	1700
6.96-11.05	2.3	0.05	0.04	0.2

3. Numerical experiments and results

With the assumptions described above, we carried out numerical simulations of the collisional evolution process spanning 100 Myr and obtained the production rate of collisional fragments in the three size bins centered at 86 m, 0.87 km, and 8.8 km. The average fragment production rates obtained from the "mean evolution" algorithm defined in Sec. 2 (no Poisson–distributed random numbers in deriving how many impacts take place in a time step) are listed in the second column of Table I. The third column of the Table gives the corresponding "mean" fragment yields to the resonances, assuming a size–independent delivery efficiency of 2% (Farinella et al. 1993a; note that in reality this efficiency is likely to be somewhat higher for smaller fragments, characterized by higher average ejection speeds from their parent bodies). It is interesting to note that Farinella's et al. (1993a) value of $\approx$ 100 fragments larger than 1 km per Myr injected into the resonances, derived from approximate estimates of asteroid collisional lifetimes, appears to be confirmed in order of magnitude by the current results of the numerical simulations. We shall discuss the last two columns of the Table, giving the steady–state abundances of NEAs corresponding to the supply flux derived from the "mean evolution", in Sec. 4.

As far as the time variability of the fragment yield is concerned, we were interested in time scales ranging from about 0.1 to 10 Myr, corresponding to different dynamical/collisional time scales for the evolution of the fragments into Earth–crossing orbits and their final loss by impact into the Sun or the planets, or hyperbolic ejection. Actually, the 10^5 yr time scale is probably relevant for bodies in comet–like orbits undergoing close approaches to Jupiter (this appears to be the most likely fate for fragments injected in the resonances located in the outer part of the belt, such as the 5/2 and 2/1 mean motion resonances with Jupiter; see e.g. Hahn et al. 1991); the 1–Myr time scale applies to "fast–track" NEAs hitting the Sun or ejected on hyperbolic orbits after their eccentricity has been drastically pumped up by the 3/1 or ν_6 resonances (Farinella et al. 1994b, Valsecchi et al. 1994); and the 10^7 yr time scale is more typical of "slow–track" non–resonant NEAs evolving mainly under the influence of close encounters with the inner planets (Milani et al. 1989; Michel et al. 1994). Thus we chose a time step of 0.1 Myr in computing the fragment production rates, and in order to estimate the fluctuations over the longer time scales, we simply averaged over 10 or 100 such steps run with the same initial conditions (the quasi–stationary population described earlier). Of course this procedure neglects the long–term variations of the asteroid size distribution, but these have been certainly minor over the last 100 Myr ($\approx 2\%$ of the age of the solar system). Our plots have the abscissae ranging from 4.5 to 4.6 Byr after the origin of the solar system and labelled "fictitious time", to remind of the way the fragment production rates have been computed.

Another important point concerning the random fluctuations of the fragment supply has to do with the way the corresponding yield to the resonance zones should be evaluated. In the "mean evolution" approach whose results were discussed above, it is sufficient to multiply the fragment production rates (given in the first column of Table I) times the 0.02 delivery efficiency factor. But the situation is different when the random fluctuations about the "mean yields" are considered. Using also for them the 0.02 efficiency factor would clearly be too conservative: when a rare event (as defined in Sec. 2) occurs near a resonance, a fraction of the created fragments much larger than 2%, and possibly not much smaller than unity, would be injected into the chaotic zone (see e.g. the fragment delivery efficiencies of several real asteroids close to resonances computed by Farinella et al. 1993a, and Morbidelli et al. 1994). On the other hand, the full fluctuations obtained by the probabilistic collision evolution codes refer to the total fragment production rate in the main belt, including parent bodies far from the resonances and therefore inefficient as NEA deliverers. To solve this dilemma, we decided to proceed in the following way: the (real) number x_{ij} of collisional events expected in each time step for every bin pair was multiplied times the 0.02 factor (roughly giving the effective number of events occurring near the resonances), and only afterwards the corresponding random integer number (to be multiplied times C_{ijk}) was computed by the Poisson probability routine, from a distribution having $0.02\,x_{ij}$ as a mean value. The figures discussed below show how the fluctuations derived in this way (referred to as "fluctuations of resonant fragments") compare with the more conservative estimates ("normalized fluctuations") which are obtained simply by multiplying the overall fragment production rates times 0.02, as explained in Sec. 2. In all cases the fluctuations are referred to the "mean evolution" values discussed earlier, which correspond to the zeros of the vertical axes.

Figs. 1 and 2 show the results on the fluctuations in the production of fragments in the 100 m bin (for the exact bin limits, see Table I). Parts (a), (b) and (c) refer to the 0.1, 1 and 10 Myr time scales, respectively. The largest fluctuations showing up in the 100 Myr time span of the simulations (some 10^6 fragments about 100 m across produced by three discrete, "rare" collisional events) are only partially smoothed out over the longest time scales. Note that, by coincidence, the number of fragments supplied to resonances in the largest fluctuations is about the same as the overall number of fragments created in the main belt over 0.1 Myr, on the average (i.e., in the "mean evolution" scenario; see Table I). Therefore, chance collisional events can increase the abundance of 100–m NEAs with lifetimes of the order of 0.1 Myr by a factor of the order of 50. This is reduced to a factor ≈ 5 and to a $\approx 50\%$ increase when 1-Myr and 10-Myr evolution time scales are considered, respectively. The corresponding "normalized fluctuations" are shown in Figs. 2: as expected, a much larger number of discrete events is recorded, but each of them gives rise to a smaller fluctuation in the fragment supply. The reduction factor with respect to the previous scenario is about 20, that is somewhat smaller than 50, since the largest "rare" event in the whole belt creates more fragments than the largest one in the reduced sample of "resonance–bordering" asteroids.

Figs. 3 and 4 refer to 1–km sized bodies, Figs. 5 to 10–km bodies. The same qualitative features as in Figs. 1 and 2 are apparent again, but the absolute numbers of fragments are reduced by factors of about 10^3 and 10^6, respectively. Note that these factors correspond to having the same total mass in the three size bins, so we can conclude that the fragments injected into to NEA orbits have approximately a power–law size distribution with the number of bodies in any interval $[D, D + \delta D]$ proportional to D^{-4} (and the number of bodies larger than D proportional to D^{-3}). These distributions are somewhat steeper than those inferred from the observed NEA population or from the lunar and terrestrial cratering record; this may be due either to problems with our collisional physics, or to preferential collisional elimination of smaller NEAs during their dynamical lifetimes. The latter explanation would be consistent with the results of Bottke et al. (1994) on the (fairly strong) size dependence of the collisional lifetime of NEAs.

The largest "rare" random events create about 700 1–km fragments, with the corresponding fluctuations surviving over the 1–Myr time scale; comparing these values with those listed in Table I, fluctuations by factors ≈ 20, 2 and 1.2 apply to the 0.1, 1 and 10 Myr time scales, respectively. Note that for 1–km fragments, the reduction in passing to the more conservative "normalized fluctuations" does not exceed a factor ≈ 6.

As regards the 10–km fragments, Figs. 5 show that over the 100–Myr time span only two "rare" events create one of these bodies each. For these large fragments, the effects of the "rare" events persist over 1 Myr, and

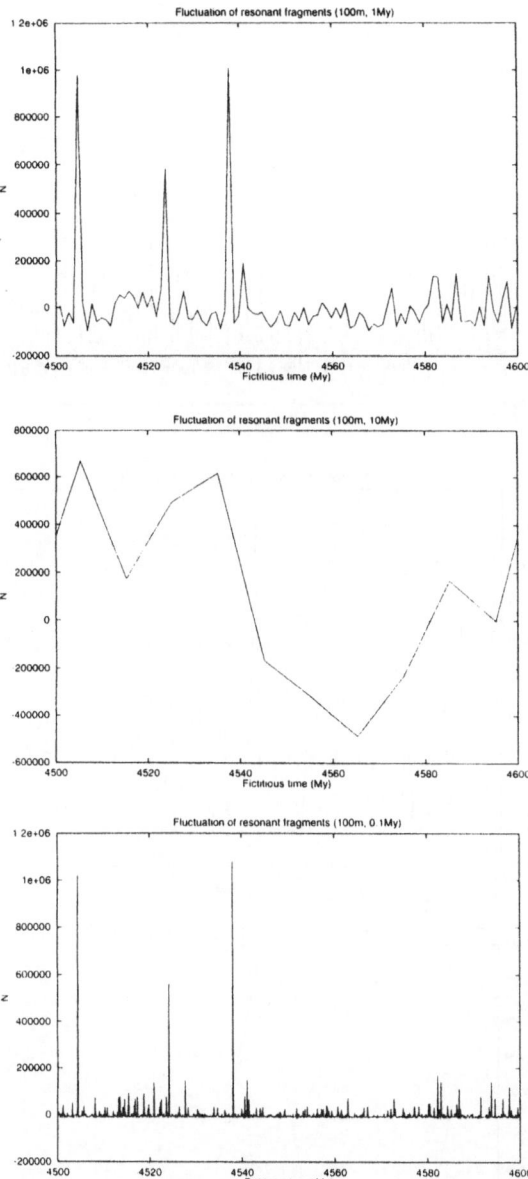

Fig. 1. The fluctuations in the number of fragments formed close to the resonances, referred to the mean evolution" case described in the text (dashed line at $N = 0$). Here fragments about 100 m in diameter are considered. The abscissae mark the "fictitious time", obtained from a sequence of 1000 equal time steps of 0.1 Myr starting from the same quasi–stationary main–belt population (see text). Since this population is close to the real one of the current asteroid belt, "fictitious times" between 4.5 and 4.6 Byr after the origin of the solar system have been used along the horizontal axis. In Fig. 1a fluctuations appearing in the original 0.1 Myr time steps have been plotted; Figs. 1b and 1c show the same simulation, but with fluctuations referred to longer time steps of 1 and 10 Myr, and obtained simply by averaging over 10 or 100 consecutive 0.1 Myr steps.

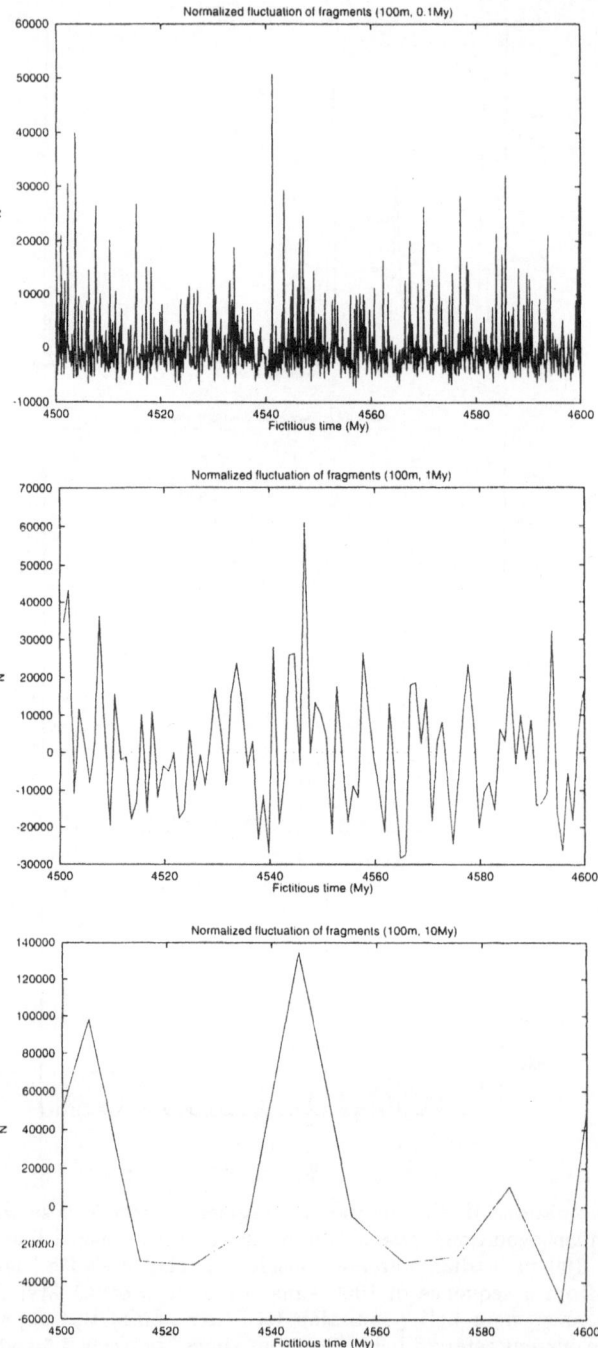

Fig. 2.   The "normalized fluctuations" in the number of fragments created in the whole belt, multiplied times the delivery–efficiency factor 0.02 (see text). As in Fig. 1, the 100–m size bin and three different time scales (0.1, 1 and 10 Myr in Figs. 2a, 2b and 2c, respectively) are considered.

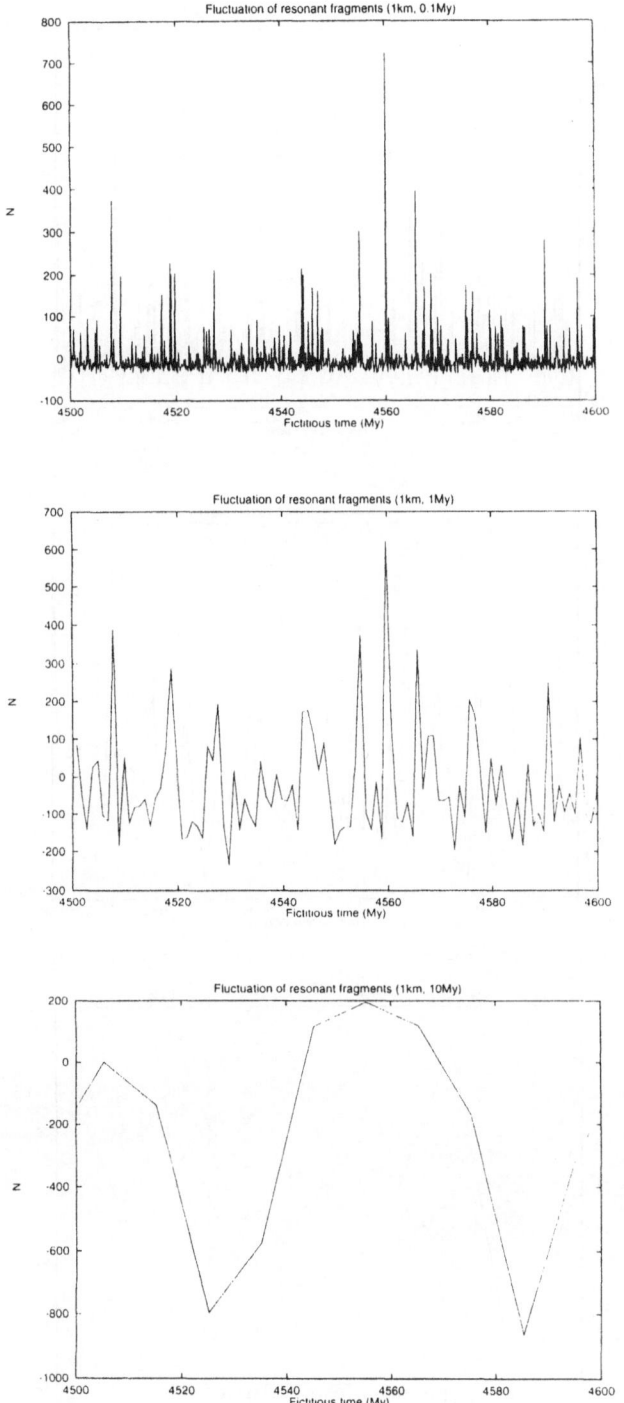

Fig. 3.   The same as Fig. 1 but for fragments in the 1–km size bin.

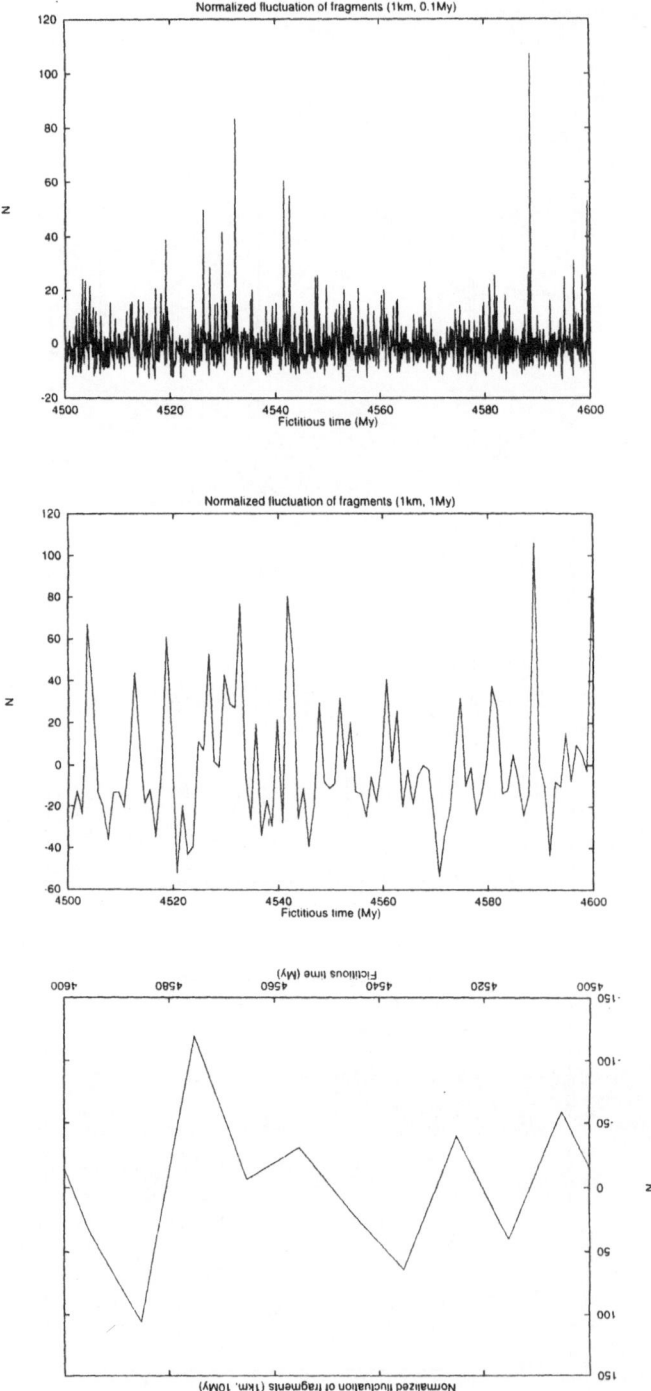

Fig. 4.   The same as Fig. 2 but for fragments in the 1–km size bin.

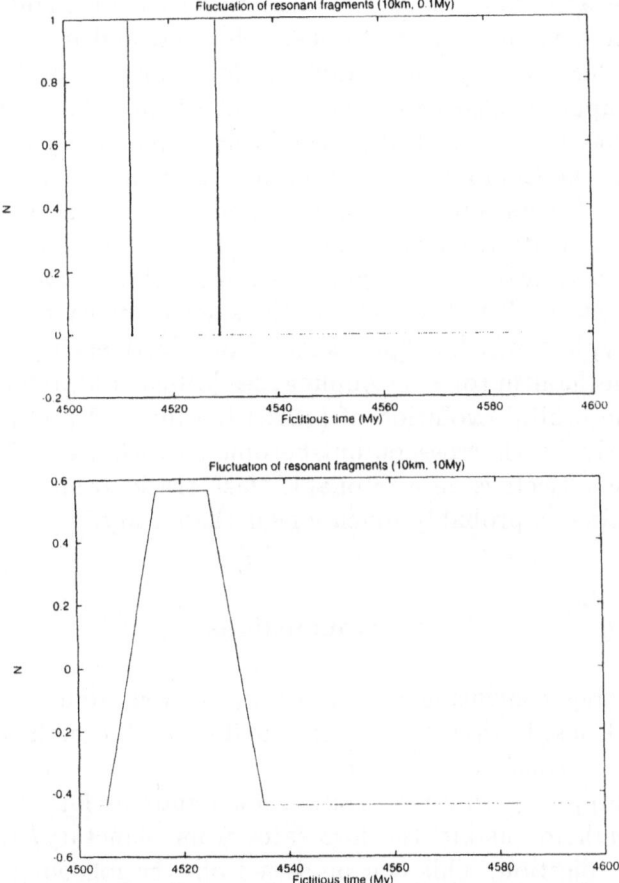

Fig. 5.   The same as Fig. 1 but for fragments in the 10–km size bin. Only the 0.1 and 10 Myr time scales have been represented in Figs. 5a and 5b, respectively. The 1–Myr plot provides no new information with respect to Fig. 5a, therefore it has been omitted.

are partially smoothed out only over 10 Myr (Fig. 5b). The "normalized fluctuations" are not meaningful, as they correspond always to values much smaller than unity. It is interesting to note that in the real NEA population several sizeable bodies exist: (1036) Ganymed, with a diameter exceeding 35 km; (433) Eros and (3552) Don Quixote, in the 20 km class; (1580) Betulia, (1627) Ivar, (1866) Sisyphus and (3200) Phaethon, all 7 to 8 km across (see data in McFadden et al. 1989). If our collisional physics is correct, we have to expect that either these NEAs are located in very long–lived, "slow–track" orbits, or they must come from a different source (i.e., comets). The latter explanation appears certainly plausible for (3552) Don Quixote, a D–type NEA with a Jupiter–crossing orbit similar to those of short–period comets, whose dynamical lifetime is only of the order of 10^5 yr (Milani et al. 1989).

Our model indicates that it is very unlikely that such a big and young object has been recently supplied from the asteroid belt, and its taxonomic type confirms that this is a very good candidate for a dormant/extinct cometary nucleus. This appears also to be the case for (3200) Phaethon, an F–type object which has been identified as the likely parent body of a prominent meteor stream (the Geminids). The situation is different for objects such as the two largest Amors, (1036) Ganymed and (433) Eros, which are likely to be "old" fragments from very rare disruption events in the main belt, as their orbits currently do not approach the Earth and evolve very slowly owing to encounters with Mars only; to the same category probably belongs also (1866) Sisyphus, which is "protected" from Earth encounters by the so-called Kozai mechanism (or e–ω coupling, see Milani et al. 1989). More rapid is probably the orbital evolution of (1580) Betulia and (1627) Ivar, which are not currently Earth–crossers, but become so within $\approx 10^5$ yr; however, neither of these objects is on a resonant, "fast–track" orbital *route*, so their dynamical lifetime is probably much longer than 1 Myr.

4. Conclusions

A first interesting conclusion of this work, derived from our "mean evolution" simulations, is that collisional evolution in the main asteroid belt, modelled in a way consistent with the evidence from laboratory experiments, is capable of supplying to NEA orbits a few hundreds km–sized fragments per Myr, enough to sustain the loss rates from planetary/solar collisions and hyperbolic ejection. This can be seen from the following simple argument. Let us we assume that: (i) the NEA population is characterized by two typical lifetimes, ≈ 1 Myr ("fast–track" objects) and ≈ 30 Myr ("slow–track objects"), as indicated by numerical integrations; (ii) some 20% and 80% of the current NEA population are accounted by fast–track and slow–track objects, respectively, again in agreement with recent dynamical work (Farinella et al. 1994b, Froeschlé et al. 1995); (iii) the steady–state abundance of both fast– and slow–track NEAs is simply given by their supply rate times their lifetime. Then, from the current proportions of NEAs of the two types, we can infer that about 15% of the fragments initially supplied to resonances end up "parked" into slow–track orbits, whereas the remaining 85% stays on short–lived fast–track orbits. Using the resonance yields given in the third column of Table I, the steady–state populations listed in the fourth and fifth columns can be derived. We obtain an overall population of about 2000 km–sized objects and more than one million of 100–m sized bodies. The former number is in excellent agreement with the observations, although we stress that uncertainties in both observations and our model imply that this may be at least in part coincidental.

Big NEAs, $\approx$ 10 or more km in diameter, appear to be overabundant in the real Apollo–Amor population with respect to the predictions of our model. Although this may be due to problems with our collisional physics and/or to random fluctuations coupled with small–number statistics, the dynamical properties of the observed sizeable NEAs rather suggest that these bodies either have a different (i.e., cometary) source, or have settled onto comparatively long–lived orbits.

Our simulations have allowed us to address for the first time in a quantitative way the important issue of the time variability of the fragment supply from the main belt to the near–Earth environment. The fast–track component of the NEA population is variable up to tens of times its average abundance, and such fluctuations are strongly enhanced for smaller bodies and faster evolutionary time scales. Although at most times (including the present one) the short–lifetime objects probably account only for a minority of the existing population, our results suggest that from time to time they may become dominant, with an important fraction of them generated from a single chance collision occurred in the main belt close to a resonance. This may be tested by physical observations of NEAs, in particular those on fast–track orbits. Relevant insight in this context may be also provided by evidence from meteorite types, thermal histories, Argon–Argon and cosmic–rays exposure ages, as well as from comparisons between the Antarctic and non–Antarctic collections (which record different average fall times). For instance, Benoit and Sears (1993) have recently found indications that the flux of H–chondrites has significantly changed in the last several hundred thousand years. As for the cratering fluxes onto the Earth, the other inner planets and the Moon, our results indicate that significant (up to an order of magnitude) peaks over the 0.1 to 1 Myr time scales are possible from time to time. Even with these sporadic spikes, however, when averaged over time scales of $\approx$ 10 Myr and longer, the overall population of NEAs down to bodies in the 100 m size range, as well as the corresponding cratering rates, are unlikely to vary by more than a factor two.

Acknowledgements

We are grateful to W. Bottke, D.R. Davis, R. Greenberg and A.W. Harris for useful discussions and comments on the subject of this paper. Inputs from C. Froeschlé, Ch. Froeschlé, P. Michel, A. Morbidelli and G. Valsecchi on dynamical issues and from A. Campo Bagatin, A. Cellino and D.R. Davis on collisional physics are also gratefully acknowledged. P.F. started to work on this project while staying at the Observatoire de la Côte d'Azur (Nice, France), thanks to the "G. Colombo" fellowship of the European Space Agency.

References

Benoit, P.H. and Sears, D.W.G.: 1993, 'Breakup and structure of an H–chondrite parent body: The H–chondrite flux over the last million years', *Icarus* **101**, pp. 188–200.

Bottke, W.F., Nolan, M.C., Greenberg, R. and Kolvoord, R.A.: 1994, 'Collisional lifetimes and impact statistics of near–Earth asteroids', in *Hazards Due to Comets and Asteroids*, Univ. of Arizona Press, Tucson, in press.

Campo Bagatin, A.: 1993, Thesis, Pisa University (unpublished).

Campo Bagatin, A., Farinella, P. and Petit, J.–M.: 1994a, 'Fragment ejection velocities and the collisional evolution of asteroids', *Planet. Space Sci.*, in press.

Campo Bagatin, A., Cellino, A., Davis, D.R., Farinella, P. and Paolicchi, P.: 1994b, 'Wavy size distributions for collisional systems with a small–size cutoff', *Planet. Space Sci.*, in press.

Cellino, A., Zappalà, V. and Farinella, P.: 1991, 'The asteroid size distribution from IRAS data', *Mon. Not. R. astr. Soc.* **253**, 561–574.

Davis, D.R., Chapman, C.R., Weidenschilling, S.J. and Greenberg, R.: 1985, 'Collisional history of asteroids: Evidence from Vesta and the Hirayama families', *Icarus* **62**, 30–53.

Davis, D.R., Farinella, P., Paolicchi, P., Weidenschilling, S.J. and Binzel, R.P.: 1989, 'Asteroid collisional history: Effects on sizes and spins', in *Asteroids II* (T. Gehrels, Ed.), pp. 805–826, Univ. of Arizona Press, Tucson.

Davis, D.R., Ryan, E.V. and Farinella, P.: 1994, ' Asteroid collisional evolution: Results from current scaling algorithms', *Planet. Space Sci.*, in press.

Dohnanyi, J.W.: 1969, 'Collisional model of Asteroids and their debris', *J. Geophys. Res.* **74**, 2531–2554.

Farinella, P. and Davis, D.R.: 1992, 'Collision rates and impact velocities in the main asteroid belt', *Icarus* **97**, 111–123.

Farinella, P. and Davis, D.R.: 1994, 'Will the real asteroid size distribution please step forward?', *Lunar and Planetary Science* **XXV**, Part 1, 365–366.

Farinella, P., Davis, D.R., Cellino, A. and Zappalà, V.: 1992a, 'The collisional lifetime of asteroid 951 Gaspra', *Astron. Astrophys.* **257**, 329–330.

Farinella, P., Davis, D.R., Paolicchi, P., Cellino, A. and Zappalà, V.: 1992b, 'Asteroid collisional evolution: An integrated model for the evolution of asteroid rotation rates', *Astron. Astrophys.* **253**, 604–614.

Farinella, P., R. Gonczi, Ch. Froeschlé, and C. Froeschlé: 1993a, 'The injection of asteroid fragments into resonances', *Icarus* **101**, 174–187.

Farinella, P., Froeschlé, Ch. and Gonczi, R.: 1993b, 'Meteorites from the asteroid 6 Hebe', *Celest. Mech.* **56**, 287–305.

Farinella, P., Froeschlé, C. and Gonczi, R.: 1994a, 'Meteorite delivery and transport', in *Asteroids, Comets, Meteors 1993* (A. Milani, M. Di Martino & A. Cellino, eds.), Proc. IAU Symp. 160, pp. 205–222, Kluwer, Dordrecht.

Farinella, P., Froeschlé, Ch., Froeschlé, C., Gonczi, R., Hahn, G., Morbidelli, A. and Valsecchi, G.B.: 1994b, 'Asteroids falling into the Sun', *Nature* **371**, 314–317.

Froeschlé, Ch., Hahn, G., Gonczi, R., Morbidelli, A. and Farinella, P.: 1995, 'Secular resonances and the dynamics of Mars–crossing and near–Earth asteroids', In preparation.

Giblin, I., Martelli, G., Smith, P.N., Cellino, A., Di Martino, M., Zappalà, V., Farinella, P. and Paolicchi, P.: 1994, 'Field fragmentation of macroscopic targets simulating asteroidal catastrophic collisions', *Icarus* **110**, 203–224.

Hahn, G., Lagerkvist, C.–I., Lindgren, M. and Dahlgren, M.: 1991, 'Orbital evolution studies of asteroids near the 5/2 mean motion resonance with Jupiter', *Astron. Astrophys.* **246**, 603–618.

McFadden, L.–A., Tholen, D.J. and Veeder, G.J.: 1989, 'Physical properties of Aten, Apollo and Amor Asteroids', in *Asteroids II* (T. Gehrels, Ed.), pp. 442–467, Univ. of Arizona Press, Tucson.

Michel, P., Froeschlé, Ch. and Farinella, P.: 1994, 'Dynamical evolution of NEAs: Close encounters, secular perturbations and resonances', this volume.

Milani, A., Carpino, M., Hahn, G. and Nobili, A.M.: 1989, 'Dynamics of planet–crossing asteroids: Classes of orbital behavior — Project SPACEGUARD', *Icarus* **78**, 212–269.

Morbidelli, A., Gonczi, R., Froeschlé, Ch. and P. Farinella, P.: 1994, 'Delivery of meteorites through the ν_6 secular resonance', *Astron. Astrophys.* **282**, 955-979.

Paolicchi, P.: 1994, 'Rushing to equilibrium: A simple model for the collisional evolution of asteroids', *Planet. Space .Sci.* **42(3)**, 207–212.

Petit, J.–M. and Farinella, P.: 1993, 'Modelling the outcomes of high–velocity impacts between small solar system bodies', *Celest. Mech.* **57**, 1–28.

Valsecchi, G.B., Morbidelli, A., Gonczi, R., Farinella, P., Froeschlé, Ch. and Froeschlé, C.: 1994, 'Dynamical evolution of objects in orbits resembling that of P/Encke', *Icarus*, submitted.

Wetherill, G.W.: 1967, 'Collisions in the asteroid belt', *J. Geophys. Res.* **72**, 2429-2444.

DYNAMICAL EVOLUTION OF NEAS: CLOSE ENCOUNTERS, SECULAR PERTURBATIONS AND RESONANCES

PATRICK MICHEL, CHRISTIANE FROESCHLÉ and PAOLO FARINELLA*

Observatoire de la Côte d'Azur, Dept. Cassini, B.P. 229, 06304 Nice Cedex 4, France.

Abstract. We discuss the main mechanisms affecting the dynamical evolution of Near–Earth Asteroids (NEAs) by analyzing the results of three numerical integrations over 1 Myr of the NEA (4179) Toutatis. In the first integration the only perturbing planet is the Earth. So the evolution is dominated by close encounters and looks like a random walk in semimajor axis and a correlated random walk in eccentricity, keeping almost constant the perihelion distance and the Tisserand invariant. In the second integration Jupiter and Saturn are present instead of the Earth, and the 3/1 (mean motion) and ν_6 (secular) resonances substantially change the eccentricity but not the semimajor axis. The third, most realistic, integration including all the three planets together shows a complex interplay of effects, with close encounters switching the orbit between different resonant states and no approximate conservation of the Tisserand invariant. This shows that simplified 3–body or 4–body models cannot be used to predict the typical evolution patterns and time scales of NEAs, and in particular that resonances provide some "fast–track" dynamical *routes* from low–eccentricity to very eccentric, planet–crossing orbits.

Key words: Near–Earth asteroids, Resonances, Close encounters

1. Introduction

Near-Earth asteroids (NEAs) are widely believed to be continuously injected into Earth-approaching orbits through a few different resonant channels, which collect fragments randomly ejected from main-belt asteroids as a consequence of energetic interasteroidal collisions (see e.g. Wetherill, 1985, 1987; Farinella *et al.*, 1993, 1994). Subsequently, these fragments undergo a fairly complex orbital evolution process, driven by (mean motion and secular) resonances, by non–resonant secular perturbations, and by a sequence of close encounters with the inner planets. This process has been and is being studied by numerical techniques (Milani *et al.*, 1989; Farinella *et al.*, 1994; Froeschlé *et al.*, 1995; Valsecchi *et al.*, 1995), with the integrations showing a puzzling variety of phenomena and behaviours. However, a basic qualitative understanding of the main mechanisms at work would also be important, both to interpret and classify the integration outputs, and to devise some simplified, statistical model of the dynamical evolution process to be applied not to single objects, but to entire populations.

* On leave from the Department of Mathematics, University of Pisa, Via Buonarroti 2, 56127 Pisa, Italy, thanks to the "G. Colombo" fellowship of the European Space Agency.

Earth, Moon, and Planets **72**: 151-164, 1996.

So far, such statistical models have been developed (Wetherill 1985, 1987; Melosh and Tonks, 1993; Bottke et al., 1994, private communication) under the following simplifying assumptions: (i) a Monte Carlo algorithm together with Öpik's (1976) analytical theory is used to assess the occurrence of planetary encounters and predict their outcomes; (ii) secular perturbations are usually taken into account only by assuming uniformly precessing apses and nodes; and (iii) resonant effects are treated as occurring at fixed values of the semimajor axis (2.5 AU for the 3/1 mean motion resonance with Jupiter, 2.05 AU for the ν_6 secular resonance, etc.) and causing essentially rapid jumps in eccentricity and/or inclination. Thus, whenever resonances are not at work, the orbital evolution is modelled basically as an encounter-driven random walk in the semimajor axis–eccentricity–inclination (a–e–I) space; if encounters with only one planet are possible (or if a planet plays a dominant role owing to its larger mass, e.g. the Earth vs. Mars), this random walk is constrained to occur near a T = constant surface, where T is the Tisserand invariant relative to the dominant planet (assumed to have a quasi–circular orbit). For a detailed discussion of the corresponding evolutionary patterns, we refer to the recent review by Greenberg and Nolan (1993).

In this paper we are going to argue that these models, albeit providing useful qualitative insights, do not account for all the main features of the orbital evolution of NEAs. The main reason is that secular perturbations play a role at least as important as that of encounters in guiding the orbital evolution in the a–e–I space, and that resonances need to be modelled in a more complex way to obtain realistic predictions of their effects. An example of this is the recent finding that resonant effects can pump up the eccentricity to almost unity on a time scale $< 10^6$ yr, thus causing a significant fraction of NEAs to end up hitting the Sun (Farinella et al., 1994; Froeschlé et al., 1995).

We shall discuss some of these problems by using as a typical example of NEA evolution the numerically integrated orbit of asteroid (4179) Toutatis. Our integrations have spanned a time scale of 1 Myr (such as required to detect and interpret most secular effects), and used several different dynamical models to point out the main mechanisms at work. As discussed by Whipple and Shelus (1993) and by Benest et al. (1994), who integrated the same orbit over shorter time spans (up to $\approx 10^5$ yr), the orbit of Toutatis is extremely chaotic, with a Lyapounov characteristic time of the order of 100 yr due to frequent encounters with the inner planets, and with the 3/1 Jovian resonance affecting the orbit on a longer time scale. Here we shall extend the afore-mentioned studies with a particular emphasis on the aspects important not just for Toutatis itself, but rather for the long–term evolution of the entire NEA population. In Sec. 2 we shall describe our integrations and discuss the most interesting results, whereas in Sec. 3 we shall elaborate on the significance and implications of these results.

2. Numerical Experiments on Toutatis

2.1. INTEGRATION METHOD AND INITIAL CONDITIONS

Given that close approaches with planets cause fast variations of the orbital elements of NEAs, it is necessary to use an accurate method to integrate the equations of motion. However, whereas a small time step is required when the integrated object is close to a planet (or the Sun) and moves very fast relative to it, a much longer time step can be used when the NEA is far from the planets. Therefore, an integration method using a variable stepsize turns out to be the optimal choice. We adopted the Bulirsch–Stoer extrapolation method ((Stoer and Bulirsch, 1980)) which, through a controlled variable stepsize, allows to handle close approaches with planets much more accurately than with a fixed stepsize. After a number of tests, we chose a restrictive value (10^{-12} instead of the usual 10^{-8}) of the convergence parameter ϵ, which determines the difference between two successive estimates in the iteration.

It is important to emphasize that, even if the integration method is very accurate, for such strongly chaotic orbits the results cannot be deterministic over a time span much longer than the Lyapounov time, and the integrations can provide only qualitative information on the long–term orbital evolution. The reasons are the following. First, the assumed dynamical model always neglects some real small perturbations, whose effects get amplified in a stochastic fashion. Second, owing to the exponential increase of small numerical errors, two different integration methods applied to the same dynamical model and the same initial conditions will give two different orbital evolutions (with similar qualitative behaviors). Likewise, the same integration method used on two different computers will yield quantitatively different results, due to different round–off errors. Thus, while for the sake of simplicity we will talk about the real NEA Toutatis, the orbital evolutions computed with even the most accurate method and the most realistic dynamical model are not necessarily predictive of the behavior of the real asteroid.

Initial conditions for the planets and Toutatis, for the epoch 1993 January 13.0 (JD = 244900.5), were kindly provided to us by G. Hahn (1994, personal communication). The orbital angles are referred to the J2000 equator and equinox. Table I contains the initial conditions of Toutatis and Table II gives the masses of the planets taken from the JPL ephemeris DE200. Note that we combined the Earth and the Moon into a single body located at the

Earth–Moon barycenter, and added the mass of Mercury to that of the Sun.
Toutatis is assumed to be massless.

TABLE I

Osculating   Orbital   Elements
of Toutatis, Referred to the J2000 Ecliptic and
Equinox

Epoch 1993 January 13.0	(JD= 244900.5)
$M = 15°.05349$	$a = 2.5054645 AU$
$e = 0.6398550$	$i = 0°.46674$
$\omega = 276°.28112$	$\Omega = 126°.48206$

2.2. THE THREE DYNAMICAL MODELS

In order to discriminate and assess the significance of each of the mechanisms which affect the orbit of Toutatis, we used three different dynamical
models:

(1) In the first model (Sun–Earth–asteroid), the perturbations are mainly due to close approaches with the Earth, located between the Sun and
Toutatis, almost in the same plane.

(2) In the second model (Sun–Jupiter–Saturn–asteroid), the orbit of Toutatis is mainly affected by resonance mechanisms with the outer planets, as
no close approaches occur.

(3) The third model (Sun–Earth–Jupiter–Saturn–asteroid) takes into account
all the main perturbation forces that affect the orbit of the real Toutatis,
and is therefore the most realistic (and complex) one.

TABLE II

Masses of Planets

Planets	Masses ($M_\odot = 1$)
Sun+Mercury	1.000000166013679527193
Earth+Moon	0.0000304043273871084
Jupiter	0.000954786104043042
Saturn	0.000285877644368210

For all the three models, we numerically integrated the Newtonian equations of motion of the planets along with Toutatis over a time span of 1 Myr.

2.3. RESULTS

2.3.1. *The Sun–Earth–asteroid model*

The evolution of Toutatis in this model is characterized by a random walk of the semimajor axis a and of the eccentricity e of its orbit due to close approaches with the Earth (see Fig. 1). Actually, the small inclination of Toutatis increases significantly the frequency of these encounters, as they can occur also relatively far from the mutual nodes. As a consequence, the random walk is fairly effective, with the semimajor axis changing by about 1 AU over the integration time span. Thus for this type of orbits, Earth encounters are sufficient to cause a comparatively fast evolution in the orbital element space and in particular (as we shall see later on) to jump frequently between different resonant states.

In Fig. 1 a strong correlation is also apparent between the semimajor axis and the eccentricity of the asteroid, with the perihelion distance staying almost constant. This fact can be explained in a simple way by using Gauss' perturbation formulae (see e.g. Bertotti and Farinella, 1990, Ch. 11). If we treat close approaches as causing fast, quasi–impulsive changes of the osculating Keplerian elements, these changes can be expressed as a function of the impulsional velocity increment δV. We have:

$$\delta a = \frac{2}{n\sqrt{1 - e^2}} [\delta V_1 + e(\delta V_1 \cos f + \delta V_2 \sin f)] \tag{1}$$

$$\delta e = \frac{\sqrt{1 - e^2}}{na} [\delta V_2 \sin f + \delta V_1 (\cos f + \frac{e + \cos f}{1 + e \cos f}] \tag{2}$$

where f is the asteroid's true anomaly at the time of the encounter and n is its mean motion. δV_1 and δV_2 are the transverse and radial components of the impulsional velocity increment. Assuming that the geometry of Toutatis' orbit is such that the approaches with the Earth always occur near perihelion (a good approximation, as the perihelion distance keeps close to 1 AU), one can approximate those equations by taking $f \approx 0$. Then, from Eq. (1), we have:

$$\delta V_1 \approx \frac{n\delta a\sqrt{1 - e^2}}{2(1 + e)}, \tag{3}$$

and substituting this expression in Eq. (2), we obtain:

$$\delta e \approx \frac{\delta a}{a}(1 - e). \tag{4}$$

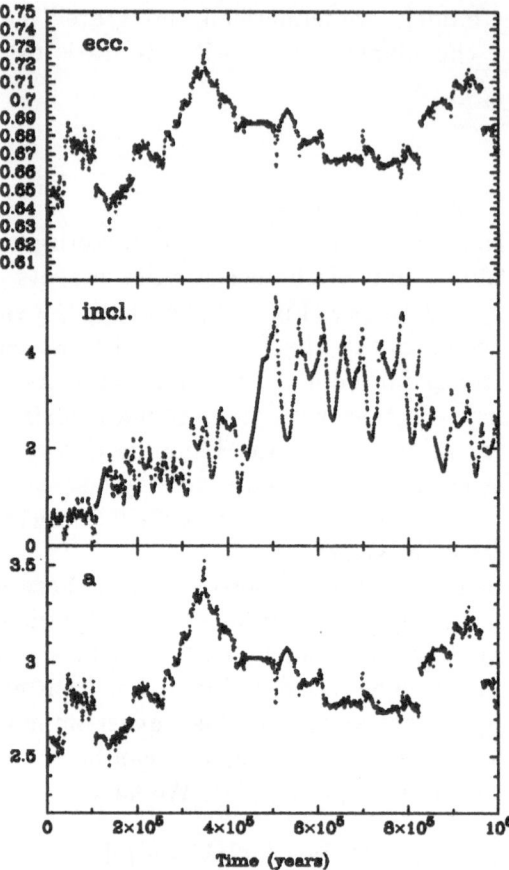

Fig. 1.   Evolution of the semimajor axis (in AU), the inclination (degrees) and the
eccentricity of Toutatis' orbit over 1 Myr in the Sun–Earth–asteroid model.

Thus δe is proportional to δa and the evolutions of e and a are well correlated. Note that Eq. (4) implies that the perihelion distance q of the orbit of
Toutatis is almost constant. This is consistent with the (approximate) conservation of the Tisserand invariant of this problem. Indeed, if we neglect
the Earth's orbital eccentricity, the model provides just a restricted three-
body problem, for which the Jacobi constant is conserved and the Tisserand
parameter

$$T = 1/a + 2[a(1 - e^2)]^{1/2} \cos I \tag{5}$$

(a being expressed in units of the perturbing planet's semimajor axis, namely, in our case, AU) is not modified by the encounters. Thus the orbit must
remain close to a surface $T = $ constant in the orbital element space. Actually,
Fig. 2 shows the contour lines for different values of the Tisserand invariant

Orbital Elements of TOUTATIS (3 Bodies Model)

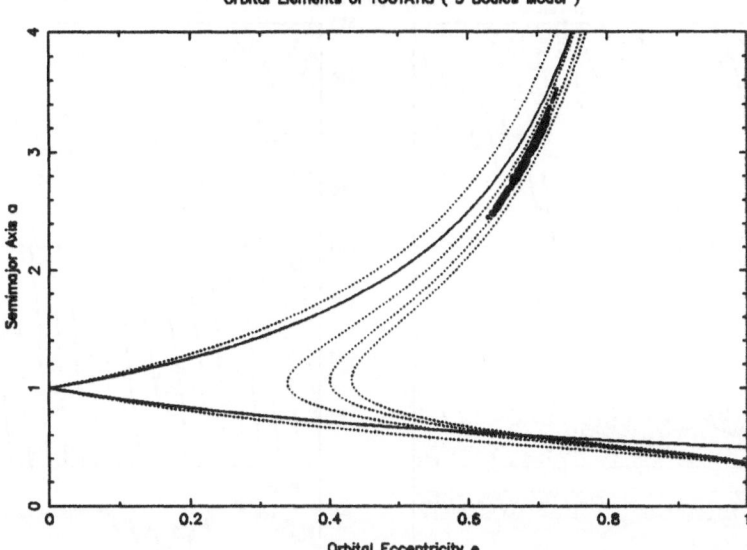

Fig. 2. The black dots show the evolution in the semimajor axis a (in AU) vs. eccentricity e plane for the numerically integrated orbit of Toutatis over 1 Myr, with the Sun–Earth–asteroid model. The two solid lines correspond to the perihelion q and aphelion Q distances equal to 1 UA. The dotted lines are the contours of the Tisserand constant T at zero inclination and correspond (from left to right) to $T = 3$, 2.88, 2.83 and 2.80.

for $I = 0$, and shows that the integrated orbit of Toutatis always stays close to the contour line $T = 2.83$, corresponding to its initial osculating elements. Note also that as $q = a(1 - e)$, another expression for T is:

$$T = \frac{1-e}{q} + 2[q(1+e)]^{1/2} \cos I .\tag{6}$$

Then, if $q = 1$ and e is small, we can expand T into a series of e:

$$T = 1 - e + 2\left[1 + \frac{1}{2}e + O(e^2)\right] \cos I ,\tag{7}$$

which, for $I = 0$, becomes:

$$T = 3 + O(e^2) .\tag{8}$$

Therefore, the orbit of a body with small inclination and for which $T \approx 3$ will evolve near the line $q = 1$, thus causing frequent Earth approaches.

2.3.2. The Sun–Jupiter–Saturn–asteroid model

The evolution of the orbital elements of Toutatis (see Fig. 3) shows that the asteroid is locked in the 3/1 mean motion resonance with Jupiter, near

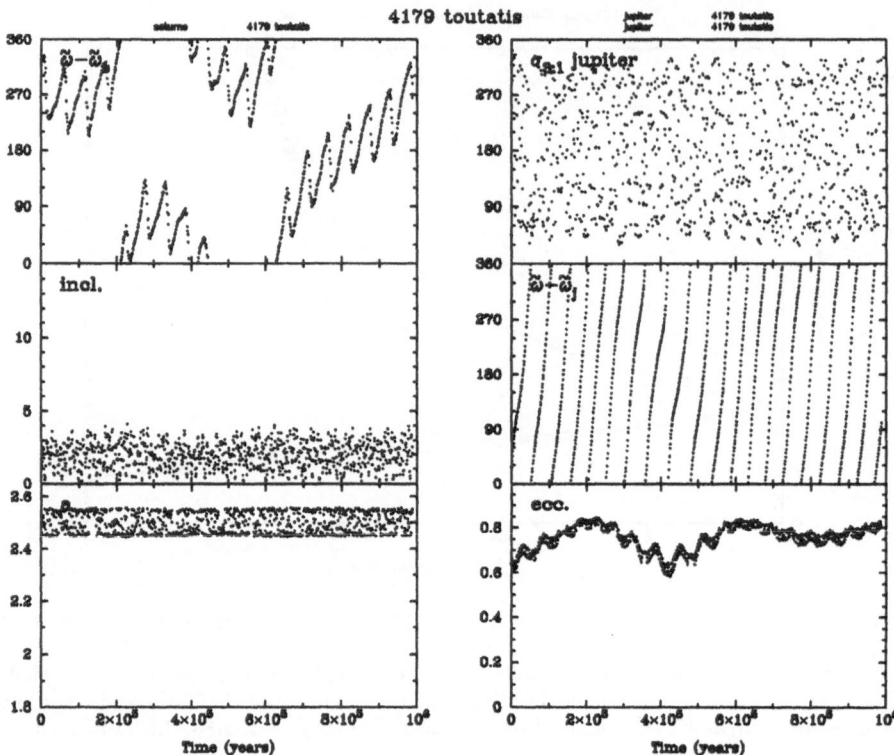

Fig. 3.   Evolution of the orbital elements of Toutatis over 1 Myr in the
Sun–Jupiter–Saturn–asteroid model. Besides semimajor axis, eccentricity and inclination,
the figure shows also three critical arguments: $\sigma_{3:1}$ of the 3/1 mean motion resonance with
Jupiter [see Eq. (9)]; $\varpi - \varpi_j$ and $\varpi - \varpi_s$ of the ν_5 and ν_6 secular resonances (ϖ_j and ϖ_s
being the perihelion longitudes of Jupiter and Saturn, respectively).

$a = 2.5$ AU, during the whole integration span. This is confirmed by the
fact that the critical argument $\sigma_{3:1}$, defined by:

$$\sigma_{i:k} = k\lambda - i\lambda_j + (i - k)\varpi, \qquad\qquad (9)$$

with $i = 3$, $k = 1$, always librates (with a large amplitude) around $180°$. Here
λ is the mean longitude and ϖ is the longitude of perihelion of Toutatis, while
λ_j is the mean longitude of Jupiter. It is worth noting that Toutatis' orbit
is also inside the secular resonance ν_6 with Saturn (perturbed by Jupiter)
during the first 6×10^5 yr of the integration, as shown by the libration of
the secular critical argument $\varpi - \varpi_s$ around $0°$. When the orbit is locked
in the ν_6 resonance, wide oscillations of the eccentricity occur. This is consistent with the findings of Morbidelli and Moons (1993) and Moons and
Morbidelli (1995) on the behavior of secular resonances inside mean motion
resonances.

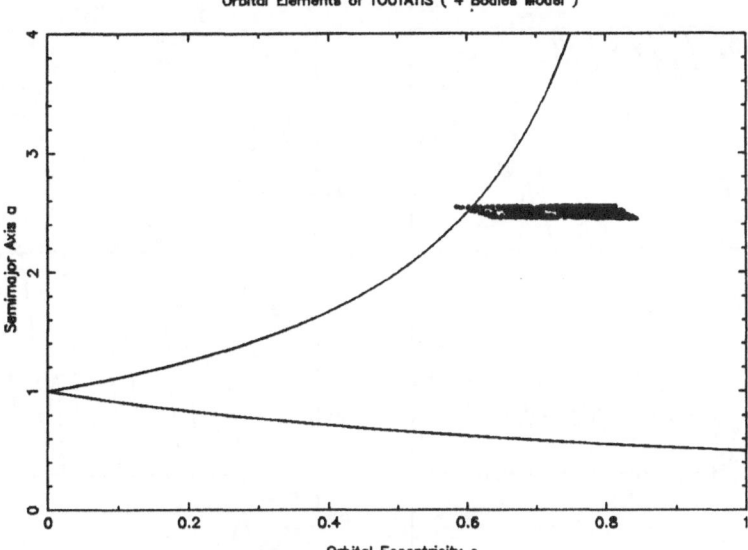

Fig. 4.   A plot of the semi-major axis a vs. eccentricity e for the numerically integrated orbit of Toutatis over 1 Myr with the Sun–Jupiter–Saturn–asteroid model. The lines $q = 1$ AU and $Q = 1$ AU are also shown.

As the Earth is not present in this model, however, there are no close approaches and no random walk of the semimajor axis during the integration time. This confirms that the stochasticity of the evolution is mostly due to encounters with the inner planets (Whipple and Shelus, 1993; Benest *et al.*, 1994). In the a vs. e plane (see Fig. 4), the motion occurs along a narrow horizontal strip, which crosses the $q = 1$ AU line.

2.3.3.   *The Sun–Earth–Jupiter–Saturn–asteroid model*
This model is the most realistic one. As many dynamical mechanisms are at work at the same time, Toutatis has a complex behavior (see Fig. 5), switching between several mean motion and secular resonances. During the first 9×10^4 yr, the orbit is locked in the 3/1 resonance with Jupiter, with the semimajor axis staying around 2.5 AU. Then, due to a close approach with the Earth, it is ejected from this resonance, with the semi-major axis jumping to lower values. But soon it enters in ν_6 secular resonance (as shown by the librations of the corresponding critical argument). Subsequently, for a fairly long time (between about 4×10^5 and 7.5×10^5 yr), the semimajor axis stays around 2.06 AU, corresponding to the 4/1 mean motion resonance with Jupiter. Actually, during this interval the corresponding critical argument $\sigma_{4:1}$ alternates between libration around $0°$ and circulation (see Fig. 6). In the libration intervals, the orbit also gets locked in the ν_5 secular resonance

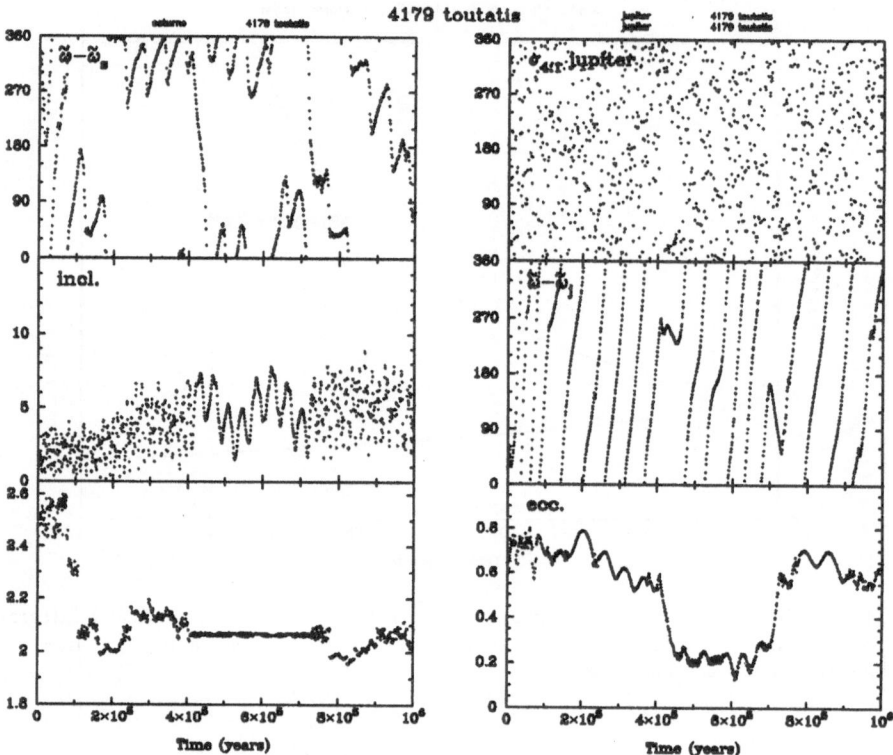

Fig. 5.   The same as Fig. 3 but for the Sun–Earth–Jupiter–Saturn–asteroid model and with the critical argument $\sigma_{4:1}$ instead of $\sigma_{3:1}$.

and the eccentricity undergoes significant changes (Moons and Morbidelli, 1995, and Yoshikawa, 1989).

Fig. 7 shows that when the outer planets are taken into account in the integration, the orbit of Toutatis does not random walk any more along the contour lines of the Tisserand invariant relative to the Earth. The corresponding path is quite complex, with vertical (semimajor axis) changes caused by encounters and large horizontal strips due to both mean motion and secular resonance effects, with superimposed oscillations due to non-resonant secular perturbations. The inclination also grows up to almost 10°. Therefore, even if the orbit's stochasticity is still dominated by close approaches with the Earth, any dynamical model considering only the effects of the encounters (i.e., conserving approximately the Tisserand invariant) cannot reproduce the trends and time scales of the real evolution. Therefore, using such models to predict the evolution of the NEA population is likely to yield misleading results.

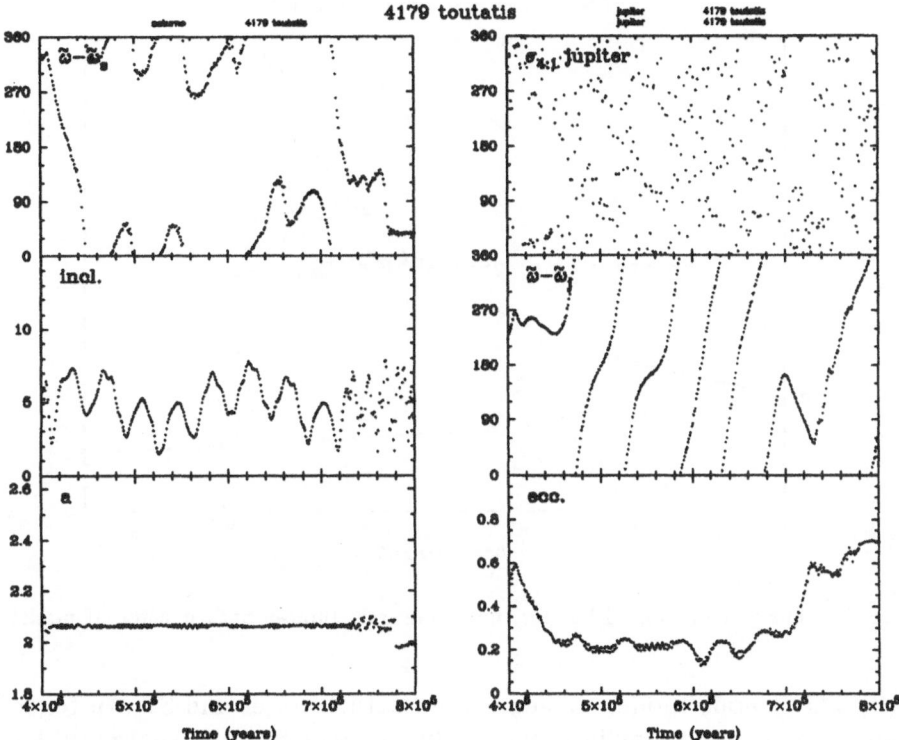

Fig. 6. The same as Fig. 5 but with the horizontal scale enlarged to better show the time interval between 4×10^5 and 8×10^5 yr.

3. Conclusions

The main conclusions of the work described above can be summarized as follows:

(1) When only the Earth is included in the dynamical model, the evolution of Toutatis is dominated by close encounters with it. The orbit is strongly chaotic and the changes of a and e have the typical features of a random walk (albeit with a strong correlation between the two elements). However, the perihelion distance and the Tisserand parameter stay almost constant, as predicted by simple analytical arguments based on the restricted three–body problem.

(2) On the other hand, the presence of the outer planets in the dynamical model causes Toutatis' orbit to get locked into resonances, in particular the 3/1 mean motion resonance with Jupiter and the ν_6 secular resonance with Saturn's longitude of perihelion (perturbed by Jupiter). In order to get the latter resonance, both outer planets have to be present in the model. Resonances cause significant variations of e but keep a almost constant.

Orbital Elements of TOUTATIS (5 Bodies Model)

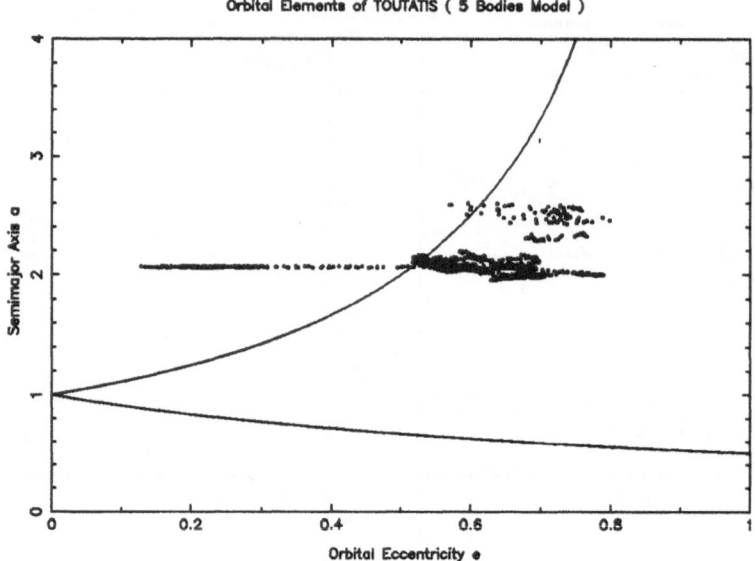

Fig. 7.   The same as Fig. 4 but for the Sun–Earth–Jupiter–Saturn–asteroid model.

(3) The 5–body model including the Earth, Jupiter and Saturn together results in a very complex interplay of dynamical effects. In summary, Earth encounters switch Toutatis' orbit between several different resonant states, including the 3/1 and 4/1 mean motion and the ν_5 and ν_6 secular resonances. The eccentricity changes are dominated by resonance effects, whereas in a mainly encounter–related variations are apparent. None of the previous two models has even a qualitative resemblance to the evolution pattern resulting in this case. In particular, the Tisserand invariant relative to the Earth is not conserved at all, as the evolution of the eccentricity is mainly controlled by the outer planets. Indeed, strong and fast eccentricity changes appear to provide a "fast–track" dynamical *route* between low– and high–eccentricity asteroid orbits.

We stress that even the third, most realistic dynamical model does not provide a quantitatively predictive description of the behavior of the real asteroid, for at least two reasons: (i) the strong stochasticity of the orbit, mainly related to Earth encounters; (ii) and the fact that the model is still an approximate one, with a number of missing perturbation effects. For instance, another numerical integration of Toutatis with a model including all the planets from Venus to Neptune (Farinella *et al.*, 1994; Valsecchi *et al.*, 1995) shows a variety of dynamical mechanisms at work as in our third model, but the evolution of the orbital elements is markedly different. Actually, in this integration, Toutatis gets ejected from the solar system on a comet-like hyperbolic orbit some 6×10^5 yr in the future, after an encounter with

Jupiter! Of course, this just shows that for such strongly chaotic, fast–track evolving orbits, very different final fates are possible, including a hyperbolic ejection, a collision with the Sun or with a planet.

Another important point to be stressed is that not all NEAs are currently evolving along the fast–track resonant routes of the type we have discussed for Toutatis. The evidence from the recent numerical work quoted above is that probably only a minor fraction (possibly some 20%) of the existing NEAs at any given time is on fast–track orbits. Other orbits (in particular those classified in the Geographos class by Milani *et al.* (1989)) actually evolve in a slower, random–walk fashion, with Earth encounters playing the dominant role, in a way qualitatively similar to that shown in Fig. 1. This is also the case for many Amor objects, including the largest NEAs, (1033) Ganymed and (433) Eros: in these cases, the evolution is still slower, as only Mars encounters are possible and the Martian mass is an order of magnitude smaller than that of the Earth. However, from the point of view of the "demography" of NEAs, it is likely that the slow–track objects represent more the exception than the rule, and that they are over–represented in the existing population just because of their much slower evolution and longer lifetime. Also, since all NEAs probably start their independent life (after collisional ejection from main–belt parent asteroids) inside Toutatis–like resonant channels, slow–track nonresonant bodies probably are the outcome of "lucky" encounters with Mars or the Earth, removing them from the resonances and putting them into long–lived "parking zones". Thus, the conclusion appears likely that there is no *typical* dynamical evolution or even lifetime for all NEAs, but that the transfer of bodies from the main asteroid belt to the Earth–crossing zone is a complex process, with a variety of time scales, dynamical mechanisms, and exchanges between different classes of objects. This scenario is further discussed by Froeschlé *et al.* (1995) and Menichella *et al.* (1995).

Much more numerical and modelling work appears needed to understand the remaining open problems. For instance: how many different "dynamical classes" are required to correctly classify the orbital evolution patterns over time spans of 10^7 to 10^8 yr? How frequently do exchanges occur among different classes, and more in general between fast–track and slow–track orbits? What are the locations of secular resonances in the little–known region with $a < 2$ AU, where the secular perturbations of the terrestrial planets need to be taken into account? And how effective are the mean motion resonances with the inner planets to protect NEAs from encounters, as in the case of the Toro–class objects of Milani *et al.* (1989)? We plan to deal with these problems in the future to obtain a plausible, self–consistent evolutionary scenario for this important and intriguing population of near–Earth interplanetary bodies.

References

Benest, D., Froeschlé, C. and Gonczi, R.: 1994, in *Seventy-five Years of Hirayama Asteroid Families: The role of catastrophic collisions in the solar system history*, Y. Kozai, R.P. Binzel and T. Hirayama (eds.), Astron. Soc. Pac. Conf. Ser. **63**, pp. 7–14.

Bertotti, B. and Farinella, P.: 1990, *Physics of the Earth and the Solar System*, Kluwer Academic Publishers, Dordrecht/ Boston/ London, pp. 219–239.

Farinella, P., Gonczi, R., Froeschlé, Ch. and Froeschlé, C.: 1993, *Icarus* **101**, 174–187.

Farinella, P., Froeschlé, Ch., Froeschlé, C., Gonczi, R., Hahn, G., Morbidelli, A. and Valsecchi, G.B.: 1994, *Nature* **371**, 314–317.

Froeschlé, Ch., Hahn, G., Gonczi, R., Morbidelli, A. and Farinella, P.: 1995, submitted to *Icarus*.

Greenberg, R. and Nolan, M.C.: 1993, in *Resources of Near-Earth Space*, J.S. Lewis, M.S. Matthews, M.L. Guerrieri (eds.), The University of Arizona Press (Tucson & London), pp. 473–492.

Melosh, H.J. and Tonks, W.B.: 1993, *Meteoritics* **Vol. 28**, Number 3, p. 398 (abstract).

Menichella, M., Paolicchi, P. and Farinella, P.: 1995, this volume.

Milani, A., Carpino, M., Hahn, G. and Nobili, A.M.: 1989, *Icarus* **78**, 212–269.

Moons, M. and Morbidelli, A.: 1995, *Icarus*, in press.

Morbidelli, A. and Moons, M.: 1993, *Icarus* **103**, 99–108.

Öpik, E.J.: 1976, *Interplanetary Encounters*, Elsevier Scientific Publishing Company, Amsterdam/ Oxford/ New-York.

Stoer, J. and Bulirsch, R.: 1980, *Introduction to Numerical Analysis*, Springer, New-York.

Valsecchi, G.B., Morbidelli, A., Gonczi, R., Farinella, P., Froeschlé, Ch. and Froeschlé, C.: 1995, submitted to *Icarus*.

Wetherill, G.W.: 1985, *Meteoritics* **20**, 1–21.

Wetherill, G.W.: 1987, *Phil. Trans. R. Soc. Lond. A* **323**, 323–337.

Whipple, A.L., Shelus, P.J.: 1993, *Icarus* **105**, 408–419.

Yoshikawa, M.: 1989, *Asron. Astrophys.* **213**, 436–458.

LONG-TERM BEHAVIOR OF THE MOTION OF PLUTO OVER 5.5 BILLION YEARS

HIROSHI KINOSHITA AND HIROSHI NAKAI
National Astronomical Observatory
2-21-1 Osawa, Mitaka, Tokyo, Japan

Abstract. The motion of Pluto is said to be chaotic in the sense that the maximum Lyapunov exponent is positive: the Lyapunov time (the inverse of the Lyapunov exponent) is about 20 million years. So far the longest integration up to now, over 845 million years (42 Lyapunov times), does not show any indication of a gross instability in the motion of Pluto. We carried out the numerical integration of Pluto over the age of the solar system (5.5 billion years ≈ 280 Lyapunov times). This integration also did not give any indication of chaotic evolution of Pluto. The divergences of Keplerian elements of a nearby trajectory at first grow linearly with the time and then start to increase exponentially. The exponential divergences stop at about 420 million years. The divergences in the semi-major axis and the mean anomaly (equivalently the longitude and the distance) saturate. The divergences of the other four elements, the eccentricity, the inclination, the argument of perihelion, and the longitude of node still grow slowly after the stop of the exponential increase and finally saturate.

1. Introduction

Sussman and Wisdom (1988) carried out the numerical integration of Pluto for 845 million years (Myr hereafter) with the Digital Orrery, a special purpose computer. This integration took about three months. In this computation Pluto was a massless particle disturbed by other giant outer planets, Jupiter, Saturn, Uranus, and Neptune. Their integration indicates that the long-term motion of Pluto is chaotic and the largest Lyapunov exponent is about $10^{-7.3}$ year^{-1} (the Lyapunov time is about 20 Myr). Laskar (1990) numerically integrated for 100 Myr the secular Hamiltonian system of the

Earth, Moon, and Planets **72**: 165-173, 1996.
© 1996 *Kluwer Academic Publishers.*

whole solar system (excluding Pluto) that is obtained after elimination of short periodic terms and is second-order with respect the disturbing masses. His integration showed that the solar system is chaotic with the Lyapunov time of only 4 Myr. Sussman and Wisdom (1992) carried out an integration of the whole solar system including Pluto with the Toolkit, another special purpose parallel computer, which took about 1000 hours. Their computation confirmed Laskar's results in direct integration instead of averaged equations. The evolution of Pluto in this 100 Myr integration is similar to that of Pluto found in 845 Myr integration of the outer planets system (Sussman and Wisdom 1988). Although the motion of Pluto is chaotic in the sense that the largest Lyapunov exponent is positive, the integration does not give any indication of a gross instability in the motion of Pluto. According to Wisdom (1992), "Since the system is apparently chaotic, we cannot rule out the possibility of gross instability. Recall some chaotic asteroid trajectories have been seen evolve chaotically for 100 Lyapunov times at low eccentricity and then suddenly jump to large eccentricity. It will be very interesting to see a number of integrations of the whole solar system for the age of the solar system and longer." We carried out the integration of the outer planets system over 5.5 billion years which is about 280 Lyapunov times. In this computation Pluto is treated as a massless particle in order to compare our results with those of Sussman and Wisdom (1988). The integrator is a 12th-order linear symmetric multistep method (Quinlan and Tremaine 1990). The error estimate of our integration is given in section 2 and preliminary results are then presented.

2. Method of Numerical Integration and its Accuracy

Pluto is integrated as a massless particle that is disturbed by four giants planets (Jupiter, Saturn, Uranus, and Neptune). The masses of the inner planets are added to the Sun. The planetary masses and their initial conditions are taken from DE245, which is the most recent planetary ephemerides developed at JPL.

As an integration formula, we adopt a linear symmetric multistep integrator (LSMI), which is one of linear multistep integrators whose coefficient in the formula are symmetric (Quinlan and Tremaine 1990). One of the great merits of LSMI is that the truncation (discretization) errors do not produce secular errors in the energy and the angular momentum, in other words no secular errors in the semi-major axis, eccentricity, and inclination.

The numerical computation was carried out on an FMR70-HX3 with an accelerator board. FMR70-HX3 is a personal computer whose cpu is Intel's i386 with clock speed of 25MHz and which does the job of input and output of data between FMR and the accelerator board. The accelerator board uses Intel's i860 cpu with clock speed of 40MHz and does only numerical integrations of equations of motion.

In order to reduce round-off errors, we evaluated the 12 th-order LSMI (Quinlan and Tremaine 1990) in the following form:

$$x_{n+k} = -\alpha_{k-1} \otimes x_{n+k-1} \ominus \cdots \ominus \alpha_0 \otimes x_n \oplus h^2 \sum_{j=0}^{k-1} \beta_j f_{n+j}, \qquad (1)$$

where $\alpha_i = \alpha_{k-i}, \beta_i = \beta_{k-i}, i = 0, \cdots, k = 12$, and $\alpha_{12} = 1$.

The $\otimes, \oplus, \ominus$ symbols mean multiplications, additions, and subtractions in quadruple precision. The quadruple arithmetic operations are carried out by software which is written in assembler. Other operations (the evaluation of the force and the last part of the right-hand side of (1)) are done with double precision.

Because of the symmetry in the coefficients of LSMI, the result by LSMI has a time-reversal nature. Therefore the time reversal test (integrating forward and then backward) does not give any information on the accumulation of the truncation errors. For checking the accuracy of orbits obtained by the 12 th-order LSMI, we made reference orbits of one Myr integrated by the extrapolation method developed by Gragg(1965), which are computed in quadruple precision. The accuracy of the reference orbits themselves are examined by the time reversal test. Assuming that the longitude error increases proportional to square of time, the longitude error of Jupiter is about 0.016 arcsecond after 5.5×10^9 years, which is precise enough to test the accuracy of orbits computed in lower precision. The difference in the longitude of Jupiter after one Myr between the orbits obtained with 12-th LSMI and the reference orbits is $1.°6 \times 10^{-5}$. Since the longitude error due to the truncation grows linearly for the LSMI and the longitude error due to round-off errors increases with the power of $3/2$ of the time, the round-off errors become dominant soon. Extrapolation by the power of $3/2$ of the longitude error of Jupiter over 5.5 billion years is about $6°.5$. Similarly the longitude error of Pluto is about $0.°3$ after 5.5 billion years.

The timespan of one run of our integration was 4×10^{10} days ≈ 110 Myr, and one run took 53 hours using the FMR with the accelerator board. We carried out 50 runs of this computation whose total time was 110 days. We made output of the positions and velocities of the 5 outer planets and a nearby orbit of Pluto in double precision for every 2 million days ≈ 5500 years, whose total amount is 296 mega bytes. These data are available on request.

3. Results

Figure 1 shows four Keplerian elements of Pluto (the eccentricity, the argument of perihelion, the inclination, and the longitude of node referred to the longitude of Neptune's node) for the first 100 Myr of 5.5 billion years integration, and Figure 2 exhibits those over the last 100 Myr. There is no indication of global instability of Pluto's motion.

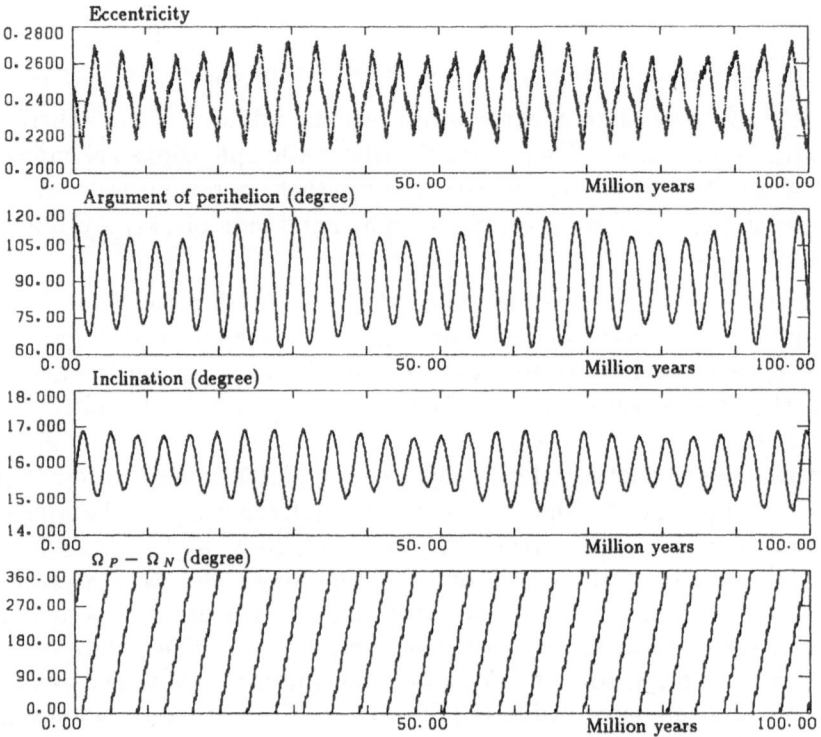

Figure 1.   Keplerian elements of Pluto (the first 100 million years).

In the motion of Pluto three resonances are found in the past works. 1) Pluto is in the mean motion resonance with Neptune. The critical argument $\theta_1 = 3\lambda_P - 2\lambda_N - \varpi_P$ of the 3:2 mean motion resonance librates around 180 degrees with the amplitude 81.2 degrees and the libration period is 2.0×10^4 years. The dominant periodic component in the variation of the semi-major axis is the libration of the critical argument θ_1. The amplitude in the semi-major axis is 0.15AU 2) The argument of perihelion $\theta_2 = \varpi_P - \Omega_P$ librates around 90 degrees and its period is 3.8 Myr. The dominant periodic variations of the eccentricity

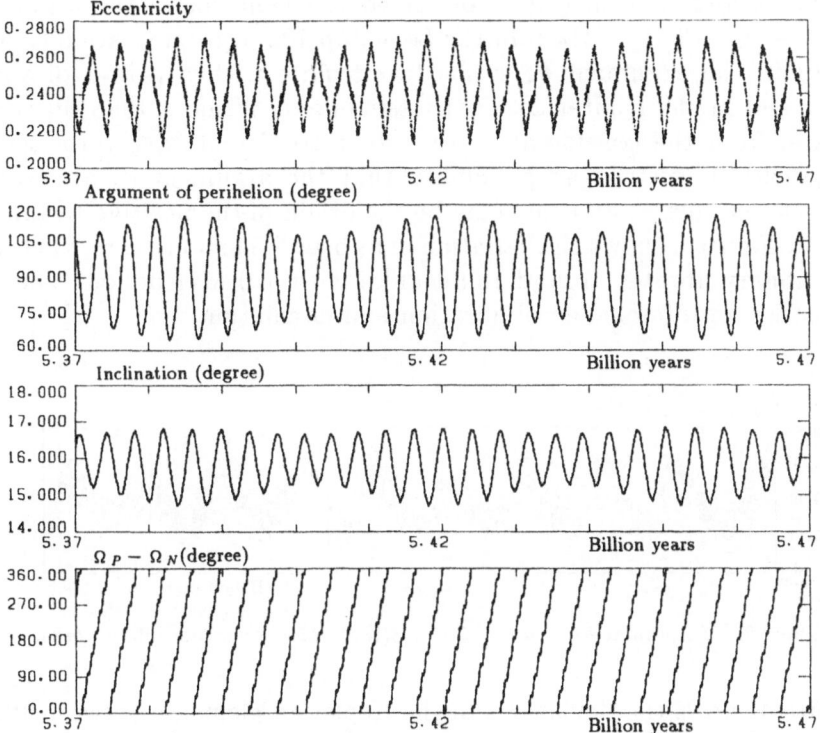

Figure 2. Keplerian elements of Pluto (the last 100 million years).

and the inclination are synchronized with the libration of the argument of perihelion, which is expected from the secular perturbation theory (Kozai 1962). The variations of θ_2, the eccentricity, the argument of perihelion, and the inclination are modulated with 34 Myr periodicity.

3) Moreover the longitude of Pluto's node referred to the longitude of Neptune's node, $\theta_3 = \Omega_P - \Omega_N$, circulates and the period of this circulation is equal to the period of the libration of θ_2. When θ_3 becomes zero, the inclination of Pluto referred to the invariable plane takes a maximum, the eccentricity reaches a minimum, and the argument of perihelion is 90 degrees. When θ_3 becomes 180 degrees, the inclination takes a minimum, the eccentricity reaches a maximum, and the argument of perihelion is again 90 degrees.

This new type of resonance was conjectured by Williams and Benson (1971) and was confirmed by Milani et al. (1989) who called this kind of resonance the 1:1 super resonance. This type of resonance is also called a secondary resonance. All these three resonances are well kept over 5.5 billion years.

Williams and Benson (1971) discussed the behavior of the argument $\theta_4 = \varpi_P - \varpi_N + 3(\Omega_P - \Omega_N)$. In the Longstop 100-Myr integration (Nobili et al. 1989) the argument θ_4 seems to circulate with a period of about 246 Myr and in the Digital Orrery (Sussman and Wisdom 1988) θ_4 seems to librate, from the consideration that $\dot{\theta}_4$ is consistent with zero. Milani et al. (1989) suggested the possibility that the argument θ_4 alternately librates and circulates and this may be the origin of the positive Lyapunov exponent. Figure 3 shows the variation of θ_4 over 5.5 billion years. θ_4 clearly and stationary librates around 180 degrees with 570 Myr period. There is no indication of interchange of libration and circulation.

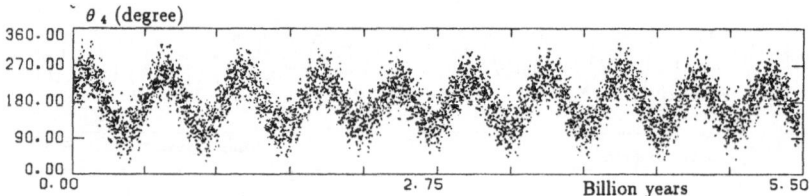

Figure 3.   Argument $\theta_4 = \varpi_P - \varpi_N + 3(\Omega_P - \Omega_N)$ over 5.5 billion years.

Figure 4 shows the deviations of the Keplerian elements between Pluto and its nearby orbit whose initial conditions are slightly different (the relative distance in the phase space is 10^{-12}). The deviations first increase proportionately with time and then from about 150 Myr start to increase exponentially. The time scale of the exponential increase is about 20 Myr, is in good agreement with the Lyapunov time (the inverse of Lyapunov exponent) of 20 Myr (Wisdom et al. 1998 and Nakai et al. 1992). The deviations of the semi-major axis and the mean anomaly saturate after 420 Myr and after do not increase. This saturation is related to the mean motion resonance. In fact the deviation of the critical argument θ_1 of the two orbits saturates at 162.4 degrees which is twice of the amplitude of the libration of the critical argument, and the deviations of the mean longitudes and the mutual distance of two orbits saturate at 70 degrees and 44 AU, respectively.

The exponential increase of the deviations of the eccentricity, inclination, argument of perihelion, and longitude of node stops after 420 Myr and then increases slowly and seems to finally saturate (see Figure 4). The final saturation after 5.5 billion years in these four elements is related to other two resonance lockings, the libration of the argument of perihelion and the secondary resonance between θ_2 and θ_3. Figure 5 shows the inclination of Pluto referred to the invariable plane versus the argument θ_3. Due to the secondary resonance, the inclination oscillates as a stationary

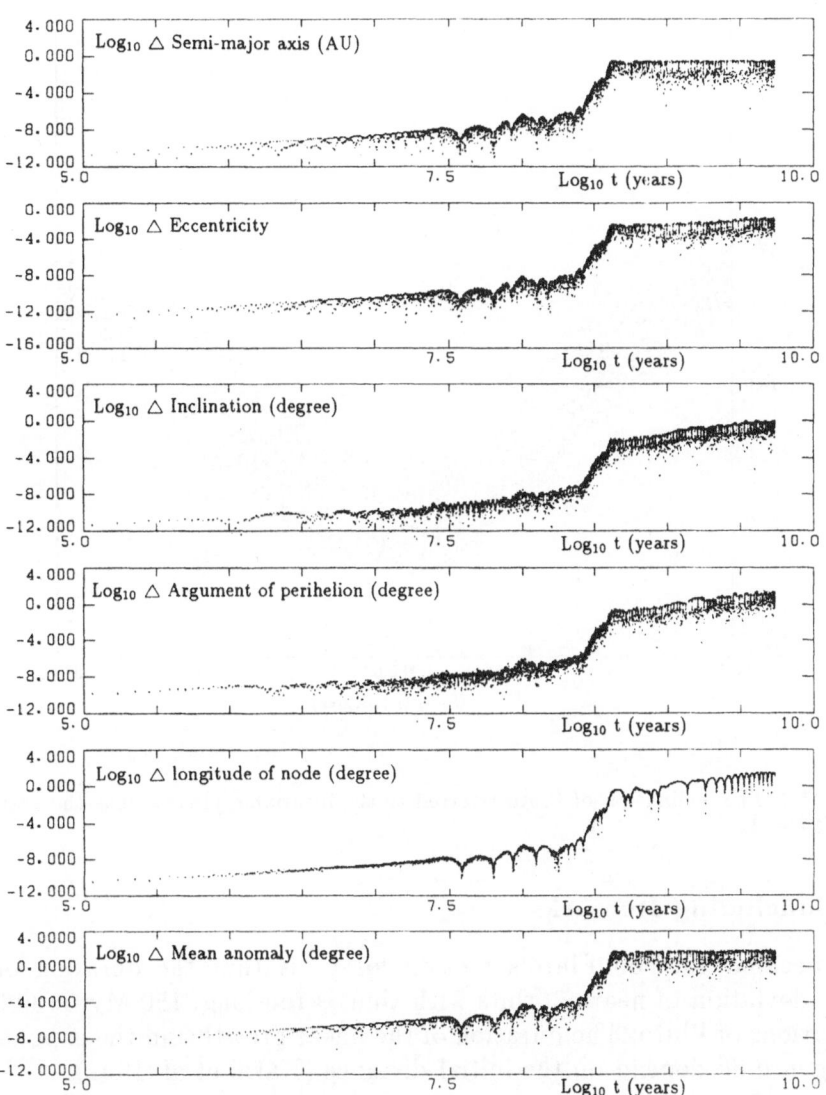

Figure 4. Deviations of the Keplerian elements between Pluto and its nearby orbit.

wave, which goes up and down with the period of 34 Myr. The width of this stationary wave is about 1.2 degree, which is equal to the saturated deviation of the inclination of two orbits. Similar discussions can be applied to the eccentricity and the argument of perihelion.

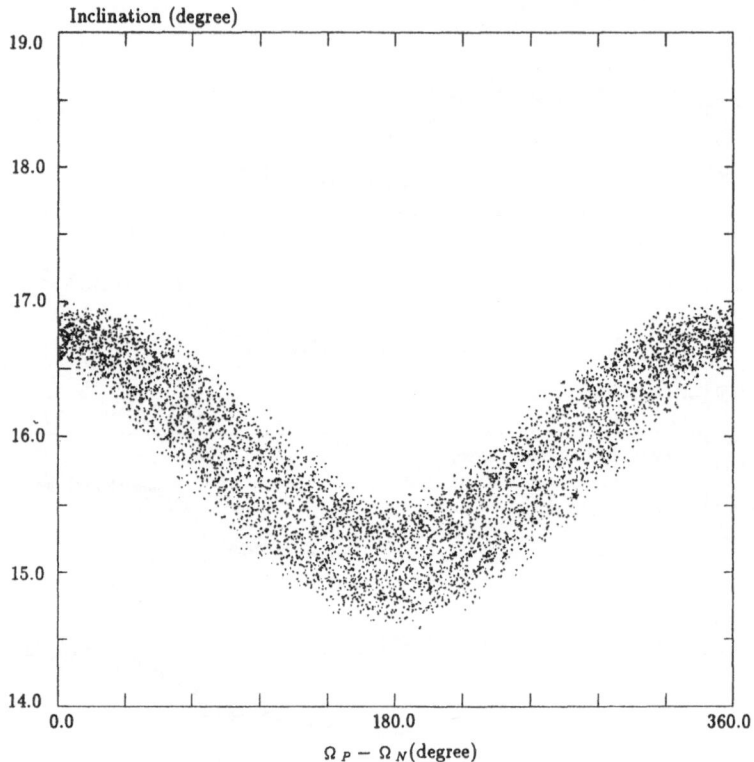

Figure 5.   The inclination of Pluto referred to the invariable plane versus the argument
$\theta_3 = \Omega_P - \Omega_n$.

4. Concluding Remarks

One peculiar fact on Pluto's motion for us is that the duration of the
linear deviation of nearby orbits with time is too long, 150 Myr $\approx$ 600,000
revolutions of Pluto. The duration of the linear growth and the exponential
divergence do depend on the initial distance (Nakai et al. 1992).

   As mentioned at the beginning of Section 3, Pluto is locked in three
resonances. Moreover the argument θ_4 librates around 180 degrees with a
570 Myr period. The behavior of θ_4 depends on the initial conditions. We
carried out several numerical integrations with different initial values of θ_1
(the critical argument of the mean motion resonance) keeping other elements unchanged (Nakai and Kinoshita 1994). When the initial value of
θ_1 is small, θ_4 circulates with prograde direction. As the initial value of θ_1
increases, the behavior of θ_4 changes from prograde circulation to libration
around 180 degrees, and then to retrograde circulation, and other orbital

elements do not show any irregular variations. From these experiments, the behavior of θ_4 does not seem to relate to the global stability of Pluto's motion. When the libration amplitude of θ_1 reaches about 90 degrees, the secondary resonance is destroyed, but orbital elements do not show any irregularities. However when the amplitude of θ_1 becomes larger than about 110 degrees, the second resonance (the libration of the argument of perihelion) is destroyed and all orbital elements show irregular changes. Even though from a very limited number of experiments we cannot derive a general conclusion of Pluto's motion, we do think the secondary resonance does not have an important role in the stabilization of Pluto's motion.

In this paper we integrated Pluto's motion towards the future. From the time reversibility of equations of motion, we do think 5.5 billion years integration of Pluto towards the past do not show any essential difference from the results of this paper. In this sense, in order to investigate how Pluto evolves to the present state of three resonance lockings, we have to take account of a non-conservative mechanism in the early stage of the solar system.

References

Gragg, W.B.: 1965, On extrapolation algorithms for ordinary initial value problems, *SIAM J. Num. Anal.* **2**, pp.384–403.

Kozai, Y.: 1962, Secular perturbation of asteroids with high inclination and eccentricity, *Astron. J.* **67**, pp.591–598.

Laskar, J.: 1989, A numerical experiment on the chaotic behavior of the solar system, *Nature* **338**, pp.237–238.

Milani, A., Nobili, A.M. and Carpino, M.: 1989, Dynamics of Pluto, *Icarus* **82**, pp.200–217.

Nakai, H., Kinoshita, H. and Yoshida, H.: 1992, Dependence on computer's arithmetic precision in calculation of Lyapunov characteristic exponent, in *Proceedings of the 25 Symposium on Celestial Mechanics*, eds. Kinoshita, H. and Nakai, H., pp.1–10.

Nakai, H., Kinoshita, H.: 1994, Stability of the orbits of Pluto, in *Proceedings of the 26 Symposium on Celestial Mechanics*, eds. Kinoshita, H. and Nakai, H., pp133-138.

Nobili, A.M., Milani, A. and Carpino, M.: 1989, Fundamental frequencies and small divisors in the orbits of the outer planets, *Astron. Astrophys* **210**, pp.313-336.

Quinlan, D. and Tremaine, S.: 1990, Symmetric multistep methods for the numerical integration of planetary orbits, *Astron. J.* **100**, pp.1694-1700.

Sussman, G.J. and Wisdom, J.: 1988, Numerical evidence that the motion of Pluto is chaotic, *Science* **241**, pp.433-437.

Sussman, G.J. and Wisdom, J.: 1992, Chaotic evolution of the solar system, *Science* **257**, pp.56-62.

Williams, J.G. and Benson, G.S.: 1971, Resonances in the Neptune-Pluto system, *Astron. J.* **76**, pp.167-177.

Wisdom, J.: 1992, Long term evolution of the solar system, in *Chaos, Resonance and Collective Dynamical Phenomena in the Solar System*, ed. S. Ferraz-Mello, Kluwer, pp.17-24.

CAPTURE OF DUST GRAINS IN EXTERIOR RESONANCES WITH PLANETS

MILOŠ ŠIDLICHOVSKÝ

Astronomical Institute, Academy of Sciences of the Czech Republic, Boční II 1401, 141 31 PRAHA 4

and

DAVID NESVORNÝ

Universidade de São Paulo, Instituto Astronômico e Geofísico, Av. Miguel Stefano 4200, Caixa Postal 9638, 01065 São Paulo, Brazil

Abstract. The temporary capture of the dust grains in the exterior resonances with planets is studied in the frames of the planar circular three-body problem with Poynting-Robertson (PR) drag. For the Earth and particles ~ 10 μm the resonances 4/5, 5/6, 6/7, 7/8 are shown to be most effective. The capture is only temporary (of order 10^5 years) and the position of resonance may be calculated from semi-analytical model using averaged disturbing function. These semi-analytical results are confirmed by numerical integration. For various planet this picture changes as with increasing planetary mass the more exterior resonances become more important. We showed that for Jupiter (at least in the space between Jupiter and Saturn) the resonance 1/2 plays the dominant role. The capture time is here several myr but again eccentricity is evolving to eccentricity $e_0 \sim 0.48$ of libration point for this resonance.
Key words: Interplanetary matter – Dust – Resonances

1. Introduction

The simplest model for studying the dust grain orbit evolution in the solar system, the two body problem (Sun + grain) with the Poynting-Robertson (PR) drag included, was solved by Wyatt and Whippple (1950). The result is well known slow decrease of the semimajor axis and eccentricity.

Here we study the effect of the planet on the spiralling grain, especially the possibility of capture into resonance with the planet. Our model is the planar circular three body problem (Sun + planet + grain) in which the grain particle is influenced by PR drag. Gonczi *et al.* (1982) showed that interior resonances are ineffective for the capture. On the other hand Jackson and Zook (1992) demonstrated in numerical examples the cases of grains in exterior resonances with the Earth. Our approach will be comparison of numerical and theoretical results. For the numerical integration we use the MSI package written by the second author. The basic method is the symmetric multistep method of the twelfth order (Quinlan and Tremaine 1990).

Earth, Moon, and Planets **72**: 175-178, 1996.

2. The capture by the Earth

This problem was studied by Šidlichovský and Nesvorný (1994). The basic equations for libration points where semimajor axis a, the eccentricity e, and the resonant argument σ

$$q\sigma = (p+q)\lambda_1 - p\lambda - q\tilde{\omega} \qquad (1)$$

are constant, were based on the method of Beaugé and Ferraz-Mello (1993) and Beaugé (1994). For exterior resonances p is negative. As usually λ is the mean longitude and $\tilde{\omega}$ the longitude of pericentre of the grain. Subscript 1 denotes the corresponding elements for the planet. The radiative force is proportional to parameter β depending on the radius and density of the particle (Burns *et al.*, 1979). Averaging over fast variables (disturbing function is averaged numerically), Šidlichovský and Nesvorný (1994) obtained three equations for a, e, σ of libration point and suggested iterative method for their solution. One of the equations depends only on eccentricity

$$(p+q)(1 + \frac{3}{2}e^2) - p(1 - e^2)^{\frac{3}{2}} = 0 \qquad (2)$$

and its solution e_0 is the universal eccentricity (compare Beaugé and Ferraz-Mello 1994). The value of e_0 is independent of the planetary mass and the β parameter and it only depends on the p, q of the resonance. Table 1 shows the value of e_0 for the first order resonances. The other two equations can be solved iteratively for a_0 and σ_0 with fixed e_0 (Šidlichovský and Nesvorný 1994). From our numerical experiments we can draw the following conclusions:

If the grain particles ($\beta = 0.01$) start on the circular orbit very close above one of the exterior resonances (we usually used 36 particles equidistantly distributed along this orbit), they are either all captured (for resonances 5/6, 6/7, 7/8, 8/9, 9/10, 10/11, 11/12 and 13/14) or none is captured (for 1/2, 2/3, 3/4, 4/5).

TABLE I

The value of e_0 for the first order resonances.

$(p+q)/p$	e_0	$(p+q)/p$	e_0	$(p+q)/p$	e_0
1/2	0.4812	5/6	0.2472	9/10	0.1878
2/3	0.3690	6/7	0.2273	10/11	0.1785
3/4	0.3108	7/8	0.2115	11/12	0.1706
4/5	0.2736	8/9	0.1986	12/13	0.1636

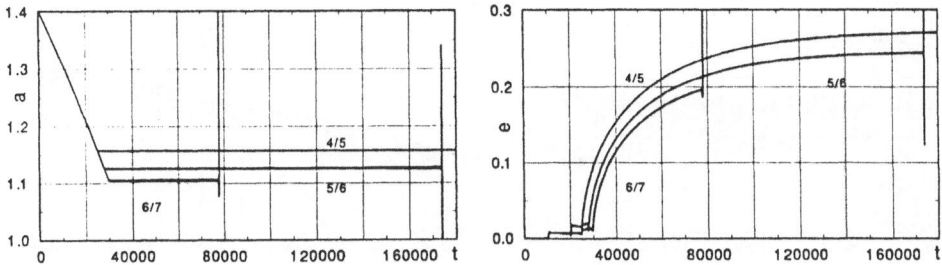

Fig. 1.   Evolution of the eccentricity and semimajor axis for three particles captured at three different resonances by the Earth. The eccentricity e increases to the values $\sim e_0$.

The boundary between capture and no capture depends on β and with decreasing β it moves from the planet. For $\beta = 0.004$ (corresponding to radius $\sim 50\,\mu$m for density 3 g/cm^3) we have no capture only for the 1/2 and 2/3 resonance and capture for 3/4, 4/5 etc.

The typical scenario for the capture is that a stops to decrease, σ starts to librate and e increases slowly to value e_0, while amplitude of σ oscillations increases. This eventually leads to close approach to planet and removal from resonance. Capture time usually of order 100 000 years.

This sharp division between capture and no capture changes if the eccentricity is non-vanishing. In one of our numerical experiments with 20 particles ($\beta = 0.01$) with initially circular orbit and $a = 1.4$AU we followed these particles for 180 000 years. We found 5 particles captured at the 4/5 resonance, 4 at the 5/6, 7 at the 6/7, 2 at 7/8 and 2 particles captured at the 8/9 resonance. Fig. 1 shows behaviour of a and e of three selected particles. Passage through the outer resonances 2/3 and 3/4 leads to increase of eccentricity to values 0.01-0.02. The capture at the 4/5 resonance is possible and on the other hand the 5/6 resonance may be passed through.

3. Jupiter

We made numerical experiment with ten particles on the planar and circular orbit at $a = 10.2$ AU ($\beta = 0.01$), for restricted circular problem with PR drag and masses of the primaries equivalent to those of the Sun and Jupiter. Even if we could afford longer time step (4 days), the calculations were time consuming, as the evolution is much slower here and we had to perform calculation for 7 Myr. We found that due to the greater Jupiter mass, the outer resonances as are now more important and all ten particles are captured at the 1/2 resonance. In the first stage a was decreasing for

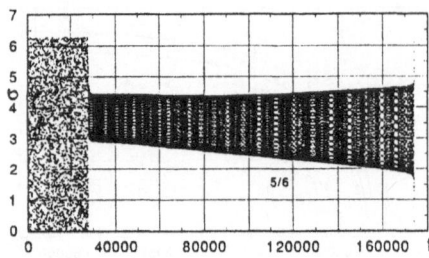 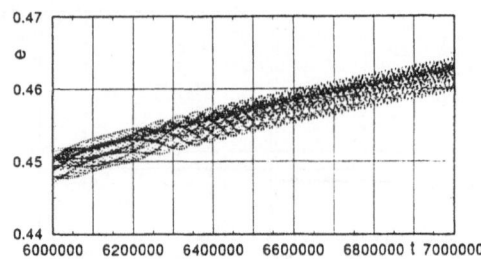

Fig. 2.   a) Behaviour of σ for the particle (one of those in Fig. 1) captured at the 5/6 resonance. b) Capture at the 1/2 resonance by Jupiter. The evolution of eccentricity for the last 1 Myr of our integration.

about 1.3 Myr when σ started to librate (a was then ~ 8.4 AU) about the value of π. At that moment e started to increase, but a was still decreasing for another 100 000 years. During this time the amplitude in σ decreased to values lower than 0.5 radians. After 1.4 Myr a stops to decrease at value $a \sim 8.24$ and at the same time the center of oscillations of σ moves sharply to values ~ 5.14 rad. After 7 Myr all particles are still captured with a oscillating about 8.24 with amplitude 0.03 AU, σ oscillates about value 5.145 with amplitude 0.035. Fig. 2 shows the behaviour of e during the last 1 Myr of integration, when it goes close to e_0 for the 1/2 resonance.

References

Beaugé, C.: 1994, 'Asymmetric Librations in Exterior Resonances', *Celest. Mech. Dyn. Astron.*, in press.

Beaugé, C. and Ferraz-Mello, S.: 1993, 'Resonance Trapping in the Primordial Solar Nebula: The Case of a Stokes Drag Dissipation', *Icarus* **103**, 301–318.

Beaugé, C. and Ferraz-Mello, S.: 1994, 'Capture in Exterior Mean-Motion Resonances Due to Poynting-Robertson Drag', *Celest. Mech. Dyn. Astron.*, in press.

Burns, J.A., Lamy, P.L., Soter, S.: 1979, 'Radiation Forces on Small Particles in the Solar System', *Icarus* **40**, 1–48.

Gonczi, R., Froeschlé, Ch., Froeschlé, Cl.: 1982, 'Poynting-Robertson Drag and Orbital Resonances', *Icarus* **51,**, 633–654.

Jackson, A.A. and Zook, H.A.: 1992, 'Orbital Evolution of Dust Particles from Comets and Asteroids', *Icarus* **97**, 70–84.

Quinlan, G.D. and Tremaine, S.: 1990, 'Symmetric Multistep Methods for the Numerical Integration of Planetary Orbits', it Astron. J. **100**, 1694–1700.

Šidlichovský, M. and Nesvorný, D.: 1994, 'Temporary Capture of Grains in Exterior Resonances with the Earth', it Astron. Astrophys., **289**, 972–982.

Wyatt, S.P. and Whipple, F.L.: 1950, 'Radiation Forces on Small Particles in the Solar System', *Astrophys. J.* **111**, 134–141.

SYMPLECTIC METHODS AND THEIR APPLICATION TO THE MOTION OF SMALL BODIES IN THE SOLAR SYSTEM

T.-Y. HUANG

Department of Astronomy, Nanjing University, Nanjing, China, 210093

K.A. INNANEN

Department of Physics & Astronomy, York University, Toronto, Canada, M3J 1P3

and

C.-B. WANG and Z.-Y. ZHAO

Purple Mountain Observatory, Academia Sinica, Nanjing, China, 210008

Abstract. Symplectic methods have been widely used in Solar System dynamics. This paper discusses both single step and multistep symplectic methods. For single step methods we point out that the modified algorithm (Wisdom et al., 1991, Kinoshita et. al., 1991) can be executed in the mass center coordinate system and in the Jacobian coordinate system. For multistep methods we describe the connections between symmetric and symplectic methods.

1. Introduction

Conservative dynamical (hamiltonian) systems are usually good approximations to old celestial systems such as the Solar System. Some classical numerical methods, such as the Störmer-Cowell integrators or the family of Runge-Kutta integrators, have been successfully applied to explore these dynamical systems quantitatively and qualitatively. But there do exist some well-known problems for these classical integrators. They introduce artificial dissipation during step-by-step integrations (Feng, 1985). They are unable to keep the Hamiltonian constant and the in-track error increases quadratically with time (Huang and Innanen, 1983). Feng (1985) pointed out that these problems arise because classical integrators do not keep the symplectic structure of a hamiltonian system during step-by-step advancement.

Symplectic methods are defined as integrators that keep the symplectic structure analytically at each step during numerical integration. Modern symplectic methods were proposed independently by Ruth (1983) and Feng (1985). Feng gave more theoretical considerations and various approaches to construct symplectic methods. Ruth succeeded in building a family of symplectic methods that are applicable to separable hamiltonian systems and that can be extended to high orders (see Yoshida, 1990).

This paper provides some discussion and elaboration on symplectic methods.

Earth, Moon, and Planets **72**: 179-183, 1996.

2. Modified symplectic methods applied to the Solar System

The Hamiltonian of the solar system in the mass center cartesian system is

$$H = \sum_{i=0}^{n} \frac{\mathbf{p}_i^2}{2m_i} - \sum_{0 \leq j < i} \frac{Gm_i m_j}{r_{ij}} \tag{1}$$

where m_i, $\mathbf{q}_i$ and $\mathbf{p}_i$ are the mass, the generalized coordinate and momentum of the ith body respectively, $r_{ij} = |\mathbf{q}_i - \mathbf{q}_j|$. Here the Sun is the body with $i = 0$.

The Hamiltonian can be separated into two parts

$$H = H_0(\mathbf{q}, \mathbf{p}) + H_1(\mathbf{q}, \mathbf{p}) \tag{2}$$

If H_0 and H_1 both represent integrable hamiltonian systems, their corresponding symplectic solution mappings, S_0 and S_1, can be associated to form a symplectic integrator (Yoshida, 1990). Wisdom and Holman (1991) and Kinoshita et al. (1991) suggested that H_0 should be chosen as Kepler's Hamiltonian and H_1 would represent the perturbation and be independent of $\mathbf{p}$. In the case of the Solar System Wisdom and Holman (1991) adopted the jacobian coordinate system and therefore succeeded in the required separation. Their Hamiltonian dropped terms smaller than m_i^3 in order to save the computation of the transformation between the jacobian and mass center coordinate systems.

We have tried another approach: separation in the mass center coordinate system:

$$H_0 = \sum_{i=1}^{n} \left(\frac{\mathbf{p}_i^2}{2m_i} - \frac{Gm_0 m_i}{r_{ij}} \right) + \frac{\mathbf{p}_0^2}{2m_0}; \qquad H_1 = \sum_{0 < i < j} \frac{Gm_i m_j}{r_{ij}} \tag{3}$$

Here H_0 represents $n + 1$ independent Kepler motions, in which the Sun moves freely with uniform velocity within every step; H_1 does not depend on the momenta and is a perturbation. The merit of this separation is that both H_0 and H_1 have no truncation. Its defect is that the Sun's Kepler motion is not modelled in H_0. To overcome this problem we improve the Sun's coordinate and momentum after each step by the integral of momentum. Several numerical experiments on the outer planets by this algorithm of the second order have shown that the same accuracy could be achieved as that in the jacobian coordinate system. Further comparisons are still in progress.

3. The connection between symmetric and symplectic methods

The symplectic methods described in the previous section are single step methods. Do there exist multistep symplectic methods? Feng (1988) provided the original idea. He pointed out that the linear multistep methods (hereafter SI)

$$\rho(E)\mathbf{z_n} = h\sigma(E)f(\mathbf{z_n}) \quad E\mathbf{z_n} = \mathbf{z_{n+1}} \tag{4}$$

with the symmetric properties

$$E^n\rho(E^{-1}) = -\rho(E) \quad E^n\sigma(E^{-1}) = \sigma(E) \tag{5}$$

are symplectic when they are applied to the linear hamiltonian system

$$\dot{\mathbf{z}} = \begin{pmatrix} 0 & 1 \\ -1 & 0 \end{pmatrix} \mathbf{z} \tag{6}$$

Later Feng's group discovered that there does not exist any linear multistep method that is symplectic for nonlinear hamiltonian systems. Feng (1990), Li (1993) and Bao and Xu (1993) gave algorithms to construct these SI methods. In the same period Eirola and Sanz-Serna (1990) independently proposed SI methods and gave a theoretical discussion. They proved that the numerical solution of (6) by SI methods does not exactly keep the symplectic structure but keeps a structure related to the coefficients of the polynomials ρ and σ. This result is a more exact description than Feng's for the question in which sense SI methods are symplectic.

The application of SI methods to harmonic oscillators has proven to be very successful: no artificial dissipation is introduced into the numerical results. Unfortunately, the methods behave badly when applied to the Kepler problem, and global errors increase rapidly in this case. This result induced us to check their numerical stability. We have found that the only stable interval is on the imaginary axis (for numerical stability background, see Lambert, 1973), and there is no absolute or relative stability region in the left half of the complex plane. Consequently, these SI methods seem to have little practical value.

Quinlan and Tremaine (1990) proposed a family of linear multistep methods called symmetric methods as well (hereafter as SII methods). They can be applied to integrate the second order newtonian equations

$$\ddot{q} = f(q) \tag{7}$$

to get the coordinates q. Their methods can be constructed by an extension of Feng's approach (1990) as follows. Let D and E be the derivative and shift operator respectively. Eq. (7) can be rewritten as

$$f_n = D^2 q_n = \frac{h^2 D^2}{h^2} q_n = \frac{\ln^2 E}{h^2} q_n \tag{8}$$

Make a rational polynomial approximation of the scalar function $\ln^2 E$

$$\ln^2 E \cong \psi(E) = \frac{\rho(E)}{\sigma(E)} \tag{9}$$

where $\rho(E)$ and $\sigma(E)$ are k-polynomials of E. Furthermore, we require that $\psi(E)$ possesses the main feature of $\ln^2 E$, that is,

$$\psi(E^{-1}) = \psi(E) \tag{10}$$

It is sufficient to require that both ρ and σ are symmetric:

$$E^k \rho(E^{-1}) = \rho(E) \quad x^E \sigma(E^{-1}) = \sigma(E) \tag{11}$$

Then we have a symmetric k-step linear method from (8)

$$\rho(E) q_n = h^2 \sigma(E) f_n \tag{12}$$

Quinlan & Tremaine (1990) have shown that the main merit of SII methods is similar to symplectic methods, that is, the total energy of the dynamical system keeps oscillating and has no secular change during step by step integration. Kinoshita & Nakai (1992) gave a proof of this feature for the Kepler problem. One would conjecture that the SII methods have close connections with symplectic methods.

The difficulties to give a strict proof of the connection are: (1) For multistep methods the solution mapping is not $q_n \mapsto q_{n+1}$ but a shift of the sequence $q_n, q_{n+1}, \cdots, q_{n+k}$, which is related to the existence of the spurious roots of a multistep method. (2) By SII methods we only get the coordinate q_n but not the momentum p_n. The question is: how to get a numerical solution p_n to make a canonical pair (q_n, p_n)? We have not succeeded in giving an answer or a proof. The following is only a tentative explanation.

For a linear oscillator

$$\ddot{q} = -\Omega^2 q \tag{13}$$

the characteristic equation of a SII method is

$$\rho(\zeta) + h^2 \Omega^2 \sigma(\zeta) = 0 \tag{14}$$

where h is the stepsize. If we neglect the effects of all the spurious roots we obtain the numerical solution as

$$q_n = c\zeta^n + d\zeta^{-n} \tag{15}$$

where ζ is the main root and c, d are constants. Within the periodic stability interval, ζ is on the unit circle. Put $\zeta = e^{\sqrt{-1}\omega h}$; then the numerical solution q_n represents an oscillator with a fixed frequency, $\omega(h)$, which is near to but different from the original frequency Ω. If p_n is computed by numerical differentiation, (q_n, p_n) would represent a harmonic oscillator with the frequency $\omega(h)$, which is a linear hamiltonian system. In this way we could consider that the SII methods are symplectic for linear hamiltonian systems.

To assure that a SII method is useful for nonlinear dynamical systems, it should have a stability region in the left half part of the complex plane. Unfortunately, SII methods have no absolute stability region besides an interval on the negative real axis because of their symmetricity, but we do find that there exist relative stability regions. This is why SII methods are useful in nonlinear hamiltonian systems, different from SI methods.

We have applied SII methods to various problems and have gained the same impressions as those in Quinlan & Tremaine (1990) and Kinoshita & Nakai (1992). In addition, they behave badly when applied to dynamical systems that explicitly depend on the velocity $\dot{q}$, such as the circular restricted three-body problem.

Acknowledgements

This work has been supported in part by the Natural Sciences and Engineering Research Council of Canada and the National Key Project Foundation of China. We are grateful to Drs. G. Quinlan, S. Mikkola and W.-Y. Li for their help and comments.

References

Bao, X.-S. and Xu, H.-Y.: 1993, *A Study on Generating Symplectic Linear Multistep Methods*, preprint.

Eirola, T. and Sans-Serna, M.: 1990, *Applied Mathematics and Computation Reports*, Report 1990/9, Universidad de Valladolid.

Feng, K.: 1985, *Proceedings of the 1984 Beijing Symposium on Differential Geometry and Differential Equations*, Feng, K. (ed.), Science Press, Beijing, 42.

Feng, K.: 1988, private communication.

Feng, K.: 1990, in *Proceedings of the Annual Meeting on Computational Mathematics in Tinjing, China*.

Huang, T.-Y. and Innanen, K.A.: 1983, *Astron. J.* **88**, 870.

Kinoshita, H., Yoshida, H. and Nakai, H.: 1991, *Celes. Mech. and Dynam. Astr.* **50**, 59.

Kinoshita, H. and Nakai, H.: 1992, in *Chaos, Resonance and Collective Dynamical Phenomena in the Solar System*, S.Ferraz-Mello (ed.), 395.

Lambert, J.D.: 1973, *Computational Methods in Ordinary Differential Equations*, John Wiley & Sons.

Li, W.-Y.: 1993, *Symplectic Multistep Methods for Linear Hamiltonian Systems*, preprint.

Quinlan, G.D. and Tremaine, S.: 1990, *Astron. J.* **100**, 1694.

Ruth, R.D.: 1983, *IEEE Trans. Nucl. Sci.* **30**, 2669.

Wisdom, J. and Holman, M.: 1991, *Astron. J.* **102**, 1528.

Yoshida, H.: 1990, *Physics Letters A* **150**, 262.

FROM COMETS TO ASTEROIDS:
WHEN HAIRY STARS GO BALD

DAVID JEWITT

Institute for Astronomy, University of Hawaii, 2680 Woodlawn Drive, Honolulu, HI.
96822 USA

Abstract. We discuss the essential differences between comets and asteroids. Ironically, with the exception of the rocky asteroids in the inner solar system, most of the objects classified as asteroids at and beyond Jupiter's orbit are likely to conceal buried volatiles, and thus are more usefully considered as comets.

Special Note

This is the written version of an invited review lecture given at the 1994 Small Bodies meeting in Mariehamn, Finland. I have retained the reductionist flavor of that lecture, with my aim being to simplify issues which are normally couched, in the literature, in more complicated terms. The style and subject matter form a sequence with two earlier reviews on cometary photometry (Jewitt 1991) and the cometary nucleus (Jewitt 1992). An extensive set of references is included to provide a relatively complete guide to the recent literature on this subject. Independent and complementary reviews of the comet-asteroid relationship have been published by Hartmann *et al.* (1987), Weissman *et al.* (1989), Luu (1994) and McFadden (1994).

1. Introduction

A troubling difference exists between the observational and physical definitions of comets and asteroids.
- Observational Definition: The presence of a spatially resolved, gravitationally unbound atmosphere ("coma") defines a comet.
- Physical Definition: The presence of bulk ice (water or other) defines a comet.

The practical problem is that the vapor pressure of water ice is an extremely strong function of the ice temperature, such that sublimation is unable to support a significant coma at temperatures $T \leq 150$ K. Therefore, even ice-rich comets (by the physical definition) fail to satisfy the observational definition when their ice is cold. The temperature of a freely sublimating ice surface can be computed from the energy balance for a sublimating volatile, which we simplify as

$$\frac{F_{sun}}{R^2}(1 - A) = \chi[\varepsilon\sigma T^4 + L(T)\frac{dm}{dt} + C(\frac{\partial T}{\partial x}) + D(\frac{\partial T}{\partial x})] \tag{1}$$

Earth, Moon, and Planets **72**: 185-201, 1996.

DAVID JEWITT

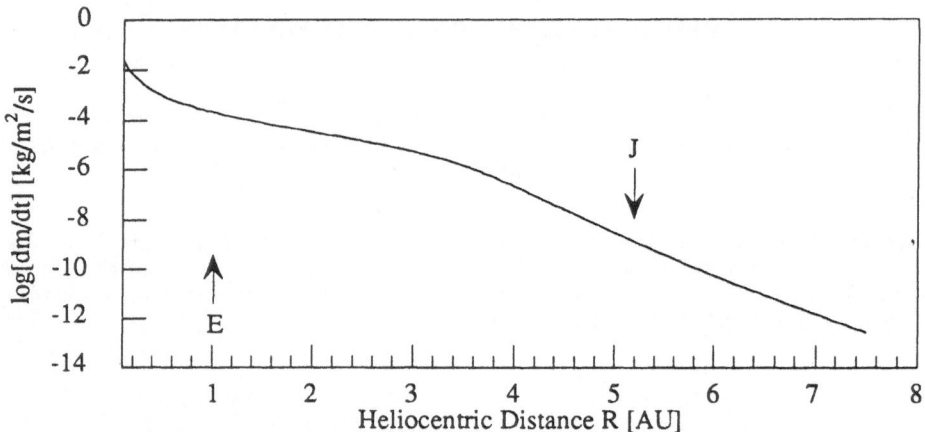

Fig. 1. Sample solution of the energy balance equation (Eq. (1)) for a sublimating water ice nucleus with $\chi = 2$, $A = 0$, $\varepsilon = 1$ and $L = 2 \times 10^6$ J kg^{-1} (see Eq. (1)). Orbits of Earth and Jupiter are marked.

Here, $F_{sun} = 1360$ W m^{-2} is the solar constant, R [AU] is the heliocentric distance, A the Bond albedo. On the right hand side, ε is the emissivity, $\sigma = 5.67 \times 10^{-8}$ W m^{-2} K^{-4} the Stephan-Boltzmann constant, T [K] the equilibrium temperature, $L(T)$ [J kg^{-1}] the latent heat of sublimation, dm/dt [kg m^{-2} s^{-1}] the specific sublimation rate and C and D represent conduction and gas-phase latent heat transfer down the temperature gradient dT/dx into the nucleus. The parameter $1 \leq \chi \leq 4$ accounts for the non-uniform distribution of solar energy across the surface of the nucleus. The term on the left represents the flux of energy absorbed from the sun. The terms on the right represent, respectively, energy lost from the nucleus surface by radiation, by latent heat of sublimation, by conduction into the interior, and by gas phase transfer into the interior.

Eq. (1) has been solved, subject to appropriate boundary conditions, in exquisite detail (e.g. Prialnik 1989). Here we present a sample solution (Fig. 1) and note a limiting case, namely that of a comet close to the sun ($R \leq 1$ AU). In this limit, the sublimation term dominates all others, and the sublimation rate can be estimated directly from

$$\frac{dm}{dt} \approx \frac{F_{sun}}{R^2 \chi L(T)} (1 - A) \qquad (2)$$

For example, with $\chi = 2$, $A = 0$, $R = 1$ AU and $L = 2 \times 10^6$ J kg^{-1} (water ice) we find $dm/dt \sim 3 \times 10^{-4}$ kg m^{-2} s^{-1}, corresponding to a flux of water molecules of 10^{22} m^{-2} s^{-1}. Bright comets (e.g. P/Halley; P/Swift-Tuttle) with peak hydroxyl production rates $Q_{OH} \sim 10^{30}$ s^{-1} must therefore be outgassing from areas ~ 100 km^2. Marginally detectable outgassing ($Q_{OH} \sim$

10^{26} s^{-1}) corresponds to exposed sublimating areas of only 10^4 m^2, or surface areas ~ 100 m on a side.

However, mass loss can be hindered or prevented altogether by an insulating, refractory mantle (as in the near-Earth asteroids (NEAs)) or by having a large heliocentric distance (as in the Jovian Trojans, the Centaurs, and the trans-Neptunian Objects, all of which are presumed to possess ice-rich interiors). Accordingly, to understand the differences between comets and asteroids, we need to consider mechanisms of heat transfer and mantle formation.

2. Heat Transfer

Heat transfer in small bodies proceeds primarily by thermal conduction and, when volatiles are present, by latent-heat effects due to sublimation and condensation. For simplicity, we first consider heat transfer in a non-volatile body. The distribution of temperature with depth is governed by the diffusion equation

$$k\nabla^2 T = \rho c_p \frac{\partial T}{\partial t} - \rho H \tag{3}$$

in which T [K] is the temperature, t [s] is time, k [W m^{-1} K^{-1}] is the thermal conductivity, ρ [kg m^{-3}] the density, c_p [J kg^{-1} K^{-1}] the specific heat capacity and H [W kg^{-1}] is the specific power production in the material (due, for example, to amorphous-crystalline phase changes (e.g. Klinger 1980), or to radioactive elements). Detailed analytic solutions of Eq. (3), subject to appropriate boundary conditions, are part of the classical literature (e.g. Carslaw and Jaeger 1959). For our present purposes, it is more revealing to consider order of magnitude solutions as follows. Setting $H = 0$, dimensional treatment of Eq. (3) shows that the timescale for the transport of heat by thermal conduction is of order

$$\tau \approx \frac{\ell^2}{\kappa} \tag{4}$$

where ℓ [m] is the characteristic dimension of the body and $\kappa = k/\rho c_p$ [m^2 s^{-1}] is the thermal diffusivity. Typical planetary dielectric solids have $\kappa \sim 10^{-7}$ to 10^{-6} m^2 s^{-1}, although values that are orders of magnitude smaller have been suggested for porous amorphous ices (Kouchi et al. 1992). To give a very terrestrial example, the conduction cooling time for a pea ($\ell \sim 3$ mm, $\kappa \sim 10^{-6}$ m^2 s^{-1}) is $\tau \sim 10$ s, while that for a potato ($\ell \sim 3$ cm) is $\tau \sim 1000$ s ~ 15 minutes. These timescales are in reasonable agreement with common experience, and the agreement seems enhanced when we remember

TABLE I

Cometary Timescales and Conduction Length Scales

Quantity	Timescale	Magnitude	Conduction Length Scale
Dynamical Lifetime	t_{dyn}	4×10^5 yr	1000 m
Orbital Period	t_{orb}	10 yr	5 m
Rotational Period	t_{rot}	10 hr	5 cm

that Eq. (4) approximates the e-folding time for cooling, and that several e-folding times must elapse before the internal heat of the pea or the potato is lost. Of course, these examples are intentionally frivolous and simplistic (e.g. we have neglected internal transport of heat by steam). But Eq. (4) also shows that the largest body able to cool by conduction in the age of the solar system ($\tau = 4.5$ Gyr $\sim 1.4 \times 10^{17}$ s) is of scale $\ell \sim (\tau \kappa)^{1/2} \sim 100$ km. Smaller bodies must have lost their initial heat and can retain no thermal memory of their formation. Therefore, all but the largest nuclei of short-period comets have interiors that have cooled by conduction of internal heat to the surface followed by (nearly instantaneous) radiation into space.

Table I lists timescales relevant to the propagation of heat in a cometary nucleus, and gives corresponding conduction length scales derived from them using Eq. (4). Several deductions about the internal thermal character of comets may be reached directly from the Table.

• First, the dynamical lifetime of short-period comets against gravitational ejection by the planets is $t_{dyn} \sim 4 \times 10^5$ yr (Levison and Duncan 1994). On this timescale, heat conducts into the nucleus by a distance $\ell_{dyn} \sim 1$ km. Since most well-studied cometary nuclei are larger than 1 km (Table II), we must conclude that these bodies are perpetually out of internal thermal equilibrium, and that deeply buried volatiles may survive even in the old, sun-baked short-period comets typified by P/Encke, P/Arend-Rigaux, and P/Neujmin 1.

• Second, in the $t_{orb} \sim 10$ yr orbital period of a Jupiter-family comet, solar heating of the surface drives a thermal wave of vertical scale-length $\ell_{orb} \sim 5$ m into the nucleus. Essentially all consequences of the annual solar heating.

• Third, the diurnal thermal skin depth (due to axial rotation of the nucleus with a period $t_{rot} \sim 10$ hr (Table II)) is $\ell_{rot} \sim 5$ cm. The presence of strong sunward emission of gas and dust from Halley (Keller et al. 1987) and other comets (e.g. Sekanina 1990; Jewitt 1991) shows that cometary outgassing occurs from the diurnal thermal skin and not from deeper layers.

Strong thermally induced stresses are also confined to a layer of thickness $\sim \ell_{rot}$. Several authors (e.g. Tauber and Kührt 1987) have suggested that thermal fracture may occur in these upper layers.

Heat may also be transported by vaporization and condensation of volatiles, assuming that sufficient porosity exists. Detailed models of heat transport have been described by Fanale and Salvail (1984) and Prialnik (1989), among others. The timescale for depletion of volatiles from a sufficiently porous body of radius r_n and density ρ_n is

$$t_{dv} \sim \frac{\rho_n r_n}{dm/dt} \tag{5}$$

With $dm/dt \sim 10^{-4} \, \mathrm{kg \, m^{-2} \, s^{-1}}$ at $R = 1$ AU, $t_{dv} \sim 10^3 \times 10^3/10^{-4} \sim 10^{10}$ s $\sim 10^3$ yr. Notice that $t_{dv} \ll t_{dyn}$, suggesting that km-sized comets should lose their volatiles before completing their dynamical evolution (in other words, that many NEAs could be completely devolatilised comets). This would be true but for the formation of surface mantles.

3. Mantles

In comets, the ice and dust are intimately mixed. Sublimation at rate dm/dt leads to a recession of the sublimating surface at rate $dl/dt = \rho^{-1} dm/dt$. For example, at 1 AU, water ice sublimates at $dm/dt \sim 10^{-4} \, \mathrm{kg \, m^{-2} \, s^{-1}}$, and with $\rho = 10^3 \, \mathrm{kg \, m^{-3}}$ the surface shrinks at rate $dl/dt \sim 10^{-7} \, \mathrm{m \, s^{-1}}$. Progressive loss of volatiles from the heated surface of a nucleus may leave behind a lag-deposit or cohesionless "rubble mantle", consisting of particles of refractory debris that are too large to be ejected against local gas drag (Whipple 1950; 1951). Balancing the gas drag force against the local gravitational acceleration towards the nucleus, one obtains a critical radius

$$a_c \sim \frac{9 C_D \dot{m} v_{th}}{16 \pi G \rho \rho_n r_n} \tag{6}$$

above which gas drag cannot eject cometary debris. Here, $C_D \sim 1$ is the dimensionless drag coefficient, dm/dt [$\mathrm{kg \, m^{-2} \, s^{-1}}$] is the specific mass loss rate, v_{th} [$\mathrm{m \, s^{-1}}$] is the speed of the escaping gas molecules, $G = 6.67 \times 10^{-11}$ [$\mathrm{N \, kg^{-2} \, m^2}$] is the Gravitational Constant, ρ and ρ_n [$\mathrm{kg \, m^{-3}}$] are the densities of the dust grain and nucleus, respectively, and r_n [m] is the nucleus radius. The multiplier $9/16 \sim 1$ is a function of the shape of the nucleus. Figure 2 shows the critical radius for ejection from spherical nuclei of radii 1 km and 5 km as a function of heliocentric distance (corresponding curves for other volatiles are presented in Luu and Jewitt 1990b). Note that optically dominant μm-sized grains can be ejected out to about the orbit

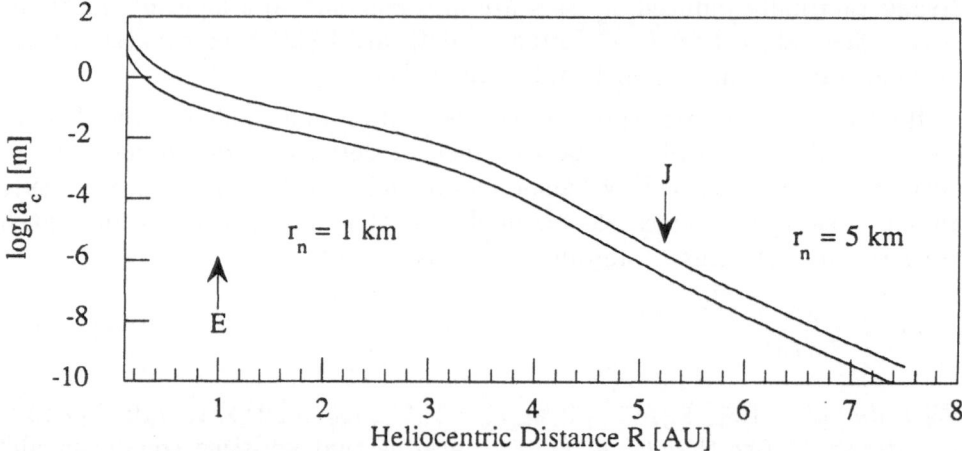

Fig. 2. Maximum ejectable grain size computed from Eq. (6) as a function of heliocentric distance, for a water ice nucleus sublimating in equilibrium with sunlight. The upper and lower curves show the effect of changing the nucleus radius by a factor of 5. Orbits of Earth and Jupiter are marked. Nuclear rotation is neglected. Eq. (6) is strictly invalid for small R, where the particle size exceeds the mean free path for molecular collisions. However, the error amounts to only about 0.6 in $\log(a_c)$, and can be ignored for the purposes of this review.

of Jupiter, while comets active much beyond 5 or 6 AU must be driven by another volatile (e.g. CO; Senay and Jewitt 1994) or another process (e.g. electrostatic charging by the solar wind). At 1 AU, a Halley-sized (5 km) nucleus can retain debris larger than $a_c \sim 5$ cm, while larger particles can be ejected if nuclear rotation creates a lower effective gravitational acceleration. Such particles are thought to clog the surfaces of comets, forming nearly continuous surface mantles of characteristic thickness a few times a_c.

Compelling evidence exists that the surfaces of short-period comets are mantled (but there is no significant evidence for or against mantles on the nuclei of long-period comets). Images of the surface of P/Halley show a dark, inert mantle punctured by regions of strong outgassing (Keller *et al.* 1987). Images of other short period comets show that sublimation is largely confined to active areas (or "vents") which have fractional coverage of the nucleus surface (Table II) $f \sim 10^{-3}$ (comets P/Encke, P/Neujmin 1, P/Tempel 2) to $f \sim 10^{-1}$ (P/Halley). These vents appear to be stable on timescales that are comparable to or longer than the orbit period (Sekanina 1990).

For a nucleus in which solids of size $a \geq a_c$ are common, the rubble mantle growth time is crudely given by $t_m \sim a_c/(dl/dt)$, or

$$t_m \sim \frac{9C_D v_{th}}{16\pi G\rho_n r_n} \tag{7}$$

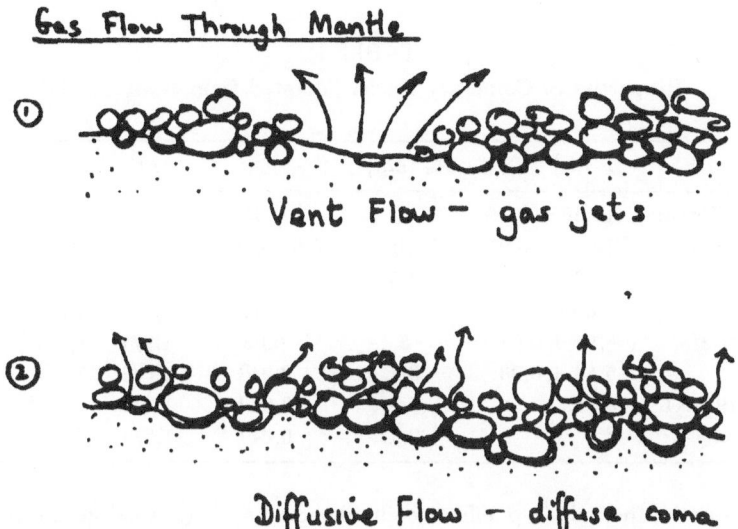

Fig. 3.   Modes of gas flow through the cometary surface. 1) vent flow through an impermeable mantle and 2) diffusive flow through a permeable mantle. Gas and dust jets in cometary comae show that vent flow is common. Diffusive flow may also occur but is observationally not well constrained.

For a 1 km radius nucleus, $t_m \sim (9 \times 1 \times 10^3)/(16\pi G 10^3 10^3) \sim 3 \times 10^6$ s ~ 0.1 yr. Two properties of Eq. (7) are worthy of note:

- t_m is independent of heliocentric distance, to first order
- $t_m < t_{orb}$.

Thus, the rubble mantle should be considered as an actively regenerated surface feature, that can grow during a single orbit of an active comet. Orbital evolution of short period comets towards smaller perihelion distances (c.f. orbital integrations in Levison and Duncan 1994) will cause repeated disruption and healing of the rubble mantle (Rickman *et al.* 1990, 1991; Rickman 1992). Rapidly growing rubble mantles choke the flow of gases from the icy interiors of comets and produce the collimated jets recorded in the comae of many comets (e.g. Jewitt 1991; Keller *et al.* 1994). Diffusive flow through weakly permeable mantles is also possible (see Figure 3).

Strengthless rubble mantles are weakly stable, and can be locally disrupted by changes in the insolation (due to orbital evolution) or nuclear spin (due to outgassing torques). For example, the nuclei of many short period comets are rotating prolate spheroids. Centripetal reduction of the local gravity might favor mantle-free "bald spots" at the sharp ends of a prolate spheroid, leading to large outgassing torques, the excitation of precession, further spin-up, possible splitting (Chen and Jewitt 1994), shape evolution and mantle disruption. Such complex cycles of feedback have yet

TABLE II

Properties of Cometary Nuclei (adapted from Jewitt 1991)

Nucleus	T [hr]$^\alpha$	R_e [km]$^\beta$	p_V^χ	a/b^δ	f^ϵ	Reference
P/Arend-Rigaux	13.56 ± 0.16	5	0.03	1.9/1	0.1-1	1,2
P/Neujmin 1	12.67 ± 0.05	10	0.03 ± 0.01	1.6/1	0.1-1	3,4
P/Encke	15.08 ± 0.08	3.5	0.04^γ	3.5/1	0.2	5
P/Halley	7 days?	5	0.04	2/1	10	6,7
P/Tempel 2	8.95 ± 0.01	5	0.02	1.9/1	0.1-1	4,8,9
P/SW2	5.58 ± 0.03	3.1	0.04^γ	1.6/1	?	10
P/Levy 1991XI	8.34	5.8	0.04^γ	1.3/1	?	11
P/Faye	?	2.4	0.04^γ	1.3/1	?	12

α–nuclear rotation period β–effective circular radius χ–visual geometric albedo
δ–projected axis ratio ϵ–active fraction*100 γ–albedo assumed

1=Jewitt & Meech 1985; 2=Millis *et al.* 1988; 3=Campins *et al.* 1987; 4=Jewitt & Meech 1988; 5=Luu & Jewitt 1990; 6=Jewitt & Danielson 1984; 7=Keller *et al.* 1987; 8=Jewitt & Luu 1989; 9=A'Hearn *et al.* 1989; 10=Luu & Jewitt 1992b; 11=Fitzsimmons & Williams 1994; 12=Lamy & Toth 1994

to be considered in published models of the evolution of the cometary nucleus (Rickman 1992) but they are probably important in real comets. Intergrain cohesion might offset some of the short-term instabilities present in purely gravitationally bound mantles but has typically been neglected in the literature (see Kührt and Keller 1994 for a counter-example). Laboratory simulations of cometary outgassing confirm the basic features of mantle growth outlined here (Grün *et al.* 1991).

It is natural to ask whether mantles could stifle outgassing so efficiently that ice-rich comets might be hidden among the near-Earth asteroids. This question has received much attention from investigators using "proxy indicators" such as the surface colors, body shapes, rotation period distributions and orbital parameters to compare the comets and NEAs (e.g. Hartmann *et al.* 1987; Jewitt and Meech 1988; Weissman *et al.* 1989; Luu and Jewitt 1990c; Binzel *et al.* 1992; McFadden 1994). Unfortunately, the properties of cometary nuclei exhibit wide diversity (Luu 1993), so that proxy indicators are difficult to use, even in a statistical sense. A more satisfactory approach is to search for outgassing directly, and two methods have been tried. First, spectral observations have failed to detect resonance fluorescence lines from cometary molecules (CN, C_2 etc) in NEAs (Cochran *et al.* 1986). A reported detection of the OH 3080 Å band in main-belt asteroid 1 Ceres (A'Hearn and Feldman 1993) awaits confirmation. It would correspond to a water source of order 1 to 10 kg s^{-1}. Second, high resolution measurements of the surface brightness profiles of NEAs place limits on the outgassing about an order of magnitude smaller than outgassing from feeble comets like P/Encke and P/Arend-Rigaux (Luu and Jewitt 1992a). However, most NEAs in the observational sample are smaller than the comets of Table II. The implied limiting active fractions, $f \sim 10^{-3}$ to 10^{-4}, are similar to f for the most feeble comets. Hence, observationally, it appears possible that some of the NEAs are mantled comets. Secular evolution of NEAs to smaller perihelion distances (e.g. Farinella *et al.* 1994) might lead to intermittent mantle disruption and outgassing. Deactivation appears to have occurred in the case of former comet P/Wilson-Harrington (1949 III) (now known as asteroid 1979 VA; Bowell 1992).

4. Trojans

The trojan asteroids of Jupiter librate around the L4 (leading) and L5 (trailing) Lagrangian stable points of that planet, at $R \sim 5.2$ AU. About 200 trojans are presently known; several thousand are thought to exist with diameters ≥ 15 km (Shoemaker *et al.* 1989). Stability calculations suggest that Saturn, Uranus and Neptune might also retain sets of trojans (e.g. Holman and Wisdom 1993) but none is known. A search for these objects is underway on Mauna Kea.

While classified as asteroids on the basis of their lack of coma, the physical nature of the trojans is observationally not well constrained. The optical spectra show only featureless, red continua, with a mean slope, $S' = 10 \pm 4$ %/1000 Å, that is statistically consistent with the mean slope of the optical spectra of cometary nuclei, $S' = 14 \pm 5$ %/1000 Å (Jewitt and Luu 1990; Fitzsimmons *et al.* 1994). The geometric albedos of trojans are 2 to 3 % (e.g. Cruikshank 1977), again comparable to the albedos of cometary nuclei (see Jewitt 1992 for a compilation). The similar optical colors and low albedos may be evidence for common organic compounds on the two classes of object (organics are favored because they provide a natural explanation for the low albedos). Unfortunately, the C-H fundamental vibration at 3.4 μm is unobservable with current technology in even the brightest trojans. Many organics show spectral features in the 1.4 μm to 2.4 μm wavelength range due to overtones and combinations of vibrations of C-H, C-O, C-N and other chemical bonds (Cloutis 1989). However, available spectra of trojans are featureless at signal-to-noise ratios ~ 20, and provide no independent evidence for the presence of organics (Luu *et al.* 1994). Perhaps carbon-rich materials on the surfaces of comets and trojans are so dehydrogenated by cosmic ray bombardment that hydrocarbon features are lost.

The bulk compositions of the trojans are unknown. Unlike the C-type asteroids, the D-types show no evidence for the 3 μm signature of water of hydration (Jones *et al.* 1990). Suggested interpretations are that the D's formed in the absence of appreciable water, or more likely, that they formed at temperatures too low for silicate hydration reactions to proceed. In the latter case, water could be incorporated as bulk ice, especially if the trojans were formed beyond the orbit of Jupiter. According to Fig. (1), low albedo water ice at $R = 5.2$ AU sublimates at $dm/dt \sim 10^{-9}$ kg m^{-2} s^{-1} corresponding to a surface recession rate $(dm/dt/r) \sim 10^{-12}$ m s$^{-1} \sim 30$ m Myr^{-1}, where $\rho = 10^3$ kg m^{-3} is the density of water ice. This is too small to sustain an observable coma, but sufficient to cause substantial topographical modification if unchecked. Note that the water ice sublimation rate at large R is highly sensitive to albedo and to heat transport by conduction, and our neglect of conduction in Fig. 1 leads to an over-estimate of dm/dt. Even so, we suspect that the dark, reddish surfaces of the trojans are thin rubble mantles, shielding buried ice perhaps just a few centimeters beneath the surface. The detection of outgassing from Jovian trojans represents a formidable observational challenge. Perhaps the best chance for success would occur following mantle disruption by collision with another body, but such events are exceedingly rare.

The origin of the trojan asteroids is a puzzle. Non-gravitational forces due to aspherical outgassing have been suggested as an agent of capture (Rabe 1972; Yoder 1979). However, sublimation of water at 6 AU is very slow and this explanation hardly seems credible for the larger (100 km size) Trojans even if more volatile ices (Senay and Jewitt 1994) were once present at the surface. The "collisional capture" hypothesis of Shoemaker *et al.* (1989) postulates the fragmentation and capture into the 1:1 resonance of precursor Jupiter planetesimals.

5. Centaurs

The orbits of Centaurs cross the orbits of gas-giant planets and are thus chaotic and short-lived (lifetime $\sim 10^6$ yr; Hahn and Bailey 1990; Dones *et al.* 1994). The three well established examples are 2060 Chiron (the only Centaur to display cometary activity; e.g. Hartmann *et al.* 1990; Luu and Jewitt 1990b), 5145 Pholus and 1993 HA$_2$. The three other Centaurs, 1994 TA, 1995 DW$_2$ and 1995 GO are recent discoveries awaiting detailed observational characterisation.

The known Centaurs (Table III) are most likely bright members of a vast population of unstable, outer-planet crossing bodies. For example, the 22nd magnitude object 1994 TA was discovered in a Mauna Kea ecliptic survey of about 3 sq. degrees. Given that the area of the ecliptic band is roughly 10,000

TABLE III

The Currently Known Centaurs*

Object	a [AU]	e	i [deg]	q [AU]	Q [AU]
2060 Chiron	13.74	0.38	6.9	8.52	18.96
5145 Pholus	20.39	0.57	24.7	8.77	32.01
1993 HA$_2$	24.80	0.52	15.6	11.90	37.70
1994 TA	17.47	0.39	5.4	10.66	24.28
1995 DW$_2$	24.2	0.22	4.2	18.9	29.5
1995 GO	14.1	0.53	19.1	6.8	21.6

*–elements compiled from the Minor Planet Electronic Circulars produced by Brian Marsden.

sq. deg., we predict that $\sim$ 3000 objects similar to 1994 TA (approximate diameter is 70 km) await discovery by future surveys. Smaller Centaurs should be even more abundant, with perhaps 10^5 to 10^6 present down to km-size. The source of these short-lived objects is plausibly identified with the trans-Neptunian Kuiper Belt (c.f. Fernández 1980; Duncan *et al.* 1988; Bailey 1994; §6).

Curiously, while the Centaurs are *dynamically* similar, their surfaces show dramatic spectral differences. For example, Chiron has a neutral visible spectrum (Hartmann *et al.* 1990; Luu and Jewitt 1990) while Pholus (Fink *et al.* 1992; Mueller *et al.* 1992; Luu 1993) and 1993 HA$_2$ (Tholen and Senay 1993; Luu 1993) are extremely red. These differences are echoed in the corresponding near infrared spectra. The 2 μm spectrum of Pholus shows deep absorptions (Fig. 4; Davies *et al.* 1993; Luu *et al.* 1994) which are absent in Chiron. The red visual slope and the near infrared features are typically interpreted in terms of a chemically complex "irradiation mantle" that has been processed by long-term exposure to cosmic rays (Andronico *et al.* 1987; Johnson *et al.* 1987; Johnson 1991). Unfortunately, no completely convincing spectral match has been achieved, and it is not even clear whether the spectral features are due to vibrations in the C-H or N-H bonds (Cruikshank *et al.* 1993; Luu *et al.* 1994), or an indeterminate mix of the two (Wilson *et al.* 1994). The neutral, featureless spectrum of Chiron might indicate burial of the irradiation mantle by sub-surface debris excavated by outgassed volatiles.

6. Trans-Neptunians

Twenty eight trans-Neptunians are known at present (August 1995). These objects have heliocentric distances $31 \leq R \leq 46$ AU, apparent red magni-

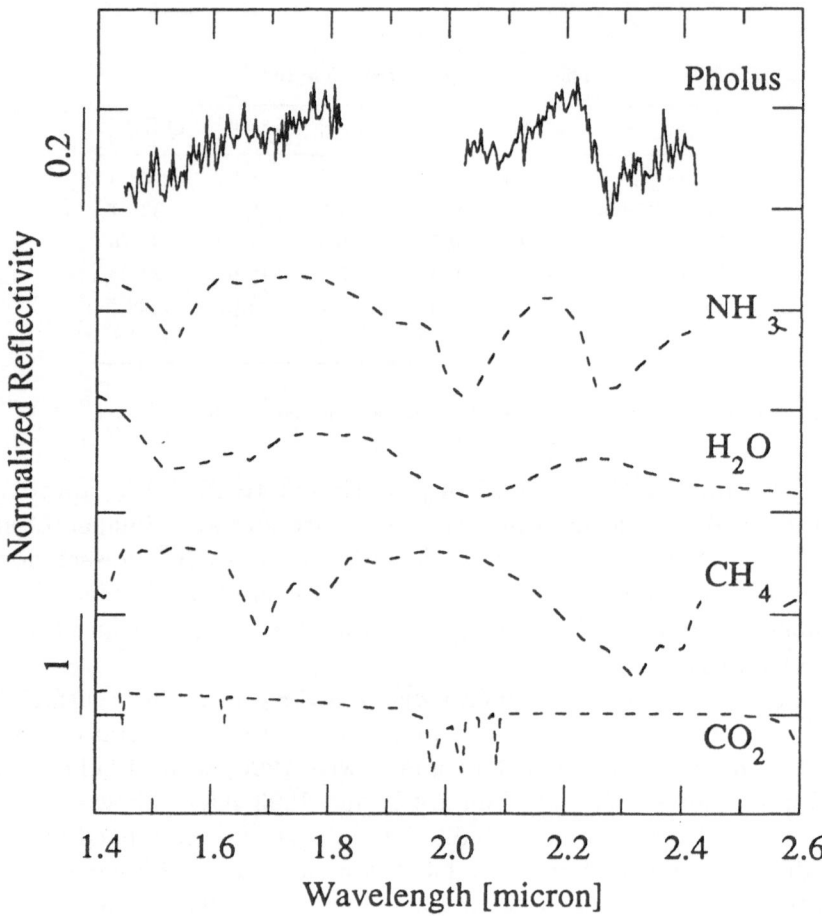

Fig. 4. Near-infrared spectrum of Centaur 5145 Pholus is compared with the reflection spectra of several common ices. A similarity with vibrational overtone and combination bands in the spectrum of NH_3 is evident, but significant differences exist (e.g. at 1.5 μm). Ammonia is not suspected, but another molecule incorporating the N-H bond might be present. Figure from Luu *et al.* (1994).

tudes $21.7 \leq m_R \leq 24.5$ and diameters (computed using an assumed 0.04 geometric albedo) $100 \leq D \leq 380$ km (Jewitt and Luu 1993, 1995). Because of their distance, it is very likely that these objects incorporated water as ice at formation; they are comets according to our physical definition. However, they are sufficiently cold that they sustain no observable comae. To date, only a very tiny fraction of the ecliptic has been searched for trans-Neptunians. The inferred total number of objects having diameters $D \geq 100$ km is $N(D \geq 100) \sim 35,000$, in the heliocentric distance range $30 \leq R \leq 50$ AU. This is several hundred times the population of comparably sized main-belt asteroids, providing a measure of the vastness of this newly discovered population. The number of trans-Neptunians down to 1 km size is probably in the 1 to 10 billion range.

The trans-Neptunians are thought to be ice-rich remnants from the outer edge of the pre-planetary disk out of which the planets accreted 4.5 Gyr ago. Those with semi-major axis $a \geq 40$ AU are probably dynamically long-lived and constitute direct evidence for the existence of Kuiper's Belt (Kuiper 1951; Fernández 1980; Duncan et al. 1988). Trans-Neptunians with $a < 40$ AU may be protected from Neptune ($a = 30$ AU) perturbations by the 2:3 mean-motion resonance. The latter objects are thus dynamically equivalent to Pluto. They may have been captured during radial migration of Neptune (in the first few $\sim 10^8$ yrs) caused by angular momentum exchange with surrounding cometesimals in the disk (Malhotra 1993). Fully 50% of the currently known trans-Neptunians may be "Plutinos", and the total population with $D \geq 100$ km in the resonance may number many thousands (Jewitt and Luu 1995). A readable review of the possible conditions of formation and scientific significance of the trans-Neptunians is given by Bailey (1994).

A distinguishing feature of the known trans-Neptunians is their large size, which is itself an artifact of observational selection favoring the closest, largest members of the population. The known trans-Neptunians are 10 to 40 times larger than typical short-period comet nuclei (c.f. Table II), but comparable to the largest Centaur, Chiron (radius ~ 90 km; Campins et al. 1994). A crude estimate of the effects of internal radiogenic heating can be obtained from Eq. (3) by neglecting the conductivity term. The temperature change in time Δt is $\Delta T \sim (H/c_p)\Delta t$. This temperature increase continues up to a maximum time $\Delta t \sim \tau$, after which internally liberated heat is lost as rapidly as it is created by conduction to the surface. Thus, the maximum temperature rise sustained by radiogenic heat is

$$\Delta T \sim (\frac{H}{c_p})(\frac{\ell^2}{\kappa}) \qquad (8)$$

With $H \sim 10^{-12}$ W kg^{-1} (Stacey 1969), $c_p = 100$ J kg^{-1} K^{-1}, $\kappa = 10^{-7}$ m^2 s^{-1}, we obtain $\Delta T \sim 10^{-2}(\ell/1\text{km})^2$. Sample temperature changes are listed in Table IV.

This is a highly simplistic calculation, but it serves to suggest that radiogenic heating in the larger bodies of the Kuiper Belt might have led to the mobilization of interior volatiles (CO, N_2, possibly CO_2; c.f. Whipple and Stefanik 1966; Yabushita 1993). We should not be surprised if some of the larger trans-Neptunians show geological evidence for cryogenic volcanism and comet-like outgassing, perhaps similar to the geyser-like activity found on Triton.

TABLE IV

Radiogenic Heating

Radius [km]	Example	ΔT [K]
5	Tempel 2, Halley	0.25 K
50	1993 RO, 1993 RP	25 K
100	Chiron, 1993 SB, 1994 JR$_1$	100 K

Acknowledgements

I thank Hans Rickman and Jing Li for help with Latex. I appreciate financial support from NASA's Planetary Astronomy Program.

References

A'Hearn, M. F., Campins, H., Schleicher, D., and Millis, R.: 1989, 'The Nucleus of P/Tempel 2', *Ap. J.* **347**, 1155-1166.

A'Hearn, M. F., and Feldman, P. D.: 1992, 'Water Vaporization on Ceres', *Icarus* **98**, 54-60.

Andronico, G., Baratta, G., Spinella, F., and Strazzulla, G.: 1987, 'Optical Evolution of Laboratory-Produced Organics', *Astron. Ap.* **184**, 333-336.

Bailey, M.: 1994, 'Formation of the Outer Solar System Bodies', in *Asteroids, Comets, Meteors 1993*, IAU Symp. 160, eds. A, Milani, M. Di Martino, and A. Cellino, Kluwer Academic Pub., Dordrecht, pp. 443-460.

Binzel, R., Xu, S., Bus, S. and Bowell, E.: 1992, 'Origins for the Near-Earth Asteroids', *Science* **257**, 779-782.

Bowell, E.: 1992, IAUC 5585.

Campins, H., A'Hearn, M., and McFadden, L.: 1987, 'The Bare Nucleus of Comet P/Neujmin 1', *Ap. J.* **316**, 847-857.

Campins, H., Telesco, C., Osip, D., Rieke, G., Rieke, M., and Schulz, B.: 1994, 'The Color Temperature of 2060 Chiron: A Warm and Small Nucleus', *Astron. J.* **108**, 2318-2322.

Carslaw, H. S., and Jaeger, J. C.: 1959, *Conduction of Heat in Solids*, 2nd. edition, Clarendon Press, Oxford.

Chen, J., and Jewitt, D. C.: 1994, 'On The Rate At Which Comets Split', *Icarus* **108**, 265-271.

Cloutis, E. A.: 1989, 'Spectral Reflectance Properties of Hydrocarbons: Remote Sensing Applications', *Science* **245**, 165-168.

Cochran, W., Cochran, A., and Barker, E.: 1986, in *Asteroids, Comets, Meteors 2*, eds. C.-I. Lagerkvist, B. Lindblad, H. Lundstedt and H. Rickman, Uppsala Universitet Reprocentralen, Uppsala, pp. 181-185.

Cruikshank, D. P.: 1977, 'Radii and Albedos of Four Trojan Asteroids and Jovian Satellites 6 and 7', *Icarus* **30**, 224-230.

Cruikshank, D. P., Moroz, L., Geballe, T., Pieters, C., and Bell, J. F.: 1993, 'Asphaltite-like Organics on Planetesimal 5145 Pholus', *BAAS* **25**, 1125.

Davies, J. K., Sykes, M. V., and Cruikshank D. P.: 1993, 'Near-infrared photometry and spectroscopy of the unusual minor planet 5145 Pholus (1992 AD)', *Icarus* **102**, 166-169.

Dones, L., Levison, H., and Duncan, M.: 1994, 'Long-Term Integrations of Chiron and Pholus', *BAAS* **26**, 1154.

Duncan, M., Quinn, T., and Tremaine, S.: 1988, 'The Origin of Short Period Comets', *Ap. J.* **328**, L69-L73.

Fanale, F. P., and Salvail, J. R.: 1984, 'An Idealized Short-Period Comet Model', *Icarus* **60**, 476-511.

Farinella, P., Froeschlé, C., Froeschlé, C., Gonczi, R., Hahn, G., Morbidelli, A., and Valsecchi, G.: 1994, 'Asteroids Falling into the Sun', *Nature* **371**, 314-317.

Fernández, J. A.: 1980, 'On the Existence of a Comet Belt Beyond Neptune', *MNRAS* **192**, 481-491.

Fink, U., Hoffmann, M., Grundy, W., Hicks, M., and Sears, W.: 1992, 'The Steep Red Spectrum of 1992 AD', *Icarus* **97**, 145-149.

Fitzsimmons, A., Dahlgren, M., Lagerkvist, C.-I., Magnusson, P., and Williams, I. P.: 1994, 'A Spectroscopic Survey of D-Type Asteroids', *Astron. Ap.* **282**, 634-642.

Fitzsimmons, A., and Williams, I. P.: 1994, 'The Nucleus of Comet P/Levy 1991 XI', *Astron. Ap.* **289**, 304-310.

Grün, E. *et al.*: 1991, 'Laboratory Simulation of Cometary Processes', in *Comets In The Post–Halley Era*, eds. R. Newburn, M. Neugebauer and J. Rahe, Kluwer Academic Publishers, Netherlands. pp. 277-297.

Hahn, G., and Bailey, M.: 1990, 'Rapid Dynamical Evolution of Giant Comet Chiron', *Nature* **348**, 132-136.

Hartmann, W., Tholen, D., Cruikshank, D.: 1987, 'The Relationship of Active Comets, Extinct Comets, and Dark Asteroids', *Icarus* **69**, 33-50.

Hartmann, W. K., Tholen, D. J., Meech, K., and Cruikshank, D. P.: 1990, *Icarus* **83**, 1.

Holman, M., and Wisdom, J.: 1993, *Astron. J.* **105**, 1987-1999.

Jewitt, D. C.: 1991, 'Cometary Photometry', in *Comets In The Post–Halley Era*, eds. R. Newburn, M. Neugebauer and J. Rahe, Kluwer Academic Pub., Netherlands. pp. 19-65.

Jewitt, D. C.: 1992, 'Physical Properties of Cometary Nuclei', in *Proceedings of the 30th Liege International Astrophysical Colloquium*, eds. A. Brahic, J.-C. Gerard and J. Surdej, Univ. Liège Press, Liège, pp. 85-112.

Jewitt, D. C., and Danielson, G. E.: 1984, 'CCD Photometry of P/Halley', *Icarus* **60**, 435-444.

Jewitt, D. C., and Meech, K. J.: 1985, 'Rotation of the Nucleus of P/Arend-Rigaux', *Icarus* **64**, 329-335.

Jewitt, D. C., and Meech, K. J.: 1988, 'Optical Properties of Cometary Nuclei and a Preliminary Comparison with Asteroids', *Ap. J.* **328**, 974-986.

Jewitt, D. C., and Luu, J. X.: 1989, 'A CCD Portrait of P/Tempel 2', *Astron. J.* **97**, 1766-1790.

Jewitt, D. C., and Luu, J. X.: 1990, 'CCD Spectra of Asteroids II. The Trojans as Spectral Analogues of Cometary Nuclei', *Astron. J.* **100**, 933-944.

Jewitt, D. C., and Luu, J. X.: 1993, 'Discovery of the Candidate Kuiper Belt Object 1992 QB$_1$', *Nature* **362**, 730-732.

Jewitt, D. C., and Chen, J.: 1994, MPEC 1994-T02 (October 5).

Jewitt, D. C., and Luu, J. X.: 1995, 'The Solar System Beyond Neptune', *Astron. J.* **109**, 1867-1876.

Jones, T. D., Lebofsky, L. A., Lewis, J. S., and Marley, M. S.: 1990, 'The Composition and Origin of the C, P and D Asteroids', *Icarus* **88**, 172-192.

Johnson, R., Cooper, J., Lanzerotti, L., and Strazzulla, G.: 1987, 'Radiation Formation of a Non-Volatile Comet Crust', *Astron. Ap.* **187**, 889-892.

Johnson, R.: 1991, 'Irradiation Effects in a Comet's Outer Layers', *J. Geophys. Res.* **96**, 17553-17557.

Keller, H. U. *et al.*: 1987, 'Comet Halley's Nucleus and its Activity', *Astron. Ap.* **187**, 807-823.

Keller, H. U., Knollenberg, J., and Markiewicz, W. J.: 1994, 'Collimation of Cometary Dust Jets and Filaments', *Planet. Space Sci.* **42**, 367-382.

Klinger, J.: 1980, 'Influence of a Phase Transition of Ice on the Heat Balance of Comet Nuclei', *Science* **209**, 271.

Kouchi, A., Greenberg, J., Yamamoto, T., and Mukai, T.: 1992, 'Extremely Low Thermal Conductivity of Amorphous Ice', *Ap. J. Lett.* **388**, L73.

Kührt, E., and Keller, H. U.: 1994, 'The Formation of Cometary Surface Crusts', *Icarus* **109**, 121-132.

Kuiper, G. P.: 1951, 'On the Origin of the Solar System', in *Astrophysics*, ed. J. A. Hynek, McGraw Hill, New York, pp. 357-424.

Lamy, P. L., and Toth, I.: 1994, *Astron. Ap.*, submitted.

Levison, H., and Duncan, M.: 1994, *Icarus* **108**, 18-36.

Luu, J. X.: 1993, 'Spectral Diversity Among the Nuclei of Comets', *Icarus* **104**, 138-148.

Luu, J.: 1994, 'Comets Disguised as Asteroids', *P.A.S.P.* **106**, 425-435.

Luu, J. X., and Jewitt, D. C.: 1990a, *Icarus* **86**, 69.

Luu, J. X., and Jewitt, D. C.: 1990b, 'Cometary Activity in 2060 Chiron', *Astron. J* **100**, 913-932.

Luu, J. X., and Jewitt, D. C.: 1990c, 'CCD Spectra of Asteroids I. Near-Earth and 3:1 Resonance Asteroids', *Astron. J.* **99**, 1985-2011.

Luu, J. X., and Jewitt, D. C.: 1992a, 'High Resolution Surface Brightness Profiles of Near-Earth Asteroids', *Icarus* **97**, 276-287.

Luu, J. X., and Jewitt, D.: 1992b, 'Near-Aphelion CCD Photometry of P/Schwassmann-Wachmann 2', *Astron. J.* **104**, 2243-2249.

Luu, J. X., Jewitt, D. C., and Cloutis, E.: 1994, 'Near Infrared Spectroscopy of Primitive Solar System Objects', *Icarus* **109**, 133-144.

Malhotra, R.: 1993, *Nature* **365**, 819.

McFadden, L.: 1994, 'The Comet-Asteroid Transition: Recent Telescopic Observations', in *Asteroids, Comets, Meteors 1993*, IAU Symp. 160, eds. A, Milani, M. Di Martino, and A. Cellino, Kluwer Academic Pub., Dordrecht, pp. 95-110.

Millis, R., A'Hearn, M., and Campins, H.: 1988, 'An Investigation of the Nucleus and Coma of P/Arend-Rigaux', *Ap. J.* **324**, 1194-1209.

Mueller, B., Tholen, D., Hartmann, W., and Cruikshank, D.: 1992, 'Extraordinary Colors of Asteroidal Object (5145) 1992 AD', *Icarus* **97**, 150-154.

Prialnik, D.: 1989, 'Thermal Evolution of Cometary Nuclei', *Adv. Sp. Research* **9**, 25-40.

Rabe, E.: 1972, 'Orbital Characteristics of Comets Passing Through the 1:1 Commensurability with Jupiter', in *Motion, Evolution of Orbits and Origin of Comets*, ed. E. I. Chebotarev, Proc. IAU Symp. 45, Leningrad, Springer Verlag, New York, pp. 55-60.

Rickman, H.: 1992, 'Physico-Dynamical Evolution of Aging Comets', in *Interrelations Between Physics and Dynamics for Minor Bodies in the Solar System*, eds. D. Benest, C. Froeschlé, Editions Frontières, Gif-sur-Yvette, France, pp. 197-263.

Rickman, H., Fernández, J. A., and Gustafson, B. Å. S.: 1990, 'Formation of Stable Dust Mantles on Short Period Comet Nuclei', *Astron. Ap.* **237**, 524-535.

Rickman, H., Froeschlé, C., Kamél, L., and Festou, M. C.: 1991, 'Nongravitational Effects and the Aging of Periodic Comets', *A. J.* **102**, 1446-1463.

Senay, M., and Jewitt, D. C.: 1994, 'Coma Formation Driven by CO Release from Comet Schwassmann-Wachmann 1', *Nature* **371**, 229-231.

Sekanina, Z.: 1990, 'Gas and Dust Emission from Comets and Lifespans of Active Areas on their Rotating Nuclei', *Astron. J.* **100**, 1293-1314.

Shoemaker, E. M., Shoemaker, C. S., and Wolfe, R. F.: 1989, 'Trojan Asteroids: Populations, Dynamical Structure and Origin of the L4 and L5 Swarms', in *Asteroids II*, eds. R. P. Binzel, T. Gehrels and M. S. Matthews, University of Arizona Press, Tucson, pp. 487-523.

Stacey, F. D.: 1969, *Physics of the Earth*, Wiley Press, New York.

Tauber, F., and Kührt, E.: 1987, 'Thermal Stresses in Cometary Nuclei', *Icarus* **69**, 83-90.

Tholen, D., and Senay, M.: 1993, *B.A.A.S.* **25**, 1126.

Weissman, P. R., A'Hearn, M. F., McFadden, L. A., and Rickman, H.: 1989, 'Evolution of Comets into Asteroids', in *Asteroids II*, eds. R. Binzel, T. Gehrels, and M. Matthews, Univ. Arizona Press, Tucson, pp. 880-920.

Whipple, F. L.: 1950, 'A Comet Model I. The Acceleration of Comet Encke', *Ap. J.* **111**, 375-394.

Whipple, F. L.: 1951, 'A Comet Model II. Physical Relations for Comets and Meteors', *Ap. J.* **113**, 464-474.

Whipple, F., and Stefanik, R.: 1966, 'On the Physics and Splitting of Cometary Nuclei', *Mém. Soc. Roy. Sci. Liège* **12**, 33-52.

Wilson, P., Sagan, C., and Thompson, W.: 1994, 'The Organic Surface of 5145 Pholus', *Icarus* **107**, 288-303.

Yabushita, S.: 1993, 'Thermal Evolution of Cometary Nuclei by Radioactive Heating', *M.N.R.A.S.* **260**, 819-825.

Yoder, C.: 1979, 'Notes on the Origin of the Trojan Asteroids', *Icarus* **40**, 341-344.

MODELING OF COMETARY EVOLUTION BY KINETIC THEORY: METHOD AND FIRST RESULTS

MAREK BANASZKIEWICZ

Space Research Centre, Bartycka 18ª, 00–716 Warszawa, Poland

and

HANS RICKMAN

Astronomical Observatory, S-75120 Uppsala, Sweden

Abstract. Physical evolution of Jupiter family (JF) comets is considered as a simultaneous process of erosion and fading. Dynamical effects are limited to discrete changes of the perihelion distance, that result in changes of the evaporation rate. Assuming that the JF comet population is in a steady state, a distribution function of this population in the two dimensional phase space consisting of radius and active fraction of the nucleus surface is found as the solution of a set of kinetic equations, each one of them for a different perihelion distance. With use of the distribution function some statistical properties of the comet population, like the total number of comets in the considered region of the phase space, the number of objects that evaporate or get dormant per unit time, etc., are obtained. The cumulative distribution function with respect to the absolute brightness is calculated and compared with the observed one as a check on the considered models.

1. Introduction

The physical evolution of Jupiter family comets (JF comets; orbital periods typically < 20 yr) is determined by two processes: (i) evaporation of volatiles from the surface followed by mass (dust and gas) outflow and gradual shrinkage of the nucleus (erosion), (ii) mantle or crust formation on the surface which limits the activity of the nucleus to a few bright areas, and finally results in a totally mantled, inactive body (Weissman et al., 1989). The first process is definitely irreversible – the lost mass cannot be reaccumulated. A simple estimate of the erosion rate (radius decrease) for a comet with perihelion distance 1 AU and a period of 6 yr gives about 2 m/yr for an assumed density of 0.3 g/cm^3 (Rickman, 1992). As to the second one, typical scenarios of mantle formation predict either periodic build-up of a mantle in the outer part of the orbit interchanged with removal of the mantle at perihelion (Brin and Mendis, 1979), or a secular increase of the mantle thickness with periodic variation during one orbital revolution (Rickman et al., 1990). In the latter case the mantle can be broken and removed only as a result of a sudden decrease of the perihelion distance or a change of the spin axis orientation. In spite of differences, most models show a trend in the mantling process, from more to less active nuclei.

For any single comet both processes, erosion and mantling, most likely act at the same time; the end point of the evolution – complete evaporation

Earth, Moon, and Planets **72**: 203–210, 1996.

or transition to a dormant state – is determined by the initial values of the size and active fraction of the nucleus, as well as by the erosion and mantling rates. Unfortunately, the time scales involved exclude, at present, an evolutionary study of single objects. We can hardly gather the brightness from more than 10 apparitions of JF comets, and the short-term brightness changes of individual comets may result from effects outside the scope of our model.

One way of extracting some evolutionary information from the present population of JF comets is to accept a hypothesis of the 'ergodic' kind – we observe a steady-state ensemble of comets at different stages of their evolution. Therefore, assuming that all the comets evolve according to similar and predictable laws we can first obtain their distribution function, and then make use of statistical theory to infer the observable quantities from the model, e.g. the magnitude distribution. A comparison with the measured values and functions should prove or disprove the validity of the model, in general, and the correctness of free parameters used, in particular. In this paper, which is a pilot study for a following extensive work (Banaszkiewicz and Rickman, in preparation), we show how such a theoretical program could be accomplished and we present the first results of calculations as an illustration of how the method is applied.

2. The model

We choose the radius R and the active fraction f of the nuclear surface as the two parameters that describe the physical state of a comet. The physical evolution of the comet is determined by the rates of change of these parameters. Since at least the erosion rate $\dot{R}$ depends on the perihelion distance q, it is clear that the physical evolution is coupled to the dynamical one. Close encounters with Jupiter, on a time scale of $10^3 - 10^4$ yr (Duncan et al., 1988; Rickman et al., 1992) can change the orbit of a comet appreciably, thus resulting in a different evolutionary path in the physical phase space. The perihelion distance seems to be by far the most important dynamical parameter for the physical evolution – therefore we limit our dynamical space to q alone. We choose a number of bins in q-space, each of them representing a subpopulation of comets with similar q. Now the evolution of comets can be envisaged as a process in which an object is brought to the population from some external source with a rate $S_{q_i}(R, f)$, spend some time in a bin q_i evolving with rates $(\dot{R}_{q_i}, \dot{f}_{q_i})$, and then, as the result of a Jupiter encounter, jumps to another bin q_j, where the evolution proceeds with different rates $(\dot{R}_{q_j}, \dot{f}_{q_j})$. After several such stages it is either ejected from the Jupiter family with a rate $l_{ko} = l_{q_k}$, or physically lost (collisions, splitting + disintegration), or, finally, reaches one of the end-states of the physical evolution: (i) evaporation to a very small size R_{min}, or (ii) decrease of the active fraction to the marginal value f_{min}.

The mathematical form of this complicated scenario applied to the whole population of comets is a set of continuity equations for the distribution functions $n_i(R, f)$ – the number of comets in the bin q_i per unit intervals of size and active fraction:

$$\dot{R}_i \frac{\partial n_i}{\partial R} + \dot{f}_i \frac{\partial n_i}{\partial f} = S_{oi} + \sum_{j \neq i} p_{ji} n_j - \left[\sum_{j \neq i} p_{ij} + l_{io} + l_i^{ph} + \frac{\partial \dot{f}_i}{\partial f} \right] n_i \qquad (1)$$

for $i = 1, \ldots, I$. Here each member expresses the co-moving derivative of the distribution function n_i (i.e., as experienced by any comet during its physical evolution). The first term in the right-hand member is the source function of injected comets, the following sum represents the comets perturbed into the i:th q-interval from other intervals, and the last term, with n_i as a common factor, represents in turn the ejections into other q-intervals, ejections back into the source, physical losses, and finally, $\partial \dot{f}_i / \partial f$ comes from the divergence term $\nabla(\dot{f}_i n)$ in the continuity equation.

Eqs. (1) are solved by the method of characteristics for $I = 5$ perihelion bins centered around $q = 0.25, 0.75, 1.25, 1.75$, and 2.25 AU. The physical phase-space is a rectangle $[R_{min}, R_{max}] \times [f_{min}, f_{max}]$, where $R_{max} = 10$ km is the radius of the largest JF comet considered, $R_{min} = 10$ m, $f_{max} = 1$, and $f_{min} = 0.001$. Objects with $R \leq R_{min}$ are no longer counted as comets but rather as icy interplanetary boulders disintegrating practically instantaneously by evaporation; comets with $f \leq f_{min}$ are probably in a dormant state. The erosion rate is $\dot{R}_i = -A(q_i) f$ with $A(q)$ chosen to be consistent with thermal model results (Rickman, 1992). For the mantling rate we arbitrarily choose the formula $\dot{f} = -(a + b R) f^k$, $k = 0, 1, 2$. The source function $S_{oi} = S_{q_i}(R, f) = q S_1(R) S_2(f)$, assumed proportional to the perihelion distance, is represented as a product of a piecewise power-law function $S_1(R) = C R^p$ and a piecewise linear function $S_2(f) = c + d f$. The jump probabilities among the bins p_{ij}, taken from the Monte Carlo study by Rickman et al. (1992), are of the order of 0.001 yr^{-1} for transits between neighbouring bins. The dynamical loss function l_{io} is, on the average, an order of magnitude smaller. For the physical loss rate l_i^{ph} we have conservatively assumed a value of 10^{-4} yr^{-1}. The main free parameters in our model are the source functions and the mantling rate. The choice of limits for the physical phase-space domain is not free but has to satisfy some accuracy criteria, i.e., a comet with $f < f_{min}$ should not be observable as an active comet, and a comet with $R < R_{min}$ must disintegrate on a very short time scale. Our present choices are conservative and might not be optimal.

Solutions of the characteristic equations for R and f give the trajectories of evolving nuclei in the (R, f) phase space (Fig. 1). The trajectories start

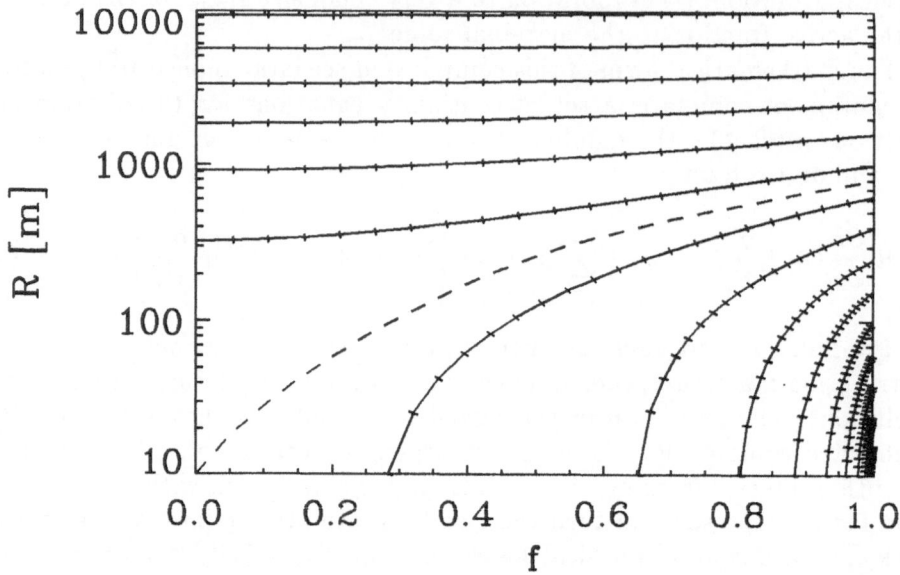

Fig. 1.   Phase space trajectories of evolving cometary nuclei with $q = 1$ AU for the standard model with the mantling rate $\dot{f} = 0.001$ yr^{-1}. Comets start from $f = 1$ and move toward smaller sizes and less active states. Ticks on the curves are equidistant in time, but the time steps are different for each curve. The dashed line is the boundary between comets that eventually become dormant (above) or evaporate (below).

at one of the boundaries $f = f_{max}$ or $R = R_{max}$ and go towards smaller sizes and less active comets. Initially large objects, i.e., in general km-sized, end their evolution at f_{min} as dormant comets. Small-size objects reach, in the end, the R_{min} limit and supply the interplanetary dust complex more efficiently. The position of the boundary line between these two evolutionary scenarios depends on the perihelion distance (bin); for smaller q, at which even relatively massive nuclei can evaporate completely during their lifetime, it is shifted towards larger R. The remaining characteristic equations provide as solutions the distribution functions $n_q(R, f)$. These functions contain all the information necessary to calculate the statistical properties of JF comets related to their physical evolution, e.g., the total number of comets:

$$N_{tot} = \sum_q \int_{R_{min}}^{R_{max}} \int_{f_{min}}^{f_{max}} n_q(R, f)\, dR\, df\, \Delta q \quad , \tag{2}$$

the number of comets per year that evaporate:

$$\dot{N}_{eva} = -\sum_q \int_{f_{min}}^{f_{max}} \dot{R}_q n_q(R_{min}, f)\, df\, \Delta q \quad , \tag{3}$$

or get dormant:

$$\dot{N}_{dor} = - \sum_q \int_{R_{min}}^{R_{max}} \dot{f} n_q(R, f_{min}) \, dR \, \Delta q \tag{4}$$

Similarly we may calculate the dust production rate $\dot{D}$, using the erosion and evaporation rates $\dot{R}$ and $\dot{N}_{eva}$, and assuming a dust/gas ratio equal to unity. Finally, the cumulative distribution of comets as a function of the absolute magnitude H can be obtained, using a model brightness function $H(R, f)$. Obviously, all statistical properties can also be calculated separately for each perihelion distance bin.

3. Results

The results of more than a dozen models will be extensively discussed in a forthcoming paper. Here we present only a small sample. In our standard model 1 it is assumed that $S_1(R) = 5 \times 10^3 R^{-2.8}$ is a simple power-law function, and $S_2(f) \equiv 1$. The mantling rate is constant ($k = 0$) and taken as $\dot{f} = -0.001$ yr^{-1}. In model 2 only the source function is changed to a steeper one: $S_1(R) = 5 \times 10^7 R^{-3.8}$. In model 3 the exponential decrease of f is considered ($k = 1$), while in model 4 a different (hat-like) source function $S_2(f)$ is introduced: $S_2(f) = 0.2$ for $0 \le f \le 0.33$ and for $0.66 \le f \le 1$, $S_2(f)$ linearly increasing to 5 for $0.33 < f \le 0.5$ and then decreasing to 0.2 at $f = 0.66$. The rest of the input parameters remain the same as in model 1. Finally, in model 5 we consider a piece-wise continuous source function, very steep for large nuclei: $S_1(R) = 5 \times 10^3 R^{-4}$ for 4 km $\le R \le$ 10 km, less steep: $\propto R^{-2.1}$ for 400 m $\le R <$ 4 km, and quite flat: $\propto R^{0.1}$ for $R <$ 400 m.

The results of our calculations are presented in Table I. The number of comets may vary appreciably between models, but it scales simply with the coefficient C in the $S_1(R)$ source function. By varying C, also the rates listed in the three rightmost columns would scale accordingly. Since the source functions in models 1-4 decrease with size and the size evolves from large to small, the distribution functions are strongly biased toward objects smaller than 100 m. This is seen from the average radii $\langle R \rangle$ of the comets. On the other hand, in model 5 where the source function has a maximum at 400 m and then rapidly falls off with size, the average radius is much larger. The total number of comets is substantially smaller in this model than in the others even though its dust production rate is the largest. The average active fraction of a nucleus $\langle f \rangle$ is typically of the order of 0.3 – larger than the observed values for short-period comets, e.g. $f \simeq 0.2$ for P/Halley, $f \simeq 0.03$ for P/Encke (Weissman et al., 1989). This discrepancy results from the assumption about uniformity of $S_2(f)$, which may overestimate

TABLE I

Results of five sample models

model	N_{Tot}	$\langle R \rangle$ [m]	$\langle f \rangle$	$\dot{N}_{ev}$ [yr^{-1}]	$\dot{N}_f$ [yr^{-1}]	$\dot{D}$ [kg yr^{-1}]
1	3040	49	0.32	87	12	1.05×10^{11}
2	216000	20	0.28	10900	1120	1.97×10^{11}
3	6950	73	0.16	97	8	1.60×10^{11}
4	1880	54	0.38	64	3	4.25×10^{10}
5	172	676	0.40	0.15	0.29	2.80×10^{11}

the number of source comets with $f > 0.5$ (even newly captured comets may be heavily mantled), but is also a consequence of the constant mantling rate. For model 3 with an exponential decay of activity ($\dot{f} \propto -f$) there is a build-up of number density $n_q(R, f)$ for smaller f and, hence, the average active fraction $\langle f \rangle$ is smaller. Since most of the dust is produced by large objects and all models show similar numbers of these, the dust production rates are comparable and $\sim 10^{11}$ kg/yr. The number of comets per year that evaporate or get dormant is strongly model-dependent.

The cumulative magnitude distributions are presented in Fig. 2. An often used reference number found from the observed distribution of JF comets is $N_{obs}(H < 10.8) = 70$ (Hughes, 1987), basically limited to $q \lesssim 2$ AU for reasonable completeness. The slope of the observed distribution is in the range 0.3 (Hughes, 1987) – 0.4 (Fernández et al., 1992). Our C values have been chosen for reasonable agreement with the observed number, and Table II shows that our slopes are also in reasonable agreement with the observed values, basically due to our choice of the exponent of $S_1(R)$.

4. Discussion

We have applied kinetic theory to descibe the average physical properties of the JF comet population. All the information necessary to obtain statistical moments of any function depending on the cometary radius and active fraction is contained in the distribution function $n_q(R, f)$. This function depends on one dynamical parameter, the perihelion distance, thus providing the coupling between physical and dynamical changes of the state of the comets. The physical evolution is determined by the erosion and mantling rates, and the dynamical one by the frequency of perihelion distance jumps

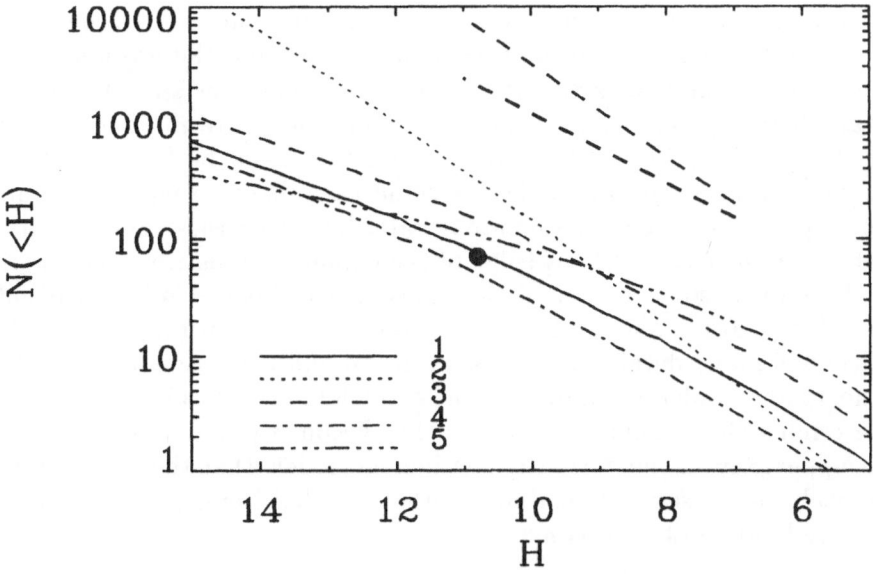

Fig. 2. The cumulative brightness distributions of comets for the five models of Table I. The black dot corresponds to the observed number of comets for $H < 10.8$. The dashed lines show the typical slopes of the $(\log(N(< H)), H)$ relation derived from the observed distribution: 0.3 (lower) and 0.4 (upper).

due to close encounters with Jupiter. The distribution functions for different q are the solutions of the kinetic equations in the (R, f) phase space. We assume that the population of JF comets is in a steady state. This is justified only if the source function, which may e.g. depend on the flux of long-period comets, does not change considerably over the time scales involved – about 10^4 years. We have also assumed that comets with perihelion distance larger than 2.5 AU are physically decoupled from the cometary population under consideration. The former group serves as a source and a sink for the latter, but once the comet's perihelion distance increases above 2.5 AU the memory about the previous physical state is lost – during the next visits to the region with $q < 2.5$ AU the physical parameters are determined by the source functions $S_1(R)$ and $S_2(f)$. This problem can be resolved by improving the source function iteratively. In the present model, the changes of the perihelion distance do not result in changes of the physical parameters. We may relax this assumption at the cost of introducing new free parameters. The splitting of comets may also be taken into account, but the number of subnuclei, their sizes and active areas are random variables for which additional information is needed.

In spite of all mentioned simplifications, the present model appears general enough to describe the observed properties of comets, provided that the input parameters are chosen appropriately. At present, we can determine with sufficient accuracy the jump probabilities p_{ij} and the erosion rate $\dot{R}$. Less certain is our knowledge about the source function and the physical loss rate. The least known parameter is the mantling rate $\dot{f}$. The uncertainty in the determination of the input parameters could be limited if better observational data were available. We could then place stronger constraints on some parameters, e.g. the mantling rate, by fitting the model results to the observed values. The knowledge of the cumulative distribution function $N(> R, > f)$ in some region of the phase space, instead of the currently observed $N(< H)$ might resolve most of the problems with input values.

Although many details should be improved and treated more accurately, the theory and models seem to be sound enough to provide a basic framework for describing the principal features of the JF comet population. In particular, it appears from the currently considered models that the mantling rate more likely decreases exponentially with time than remains constant over the active lifetime of a JF comet.

Acknowledgements

We thank Paul Weissman for many useful suggestions. Marek Banaszkiewicz is very grateful to the organizers of the Mariehamn Conference for the financial support.

References

Brin, G.D. and Mendis, D.A.: 1979, *Astrophys. J.* **229**, 422.
Duncan, M., Quinn, T. and Tremaine, S.: 1988, *Astrophys. J.* **328**, L69.
Fernández, J.A., Rickman, H. and Kamél, L.: 1992, in *Proc. of the International Workshop on Periodic Comets*, J. Fernández and H. Rickman (eds.), Montevideo, Uruguay, p. 143.
Hughes, D.W.: 1987, *Nature* **325**, 231.
Kresák, L. and Kresáková, M.: 1990, *Bull. Astron. Inst. Czechosl.* **41**, 1.
Rickman, H.: 1992, in *Interrelations between Physics and Dynamics for Minor Bodies in the Solar System*, D. Benest and C. Froeschlé (eds.), Gif-sur-Yvette, France, p. 197.
Rickman, H., Fernández, J.A. and Gustafson, B.Å.S.: 1990, *Astron. Astrophys.* **237**, 524.
Rickman, H., Froeschlé, C., Kamél, L. and Festou, M.C.: 1991, *Astron. J.* **102**, 1446.
Rickman, H., Bailey, M.E., Hahn, G. and Tancredi G.: 1992, in *Proc. of the International Workshop on Periodic Comets*, J. Fernández and H. Rickman (eds.), Montevideo, Uruguay, p. 55.
Weissman, P.R., A'Hearn, M.F., McFadden, L.A. and Rickman, H.: 1989, in *Asteroids II*, R.P. Binzel, T. Gehrels, and M.S. Matthews (eds.), Tucson, Arizona, p. 830.

MIGRATION OF SMALL BODIES IN THE SOLAR SYSTEM

S.I. IPATOV

Institute of Applied Mathematics, Miusskaya Sq. 4, Moscow, Russia

Abstract. We investigate several parts of the process of migration of small bodies to the Earth from the asteroid and transneptunian belts. The obtained characteristic times up to collisions of near-Earth objects with the Earth are less than those obtained by other scientists.

1. Time until Collision of Two Celestial Bodies

For the investigations of lifetimes of the bodies, Bottke *et al.* (1994), Farinella and Davis (1992), and others used modifications of the formula obtained by Öpik (1951). In the present paper for the characteristic time T until a collision of two bodies circling the Sun, we use another formula obtained by Ipatov (1988a) for the case when semimajor axes a, eccentricities e, and the angle between the orbital planes, Δi, do not change before the collision:

$$T \approx 2\,\pi^2\,\xi\,\sin\Delta i\,T_s\,(R/r_\Sigma)^2/(k_\varphi\,k_\Theta), \tag{1}$$

where $k_\Theta = 1 + (v_p/v_r)^2$, $k_\varphi = \Delta\varphi/r_s^*$, $r_s^* = r_s/R$, $\Delta\varphi$ is the sum of angles (in radians) with apices in the Sun, within which the distance between orbits is less than r_s, for the planar model; T_s is the synodic period of revolution, r_Σ is the sum of bodies' radii, v_p/v_r is the ratio of parabolic and relative velocities of the bodies, $r_s \approx R'\,\mu^{2/5}$ is the radius of the Tisserand sphere, R is the bodies' distance to the Sun, μ is the ratio of the total mass of bodies to the mass of the Sun, and ξ may vary between 0.5 and 1 depending on the considered model. We assume that ω, the argument of perihelion, and Ω, the longitude of ascending node, change considerably before a collision. In contrast to Öpik's formula, T in Eq. (1) depends on T_s and k_φ.

Rabinowitz *et al.* (1993) investigated the orbits of 15 near-Earth objects (NEOs) with diameters $d \le 50$ m and found that nearly half of these objects have $e < 0.2$ and for 11 NEOs $0.93 \le q = a(1-e) \le 1.02$ AU. They suggested that there is a belt of small asteroids near the Earth. The values of k_φ are greater for smaller eccentricities and for q closer to 1 AU. Let us assume r_s to be equal to the maximum asteroid distance to the Earth, where we can watch small asteroids. Then we obtain that all the above distributions of orbital elements of the watched small NEOs may be explained basing on the above dependencies of k_φ, and the orbital distribution of small NEOs may be the same as that for greater NEOs.

For real asteroids, Δi usually varies before a collision due to variations in i, ω, and Ω caused by gravitational influence of planets. If Δi varies between

Earth, Moon, and Planets **72**: 211–214, 1996.
© 1996 *Kluwer Academic Publishers.*

0 and $\Delta i_{\max}$ before encounters up to r_s and $\sin \Delta i \approx \Delta i$, then Ipatov (1988a) obtained the formula that use $\Delta i_{\max}/\eta$ instead of $\sin \Delta i$ in formula (1), where $\eta = 0.5 + \ln(\Delta i_{\max}/r_s^*)$. We denote by T^* the values of T for such model. We think that it is better to use T^* than T, if we consider collisions of asteroids with other asteroids and with planets. If $\Delta i_{\max} = 2\Delta i$, then $\eta/2 = 2.5$ for NEOs crossing the Earth at $\Delta i = 15°$, and at $\Delta i = 10°$ we have $\eta/2 \approx 4.5$ and $\eta/2 \approx 11.5$ for mutual collisions of main-belt asteroids with $d = 100$ km and $d = 1$ km, respectively.

2. Migration of Bodies to Orbits of Planets

A large number of papers, devoted to the problem of migration of bodies to the Earth, was published during last years [see review by Ipatov (1995b)]. It was found by some authors that bodies from the 3:1, 5:2, and 2:1 Kirkwood gaps and ν_6, ν_5, and ν_{16} secular resonances can migrate to the terrestrial planets. Ipatov (1988b, 1995a) showed that the gravitational influence of the largest objects of the transneptunian belt may be one of the reasons of the migration of bodies from this belt to the orbit of Neptune.

We found that asteroids, which migrate to the Kirkwood gaps due to the gravitational influence of asteroids, can cause only several percent of NEOs. A larger number of bodies can migrate to the gaps due to mutual collisions of asteroids. Farinella and Davis (1992) and Bottke *et al.* (1994) considered collisions between main-belt asteroids larger than 50 km and obtained the mean intrinsic collision probability $P = 2.9 \cdot 10^{-18}$ km^{-2} yr^{-1}. At $R = 2.8$ AU, $\Delta i = 10°$, $T_s = 10$ yr, $k_\varphi \approx 5$, $k_\Theta \approx 1$, and $\xi = 1$, we have $P = 1/T = 8.3 \cdot 10^{-19}$ km^{-2} yr^{-1} and $P^* = 1/T^*$ is equal to $6 \cdot 10^{-18}$ and $3.7 \cdot 10^{-18}$ km^{-2} yr^{-1} for 1 and 100 km asteroids, respectively. If we take into account variations in e, and distributions in a, then we obtain larger values of P^*. Basing on the values of T^*, we obtained, as Farinella *et al.* (1993), that for $d < 100$ km the average lifetime of main-belt asteroids is less than the age of the solar system. The average time until the moment of the first collision of some asteroid of diameter $d = 1$ or $d = 10$ km with some asteroid of diameter $d' \geq 0.1d$ is less than 10^4 or 10^6 yr, respectively. Therefore, small debris can frequently be ejected into the Kirkwood gaps.

3. Computer Simulation of Migration of Bodies to Earth's Orbit

We investigated the evolution of three-dimensional (spatial) disks initially consisting of all planets (excluding Pluto) and $N_o = 500$ bodies, moving round the Sun. The used algorithm was presented by Ipatov (1991). The gravitational influence of the planets was taken into account by the Tisserand spheres' method (two two-bodies' problems). Initial values of semimajor axes and eccentricities of all bodies were equal to a_o and e_o. Initial orbital orientations of bodies were different.

We found that amongst the bodies, which have come from the zone of the giant planets, the number of Earth-crossers is ten times greater than the number of only Mars-crossers and usually $e > 0.6$. Therefore, most Amor objects must have come from the asteroid belt. For $a_o = 1.7$ AU, $e_o = 0.5$ (a typical NEO), and $a_o = 2.82$ AU, $e_o = 0.7$ (a typical Earth-crossing orbit at the 5:2 resonance), the values of the time interval, t_h, during which the number of bodies, N, decreases to $N_o/2$, are hundreds of times smaller than the average time until a collision with the Earth obtained by using formula (1) for these values of a_o and e_o at $\Delta i = 10°$. When $a_o = 2.82$ AU, $e_o = 0.7$ and $N = N_o/2$, $0.1N_o$ bodies are Earth-crossers but not Jupiter-crossers and the time of further evolution is much greater than t_h.

Perihelia or aphelia of bodies colliding with the Earth were located mainly near the Earth's orbit. Many bodies which struck the Earth had aphelia close to 1 AU even for $a_o > 1$ AU. Their eccentricities usually were not large. The part of bodies colliding with the Earth was found to be about 6 % and 1 % for the pairs (2.5 AU, 0.4, i.e. a typical Mars-crossing orbit at the 3:1 resonance) and (2.82 AU, 0.7), respectively. A large amount of water could be delivered to the Earth during the accumulation of Uranus and Neptune.

During the last stages of the evolution of all considered disks, the orbits of some bodies were located inside the orbits of Earth and Venus. The number of such bodies in the solar system can be large and some of these bodies can migrate to the orbit of the Earth. At the late stages of disk evolution for $a_o = 25$ AU and $e_o = 0.5$, we obtained gaps in the distribution of perihelia of bodies near the orbits of the giant planets.

4. Characteristic Times until Collisions of NEOs with the Earth

The orbital elements of real NEOs can change significantly before their collisions with planets. Therefore, the values of the time until these collisions may differ by a factor of several from the values of T obtained for fixed orbital elements, and the lifetime of a NEO does not depend much on its initial orbit. We found the average time until a collision of an Earth-crossing object (ECO) with the Earth $T_* = N/\sum(1/T_k) = 75$ Myr at $\xi = 1$, where T_k is the value of T for the k-th ECO and the sum is taken for known ECOs. The value of T_* is 5 times less than the value of T obtained for average values of e and Δi for ECOs and is about two times less than the value obtained by Bottke et al. (1994). Let us consider that N_* objects become new ECOs at some moment of time. After time t there will be $N = N_* \cdot e^{-t/T_d}$ of these ECOs, where $T_d = T_*/(k + 1)$ is the dynamical lifetime of a ECO,

$1/k$ is the ratio of the number of objects hitting the Earth to the number of objects ejected into hyperbolic orbits or colliding with other planets and the Sun. For $1 + k = 10$ at N/N_* equal to 0.5, 0.1, and $4.54 \cdot 10^{-5}$, we have $t = T_{**} \approx 0.07T_*$, $t = 0.23T_*$, and $t = T_*$, respectively. Therefore, half the ECOs that hit the Earth did so in less than 5 Myr after these objects became ECOs. The collisional lifetime of 1 m NEO was found to be several times less than T_{**}. This result agrees with the fact that stony meteorites are usually the result of several destructions (Chapman 1990).

A little more than half the known NEOs (with $q \leq 1.3$ AU) are ECOs. Therefore, the average time up to a collision of a NEO with the Earth is less than 150 Myr. The total number of NEOs with diameter $d \geq 1$ km may reach 2000, so the average time between impacts of such NEOs with the Earth may not exceed 100,000 yr. H-chondrites probably come to the Earth, passing through the resonances in the asteroid belt. The small number of LL-chondrites with ages $t < 8$ Myr may be due to the long way of LL-chondrites from the transneptunian belt. The large age of iron meteorites may be caused by the large time interval between collisions of large asteroids.

Acknowledgements

This work was supported by the Russian Fund of Fundamental Investigations under Grant 93-02-17035.

References

Bottke, W.F., Nolan, M.C., Greenberg, R., Kolvoord, R.A.: 1994, "Collisional lifetimes and impact statistics of near-Earth asteroids." In *Hazards due to comets and asteroids*, University of Arizona Press.
Chapman, C.R.: 1990, "Meteorite parent bodies." *Nature*, **344**, 813-814.
Farinella, P. and Davis, D.R.: 1992, "Collision rates and impact velocities in the main asteroid belt." *Icarus*, **97**, 111-123.
Farinella, P., Gonczi, R., Froeschlé, Ch., Froeschlé, C.: 1993, "The injection of asteroid fragments into resonances." *Icarus*, **101**, 174-187.
Ipatov, S.I.: 1988a, "Evolution times for disks of planetesimals." *Sov. Astron.*, **32(65)**, N 5, 560-566.
Ipatov, S.I.: 1988b, "Computer simulation of the possible evolution of the orbits of Pluto and bodies of the trans-Neptune belt." *Kinematics Phys. Celest. Bodies*, **4**, N 6, 76-82.
Ipatov, S.I.: 1991, "Evolution of initially highly eccentric orbits of the growing nuclei of the giant planets." *Sov. Astron. Lett.*, **17**, N 2(3), 113-119.
Ipatov, S.I.: 1995a, "Gravitational interaction of objects moving in crossing orbits." *Solar System Research*, **29**, N 1 (p. 11-23 in Russian edition).
Ipatov, S.I.: 1995b, "Migration of small bodies to the Earth." *Solar System Research*, **29**, N 4 (in press).
Öpik, E.J.: 1951, "Collisional probabilities with the planets and the distribution of interplanetary matter." *Proc. Roy. Irish. Acad.*, **A54**, 165-199.
Rabinowitz, D.L., et al.: 1993, "Evidence for a near-Earth asteroid belt." *Nature*, **363**, 704-706.

NEAR-EARTH ASTEROIDS: SURFACE STRUCTURE AND SHAPES

I.N. BELSKAYA and D.F. LUPISHKO
Astronomical Observatory of Kharkov University, Sumskaya str. 35, Kharkov 310022, Ukraine

and

A.N. DOVGOPOL
Main Astronomical Observatory of Ukrainian Academy of Sciences, Goloseevo, Kiev 252157, Ukraine

Abstract. The available data characterizing surface properties of near-Earth objects have been analyzed and compared with the data of large main belt asteroids. There are no evidences of the existence of solid rock surfaces among near-Earth objects.

1. Introduction

Surface structure and shapes of near-Earth asteroids (NEAs) are expected to be different from large main-belt asteroids (MBAs) because of the difference in their location and gravity (e.g. see McKay *et al.*, 1989). We have compared the mean values of various physical parameters characterizing surface properties of NEAs and MBAs. Consideration has been restricted to the S-type asteroids which are well-representative among NEAs. These two asteroid groups probably have similar composition and hence internal strength of surface material which is critical for regolith formation. Models of asteroidal regoliths have predicted a thick regolith layer for large "strong" asteroids (> 100 km in diameter) and bare rock surfaces for "strong" asteroids with diameters smaller than 10 km (see McKay *et al.*, 1989).

2. Results

The mean value, standard deviation and sample size of each group of asteroids are given in Table I for the following parameters: diameter, UBV-colours, phase coefficient β_v of the asteroid magnitude-phase dependence; the negative polarization P_{min}, inversion angle α_{inv} and slope h of the polarization-phase dependence; polarimetric and radiometric albedos, p_v, radar albedo σ_{oc} and polarization ratio μ_c, and lightcurve amplitude. The data were taken from the available asteroid data-base (Lagerkvist *et al.*, 1989; Morrison and Zellner, 1979; Tedesco, 1989) and supplemented with recently published data for NEAs. The mean diameters of the NEA and MBA samples differ by nearly 20 times. Nevertheless, the mean values of

Earth, Moon, and Planets **72**: 215–218, 1996.

TABLE I

Mean values of the parameters for near-Earth and main belt asteroids

Parameter	NEAs, S-type		MBAs, S-type	
D, km	8 ± 8	11	140 ± 36	32
$U - B$, mag	0.45 ± 0.05	16	0.45 ± 0.04	33
$B - V$, mag	0.86 ± 0.05	16	0.85 ± 0.03	33
β_v	0.029 ± 0.006	9	0.030 ± 0.006	16
P_{min}, %	0.77 ± 0.07	3	0.76 ± 0.08	19
α_{inv}, deg	20.2 ± 0.7	5	19.9 ± 1.1	19
h	0.110 ± 0.013	9	0.112 ± 0.015	19
p_v, pol	0.16 ± 0.03	9	0.15 ± 0.02	19
p_v, rad	0.21 ± 0.06	10	0.20 ± 0.04	29
σ_{oc}	0.17 ± 0.09	10	0.16 ± 0.04	6
μ_c	0.26 ± 0.09	9	0.13 ± 0.07	6
Amplitude	0.72 ± 0.56	16	0.26 ± 0.17	31

most of the other parameters are almost the same for these asteroid groups. Below we discuss implications from the various parameter comparisons.

2.1. PHASE COEFFICIENT

For objects of similar surface mineralogy a difference in phase coefficient values is evidence of differences in roughness and porosity of their surface layers (e.g. Helfenstein and Veverka, 1989). To accurately compare phase coefficients of NEAs and MBAs we should consider the dependence on geometry of observations. This is difficult because of poor knowledge of asteroid shape and pole coordinates. We have estimated the influence of aspect changes on phase coefficient value from an ellipsoidal model assuming the same rate of aspect and phase angle changes. The coefficient β_v was calculated for the phase dependence of lightcurve maximum at various axis ratios and pole positions of the model. For a ratio $b/c = 1.2$ the dispersion in phase coefficient due to aspect is 0.003 mag/deg while for $b/c = 1.4$ it becomes 0.006 mag/deg. Thus, the phase coefficients of individual near-Earth asteroids should not be interpreted without careful aspect correction. Taking into account various aspect changes both increasing and decreasing the phase slope, we can consider the mean value of the phase coefficient of NEAs as some average characteristic of the asteroid surface layer. The coincidence of the mean values for NEAs and MBAs suggests a similarity of their surface layers at the submicron scale.

2.2. POLARIZATION PROPERTIES

Polarimetric properties are widely considered to be very sensitive to surface structure. According to the numerous laboratory data (Dollfus et al., 1977; Shkuratov, 1994) the inversion angle is noticeably smaller for bare rock and increases with decreasing particle size. Within the data set considered, we see no differences in the values of minimum polarization and inversion angle of small NEAs and large MBAs. This implies that NEAs have dusty surfaces similar to those of MBAs.

2.3. RADIOMETRIC DATA

There is convincing evidence that some NEAs do not satisfy the standard thermophysical model. These objects probably have thin (if any) regolith layer while others have the same thermal properties as MBAs (Veeder et al., 1989). We do not see a systematic difference between the mean albedos of NEAs and MBAs in either radiometric or polarimetric values.

2.4. RADAR DATA

Mean values of radar parameters were calculated from data taken from Ostro et al.(1991). The mean circular polarization ratio μ_c which characterizes surface roughness is two times larger for NEAs than for MBAs (see Table). Ostro (1989) concluded that surfaces of NEAs tend to be much rougher at decimeter scales compared to MBAs. Radar data have also proved more irregular global shapes of NEAs with an abundance of exotically shaped objects (Ostro, 1989).

2.5. LIGHTCURVE AMPLITUDE

The mean amplitude of asteroid lightcurves, which characterize to some extent the overall shape of asteroids, is larger for NEAs compared to MBAs. Partly the difference in amplitude may be caused by geometry of NEA observations (large phase angle and rapid changes of aspect). It is difficult to correct amplitudes in a proper way because of many unknowns. The great dispersion of lightcurve amplitudes of NEAs confirms the diversity of their shapes.

3. Conclusions

We have compared available physical properties of small and large asteroids which characterize surface structure and shapes. All asteroids considered belong to the S-type, implying a similar composition and internal strength.

On the average we do not see a difference in their surface layers at the optical wavelength scale. However, there are differences in surface roughness at centimeter to meter scales and in overall asteroid shapes. Most likely we have not observed bare rock surfaces among km-size S-type NEAs.

Acknowledgements

We thank A.W. Harris for useful comments. This work was partly supported by the Russian Fund of Fundamental Exploration.

References

Dollfus A., Geake J.E., Mandeville J.C., Zellner B.: 1977, in *Comets, Asteroids, Meteorites*, A.H. Delsemme (ed.), Univ. of Toledo Press, p. 243

Helfenstein P., Veverka J.: 1989, in *Asteroids II*, R.P. Binzel, T. Gehrels, M.S. Matthews (eds.), Univ. of Arizona Press, p. 557

Lagerkvist C.-I., Harris A.W., Zappala V.: 1989, in *Asteroids II*, R.P. Binzel, T. Gehrels, M.S. Matthews (eds.), Univ. of Arizona Press, p. 1162

McKay D.C., Swindle T.D., Greenberg R.: 1989, in *Asteroids II*, R.P. Binzel, T. Gehrels, M.S. Matthews (eds.), Univ. of Arizona Press, p. 617

Morrison D., Zellner B.: 1979, in *Asteroids*, T. Gehrels (ed.), Univ. of Arizona Press, p. 1090

Ostro S.J.: 1989, in *Asteroids II*, R.P. Binzel, T. Gehrels, M.S. Matthews (eds.), Univ. of Arizona Press, p. 192

Ostro S.J., Campbell D.B., Chandler J.F., Hine A.A., Hudson R.S., Rosema K.D., Shapiro I.I.: 1991, *Science* **252**, 1399

Shkuratov Yu.G.: 1994, *Astron. Vestnik* **28**, N 3, 23 (in Russian)

Tedesco E.F.: 1989, in *Asteroids II*, R.P. Binzel, T. Gehrels, M.S. Matthews (eds.), Univ. of Arizona Press, p. 1090

Veeder G.J., Hanner M.S., Matson D.L., Tedesco E.F., Lebofsky L.A., Tokunaga A.T.: 1989, *Astron. J.* **97**, 1211

SPIN RATES OF ASTEROIDS

C.-I. LAGERKVIST and Å. CLAESSON

Astronomiska observatoriet, Box 515, S-751 20 Uppsala, Sweden

Abstract. The Asteroid Photometric Catalogue was used to redetermine the rotation periods of all asteroids with data in the catalogue. The quality of the period determinations was divided into five groups. The total number of asteroids studied were 710 and 225 of these were considered not to be observed enough to yield any rotation period (code 0). For 121 asteroids the uncertainty was several hours (code d) and for 180 the uncertainty was less than one hour (code c). Code a was used for asteroids with reliable pole determinations (47 asteroids) and code b was used for asteroids with very reliable synodic rotation periods (137 asteroids). Some statistic properties of the rotation periods of asteroids are presented.

1. Introduction

The Asteroid Photometric Catalogue (Lagerkvist et al., 1993) is available in digital form at the Uppsala observatory (Magnusson et al., 1994) and may be obtained via anonymous ftp. The aim of this study was to make a redetermination of asteroid rotation periods using all published observations. The period determinations were made with the software package developed by Magnusson (Magnusson et al., 1994). Unfortunately there are still data in the catalogue consisting of composite lightcurves not possible to separate into individual nights. For a few single asteroids this prevented us from making an independent determination of the rotation period of these objects. Some of these were considered reliable in spite of this and were not placed in code 0. A total of 225 asteroids had not been observed enough in our opinion, or were only available in such a form that no independent determination of the rotation period could be done (code 0). For the remaining 485 asteroids we divided the results into four groups:

a) a reliable pole solution exists (Magnusson et al., 1994) meaning that normally at least three apparations are covered (47 asteroids)

b) more or less a complete rotational cycle is covered and the result is given to two decimal places (137 asteroids)

c) the asteroid was observed less frequently and the result should be correct to within one hour (180 asteroids)

d) the result may be in error of several hours (121 asteroids)

Earth, Moon, and Planets **72**: 219–223, 1996.

In the next section the result is briefly discussed. We have in the discussion only used the data for the numbered asteroids. All the tabular data are available via internet by typing **ftp ftp.astro.uu.se**. When prompted for the username give **anonymous** and identify yourself when asked for a password. Proceed to the directory **pub/Asteroids/RotationPeriods**.

2. Results

The distributions of the spin rates for the various codes may be found in Figures 1a-1d. Codes a and b have been treated together since the errors here are practically the same (Figure 1a). In all figures the unit on the x-axis is revolutions per day. One can clearly see that there is a difference in the sense that asteroids with well determined rotation periods have faster spin than those with less well defined rotation periods. This might be natural from the observers point of view but causes that the rotation periods of asteroids will be heavily biased. For asteroids with pole determinations this means that basically no pole solutions exist for asteroids with long rotation periods. The distributions for codes c and d seems to be essentially the same (Figures 1b,1c). The gap around 12 hours might be caused by the obvious difficulty to determine rotation periods close to 24 or 12 hours. Figure 1d gives the overall distribution for the total sample of 445 numbered asteroids. The various codes have not been weighted but doing so gives no difference in the distribution. In none of Figures 1a-1d were the asteroids with perihelion distances smaller than 1.4 AU included. A comparison with Figure 4 of Binzel et al. (1989) shows that our distribution peaks at a rotation rate of 2.9 compared to 2.6 revolutions per day. The reason for this is not obvious but shows that great care has to be taken when using a heavily biased sample of asteroid rotation rates.

In Figures 2a-2d the asteroids have been divided into three diameter ranges $D \leq 60$ km (Figure 2a), $60 \leq D \leq 120$ km (Figure 2b) and $120 \leq D$ (Figure 2c). Only asteroids with diameters determined by IRAS have been considered and Figure 2d gives the distribution for all asteroids with diameters determined by IRAS. There is a difference in the sense that the small and the large asteroids seem to spin faster than those of intermediate size which also have a more concentrated distribution. Asteroids with perihelion distances smaller than 1.4 AU have not been included in the analysis since these have considerably shorter rotation periods and larger lightcurve amplitudes. However, they do not effect the overall distributions very much anyhow, since they constitute quite a small number.

A division into compositional types are shown in Figures 3a-3d where the distributions of the spin rates for the S type, M type, C type and the total population may be compared. The M-type stands out quite clearly from the

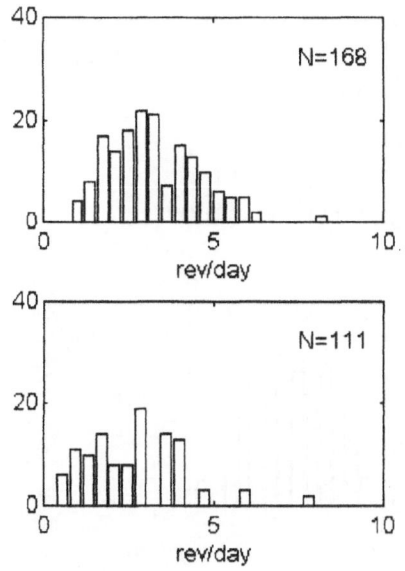
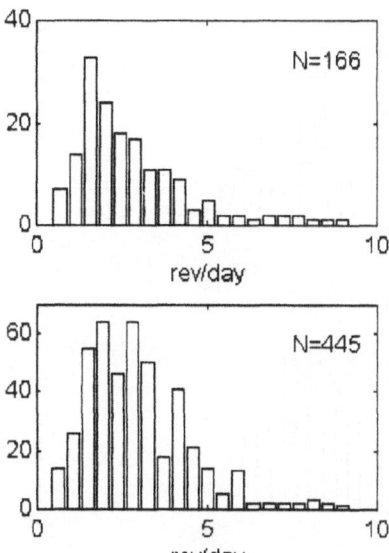
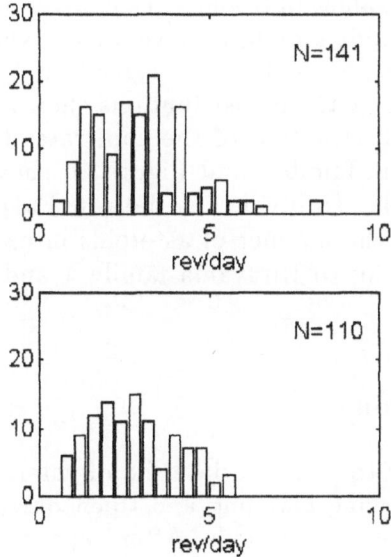
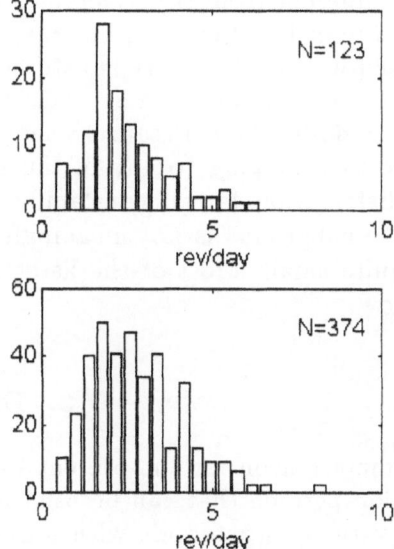

Fig. 1.   a–d

Fig. 2.   a–d

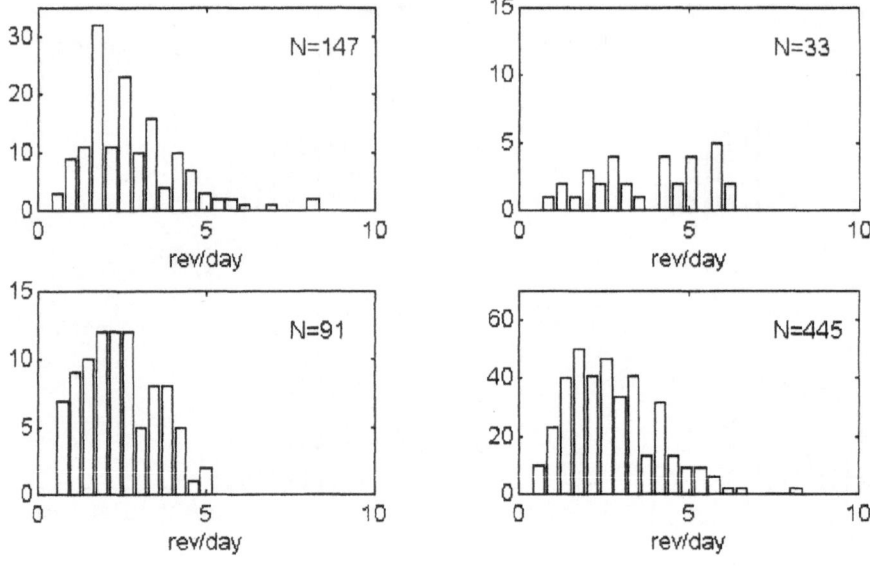

Fig. 3.   a–d

other types in the sense that the spin is faster but the distribution is also quite different from the others. The mean values of S and C type asteroids are more or less the same but S type asteroids may have have a somewhat slower rotation than C type asteroids do.

If the family asteroids are considered one gets the distributions shown in Figures 4a-b where Figure 4a shows the distribution of the spin rates for asteroids belonging to any of the Hirayama families and Figure 4b shows the distribution for all non-family asteroids. To divide the data in Figure 4a into subgroups is not meaningful since the number of asteroids in each get quite small. Most of the asteroids belong to Hirayama family 1 and 2 anyhow.

3. Discussion

The main reason for the present work was to create a data file of asteroid rotation periods that can be used for observing planning and to encourage observations of asteroids with uncertain rotation periods. After examining all these thousands of asteroid lightcurves one gets the impression that many observers have a more optimistic view of the errors in their rotation periods than is justified when the observations are compared to others observations

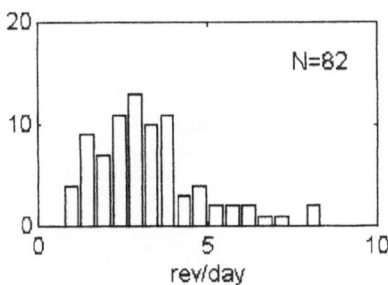 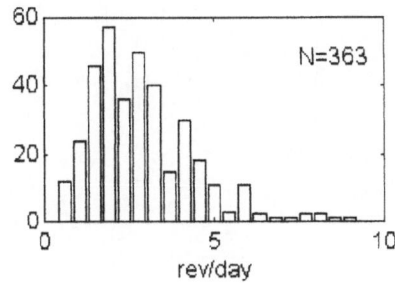

Fig. 4.   a–b

and examined in a more homogenous manner. A thorough general discussion
of spin rates may be found in Binzel et al. (1989) and references therein.

References

Binzel, R.P., Farinella, P., Zappalà, V., Cellino, A.: 1989, in *Asteroids II*, eds. R.P. Binzel,
   T. Gehrels, M.S. Matthews. Univ. of Arizona Press, pp. 416–441.
Lagerkvist, C.-I., Magnusson, P., Belskaya, I.N., Erikson, A., Dahlgren, M., Barucci, M.A.:
   1993, *Asteroid Photometric Catalogue, third update*. Reprocentralen HSC, Uppsala.
Magnusson, P., Lagerkvist, C.-I., Dahlgren, M., Erikson, A., Barucci, M.A., Belskaya, I.N.,
   Capria, M.T.: 1994, *The Uppsala Asteroid Data Base*, in *Asteroids, Comets, Meteors
   1993*, eds. A. Milani, M. Di Martino, A. Cellino. Kluwer Academic Publishers, pp.
   471–476.

MAGNETIC PROPERTIES OF ASTEROIDS FROM METEORITE DATA –
IMPLICATIONS FOR MAGNETIC ANOMALY DETECTIONS

MAURI TERHO, LAURI J. PESONEN and ILMO T. KUKKONEN

Geophysics Department, Geological Survey of Finland, FIN-02150 Espoo, Finland

Abstract. Magnetic measurements of meteorites suggest that small bodies (e.g. asteroids) in the Solar System have small but distinct magnetic fields produced by the bulk remanent magnetisation (NRM) of the body. Here we report calculations of magnetic fields of small bodies, assuming that they can be approximated as homogeneously magnetised spheres with dipole moments derived from NRM data on known meteorites. The magnetic fields are compared with the field of the asteroid 951 Gaspra measured by spacecraft Galileo in 1991 (Kivelson et al., 1993). The result of this comparison suggests that the field of Gaspra could be caused by an L-, H- or E-chondritic or a pallasite body. The spectral reflectance data on Gaspra suggest, however, that it is a basaltic achondrite. The problem can be resolved if Gaspra is a differentiated body, its surface material being closer to that of basaltic achondrites, and the bulk closer to ordinary chondrites or pallasites. We also present magnetic anomaly profiles along the surface of Mars such as would be measured with a magnetometer installed on a Rover-type vehicle by assuming that the main sources of the surface anomalies are the NRMs of the boulders on the Martian surface. The NRM values are taken from the data measured on SNC meteorites. The results suggest large oscillations in magnetic field intensity at the Martian surface.

1. Introduction

Meteorites are unique material of the Solar System in their ability to provide tangible information on the chemical composition, physical properties and evolution of their parent bodies. During the last twenty years, the visual and infrared reflectance spectra of a number of asteroids and meteorites have been measured telescopically and in laboratories. Correlation of the results has led to the identification of the surface materials of several asteroids, e.g., the asteroid 2 Pallas, which is like a carbonaceous chondrite, 4 Vesta, which is similar to a basaltic achondrite, and 230 Athamantis, which resembles a mesosiderite (e.g. Chapman, 1976; Wood, 1988; Hiroi *et al.*, 1993; Kivelson *et al.*, 1993; Pieters and McFadden, 1994).

A meteorite database compiled at the Geological Survey of Finland (GSF) contains petrophysical data (such as bulk dry density, porosity, bulk susceptibility and intensity of the natural remanent magnetisation (NRM)) on more than 400 meteorites and the magnetic hysteresis data on more than 50 meteorites measured during 1979 – 1994. The measurements and interpretations are reported in detail in Kukkonen and Pesonen (1983), Pesonen *et al.*, (1993) and Terho *et al.*, (1993). The petrophysical data on meteorites have

Earth, Moon, and Planets **72**: 225-231, 1996.

been analysed with the aid of bivariate, e.g. susceptibility vs. density, diagrams (Terho *et al.*, 1993). Because of systematic variations in FeNi content between achondrites and irons, and owing to variations in the grain size and shape of the ferromagnetic minerals, the meteorite classes plot into distinct clusters in these diagrams, thus giving us a tool for classifying meteorites and for rapidly discriminating a meteorite from terrestrial rocks or artefacts. The mean values for the petrophysical properties of the main meteorite classes provide a unique opportunity for calculating the geophysical anomalies of small bodies (asteroids) by assuming that the physical properties of these bodies can be represented by those of meteorites (see Terho *et al.*, 1993).

We present here some applications of the meteorite petrophysics database, in which the magnetic fields of small solar system bodies are calculated using the NRM data on meteorites measured in the laboratory. We demonstrate how this technique can be applied to recent magnetic intensity data obtained during flyby measurements of the asteroid 951 Gaspra (Kivelson *et al.*, 1993; Baumgärtel *et al.*, 1994). We also estimate the amplitude of local magnetic anomalies caused by boulders along a simulated magnetometer profile on the surface of Mars assuming that the magnetic properties of boulders are similar to those of SNC meteorites.

2. Magnetic modelling

In modelling magnetic anomalies of small bodies we assume that the body can be treated as a homogeneously magnetised sphere (Fig. 1). We assume that the magnetic field of the body is caused by the bulk NRM and that no significant interplanetary or internal dynamo magnetic fields are present to produce induced magnetisation. We further assume that certain meteorites represent their host asteroids. The NRM of the body is taken from the NRM data on meteorites. The vector components (ΔX, ΔY, ΔZ) or the total component (ΔB_t) of the magnetic field of the small body at its surface, or at a distance r from the centre of the body (Fig. 1), is calculated (see e.g., Parasnis, 1979). Figure 1 demonstrates the technique and shows the ΔB_t of a hypothetical H chondrite asteroid, 10 km in size, having a bulk NRM of 0.5 Am^2/kg and magnetic inclination $\sim 40°$. The inclination is naturally unknown and thus the shape and amplitudes of ΔB_t will vary accordingly (Parasnis, 1979); here we demonstrate only the principle with fixed parameters. The results in Fig. 1 nevertheless show that anomalies of up to $10-20$ nT are expected during flybys at distances of less than 200 km. The magnetic field ΔB_t falls rapidly with increasing measuring distance r, as it is proportional to r^{-3}.

There are a number of imponderables in these calculations. For example, the bodies are probably not homogeneously magnetised, nor are they per-

TABLE I

Calculated magnetic fields at surfaces of achondritic, chondritic and stony-iron bodies the size of Gaspra ($D = 14$ km)

BODY (D = 14 km)	Mean NRM $(10^{-6}$ Am^2/kg)	SURFACE FIELD B (μT) min – max
BASALTIC ACHONDRITES:		
Eucrites	509	0.6 – 1.2
Howardites	2072	2.5 – 5
CHONDRITES:		
LL	675	1 – 2
L	4215	5 – 10
H	36848	55 – 110
E	43511	66 – 132
STONY-IRONS:		
Mesosiderites	206770	360 – 720
Pallasites	31219	75 – 150

BODY = type of small body, NRM = mean NRM of each meteorite group (corrected for shape demagnetisation), The NRMs were taken from the meteorite database of the Geological Survey of Finland (Terho *et al.*, 1993). SURFACE FIELD = magnetic field at surface of body, where min, max denotes field at magnetic equator and pole of body, respectively.

fectly spherical. We do not know whether the meteorite sample we used for magnetic measurements is really representative of the whole small body, even though the spectral reflectance of the body corresponds to that of meteorites, particularly if the body is differentiated or has a well-developed regolith. Moreover, the interplanetary magnetic field will cause a small component of induced magnetisation, which is not taken into account here. Secondary magnetisation events may have occurred in both the parent body and the meteorite. Components that characterise later geological events (e.g. shock remanence), but which are not necessarily related to bulk chemical composition (e.g. FeNi content), could then have been produced, thus violating our assumptions. The multicomponent nature of NRM in a meteorite can be established with various techniques (Kukkonen and Pesonen, 1983), but the ages of these events are difficult to determine.

3. Magnetic fields of the asteroid 951 Gaspra

Using the dipole equations (Parasnis, 1979) and measured NRM intensities of meteorites (Table I), we calculated conceivable magnetic fields for the

TABLE II

Petrophysical properties of SNC meteorites

Meteorite	n	NRM (10^{-6} Am^2/kg)	χ_0 (10^{-8} m^3/kg)	Q	ρ (kg/m^3)	Ref.
Chassigny	1	19	14	3.22	3319	3
Nakhla	1	13	182	0.18		1,2
Governados Valadares	1	51	167	0.76		1,2
Shergotty A-B	4	41	127	0.81		1,2
Shergotty C	1	361	112	8.1		1,2
Zagami fc	3	260	100	6.5		1,2
Zagami	2	4069	54	188	3071	3
ALHA 77005.77	2	32	645	0.12		1,2
ALHA 77005	3	23	347	0.17		4
EETA 79001 A	1	17	88	0.48		1,2
EETA 79001	3	30	47	1.6		4
EETA 79001.183	1	24	63	0.95	3124	3

A-B denote subsamples of the same parent sample, C is another parent sample. fc: fusion crust is present. n = number of specimens. **NRM** = intensity of Natural Remanent Magnetisation per unit mass. χ_0 = initial susceptibility per unit mass (NRM and χ_0 values corrected for shape demagnetisation). **Q** = Koenigsberger ratio [=NRM/$\chi_0 H$, where H was arbitrarily chosen as 39.8 A/m (= $50\mu T$)]. ρ = density (kg/m^3).
References: 1 – Cisowski (1985); 2 – Cisowski (1986); 3 – Terho et al. (1993); 4 – Collinson (1986).

asteroid Gaspra. Table I shows the magnetic fields at surfaces of chondritic and stony-iron meteorite bodies with the same diameter (14 km) as Gaspra. The minimum and maximum values correspond to the fields at the magnetic equator (min) and at the pole (max) of the body, respectively. Results from the Galileo flyby revealed that the inferred field at the surface of Gaspra ranges from 4 to 140 μT (Kivelson et al., 1993; Baumgärtel et al., 1994). Thus, according to Table I, the bulk composition of Gaspra could correspond to that of L-, H- or E-chondrites or pallasites. The spectral reflectance data on Gaspra suggest, however, that it is a basaltic achondrite (Kivelson et al., 1993). The mean density of Gaspra was estimated to be 4000 kg/m^3 (Kivelson et al., 1993), which is too high for the density ($\sim$ 3000 kg/m^3) of basaltic achondrites (Terho et al., 1993). We therefore suggest that Gaspra has a differentiated structure similar to that proposed by Delaney et al., (1981) for mesosiderites, namely, that the surface material is much lighter than that of the inner parts of the body. This model for Gaspra could explain both the spectral reflectance data and the magnetic observations (e.g. Kivelson et al., 1993).

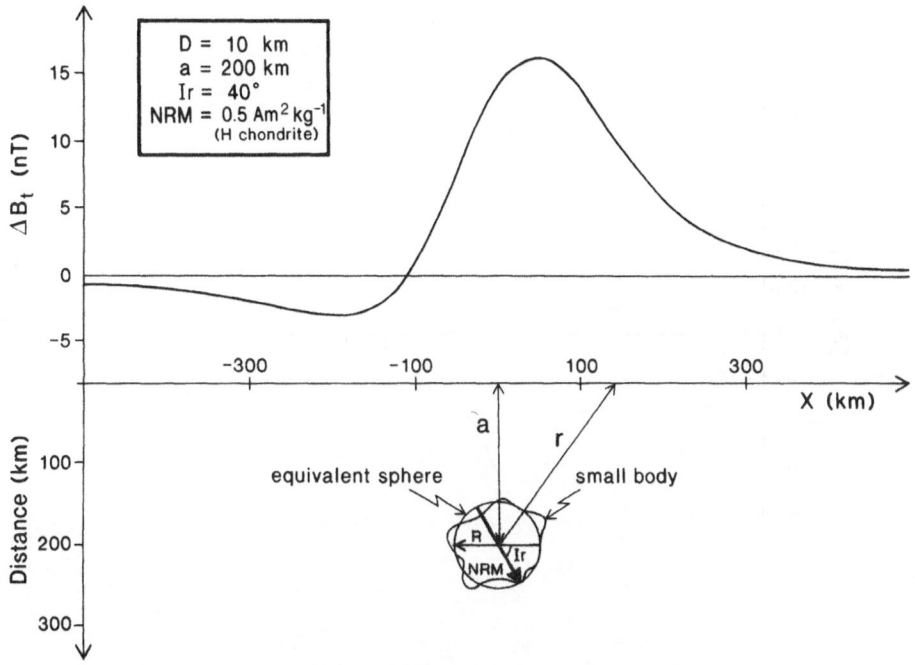

Fig. 1.   Total field magnetic anomalies caused by homogeneously magnetised boulders on the surface of Mars. The boulders are treated as equivalent spheres, and their centres lie on a same line under the magnetometer. The arrows in the boulders show the random direction of NRMs. Magnetic dipole moments for each boulder were calculated from NRM data on SNC meteorites: Zagami (NRM $= 4069 \cdot 10^{-6}$ Am^2/kg) and EETA 79001 (NRM $= 17 \cdot 10^{-6}$ Am^2/kg).

4. Local magnetic field on the Martian surface

SNC meteorites are thought to derive from the Martian surface (Laul, 1986). The magnetic properties of SNC meteorites are summarised in Table II (Cisowski, 1985, 1986; Collinson, 1986; Terho *et al.*, 1993). From Viking lander photographs we know that the Martian surface is strewn with boulders (Carr, 1990). Figure 2 gives an example of 3D-modelling of magnetic field anomalies on the surface of Mars produced by homogeneously magnetised boulders of SNC meteorites with NRMs of varying direction. The minimum distance between a boulder and a magnetometer is 20 cm. Here we have only used data on Zagami and EETA 79001 that represent both high and low NRM levels of SNCs. Figure 2 shows that considerable magnetic variations, with amplitudes of up to ±600 nT, that are produced by NRMs of SNC boulders on the Martian surface and would be detectable by a Rover-type vehicle survey. Since EETA 79001 has a much lower NRM inten-

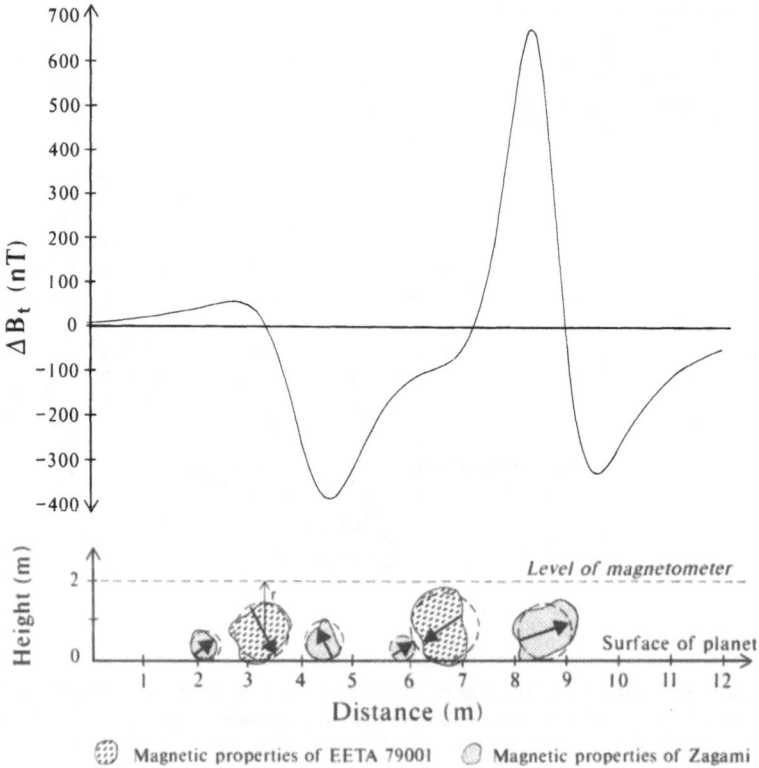

Fig. 2. The principle of calculating the total field magnetic anomaly (ΔB_t) at distances r of a small body. The body is treated as a homogeneously magnetised sphere with radius R, natural remanent magnetisation NRM and inclination I with respect to a horizontal reference frame. ΔB_t denotes the total magnetic field anomaly caused by the NRM of the body. No significant interplanetary or internal dynamo magnetic field is assumed.

sity ($24 \cdot 10^{-6}$ Am^2/kg) than Zagami ($4069 \cdot 10^{-6}$ Am^2/kg), the magnetic signal would be weaker near EETA 79001 boulders (Fig. 2).

5. Conclusions

Petrophysical data on meteorites suggest that small bodies in the solar system have small but distinct magnetic fields produced by their bulk NRM. The bulk compositions of small bodies can be estimated using petrophysical data on meteorites if the magnetic field of the body is first measured during a flyby. Further, from NRM data on SNC meteorites, we predict that considerable field oscillations will be seen in Rover-type magnetometer surveys along the Martian surface. Our findings emphasise the importance of the magnetic properties of meteorites and of taking their implications into account in planning magnetic instrumentation for use on future spacecraft missions.

References

Baumgärtel, K., Sauer, K. and Bogdanov, A.: 1994, A Magnetohydrodynamic Model of Solar Wind Interaction with Asteroid Gaspra, *Science* **263**, 653–655.

Carr, M.H.: 1990, Mars, in *The New Solar System*, Third edition, J.K. Beatty and A. Chaikin (eds.), Cambridge University Press & Sky Publishing Corporation, 53–64.

Chapman, C.R.: 1976, Asteroids as meteorite parent-bodies: the astronomical perspective, *Geochim. Cosmochim. Acta* **40**, 701–719.

Cisowski, S.M.: 1985, Magnetism of Shergottite Meteorites, in *Lunar and Planetary Science XVI*, Suppl. **A**, 9–10.

Cisowski, S.M.: 1986, Magnetic Studies on Shergotty and Other SNC Meteorites, *Geochim. Cosmochim. Acta* **50**, 1043–1048.

Collinson, D.W.: 1986, Magnetic Properties of Antarctic Shergottite Meteorites EETA 79001 and ALHA 77005: Possible Relevance to a Martian Magnetic Field, *Earth Planet. Sci. Lett.* **77**, 159–164.

Delaney, J.S., Nehru, C.E., Prinz, M. and Harlow, G.E.: 1981, Metamorphism in mesosiderites, *Proc. Lunar Planet. Sci.* **12B**, 1315–1342.

Hiroi, T., Pieters, C.M., Zolensky, M.E. and Lipschutz, M.E.: 1993: Evidence of Thermal Metamorphism on the C, G, B and F Asteroids, *Science* **261**, 1016–1018.

Kivelson, M.G., Bargatze, L.F., Khurana, K.K., Southwood, D.J., Walker, R.J. and Coleman Jr., P.J.: 1993, Magnetic Field Signatures Near Galileo's Closest Approach to Gaspra, *Science* **261**, 331–334.

Kukkonen, I.T. and Pesonen, L.J.: 1983, Classification of Meteorites by Petrophysical methods, *Bull. Geol. Soc. Finl.* **55**, 157–177.

Laul, J.C.: 1986, The Shergotty Consortium and SNC meteorites: An overview, *Geochim. Cosmochim. Acta* **50**, 875–887.

Parasnis, D.S.: 1979, *Principles of Applied Geophysics*, Third Edition, Chapman and Hall, London, New York, pp. 275.

Pesonen, L.J., Terho, M. and Kukkonen, I.T.: 1993, Physical Properties of 368 Meteorites – Implications Meteorite Magnetism and Planetary Geophysics, *Proc. NIPR Symp. Antarct. Meteorites* **6**, 401–416.

Pieters, C.M. and McFadden, L.A.: 1994, Meteorite and Astreroid Reflectance Spectroscopy: Clues to Early Solar System Processes, *Annu. Rev. Earth Planet. Sci.* **22**, 457–497.

Terho, M., Pesonen, L.J., Kukkonen, I.T. and Bukovanská, M.: 1993, The Petrophysical Classification of Meteorites, *Studia geoph. et geod.* **37**, 65–82.

Wood, J.A.: 1988, Chondritic meteorites and the solar nebula, *Annu. Rev. Earth Planet. Sci.* **16**, 53–72.

THE POPULATION OF NEAR-EARTH OBJECTS
DISCOVERED BY SPACEWATCH

T. GEHRELS AND R. JEDICKE

Lunar & Planetary Laboratory
University of Arizona, Tucson, AZ, 85721

Abstract.

In the past three years the Spacewatch program at the University of Arizona's Lunar and Planetary Laboratory has discovered ~45% of the new Earth Approaching asteroids and found evidence for an unheralded population of small (~10m) objects in the inner solar system. This success is due to the automated Moving Object Detection Program (MODP) which searches successive scans over the same region for objects showing consistent motion. Highlights of recent discoveries, an update on research, and the development and potential of the new 1.8m Spacewatch facility will be discussed.

1. Introduction

In 1983 Spacewatch began the first long term CCD-based discovery program for Near Earth Asteroids (NEA) utilizing the Steward Observatory's 0.91m $f/5$ Newtonian at Kitt Peak. The RCA SID 52612 512×320 pixel CCD used at the outset lacked the areal coverage necessary for detection of the sparsely distributed and faint NEAs. In 1988 a thick, front-illuminated, Tektronix TK2048SP 2048×2048 pixel CCD was delivered and the software required for real-time analysis of the images was developed (Rabinowitz, 1991). The first automated discovery of an NEO (1989 UP) was on the 27th of October, 1989, and since that time Spacewatch has found over 75 NEAs, two Centaur class asteroids, two comets, and recorded astrometric positions for over 50,000 main belt asteroids. Yet another twofold improvement in the NEA discovery rate was achieved in 1991 with the delivery of a thinned,

Earth, Moon, and Planets **72**: 233-242, 1996.
© 1996 *Kluwer Academic Publishers.*

back-illuminated, Tektronix TK2048EB1 2048×2048 pixel CCD with twice the quantum efficiency of the earlier Tektronix device.

Perhaps the most interesting Spacewatch discovery, and certainly the most controversial, was an indication for the existence of a NEA belt (Rabinowitz, 1993) populated by objects with orbital elements similar to the Earth's ($a \sim 1$, $e \sim 0$, and $i \sim 0$). If these objects are confirmed, they represent an unusual and unexpected addition to the inventory of the solar system.

The Spacewatch program continues in earnest with plans to improve the discovery rate through revised detection techniques, improved hardware and software, and the construction of a new $1.8m$ Spacewatch facility at Kitt Peak.

2. Near-Earth Asteroid Detection

Spacewatch operates the telescope and CCD in a *drift scan* mode where the telescope drive is turned off during an exposure. Instead of tracking the telescope with the stars, the charges generated on the CCD representing the star's image are transferred across the face of the CCD at the same rate as that of the star's image. When the star's image passes off the CCD the charge is read out and almost instantaneously displayed on a computer screen. An image of the sky is built up during a scan which has a fiducial extent of about 32′ in declination and a practicable length of up to forty-five minutes in right ascension.

Each scan is repeated three times over the same section of sky. While the image is being read from the CCD and displayed on the computer screen, MODP (developed by D.L. Rabinowitz and J.V. Scotti) searches through the image seeking out point sources and also streaks which may be consistent with an interpretation as nearby Fast Moving Objects (FMO). The streak detection mode is efficient for bright objects moving with angular rates of motion (ω) greater than about 2°/day. Streaks are identified in real-time within each of the three passes of a scan and their locations are highlighted in order to notify the observer who decides on the credibility of the object.

In the range 0.02°/day$< \omega < $2°/day MODP detects objects through their consistent motion from pass to pass. Point sources identified in the second pass are matched with sources in the first pass. Most of these are stars which align exactly with one another except for a small offset (normally less than about 10″) in right ascension and declination. The offset is determined and corrected in software. When there is no direct match in the first pass for a source in the second pass, MODP searches (normally a radius of 120 pixels) for a second unmatched point source. When an un-

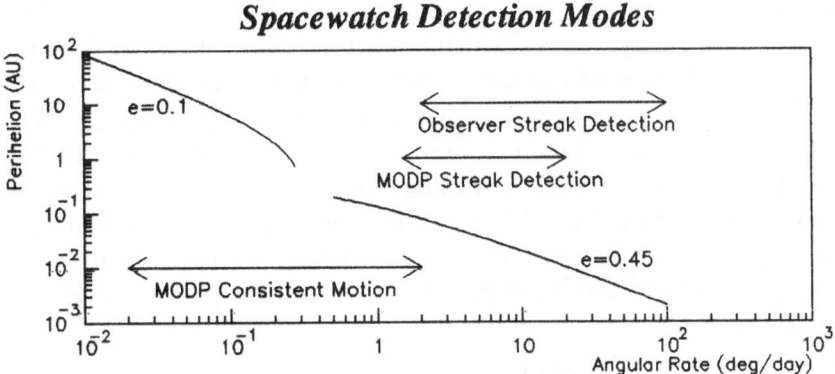

Figure 1. Spacewatch NEA detection modes as a function of rate of motion and distance. Rates were calculated for the eccentricity specified on the figure and for objects at perihelion and opposition.

matched point source is identified in both passes. MODP assumes that they are the same object, and predicts its location in the third pass assuming that it moves at a consistent rate. During the third pass, MODP matches triplets of otherwise unmatched point sources with consistent motion, and informs the observer of the location, brightness, and direction of motion of the new object.

A third mode of NEA detection relies entirely on the observer to identify faint and/or long trails missed by the automated streak detection. These Very Fast Moving Objects (VFMO) are typically within 0.2 AU of the Earth, are moving at $\omega > 2°/$day, and have signal to noise (S/N) in a single pixel between one and three. Their proximity to the Earth, small size, and extreme rate of motion make them exciting objects to follow during the few days in which they are visible.

Figure 1 indicates the modes of NEA detection used by Spacewatch as a function of the range in rate of motion and distance to the object. The distance has been calculated for objects found at opposition and at perihelion. In the range $0.01 < \omega < 0.3$ (°/day) the distance has been calculated for asteroids with $e = 0.1$ while for $0.5 < \omega < 100$ the eccentricity is set equal to the mean for the Spacewatch NEAs ($e = 0.45$). The limit at the lowest rate of motion is determined by the rate required for the object to move two pixels during the time interval between passes. The limit at the highest rate of motion is set by the ability to distinguish between meteors or faint satellites (which traverse the entire field of view in declination or about 2048 lines in right ascension) and VFMOs (which are required to begin and end within a single fiducial frame of 1900×2048 pixel area). At the highest rates of motion the system has been limited by the ability of the observer to identify the object and calculate its position at the middle of another scan though new software now automates this process as well.

The Spacewatch camera and software system enable efficient (>10%) detection of asteroids between apparent magnitudes of about 13 to 20. Brighter objects saturate the CCD image and make it difficult for the software to determine the centroid, while fainter objects are limited by the telescope, seeing, and lower limit set on S/N for identification of a source in a single pixel above the background level.

3. The Spacewatch Legacy

Since the first object was discovered in October 1989, Spacewatch has accumulated discoveries of over 75 NEAs. Figure 2 shows the distribution of eccentricity (e), semi-major axis (a), and inclination (i) of all the Spacewatch discoveries since that time. The curves in Figure 2a) indicate the $a - e$ combination which place the object's perihelion or aphelion at the mean heliocentric distance of an inner planet. An obvious clustering of the $a - e$ elements with perihelion near the Earth's orbit reflects the bias inherent to the Spacewatch search. In total, there are 31 Amor, 42 Apollo and 4 Aten asteroids. In addition, fifteen of the objects cross the orbits of Venus, Earth and Mars, while all four Atens cross the orbits of Mercury, Venus and Earth!

The raw distribution in absolute magnitude of the Spacewatch objects is shown in Figure 3a. It is bi-modal due to the different techniques, and therefore different range of apparent visual magnitude, used to identify the object. The faintest NEAs ($H > 22$) are almost entirely visual discoveries made by the observer, while the brighter objects are usually detected by MODP through their consistent motion from pass to pass within a scan. Assuming that C and S type asteroids are equally probable in the Spacewatch NEA data, and that their mean albedos are 0.050 and 0.155 respectively (Tholen *et al.*, 1989), the size distribution of these objects is shown in Figure 3b. It should be noted that since the albedo of the S type asteroids is $\sim 3\times$ greater than the albedo of the C type asteroids, any magnitude limited survey of the NEAs will be biased towards discovery of the S types (Luu & Jewitt, 1989). This bias factor is unknown for the Spacewatch data but would have the effect of shifting the size distribution to smaller objects in Figure 3b. It is clear from these distributions that the system is sensitive to asteroids over four orders of magnitude in size. One-third of the discoveries are of objects greater than about one kilometer in diameter.

Although Spacewatch is primed for the detection of NEAs it also excels

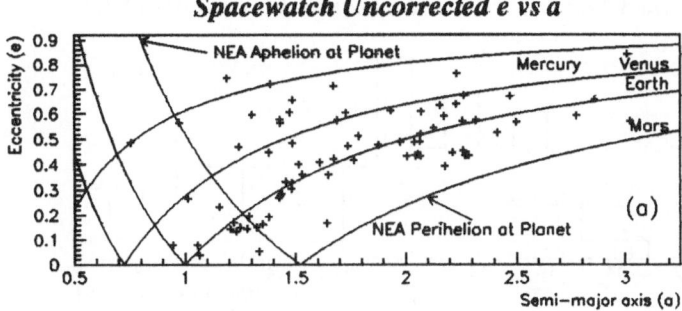

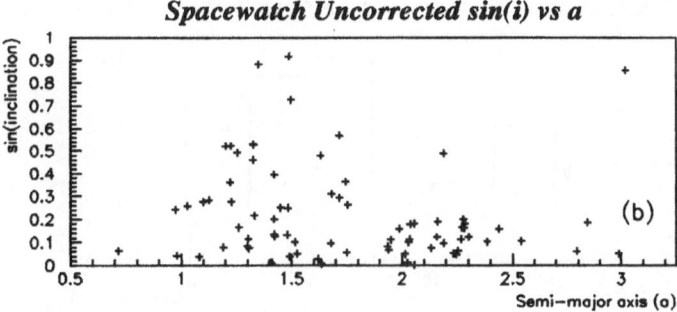

Figure 2. (a) Raw distribution of *e* vs. *a* for all Spacewatch discoveries. (b) Raw distribution of $sin(i)$ vs. *a* for all Spacewatch discoveries.

at detecting Main Belt (MB), Trojan, and other asteroids and has been a prolific provider of astrometric positions of these objects and comets to the Minor Planet Center. Most of this data (and a wealth of information on meteors) has lain unexplored even though it has the potential of providing a high statistics, faint object, single camera, multi-year, improvement in the understanding of the orbital and absolute magnitude distributions of various asteroid groups.

The data from the past two years (using the TK2K EB1) have been accumulated and circular orbits calculated for asteroid observation triplets in which the time separation between the first and third measurement is greater than one hour. Restricting the observations to within ±20° in ecliptic longitude of opposition yields orbits for over 20,000 MB objects which are good to about 5% in *a* and 25% in *i*. These fractional errors were determined by generating positions for an ensemble of MB asteroids, smearing them in two dimensions by the measured astrometric error for Spacewatch, and comparing the orbit calculated from the smeared positions to the actual

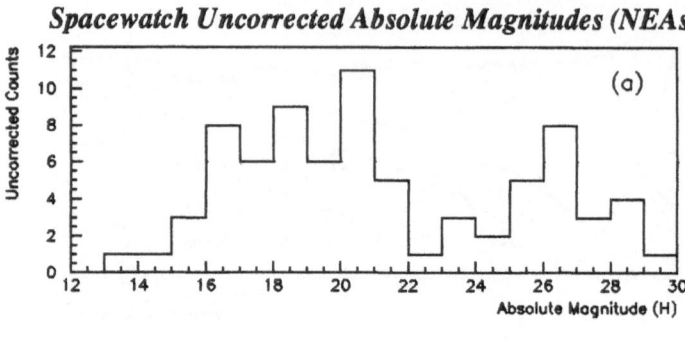

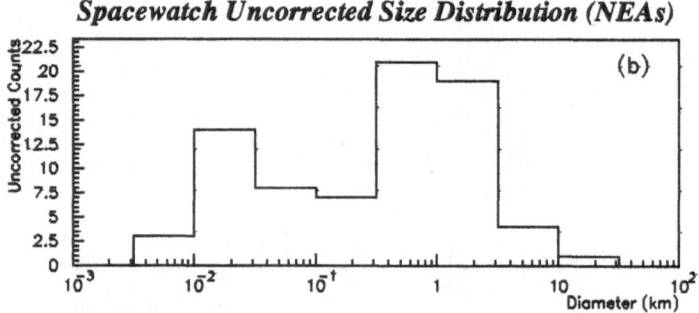

Figure 3. (a) Raw distribution of absolute magnitude (*H*) for all Spacewatch discoveries. (b) Size distribution of Spacewatch objects under assumptions outlined in text.

orbit. The contamination of these MB objects by mis-identified Trojans and NEAs is very small. Figure 4 illustrates the distribution of MB asteroids in $\sin(i)$ vs a space while Figure 5 shows the raw distribution of absolute magnitudes (*H*) for the same objects. The absolute magnitude is good to $\sim$0.7 magnitudes due to errors in Spacewatch's apparent magnitude measurement and the determination of the geocentric distance.

Despite the coarse resolution, the Eos group ($a \sim 3.01$, $\sin i \sim 0.17$) is visible and the Koronis and Themis groups are merged at low inclination ($2.83 < a < 3.24$). The ν_6 resonance is apparent as a cutoff in the distribution of asteroids in the upper left of the $\sin i$-a space. The superposed histogram of the semi-major axis distribution in Figure 4 shows the characteristic dip at the 3:7 resonance at about 2.9 AU.

The raw magnitude-frequency relation in Figure 5 shows that the Spacewatch camera is capable of detecting Main Belt asteroids over a range of twelve magnitudes corresponding to asteroids down to less than 2 km diameter in the MB. Debiasing of this data to compensate for observational

Spacewatch Uncorrected sin(i) vs Semi-major Axis (Main Belt)

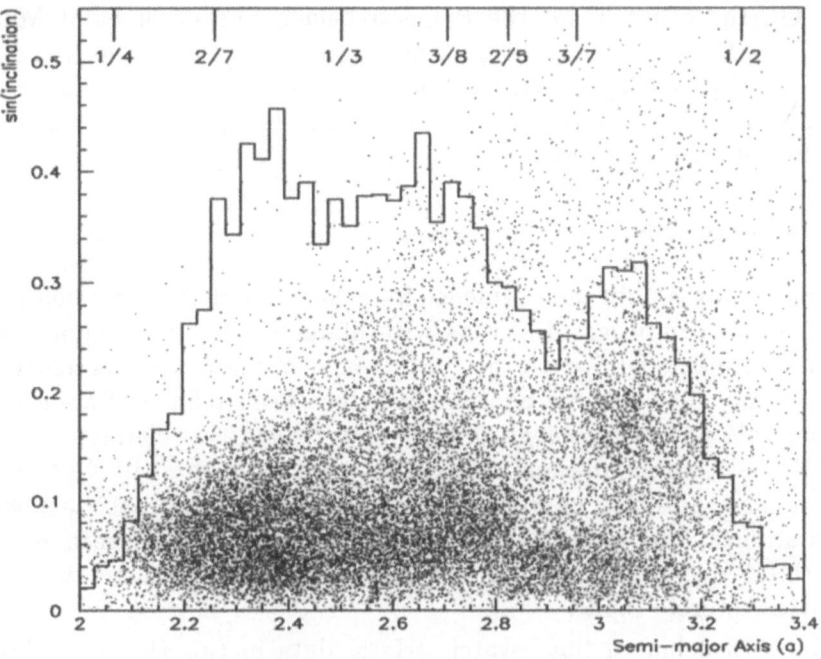

Figure 4.   Main Belt sin i vs a for Spacewatch discoveries since September 1991.

Spacewatch Uncorrected Absolute Magnitudes (Main Belt)

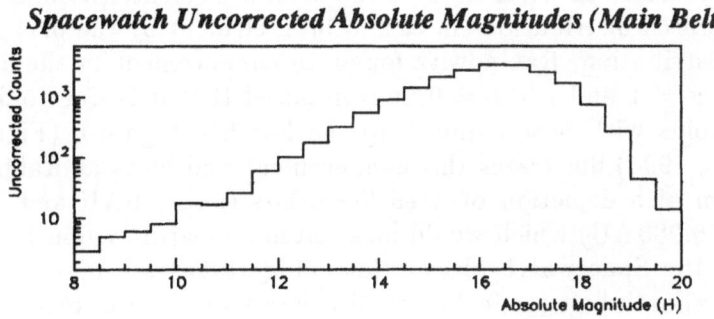

Figure 5.   Raw distribution of absolute magnitudes for Spacewatch discoveries since September 1991.

selection is currently in progress.

Magnitude-frequency relationships derived from the numbered asteroids are seriously affected by the attempt to debias the data from hundreds of observatories distributed across the world and observations which span a century. On the other hand, the Palomar-Leiden Survey of Faint Minor Planets (van Houten *et al.*, 1970) found good orbits for 1800 asteroids in about 216deg^2 of sky centred at the vernal equinox. The Spacewatch survey for NEAs has covered over 3000deg^2 and has more than ten times the number of asteroid orbits.

4. Evidence of a Near-Earth Asteroid Belt

In 1993 Spacewatch published evidence for a Near-Earth Asteroid belt (Rabinowitz, 1993) composed of objects with diameters less than about $50m$ and orbital elements similar to the Earth's in semi-major axis, eccentricity and inclination. The original analysis has since been refined (Rabinowitz, 1994) and this new work suggests that there exists two distinct populations of NEAs. One population corresponds to the 'traditional' NEAs with distributions in absolute magnitude and orbital elements consistent with previous studies. The second group of smaller asteroids are characterized by an orbital element distribution peaked in such a way as to suggest a Near Earth Asteroid 'belt'.

Rabinowitz debiased Spacewatch orbital data in (a,e,i)-space using a large, randomly generated, set of NEA-*like* orbits which were run through a software simulation of the detector system. 'Observations' of these simulated NEAs were recorded, and comparing the generated to the observed distribution in a, e or i provided a measure of the bias (efficiency) for detecting asteroids as a function of the orbital element. Dividing the actual Spacewatch distributions in each orbital element by the bias gave the debiased distribution. Rabinowitz found an enhancement in the $a - e - i$ space near $a = 1$ and $i = e = 0$ and proposed that it is due to a belt of small asteroids with orbits similar to the Earth's. Figure 6 (Figure 4 in Rabinowitz, 1994) illustrates this enhancement and hints at Rabinowitz's recent claim of a depletion of Aten-*like* orbits ($a < 1.0\,\text{AU}$ and aphelion distance $> 0.983\,\text{AU}$) which would most often appear in region D. He proposes that the Spacewatch observations (Figure 6a) are better modelled with a near-Earth belt (as in Figure 6b) than with a set of orbits statistically similar to the larger Earth approaching asteroid orbits (as in Figure 6c).

The controversial suggestion of a NEA-belt has generated interest in developing an understanding of a mechanism for producing objects in Earth-*like* orbits. Recent work by Bottke *et al.* seems to prohibit the production of Rabinowitz's small-NEAs by resonant delivery from the Main belt or through escaped ejecta from the surface of Mars. The only source of ob-

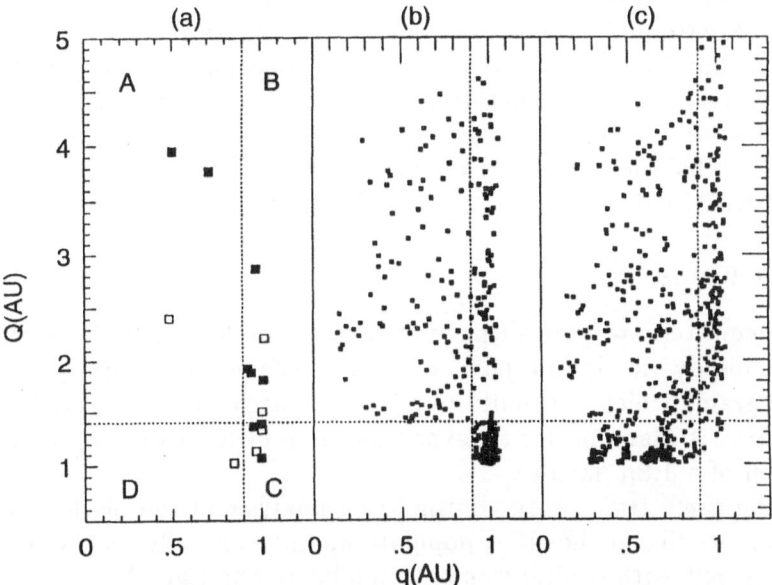

Figure 6.   Aphelion (Q) versus perihelion (q) for detected Small Earth Asteroids. Filled squares indicate asteroids which meet all the detection criteria of Rabinowitz's study. (a) Spacewatch observations (b) Simulated detections assuming a distribution of Earth-approaching orbits modified to include a near-Earth belt. (c) Simulated detections assuming the debiased orbit distribution of large Earth approachers.

jects consistent with Rabinowitz's claim is ejecta from the Earth-Moon system or from Venus, though they have not developed their model enough to determine the steady state distribution of this material as a function of the mass ejection rate from the planetary systems.

5. The Future of Spacewatch

While Spacewatch pioneered automated CCD searching for NEAs, and maintained a monopoly on the technique for many years, there are currently a number of systems in development which may supercede the existing camera within the next few years (LONEOS, GEODDS, Observatoire de la Côte d'Azur). All of these systems are based on wide field, multi-CCD image planes on telescopes of modest aperture ($\lesssim 1m$). At roughly the same time that these systems come on-line, Spacewatch will be ramping up operations on a new $1.8m$ $f/2.72$ telescope system also to be located on Kitt Peak. The smaller field-of-view of the $1.8m$ relative to the other planned systems is compensated by the increased light gathering power of the larger telescope. This will allow the added flexibility of performing deep stares in small areas or scanning (at non-sidereal rates) to cover a large area. In addition, real-time follow-up of the 1.8m telescope's discoveries will be possible with the existing 0.91m camera, and the larger telescope may be used for continued monitoring of objects which are beyond the detection limit of the other systems. Construction of the 1.8m altazimuth base, optical design and assembly, and preliminary tests of the optic path are proceeding well. A site has been selected on the Southwest Ridge (within the picnic area) of Kitt Peak.

6. Conclusion

The Spacewatch program's legacy of success is reflected in its track record of NEA discoveries in the past five years. While paving the path for future programs which will build upon Spacewatch's trials and tribulations, Spacewatch is planning for an expanding role in NEA discoveries with the operation of a 1.8m facility.

At the same time, work continues on analysis of the plethora of main belt data, studies of the NEA population, and especially on the possibility of a NEA belt with orbital elements similar to the Earth's.

References

C.J. van Houten, I. van Houten-Groeneveld, P. Herget, T. Gehrels, Astr. Astrophys. Suppl. **2**, p339-448, 1970.

D.L. Rabinowitz, Astrom. J. **101**(4), p1518-1559, April 1991.

D.L. Rabinowitz *et al.*, Nature 363, p701-706, 1993.

D.L. Rabinowitz, ICARUS **111**, p364-377, 1994.

J. Luu and D. Jewitt, Astrom. J. **98**(5), p1905-1911, November 1989.

Tholen D.J., Barucci M.A., 1989, Asteroid Taxonomy, in: Asteroids II, eds. R.P. Binzel, T. Gehrels, M.S. Matthews, The University of Arizona Press, Tucson, p. 298.

W.F. Bottke, Jr., M.C. Nolan, R. Greenberg, A.M. Vickery, J. Melosh: personal correspondence with W.F. Bottke.

ON THE EXISTENCE OF SMALL COMETS AND THEIR INTERACTIONS WITH PLANETS

J.C. BRANDT

Laboratory for Atmospheric and Space Physics, University of Colorado at Boulder

M.F. A'HEARN

Department of Astronomy, University of Maryland, College Park

C.E. RANDALL

Laboratory for Atmospheric and Space Physics, University of Colorado at Boulder

D.G. SCHLEICHER and E.M. SHOEMAKER

Lowell Observatory, Flagstaff, Arizona

and

A.I.F. STEWART

Laboratory for Atmospheric and Space Physics, University of Colorado at Boulder

Abstract. Arguments are presented for a substantial, unexplored population of comets with radii less than 1 km. Known examples confirm this population and extrapolation of any plausible size-distribution function indicates large numbers. However, their accurate numbers, orbital characteristics, and physical properties are unknown. Thus, even though the small comets may be the most frequent cometary bodies impacting the planets, a quantitative evaluation is not currently possible. We advocate an optimized, dedicated search program to characterize this population.

Key words: Comets, small; search program, planets

1. Introduction

We define small comets (SCs) as icy bodies with radii ≤ 1 km. In the past, SCs have been invoked, for example, to explain craters on the Moon and Ganymede, to supply a source of hydrogen to the interplanetary medium, and to supply volatile materials for the terrestrial planets; see Solomon (1991).

The general approach of inferring the numbers of SCs by effects ascribed to them has produced widely varying estimates, but has also served to heighten interest in SCs. The differential size-distribution function for comets is usually taken to be r^{-3}. In Figure 1, we show extrapolations to smaller sizes for r^{-2} and r^{-3}. Unless there is a real, physical cutoff in the distribution function close to the practical observational limit of $r = 1$ km, Figure 1 shows that almost any reasonable extrapolation to smaller sizes yields large numbers of SCs. As we show in the next section, there are direct observations of SCs which leave little doubt of their existence.

Earth, Moon, and Planets **72**: 243–249, 1996.

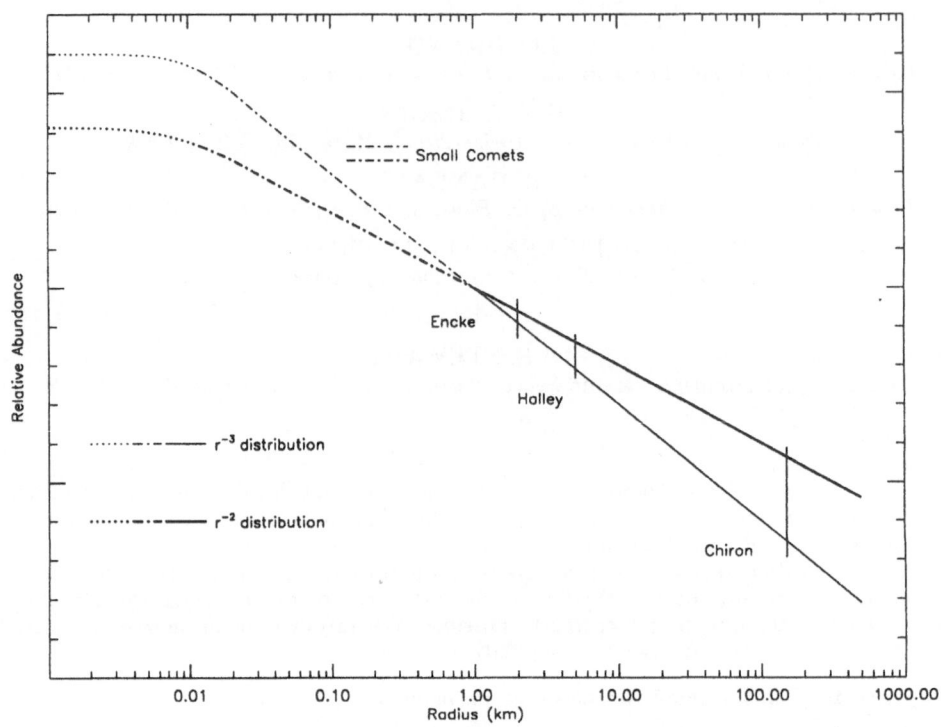

Fig. 1. Plausible Size Distribution of Comets

Our interests are in the small, icy bodies that are sublimating. Thus, we expect that SCs (as we define them) are surrounded by a coma of dust and gas, presumably produced by the sublimation of water.

2. Direct Observations of SCs

The idea of a major SC population is not new. In 1966, Roemer wrote: "The nuclei of most periodic comets appear to be less than 10 km in radius, and many may have radii of less than 1 km." This conjecture was based on a major study which Roemer (1966) carried out to determine the dimensions of cometary nuclei. One certain SC was found in this work, Comet Tuttle-Giacobini-Kresák (1962). This comet has a radius of 800 meters for an albedo of 0.02 (Roemer 1966). Occasionally SCs pass sufficiently close to earth that even smaller sizes can be determined with some accuracy. Two examples are:

(1) Comet Pons-Winnecke (1927). This comet has a radius of 310 meters or 210 meters for albedos of 0.05 or 0.10, respectively (e.g., Yeomans, 1991).

(2) Comet Sugano-Saigusa-Fujikawa (1983). Infrared, visual, and radar observations are consistent with an accurately determined radius of 370 meters (Hanner et al. 1987).

2.1. SUNGRAZERS

At times, SCs pass very near the sun and often do not survive perihelion passage. Fifteen such comets were observed by coronagraphs on the SOL-WIND and SMM spacecraft over a ten-year period. None of these comets were detected from earth while they were moving toward the sun. Marsden (1989) has noted that "...all must have brightened up tremendously as they approached the Sun, and it seems that it would have been quite impossible to detect them in a dark sky beforehand."

All of these sungrazing comets are probably members of the Kreutz sun-grazers (Marsden 1989) and this fact might suggest that, in some sense, they are not representative of SCs in general. However, work on the evolution of comet orbits indicates that Kreutz-type orbits may be a normal end state. Specifically, orbital integrations of intermediate period (or Halley-type) orbits show sungrazing to be a common end state for comets (Asher et al., 1994). Thus, the Kreutz sungrazers cited here as directly observed examples of SCs are not necessarily extraordinary.

2.2. SCs STRIKING THE EARTH

Use of meteoroids impacting the earth's atmosphere may seem to stretch the idea of direct observation a bit, but the results appear to be quite solid. Statistical data on meteoroids leads Ceplecha (1994) to conclude that "...the majority of 10m size bodies are cometary bodies of the weakest known structure." Earlier, Ceplecha (1992) had noted that "the almost unknown bodies in the 10 to 100m size range prove to be the second most important contributor to the Earth mass and their study deserves the fullest attention in the near future." This gap is beginning to be filled in for asteroids by the Spacewatch Project (Rabinowitz 1993), but perhaps not for cometary bodies.

Ceplecha's (1992, 1994) work on meteoroids immediately leads to the conclusion that 10-meter radius cometary bodies are plentiful, and Ceplecha (Mariehamn Meeting and private communications) finds that our guess on numbers of SCs is roughly consistent with the meteor data. Specifically, in his summary of "Meteoroids" for the Mariehamn Conference, Ceplecha writes "As Brandt *et al.* posed it, we should try to learn more about small comets with radii less than 1 km: their guess of 1×10^6 to 5×10^7 10-m-size comets in the shell from 1 to 2 AU is not far from the observed impacts of these bodies onto Earth, if strong ecliptic concentration and somewhat steeper size distribution are assumed".

We stress that our guess is quite crude and based on the simple assumption that the number of SCs increases by 10^3 per decade, i.e., that there would be 10^3 more 100-meter comets than 1-km comets, etc. Any detailed estimate of SC numbers would be based on an integration over a specific distribution function. In any event, these examples of direct observations of SCs and straightforward deductions from meteoroid statistics show that the existence of SCs is not in question, but their numbers are unknown.

A significant issue is whether the 10–100 meter cometary bodies are sublimating bodies. Thus, we need to consider the time scales for completely outgassing icy bodies of different sizes and the mixture of pristine and extensively processed bodies at each size. We discuss this issue below.

3. Discussion

Despite considerable evidence for substantial numbers of SCs, an argument against this position is essentially that everyone knows they don't exist. Surely, the correct question is "What are the numbers of SCs (say) in the 100-meter size range?" Even if the numbers of SCs are much lower than those estimated by us, a definitive determination of the size-distribution function for SCs would provide direct evidence for the processes producing the turnover in the distribution function.

There are some theoretical reasons for doubting the existence of SCs. On the theoretical side, for example, Whipple (1994, private communication) sees difficulties in getting the SCs out of the formation region (see also Stern 1990), hazards in the Oort cloud, and difficulties in getting them back into the inner solar system. While there certainly are processes that destroy SCs, other processes such as collisions and splitting create SCs. Thus, it is the balance that is important.

A specific example of a current creation process is the cloud fragments constituting comet P/Shoemaker-Levy 9. The crater-producing comets near Jupiter may be a useful surrogate for the global population. If so, the size distribution of craters has implications for the size distribution of comets. At the poles of Ganymede, the r^{-3} differential distribution holds down to about 2 km (Shoemaker *et al.* 1982). This implies the same distribution in the radii of the impacting cometary bodies down to 100 to 200 meters (using the usual scaling laws).

The time required for devolatilization of bodies of cometary origin with different sizes is also relevant. If the SCs lose their volatiles, search strategies based on the existence of a coma would fail. An advance in our understanding of comets since the 1986 apparition of comet Halley is the realization that usually only a small fraction of the surface of the nucleus is involved in sublimation. Presumably, this situation occurs because a dust mantle or crust commonly forms (Thiel *et al.*, 1990; Rickman *et al.*, 1990). The outgassing rate may be reduced by as much as several orders of magnitude and the time for complete loss of volatiles greatly extended over previous estimates (e.g., Weissman 1980). An important question is whether a size threshold exists below which a protective crust does not form. Observations are still needed and could advance our understanding of the devolatilization time scales for cometary nuclei.

There are also some observational reasons for doubting the existence of SCs. For example there is concern that the Spacewatch Survey should be finding the SCs if their numbers are substantial. We note that the Spacewatch program is not optimized for SCs (Morrison 1992) and in fact is not finding many comets, either large or small (Scotti 1994, Jedicke and Gehrels 1995). These remarks should not be taken as a criticism of the Spacewatch program. Their searches are optimized for objects that pose a threat to earth, not for SCs. Nuclei of 100-meter comets would have V magnitudes in the range 22.2 to 22.9. This is well outside the usual range for detection by Spacewatch and there is no conflict whatsoever. Nevertheless, understanding the reasons for Spacewatch's low cometary discovery rate is important and could help optimize strategies and techniques for detecting SCs.

A potentially strong argument has been put forward by Sekanina and Yeomans (1984). In essence, they argue that, with improved observing techniques, we should have picked up more intrinsically faint, but close-approaching comets than before. We agree with the thrust of the argument, but note that for the improved capability to be effective it must be applied in a systematic manner.

The number of SCs could be estimated from known distributions with corrections for incompleteness. A major study along these lines has been carried out by Hughes (1990). The problem is that large differences, attributed to a real physical cutoff, between the observed distribution and the distribution corrected for incompleteness occur just at the radius (≈ 1 km) corresponding to the effective observational limit; see Hughes (1990), Figure 10. The corrections are large and should be regarded with some suspicion. Hughes (1990) himself notes that "Small faint comets are difficult to see and many will pass the sun unnoticed." Also note that in Hughes' (1990) "corrected" distribution function (non-cumulative) the maximum lies below 1 km at about 300 meters. Our basic point is not, however, whether the final distribution function adopted by Hughes is essentially correct, but that the issue can be settled by observational work that could provide valuable new information on the origin of comets, aging processes, the importance of splitting and collisions, etc.

Essentially the same conclusion was reached by Parker *et al.* (1990). Their Abstract summarizes the issue succinctly. "The distribution of absolute magnitudes for comets suggests a cutoff close to the point where we would expect the numbers found to diminish due to observational effects. A number of authors consider this cutoff to be real but the data does not yet permit a definitive statement. We have embarked on a long term photographic survey which will push the observational limits fainter by several magnitudes and thus allow a resolution of the question of whether there is a sharp cutoff in cometary numbers below a particular absolute magnitude." Our approach (Brandt *et al.*, 1995) differs only in favoring a CCD survey that utilizes the properties of SCs, viz. extended source, motion, and characteristic spectrum.

4. Conclusions

Our interpretation of the evidence currently available is that SCs surely exist and that a major population is likely. We believe that the best approach to their study should emphasize direct detection, and the techniques optimized for icy, sublimating bodies. By characterizing the population of SCs – number, orbital parameters, and ultimately physical properties – we can learn a great deal about the formation and evolution of comets and, in addition, definitively evaluate the role of SCs in planetary interactions. If the numbers of SCs are as high as we suspect, the majority of planetary interactions with comets are produced by SCs. Knowledge of their present size distribution will also help us understand their aging processes and their importance in the past.

5. Acknowledgments

We thank Zdenek Ceplecha, Fred Whipple and Donald Yeomans for their comments on the subject of this paper.

References

Asher, D.J., Bailey, M.E., Hahn, G., and Steel, D.I.: 1994, *M.N.R.A.S.* **267**, pp. 26-42.
Brandt, J.C., A'Hearn, M.F., Randall, C.E., Schleicher, D.G., Shoemaker, E.M., and Stewart, A.I.F.: 1995, ASP Conference Series volume, *Completing the Inventory of the Solar System*, Bowell, E.L.G., Ed., in press.

Ceplecha, Z.: 1992, *Astron. Astrophys.* **263**, pp. 361-366.

Ceplecha, Z.: 1994, *Astron. Astrophys.* **286**, pp. 967-970.

Hanner, M.S., Newburn, R.L., Spinrad, H., and Veeder, G.J.: 1987, *Astron. J.* **94**, pp. 1081-1087.

Hughes, D.W.: 1990, in *Asteroids, Comets, Meteors III*, Lagerkvist, C.-I., Rickman, H., Lindblad, B.A., and Lindgren, M., Eds., Uppsala University, pp. 327-342.

Jedicke, R. and Gehrels, T.: 1995, *this volume.*

Marsden, B.G.: 1989, *Astron. J.* **98**, pp. 2306-2321.

Morrison, D. (Ed.): 1992, *The Spaceguard Survey*, Jet Propulsion Laboratory, Section 5.6.2.

Parker, Q.A., Hartley, M., Russell, K.S., and McNaught, R.: 1990, *Asteroids, Comets, Meteors III*, Lagerkvist, C.-I., Rickman, H., Lindblad, B.A., and Lindgren, M., Eds., Uppsala University, pp. 413-416.

Rabinowitz, D.L.: 1993, *Astrophys. J.* **407**, pp. 412-427.

Rickman, H., Fernández, J.A., and Gustafson, B.Å.S.: 1990, *Astron. Astrophys.* **237**, pp. 524-535.

Roemer, E.: 1966, *Mém. Soc. Roy. Sci. Liège* **12**, pp. 23-28.

Scotti, J.V., 1994, in *Asteroids, Comets, Meteors 1993*, Milani, A., DiMartino, M., and Cellino, A., Eds., Kluwer Academic Publishers, Dordrecht, pp. 17-30.

Sekanina, Z. and Yeomans, D.K.: 1984, *Astron. J.* **89**, 154-161.

Shoemaker, E.M., Lucchitta, B.K., Wilhelms, D.E., Plescia, J.B., and Squyres, S.W.: 1982, in *Satellites of Jupiter*, D. Morrison, Ed., University of Arizona Press, Tucson, pp. 435-520.

Solomon, S.C.: 1991, *Rev. Geophys. Supp.*, **April**, pp. 1089-1109.

Stern, S.A.: 1990, *B.A.A.S.* **22**, p. 1099.

Thiel, K., Kölzer, G., Kochan, H., Grün, E., Kohl, H., and Hellmann, H.: 1990, in *The 20th Lunar and Planetary Science Conference*, Lunar and Planetary Institute, Houston, pp. 389-399.

Weissman, P.R.: 1980, *Astron. Astrophys.* **85**, pp. 191-196.

Yeomans, D.K.: 1991, *Comets*, Wiley Science Editions, New York, p. 246.

POSSIBILITY OF SELF-SUSTAINING BOMBARDMENT
OF INNER PLANETS

E.M. DROBYSHEVSKI

A.F. Ioffe Physico-Technical Institute, The Russian Academy of Sciences, 194021
St. Petersburg, Russia; E-mail: emdrob@drob.pti.spb.su

Abstract. The great strengthening the material undergoes under high confining pressure, and jet pattern of matter outflowing from large impact craters make possible the ejection of asteroid-size bodies from the Earth into space. The ejected bodies, after gaining energy in planetary perturbations, may fall back with a velocity higher than that of their ejection. This solves, in particular, the problem of shower bombardments with ~ 25 Myr interval (Drobyshevski, Sov. Astron. Let. 16(3), 193, 1990), and a question arises whether this process could become self-sustained, like a chain reaction, when secondary impacts release an energy higher than that of primary impact. Estimates show that such a possibility could have been realized for Mercury (Drobyshevski, Lunar Planet. Sci. Conf. Abstr. 23(1), 317, 1992) due to its low escape and high orbital velocities. Self-sustained bombardment can account for the loss of the silicate mantle from Mercury. The energy and angular momentum conservation laws imply that its orbit contracted toward the Sun in the course of ejection of the mantle fragments by Mercury's perturbations beyond its orbit. Straightforward calculations show the initial orbit to have practically coincided with the Venusian orbit. This puts the old hypothesis of Mercury being a lost satellite of Venus on a solid ground and provides an explanation for many facts from the origin of the Imbrium bombardment to the observed locks in the axial and orbital rotation of Mercury, Venus, and the Earth.

1. Introduction.
On the Problem of Solid Body Ejection from Planets

The composition of eight basaltic SNC meteorites indicates that they were formed in a planet no smaller than the Moon, with high contents of volatiles (McSween, 1985). Ten Moon meteorites are also known with masses about 40 times smaller than SNCs. Mars is believed to be the most likely source of SNCs. However, Martian scenario faces a number of problems: the mass ratio SNC/Moon meteorites should be opposite to the observed, there are no young craters of needed size on Mars, etc.

Calculations of rock ejection from impact craters, entrained by gaseous products of impact heating/evaporation of both the target and the impactor, show that only fragments of $\phi = 5-10$ m could be ejected from Mars (Vickery, 1986), whereas the cosmic ray irradiation of SNC matter requires $\phi \geq 15$ m (Wetherill, 1984). The problem is mainly the low efficiency of shock heating as the shock wave carries greater part of the impact energy into the target. On the other hand, Shoemaker (1977) and O'Keefe and Ahrens (1977) noted that experiments usually give greater ejection velocities than calculated. Collision experiments at velocities of 4–5 km/s show

Earth, Moon, and Planets **72**: 251-255, 1996.

bright flashes (Kondo and Ahrens, 1983; Drobyshevski *et al*, 1990) indicating the presence of overheated gas in such regimes when, by calculations, liquid phase would just appear. The averaged hydrodynamic description obviously fails in such conditions. Moreover, going from laboratory energy scale of $\sim 0.01-1$ MJ and even nuclear explosion energies (~ 1 Mt TNT) to meteoroid impacts ($m \gtrsim 10^{15}$ g, $V \gtrsim 20$ km/s) with energies of several Tt TNT one is sure to find mechanisms that would cause rapid local dissipation leading to evaporation of matter (Drobyshevski, 1995). From this point of view all existing calculations of large-scale impacts can only provide estimate of lower limit for ejection efficiency.

2. Possibility of Asteroid-Size Body Ejection from Planets

Gasdynamic calculation of impact crater formation yields parameters of the jet flowing out of the crater. A body of diameter ϕ will be entrained by the jet with acceleration

$$a = \frac{3}{2} \frac{C_x}{\rho_r \phi} \rho_g (V_g - V_r)^2 \tag{1}$$

where C_x is the gas drag coefficient, V is velocity, ρ is density, and r and g are indices for rock and outflowing gas (e.g. Vickery, 1986). The maximum possible acceleration depends on the body size and material strength $a_{max} = \sigma_{max}/\rho_r \phi$, and effective acceleration path can be defined as $S_{eff} = V_{r\,max}^2/2a_{max}$. S_{eff} can be related to some specific length of the problem, e.g. to crater diameter D. Then the maximum size of a body accelerated to $V_{r\,max}$ is

$$\phi_{max} = \frac{2 S_{eff} \sigma_{max}}{\rho_r V_{r\,max}^2} \tag{2}$$

Vickery (1986) considered entrainment of boulders lying on the surface by gas expanding hemi-spherically from a crater. Assuming $D = 30$ km and $\sigma_{max} = 1$ kbar, she found $V_{r\,max} = 5.2$ km/s for boulders of $\phi = 5-10$ m; $S_{eff} \simeq D/10$.

Laboratory tests show that strength of rock increases linearly under confining pressure up to $p \sim 50$ kbar to the values of $\sigma_{max} \sim 30-100$ kbar (Beresnev and Trushin, 1976; Kinsland and Bassett, 1977; Meada and Jeanloz, 1990). The increased strength of fragments and jet-like outflow of compressed gas from impact craters (with the vertex angle $\sim 50°$ by Drobyshevski *et al*, 1990) allow increasing of S_{eff} to $\sim D/3$. Then, since $D = KW^{1/3.4}$ cm (where W is the impact energy in ergs and $K = 0.016$ (Shoemaker and Wolfe, 1982)), for an impact with energy $W \sim 3 \times 10^6$ Mt TNT

($D \approx 60$ km) one can find from Eq. (2) that a fragment with $\phi \simeq 1$ km can be accelerated to $V_{r\,\text{max}} \simeq 11.5$ km/s!

Thus, impacts with $W \geq 10^6$ Mt TNT are capable of ejecting asteroids from Earth-type planets (Drobyshevski, 1990)! That resolves contradictions of the Martian origin of SNCs (some of which may be from Earth, as gas contents and isotopic shifts in them are similar to those in some Earth rocks), and provides additional source for NEAs, explaining the 100-fold excess of bodies with $\phi \leq 10$ m and their small eccentricities (Rabinowitz, 1993) by a recent impact with moderate energy. One might expect to find meteorites of acid and sedimentary earthy rocks, even containing fossils (Drobyshevski, 1990).

Since fragments ejected into space can gain from planetary perturbations energy sufficient for leaving the Solar System, some of them can fall back onto the Earth having much greater energy within the following 1–3 Myr, causing new "nuclear winters", i.e. the step-wise pattern of a mass extinction following an initial great extinction due to original impact. From this point of view the 25–30 Myr intervals between extinctions are of stochastic nature (Drobyshevski, 1990) and hardly should be periodic (Yabushita, 1994).

The presence of secondary craters with $D \simeq 20-30$ km on the Moon, Mars etc also favors the hyperimpact ejection of km-size bodies (I am grateful to W.M. Napier for drawing my attention to this fact), as does the discovery by Binzel and Xu (1993) of ~ 10 km chips of Vesta. Recently Melosh (1993) and Bottke et al (1994) also argued in support of rock ejection from Earth and Venus.

3. Self-Sustaining Avalanche-Like Bombardment of Planets

As fragments falling back onto a planet after long evolution usually have energies greater than that with which they left the planet, the question is: could the back-infall energy exceed that of the primary impact, to make the impact-ejection process self-sustaining? For this to happen, at least two conditions must be satisfied (Drobyshevski, 1992a,b):

(1) The energy of secondary impact $W_{\text{sec}} = \frac{\pi}{12}\phi_{\text{max}}^3\rho_r V_{\text{thresh}}^2$ must exceed that of the original impact $W = (D/K)^{3.4}$. This means that if the largest fragment falls back with velocity

$$V \geq V_{\text{thresh}} \approx \left(\frac{81}{2\pi}\right)^{1/2} \rho_r \frac{D^{0.2}V_{\text{esc}}^2}{K^{1.7}\sigma_{\text{max}}^{3/2}} \tag{3}$$

the process can be self-sustaining. For Venus $V_{\text{thresh}} = 105$ km/s, for Mercury only 7.2 km/s. As the orbital velocity of a planet is a natural scale for meteoroid impact velocity, for Mercury Eq. (3) can be satisfied.

(2) The time τ_w needed for the fragment to gain necessary energy from planetary perturbations, which by Chandrasekhar (1942) is

$$\tau_w = \frac{V_{\text{thresh}}^3}{32\pi g N G^2 M_p^2 \ln\left(D_0 V_{\text{thresh}}^2/GM_p\right)}, \tag{4}$$

should be comparable with the time $\tau_\sigma = 1/\left(\pi R_p^2 N V_{\text{thresh}}\right)$ of its return infall. Then $\tau_\sigma/\tau_w = 8g\left(V_{\text{esc}}/V_{\text{thresh}}\right)^4 \ln\left(D_0 V_{\text{thresh}}^2/GM_p\right)$. Otherwise, if $\tau_w \gg \tau_\sigma$ fragments fall with insufficient energy, and if $\tau_w \ll \tau_\sigma$ they leave the Solar system without falling on planets. Here: N and D_0 are volumetric concentration of planets and mean distance between them; M_p and R_p are the mean planet mass and radius; G is the gravitational constant; $g \approx 0.18$.

4. Impact Self-Destruction of Mercury. Some Inferences

For Mercury, $V_{\text{esc}} = 4.25$ km/s, so $\tau_\sigma/\tau_w \simeq 16(V_{\text{esc}}/V_{\text{thresh}})^4 \sim 1$. Therefore there are grounds to believe that numerous peculiarities of the Mercury-Venus system could be explained by the hypothesis of self-sustaining impact destruction of Mercury. Briefly, the consequences of this process are (Drobyshevski, 1992a,b):

(1) Loss of Mercury's silicate mantle: now 70% of the planet's mass falls to its iron core.

(2) If the original composition of Mercury was similar to that of the Earth and Venus, then its original mass was $M_{0\text{merc}} \approx 2.25 M_{\text{merc}}$.

(3) Considering the loss of energy and angular momentum by Mercury due to its gravitational sweeping-out of all its fragments, one can show that Mercury's original orbit was very close to that of Venus!

(4) I.e., Mercury is the lost Venusian satellite! This was critically analyzed by Van Flandern and Harrington (1976) and not rejected. The only difficulty remained was the cause of Mercury's orbit contraction. The previous paragraph provides the explanation.

(5) By Van Flandern and Harrington (1976), Mercury could be lost due to tidal evolution in the first 0.5 Byr. The loss of Mercury caused its self-destruction which roughly coincides with the Imbrium bombardment of the Moon ($\simeq 3.9$ Byr ago).

(6) As Venus scattered gravitationally some of the Mercury's fragments, its orbit also slightly contracted and thus fell out of precise 2/3 lock with the Earth orbital period.

(7) The basalt crust on Venus could be the result of its acid crust loss due to bombardment by Mercury's fragments.

The proposed mechanisms of impact ejection of large rock fragments from Earth-like planets and of self-sustaining impact process for inner planets resolve a number of old problems and link many facts and phenomena. It is clear, however, that this approach requires further refinement. Some of the problems are already evident:

(1) The very mechanism of ejection of large blocks from impact craters is not yet clear. One should seek additional processes for rapid dissipation of large-scale impact energy in order to increase the efficiency of the shocked rock evaporation (Drobyshevski, 1995).

(2) The inevitable impact self-destruction of inner planets makes impossible their formation in the framework of traditional nebular hypotheses. The alternative approach considers the Jupiter-Sun system as the limiting case of a binary star (Drobyshevski, 1978).

(3) It is essential to search for sub-Aten asteroids ejected from Mercury and Venus.

References

Beresnev, B.I., and Trushin, E.V.: 1976, *Process of Hydroextrusion*, Publ. Nauka, Moscow, 200 p. (in Russian).

Binzel, R.P., and Xu, S.: 1993, *Science* **260**, 186–191.

Bottke, W.F., Jr., Nolan, M.C., Greenberg, R., Vickery, A.M., and Melosh, H.J.: 1994, in Abstracts for *Small Bodies in the Solar System and their Interactions with the Planets*, Reprocentralen HSC, Uppsala, p. 18.

Chandrasekhar, S.: 1942, *Principles of Stellar Dynamics*, Univ. of Chicago Press.

Drobyshevski, E.M.: 1978, *Moon Planets* **18**, 145–194.

Drobyshevski, E.M.: 1990, *Sov. Astron. Lett.* **16**, 193–196.

Drobyshevski, E.M.: 1992a, *23 Lunar Planet. Sci. Conf. Abstr.* **1**, 317–318.

Drobyshevski, E.M.: 1992b, *Sov. Astron.* **36**, 436–443.

Drobyshevski, E.M.: 1995, *Int. J. Impact Engng.* **17** (in press).

Drobyshevski, E.M., Zhukov, B.G., Rozov, S.I., Sokolov, V.M., Kurakin, R.O., and Savelyev, M.A.: 1990, *Sov. Tech. Phys. Lett.* **16**, 468–469.

Kinsland, G.L., and Bassett, W.A.: 1977, *J. Appl. Phys.* **48**, 978–985.

Kondo, K., and Ahrens, T.J.: 1983, *Phys. Chem. Minerals* **9**, 173–186.

McSween, Jr., H.Y.: 1985, *Revs. Geophysics* **23**, 391–416.

Meada, C., and Jeanloz, R.: 1990, *Nature* **348**, 533–535.

Melosh, H.J.: 1993, *Nature* **363**, 498–499.

O'Keefe, J.D., and Ahrens, T.J.: 1977, *Science* **198**, 1249–1251.

Rabinowitz, D.L.: 1993, *Astrophys. J.* **407**, 412–427.

Shoemaker, E.M.: 1977, in *Impact and Explosion Cratering* (eds. D.J. Roddy, P.O. Pepin, R.B. Merill), Pergamon Press, pp. 1–10.

Shoemaker, E.M., and Wolfe, R.E.: 1982, in *Satellites of Jupiter* (ed. D. Morrison), Univ. Ariz. Press, Tucson, pp. 277–339.

Van Flandern, T.C., and Harrington, R.S.: 1976, *Icarus* **28**, 435–440.

Vickery, A.M.: 1986, *J. Geophys. Res.* **91**, 14139–14160.

Wetherill, G.W.: 1984, *Meteoritics* **19**, 1–13.

Yabushita, S.: 1994, in Abstracts for *Small Bodies in the Solar System and their Interactions with the Planets*, Reprocentralen HSC, Uppsala, p. 176.

DISCOVERY OF VERY SMALL ASTEROIDS BY AUTOMATED TRAIL DETECTION

A. MILANI AND A. VILLANI

Department of Mathematics, University of Pisa, Italy

AND

M. STIAVELLI

Scuola Normale Superiore, Pisa, Italy

Abstract. Faint trails left by asteroids on CCD frames can be detected at a signal to noise ratio ≤ 1. We propose the use of a suitable algorithm, applicable to both spaceborne and groundbased searches for small bodies.

The size distribution of very small asteroids is largely unknown; on the number of small comets we have very weak constraints. In the diameter range between 10 and 300 metres only a small number of Earth–approaching objects have been discovered by the Spacewatch camera. An estimate of the population in this size range would be important to study the source of Near Earth Objects, to understand the transition between asteroids and comets at small sizes, to allow absolute datation of asteroid surfaces by crater counting, and to estimate the frequency of Tunguska class impacts. We propose a direct measurement of the number density of very small objects in the asteroid belt by using the camera system of the already approved E.S.A. mission *ROSETTA* (Bar-Nun et al., 1993). The spacecraft will cross the entire asteroid belt on its way to a comet; if one of the on-board camera systems is left in operation during the cruise phase, a large number of objects will pass in the field of view. However, an assessment of the performance of such a system is required.

Very small asteroids (e.g. in the $10\,m$ of diameter range) are discovered by their *trail*, with a length of N_p pixels, on a CCD frame. This applies both to a groundbased telescope (such as the *Spacewatch camera*) and to a spaceborne camera, e.g. from *ROSETTA* cruising across the asteroid main belt. On each CCD pixel the signal/noise is decreased (*trailing loss*) by a factor $1/N_p$. The human eye is very effective in detecting long trails down

Earth, Moon, and Planets **72**: 257-262, 1996.

to a signal/noise (on each pixel) $\simeq 1$ (Scotti, 1994), but this requires well trained and dedicated observers to be at the telescope all the time and makes more difficult to estimate the completeness of a survey. In space full automation is essential because not enough downlink bandwidth is available to transmit all the CCD frames for processing on the ground, and the main constraint is the availability of sufficient on-board computing power.

We have developed an algorithm exploiting the knowledge of the rectilinear shape of the trail to reduce the equivalent trailing loss to a factor $1/\sqrt{N_p}$. The algorithm is presented below; let us first assess how relevant is such a reduced trailing loss to the performance of a camera in detecting small asteroids. The computation of the expected number of discoveries is done on the following assumptions: average relative velocity between an asteroid and the camera platform $10\,Km/s$; pixel size 30" × 30"; limiting visual magnitude 14 for an asteroid far enough to stay on one pixel in the integration time of $100\,s$ (this is a rather pessimistic estimate, given current camera technology). From this we derive: R_1 = distance at which the average relative velocity corresponds to 1 pixel in the integration time $= 6.9\,MKm$; D_1 = diameter of an asteroid appearing of the limiting magnitude at the distance $R_1 = 1\,Km$ (assuming average albedo). If L_1 is the flux of light on one pixel from an asteroid of diameter D_1 at a distance R_1, then for an asteroid at a distance R and of diameter D the flux is:

$$L = L_1 \left(\frac{D}{D_1}\right)^2 \left(\frac{R_1}{R}\right)^2 \times (trailing\ loss)$$

If the trailing loss is $1/N_p = R/R_1$, then $L = L_1$ (detection threshold) occurs for: $D/D_1 = \sqrt{R/R_1}$. Let us suppose the longest trail we can detect has $N_p = 1024$ pixels, that is the minimum distance is $R_0 = R_1/1024$, then the minimum size is $D_0 = 31\,m$. A trail longer than 1024 pixels in most cases can not be used because it ends outside the CCD frame.

If the trailing loss is $1/(k\sqrt{N_p})$, that is trails are detected with a gaussian statistics and a safety factor k (as we will see below, $k \simeq 1.5$ can be assumed), then $L = L_1$ occurs for:

$$\frac{D}{D_1} = \left(\frac{R}{R_1}\right)^{3/4} \sqrt{k}$$

and again at $N_p = 1024$ pixels, the minimum detectable diameter is $D_0 = 6.7\,m$. To assess how relevant is this reduction in minimum size for the discovery rate we need to use a population model; a simple model could be (*number of asteroids with diameter* $\geq D$) $= n(D/D_1)^{-\alpha}$ where the number n of asteroids of more than $1\,Km$ diameter is $\simeq 2 \times 10^6$ (already highly uncertain) and α –the slope parameter to be determined by the experiment– should be in the range between 2 and 4. If the trails can be detected only at an high ($\simeq 4$) signal/noise on each pixel, integrating the number discovered between diameters D_0 and D_1:

$$N_{discovered} = F_{swept} \frac{n\alpha}{4 - \alpha} \left[1 - \left(\frac{D_0}{D_1} \right)^{4-\alpha} \right] \qquad \text{(for } \alpha \neq 4) \qquad (1)$$

where F_{swept} is the fraction of the volume of the asteroid belt swept during the experiment (realistically, $F_{swept} \simeq 10^{-4}$). On the contrary, for statistical detection at a signal/noise level on each pixel $\simeq 4k/\sqrt{N_p}$:

$$N_{discovered} = \frac{F_{swept}}{k^{4/3}} \frac{n\alpha}{\frac{8}{3} - \alpha} \left[1 - \left(\frac{D_0}{D_1} \right)^{\frac{8}{3} - \alpha} \right] \qquad \text{(for } \alpha \neq \frac{8}{3}) \qquad (2)$$

As an example, if we assume $\alpha = 2.5$ then by (2) the statistical detection of trails allows $\simeq 1000$ discoveries while (1) allows $\simeq 300$ discoveries. For $\alpha = 3$, (1) gives $\simeq 600$ discoveries, (2) gives $\simeq 4500$ discoveries. The statistical detection of trails allows a much higher rate of discoveries; moreover this method allows much higher sensitivity to the value of the slope parameter α. The above estimates have been based on a spaceborne camera (with a very small aperture), but it is easy to rescale the computations for a *Spacewatch* class groundbased system, and even for the proposed *Spaceguard* system.

Let us discuss the procedure to detect a trail at low signal/noise. The first step is the *cleaning* of the frame; by this we mean removing from the digital image all the objects which can be detected at a high signal/noise: stars, galaxies, brighter asteroids as well as cosmic rays and other imperfections. The problem is, one such "object" is not simply a bright pixel, but (depending upon the scale) can have an extended halo (whether this is a physical halo, as for the galaxies, or a seeing disk, or a diffraction pattern, does not matter for our purpose). Imperfect removal of a relatively bright object results in an enormous number of spurious trail detections: e.g. an object with a peak signal/noise above 4 can have an halo above 1, and all the segments contained in the halo have an integrated signal above the detection threshold (see an example of a galaxy with a large halo in Fig. 1). We have experimented with several different solutions for this problem; the most effective method to clean the frame proved to be the following.

We first use an adaptive contrast enhancement filter to increase the signal/noise of the "halos". The filter we have used is an adaptive wavelet transform (Lorenz and Richter, 1993). The filtered image is used to generate a *mask* which is 1 above a threshold, and 0 otherwise (see Fig. 2). Then we "erase" all the objects from the original image by setting at the mean sky background value all the pixels with 1 in the mask. The resulting "empty" frame is still not suitable for trail detection because of local inhomogeneities of the sky background: thus we apply a further filtering stage, by subtracting from each pixel value the median of a small square around it. The end product of this procedure really looks like random noise, and indeed we assume the root mean square value σ of this stage as the noise RMS (the RMS of the original frame is much larger, being dominated by the brightest objects). However, the long trails at a low signal/noise are not removed by this procedure; whenever a segment of length N_p pixels has a sum S of the values on all the pixels such that $S > k4\sigma\sqrt{N_p}$, then it can be used as a tentative trail detection. A number of false detections (e.g. a few hundreds per frame) can be accepted since each tentative detection needs anyway to be confirmed by a matching detection on frames taken shortly before (after); we have found that a threshold at $4k\sigma\sqrt{N_p} = 6\sigma\sqrt{N_p}$ gives acceptable results.

However, to compare the sum of the values along a segment to some threshold value requires to compute these sums for all the segments in the frame: if the CCD has $N \times N$ pixels, there are $N^4/2$ such segments, and for $N = 2048$ the computational load is a main concern. We have found an algorithm which tests "almost all" the segments of the frame, and which has a computational complexity of the order of N^3. The segments are tested on all the possible straight lines crossing the frame, but only if their length is 2^j pixels (for some integer j). Then a binary sum algorithm requires only of the order of N operations (and of the order of N tests) per line. The $O(N^2)$ "lines" are prepared by a Bresenham (1965) algorithm.

Fig. 3 shows the output, with all the segments above the threshold $6\sigma\sqrt{N_p}$ marked. The two trails marked with an arrow are the "real" ones (the simulated asteroid trails added to the original frame). The shorter of the two has length $N_p = 30$ and an average signal $\simeq 1.3\sigma$; it is almost invisible in the printout, but a good observer should be able to spot it on a good screen (see Scotti, 1994).The longer simulated trail had $N_p = 200$ and an average signal $\simeq 0.5\sigma$; even knowing where it is, the human eye cannot detect it on the original frame (and not even on the "cleaned" figure). The conclusion is that automated trail detection allows either to discover asteroids fainter by a factor of at least 2 (with respect to visual detection), or to use a telescope with an aperture smaller by a factor $\sqrt{2}$.

Although an N^3 computational complexity is much better than an N^4 one, the computational load is still very severe. As an example for the comparatively small ($N = 512$) frame of the figures 1–3, our software requires $\simeq 1500$ CPU seconds on a state of the art workstation (IBM RISC 6000/350). Although much work remains to be done on the optimization of the algorithm and of the software, it is clear that a dedicated Digital Signal Processor needs to be used. We estimate that for a 2000 × 2000 CCD, if the exposure time is $\simeq 2$ minutes, an effective computing speed of $\simeq 200\,MFLOPS$ will be required to be able to process in real time all the

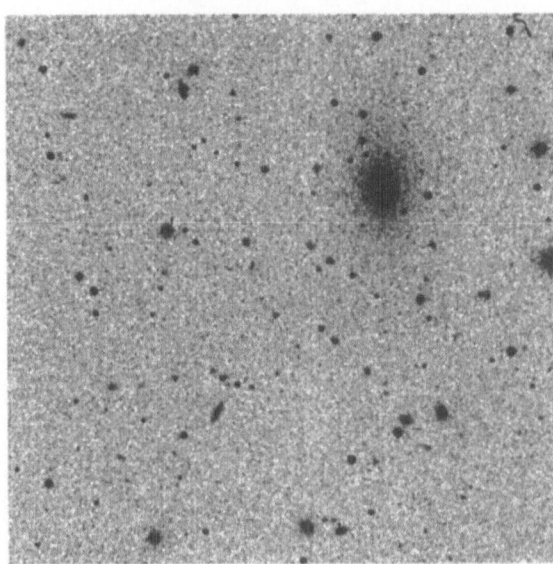

Figure 1. In this sample image, a 512 × 512 frame, there is a relatively big galaxy with a large halo. The RMS of the frame is $\simeq 560$ while the RMS of the background is $\sigma \simeq 28$. We added to the original image 2 tracks: one of length 200 and signal $\simeq 0.5\ \sigma$, and the other of length 30 and signal $\simeq 1.3\ \sigma$. A good observer could spot the latter on a good screen.

Figure 2. This is the mask obtained with the adaptive filter. The largest object is satisfactorily removed together with its halo.

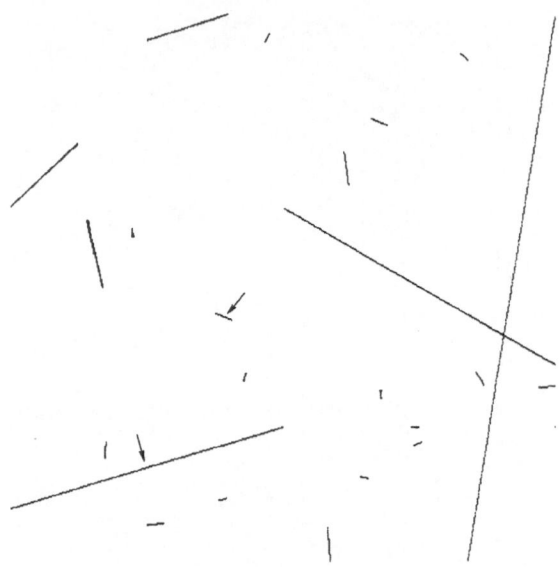

Figure 3.   The result of our algorithm: we have found the two simulated tracks and about 20 spurious tracks (for a total of 339 groups over the threshold of $6\sigma\sqrt{N_p}$).

frames. For a groundbased Spacewatch/Spaceguard observatory this is not a technological problem, also because the algorithm is easily parallelisable; we only need not to underestimate the computing power required by such a system. We hope to be able to test our algorithm under the real conditions of a Spaceguard–type survey as soon as suitable telescopes will be available.

For a spaceborne experiment, the challenge is to develop a dedicated, space qualified, DSP system. The components for such a system already exist and space qualification is under way; a preliminary study has already been performed by the avionics company LABEN. The proposal of a space experiment to study the size distribution of small asteroids/comets/meteoroids in the main asteroid belt consists of two components: the use of one of the already planned cameras on board ROSETTA during the cruise phase, and the installation of a DSP system capable of fully automated real time detection of trails of objects down to $\leq$ 10 metres in diameter.

References

Bar-Nun, A. et al.: 1993, 'ROSETTA Comet Rendezvous Mission', *ESA SCI*, **(93)7**.

Bresenham, J.E.: 1965, 'Algorithm for computer control of a digital plotter', *IBM Syst. J.*, **4(1)**, 25–30.

Lorenz, H. and Richter, G.M., 1993, 'Wavelet transform and adaptive filtering', *Proceedings of fifth ESO/ST-ECF Data Analysis Workshop, April 26-27, 1993*.

Scotti, J.V., 1994, 'Computer aided near earth object detection', *in: Asteroids Comets Meteors 1993, Milani, A. Di Martino, M. and Cellino, A. eds.*, Kluwer, 17–30.

ASTEROIDS APPROACHING THE EARTH FROM DIRECTIONS AROUND THE SUN

SYUZO ISOBE

National Astronomical Observatory, Mitaka, Tokyo 181, Japan

and

MAKOTO YOSHIKAWA

Communication Research Laboratory, Hirai, Kashima, Ibaraki 314, Japan

Abstract. We have estimated close asteroid encounters with the Earth by numerical integrations of a system with the Sun, 9 planets, and 188 near-earth-asteroids during the period 1994-4600. Asteroids approach the Earth from directions within 30° around the Sun in more than 20% of encounters with the closest distance less than 0.01 AU. Since ground-based observations cannot detect these objects, we should develop space-borne and/or lunar observatories in a short time to allow enough warning time before a catastrophic collision.

1. Introduction

There are many discussions relating to hazards caused by an asteroid collision with the Earth (for example, refer to the Space-Guard project by NASA). It would be a very difficult job to deflect an asteroid on a collision orbit with the Earth if it would be discovered only days before its collision. The asteroid 1991BA discovered in 1991 January 18.23 UT, 1991 passed at its nearest distance of 170,000 km to the Earth just half a day after its discovery. To keep enough warning time, it is the aim of the Space-Guard project that more than 99% of near-earth-asteroids with diameters larger than 1 km will be detected through a 20 year survey by the five newly built 2.5 m telescopes.

The Space-Guard project and/or the similar international projects should start first. However, even after a completion of those detection programs, there will still remain 1% of asteroids with diameters larger than 1 km and most of the asteroids with diameters smaller than 1 km that will be undetected. However if the Space-Guard system is in operation, it will detect asteroids at least several weeks before collision with the Earth.

There is a blind point in the Space-Guard system. If asteroids approach the Earth from directions around the Sun, they cannot be detected by any ground-based observations because of sky-brightness during the day-time. In this paper, we estimate the percentage of asteroids approaching the Earth from various directions to the Sun. If the number is high, we have to develop a new system other than the Space-Guard system to escape a hazardous event.

Earth, Moon, and Planets **72**: 263–266, 1996.
© 1996 *Kluwer Academic Publishers.*

2. Numerical estimates of close encounters of asteroids with the Earth

Numerical integrations for a system of the Sun, 9 planets, and 188 near-earth-asteroids have been carried out. A detailed explanation is given by Yoshikawa (1994). During the period 1994-4600, 6694 cases of close asteroid encounters with the shortest distance to the Earth within 0.1 AU are obtained. Certainly, some asteroids approach from the outside of the Earth's orbit, but some others come from inside. If an asteroid approaches from a direction quite near the Sun, one cannot detect it by ground-based observations, because of sky-brightness. As a working condition, we define this angular distance as 30° from the direction of the Sun. Table 1 shows a number distribution of close asteroid encounters with the Earth for each time period during which each asteroid is inside a 30° zone from the Sun in a period 10 days prior to the closest approach to the Earth.

Out of 6694 close encounters with the Earth within 0.1 AU, less than 5% are undetectable. However, if we limit close asteroid encounters to be within 0.01 AU, out of 94 close encounters one cannot detect asteroids in 22 cases with each time period longer than 8 days during which each asteroid is inside the 30° zone from the Sun. This fraction (over 20%) is rather large. Collisions of asteroids do not occur for all the cases, and the direction of each asteroid in ecliptic coordinates varies by a large amount at both sides of the closest approach. For a collisional case, an asteroid approaches directly to the Earth and hardly changes its direction.

Figure 1 shows 9 examples of asteroid distances from the Earth and angular distances from the solar direction, depending on the time and the asteroid motion on the celestial sphere in ecliptic coordinates. All of these asteroids cannot be detected prior to the closest approach. If one of these asteroids would be on a collisional orbit, we would have a very short warning time. In our numerical calculations we considered only 188 near-earth-asteroids with well-determined orbital elements in a time interval about 2,600 years. Since it is expected that there are 20,000 asteroids with diameters larger than 1 km and a much large number of asteroids smaller than 1 km, a number of asteroids are certainly on a collisional orbit. Yoshikawa (1994) estimated statistically that the mean collision interval of asteroids with diameter larger than 1 km is about 10^5 years and smaller asteroids have much shorter intervals. These asteroids are dangerous objects for humanity even if we complete the Space-Guard system.

TABLE I

Numbers and percentages of asteroids which are inside a 30° zone from the Sun in n days during 10 days prior to the closest approaches to the Earth.

	For cases with distances of the closest approach less than 0.1 AU		For cases with distances of the closest approach less than 0.01 AU	
Total number of encounters	6694	100%	94	100%
$n = 0$	5546	82.85	64	68.09
1	142	2.12	1	1.06
2	143	2.14	0	0.00
3	139	2.08	0	0.00
4	153	2.29	3	3.19
5	131	1.96	3	3.19
6	108	1.61	1	1.06
7	120	1.79	0	0.00
8	91	1.36	4	4.26
9	76	1.14	9	9.57
10	45	0.67	9	9.57

3. Necessity of space-borne and/or lunar base observations of near-earth objects

As shown in Sect. 2, we have a significant possibility for a collision of an asteroid approaching from a direction around the Sun. To escape this difficulty, we have two ways. One is to detect all the near-earth-asteroids and to determine their precise orbits well in advance of the collisions. However this task is very difficult even for the Space-Guard system which may miss some near-earth objects with specific orbits. To keep enough working time before collision, we need space-borne and/or lunar base observations. Although this is not an urgent task, we need these observational systems since the collision will happen within approximately 10^5 years for asteroids larger than 1 km and 10^3 years for asteroids larger than 0.1 km. This collision will produce a catastrophic event and may cause an extinction of humans. Therefore, it is very important for us to set up space-borne and/or lunar observatories as well as the Space-Guard system.

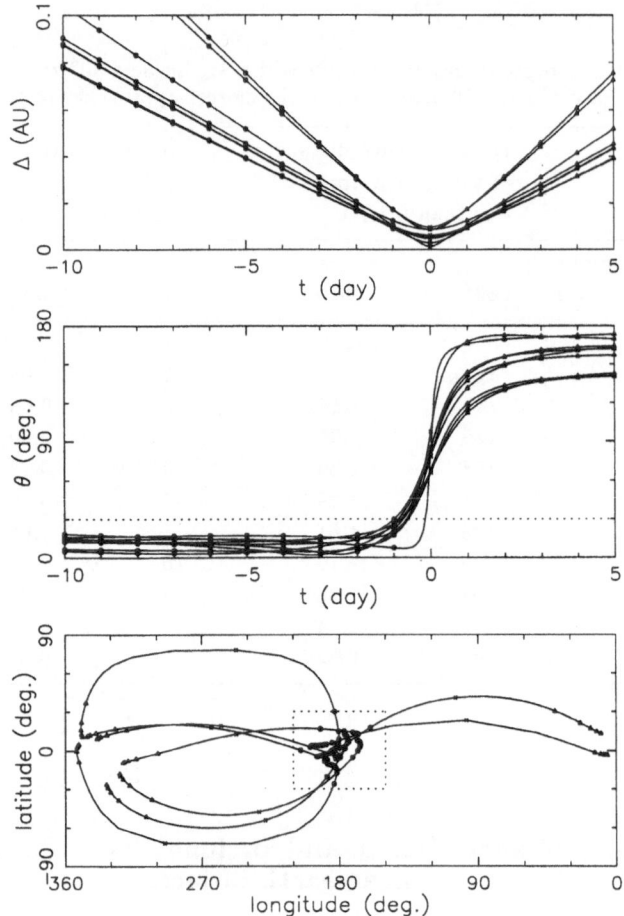

Fig. 1.   Diagrams for 9 near-earth-asteroids which are inside a 30° zone from the Sun all the 10 days prior to the closest approach of less than 0.01 AU to the Earth. Upper: Time dependence of distances from the Earth. Middle: Time dependence of each asteroid direction from the Sun. Lower: A trace of each asteroid on the celestial sphere. In all three diagrams, circles, crosses, and triangles are asteroid position before, at, and after those closest approaches.

References

Yoshikawa, M.: 1994, in *Seventy-Five Years of Hirayama Asteroids Families: The Role of Collision in the Solar System History*, eds. Y. Kozai, R. P. Binzel, and T. Hirayama (Astronomical Society of the Pacific, San Francisco), pp 28-38.

UESAC – THE UPPSALA-ESO SURVEY OF ASTEROIDS AND COMETS *

C.-I. LAGERKVIST
Astronomiska observatoriet, Box 515, S-751 20 Uppsala

O. HERNIUS
Observatory, P.O. Box 14, FIN–00014 University of Helsinki, Finland

M. LINDGREN
Astronomiska observatoriet, Box 515, S-751 20 Uppsala

and

G. TANCREDI
Depto. de Astronomia, Facultad de Ciencias, Tristan Narvaja 1674, Casilla de Correo 10773, 11200 Montevideo, Uruguay

Abstract. The Uppsala-ESO Survey of Asteroids and Comets was undertaken to find previously undetected comets in the vicinity of Jupiter.

Over 15000 positions of moving objects have been detected on 74 plates obtained from the European Southern Observatory in Chile and from the Anglo-Australian Observatory in Australia in 1992 and 1993. Two or more positions were secured for about 3300 asteroids and orbits have so far been calculated for 1944 asteroids. The main bulk of these asteroids are previously undetected.

We present absolute magnitudes and diameters for asteroids which have an accurate orbit. The magnitude and diameter distributions are compared to the results of the Palomar-Leiden Survey of Faint Minor Planets.

1. Introduction

The Uppsala-ESO Survey of Asteroids and Comets (hereafter UESAC) was started to find previously undetected comets in the vicinity of Jupiter (Tancredi and Lindgren, 1994). A more detailed report concerning the reductions can be found in Lagerkvist et al. (1995).

In this paper we present a short summary of the observations and the results with respect to the magnitude and size distribution. The present data are compared with results from the Palomar-Leiden Survey of Faint Minor Planets (van Houten et al., 1970; hereafter PLS.)

2. Observations

74 photographic plates and films were obtained from observing runs in 1992 and 1993 with the Schmidt telescopes at the European Southern observatory and the Anglo-Australian observatory.

* Based on observations collected at the European Southern Observatory, La Silla, Chile

Earth, Moon, and Planets **72**: 267–274, 1996.
© 1996 *Kluwer Academic Publishers.*

The 1992 and 1993 observing campaigns (UESAC'92 and UESAC'93) were carried out in a similar fashion. The investigated sky-field was covered with nine plate-fields. Since each plate-field covers a 5.4×5.4 square degrees large sky-field the total covered area in each campaign is approximately 16×16 square degrees. The central plate field was in both campaigns centered on Jupiter's position.

The two nights of observation scheduled for April 1993 at ESO were cancelled because of the lack of suitable plates. Some plates could, however, be taken with the Schmidt telescope at Siding Spring Observatory during April.

All plates and films were scanned manually at least three times. About 15000 positions were measured with the "Optronics" measuring system at ESO headquarters in Garching bei München. The PPM and SAO catalogues were used for the reductions.

Table I gives the results of the measurements at ESO. The mean standard deviation of the residuals in right ascension and declination for the reference stars are given for three concentric regions on each plate. Section 1 refers to the most central concentric region, section 2 to the intermediate and section 3 to the region reaching out to the edge of the plate.

TABLE I

Average quality of the plate reductions.

	$\overline{\sigma_\alpha}$	$\overline{\sigma_\delta}$	Ref. stars
1992 Section 3	0.814"	0.770"	72 - 50
1992 Section 2	0.607"	0.593"	69 - 37
1992 Section 1	0.352"	0.335"	43 - 22
1993 Section 3	0.956"	0.832"	38 - 25
1993 Section 2	0.676"	0.517"	33 - 19
1993 Section 1	0.410"	0.380"	26 - 19

2.1. LINKAGE OF OBJECTS

The linkings between the March exposures were done by trail-identification using the length, position, angle and brightness of the trails. The trails were linearly extrapolated from one exposure to the next and in most cases it was obvious which trails belonged to the same object. About 70% of the asteroids could be detected on two plates and about 45% on three plates. These figures are based on both UESAC surveys. The asteroids which were found only once or twice are generally faint and were lost mainly because of different seeing conditions during the different observing nights.

It was not possible to predict an asteroid's position on the April exposures by a linear extrapolation of the March trails. The asteroid motions in right ascension and declination can not be approximated with a straight line on longer time-scales. We added a fictitious fourth position, chosen as the continuation of the approximately linear March motion to the April date, and calculated an orbit using only the three March positions. The orbit determination program outputs residuals in right ascension and declination for the fourth fictitious position. An accurate estimate of the true April position could then be determined by subtracting these residuals from the fictitious April position. About 45% of the asteroids found on three March exposures could thus be recovered on the April plates.

3. Magnitudes

3.1. DETERMINATION AND ACCURACY OF APPARENT MAGNITUDES

The apparent magnitudes were estimated by assigning one of eight brightness classes to each object simultaneously with the measuring of the positions. The estimates were transformed to V-magnitudes by performing a linear data fit with the magnitude estimates for the observed numbered asteroids and the corresponding ephemeris V-magnitudes. The magnitudes will be improved when magnitude sequences are available in these fields.

The mean errors for the V-magnitudes are computed by comparing the magnitudes of the numbered asteroids included in the data fits with the corresponding ephemeris values. These errors are only valid for magnitudes up to V=18, since the data fits are mainly based on asteroids which have V< 18. Table II gives the mean differences (ephemeris minus UESAC) for the 1992 and 1993 observing campaigns separately. The standard deviations for the errors with regard to sign and the number of included asteroids are also given.

3.2. DETERMINATION AND ACCURACY OF ABSOLUTE MAGNITUDES

To calculate the absolute magnitude the formula

$$H = V - 5\log(r\Delta) + 2.5\log[(1-G)\phi_1 + G\phi_2] \tag{1}$$

was used (Bowell et al., 1989). r and Δ are the heliocentric and geocentric distances, respectively. V is the apparent magnitude and ϕ_1 and ϕ_2 are two phase functions. G is defined as the slope parameter, dependent on the albedo of the asteroid. For low and high albedo asteroids we have used G values of 0.09 and 0.22 respectively (Lagerkvist and Magnusson, 1990).

TABLE II

Mean apparent magnitude differences (ephemeris
minus UESAC) for the observed numbered asteroids

Plates	$\overline{|\Delta V|}$	$\overline{\Delta V}$	$\sigma_{\Delta V}$	N
March 1992	$0^m.51$	$+0^m.00$	$0^m.66$	56
March 1993	$0^m.51$	$+0^m.00$	$0^m.63$	67

The mean errors for the absolute magnitudes are determined by comparing the H-magnitudes of the observed numbered asteroids with the corresponding EMP/MPC H-magnitudes (EMP="Ephemerides of Minor Planets", MPC="Minor Planet Circulars"). Table III gives the mean differences
(ephemeris minus UESAC) for the 1992 and 1993 observing campaigns separately. The standard deviations for the errors with regard to sign and the
number of included asteroids are also given.

TABLE III

Mean absolute magnitude differences (ephemeris
minus UESAC) for the observed numbered asteroids

Plates	$\overline{|\Delta H|}$	$\overline{\Delta H}$	$\sigma_{\Delta H}$	N
March 1992	$0^m.49$	$-0^m.04$	$0^m.63$	49
March 1993	$0^m.45$	$+0^m.05$	$0^m.56$	45

3.3. STATISTICS OF ABSOLUTE MAGNITUDES

The statistics in this section are based on H-magnitudes for UESAC and
PLS asteroids for which there has been possible to calculate an orbit. For
some asteroids in UESAC'93 we used orbital elements published in Minor
Planet Circulars (Williams, 1994). In particular those then linkings to other
apparitions were made.

Figure 1 shows the logarithm of the cumulative numbers of absolute magnitudes per half magnitude interval for UESAC'92, UESAC'93 and PLS
asteroids. The PLS absolute magnitudes were taken from EMP/MPC files

and not from the values given in the paper by van Houten et al. (1970). We corrected the PLS numbers by a factor 0.9 to compensate for the difference in the sky area covered.

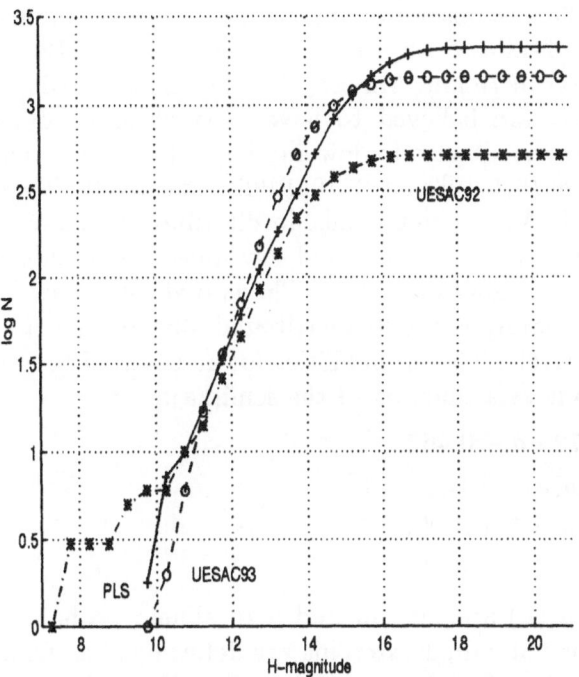

Fig. 1.   The logarithm of the cumulative number of asteroids as a function of absolute magnitude for UESAC'92, UESAC'93 and PLS (including numbered asteroids).

Note that the UESAC'93 asteroids are more numerous from about $H = 12$ to about $H = 15$. This is peculiar since the total number of asteroids is higher for PLS. This might be an artifact of the different magnitude calculation methods. It is interesting to note that if this is a true physical effect it implies that the number of observed asteroids is higher when a region close to an ecliptic longitude of 180° (the UESAC situation) is surveyed as compared to a survey concentrating on a region close to an ecliptic longitude of 0° (the PLS situation).

4. Asteroid diameters

The diameter calculation is based on, in part, randomly chosen albedos. The probability for an asteroid to have high or low geometric albedo was calculated as a function of the semimajor axis. Previous investigations of the ratio between different taxonomy classes as a function of heliocentric distance have, among others, been done by Zellner and Bowell (1977), Ishida et al. (1984) and Gradie and Tedesco (1982). We have made an independent investigation based on albedo values from the IRAS Minor Planet Survey (Tedesco et al., 1992; hereafter IRAS).

To avoid the effect of observational bias towards inner main-belt asteroids (and therefore high-albedo objects) we have only used IRAS albedo data down to a diameter of 70 km. Almost all main-belt asteroids with diameters larger than 70 km are believed to have been observed during the IRAS survey. The geometric albedos for low and high-albedo asteroids were chosen to lie in the range $[0, 0.10]$ and $[0.10, 1.00]$ respectively. These limits were chosen since the IRAS geometric albedo distribution shows a clear change at $p = 0.10$. The fraction of high and low albedo asteroids is assumed to be constant with changing diameters. The two albedo distributions have, if assuming approximately normally distributed albedos, mean values of 0.053 and 0.203, respectively. The probabilities for an asteroid to have low or high geometric albedo is as a function of the semimajor axis:

$$g_l(a) = 0.4272 * a - 0.4678 \qquad\qquad (2)$$

$$g_h(a) = 1 - g_l(a)$$

$$g_l(a) = 1.0, \ a > 3.4 \ A. \ U.$$

$$g_l(a) = 0.0, \ a < 1.1 \ A. \ U.$$

where $g_l(a)$ and $g_h(a)$ are the low and high albedo probabilities. The geometric albedos for individual asteroids are determined as follows:

1) The probability for the asteroid to have low ($p < 0.10$) geometric albedo is calculated using relation (2).

2) A random value, evenly distributed in the range $[0, 1]$, decides albedo class. If the random number is lower than the probability calculated in step 1 the low albedo class is chosen and if the random number is higher the high albedo class is chosen.

3) The geometric albedo is taken from the values given above.

Given the geometric albedo and absolute magnitude the formula:

$$\log d = 3.122 - 0.5 \log p - 0.2H \qquad\qquad (3)$$

was used to calculate the diameter d in kilometers (Bowell and Lumme, 1979).

There should not exist any large systematic errors in the diameters. However, for individual asteroids the errors can be very large due to the, in part, randomly chosen albedos.

To be able to compare our results with PLS we have calculated diameters for the PLS asteroids, using the same method as described above. Figure 2 shows the cumulative number per equal (0.1) log d interval for UESAC'92, UESAC'93 and PLS asteroids. The numbers in Figure 2 are mean values of 100 independent calculated cumulative distributions.

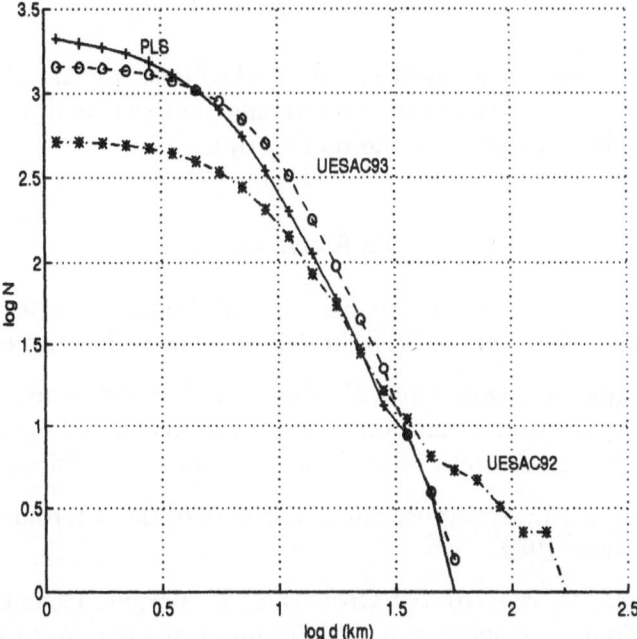

Fig. 2. The logarithm of the cumulative number of asteroids per equal (0.1) log d interval for UESAC'92, UESAC'93 and PLS (including numbered asteroids).

The difference seen for the distributions of the absolute magnitudes is also reflected in the size distributions. The cumulative numbers for the UESAC'92 asteroids are higher for diameters larger than 30 km. However, this is not significant since the total number of observed UESAC'92 asteroids larger than 30 km is only 13. The power-index change in the diameter distributions seen at around 160 km and 25 km (Ishida et al. (1984), Zellner and Bowell (1977)) can not be observed very clearly in our results. Both PLS and UESAC seems to be essentially complete down to sizes of around 15 km (based on the deflection from linearity of the logarithmic diameter distributions around this value) and our results indicate that there are no major change in the power-index between diameters of 15 and 30 km. However, both UESAC and PLS observed too few asteroids with diameters larger than 25 km to definitely confirm or rule out the 25 km index change found by Ishida et al. (1984).

Acknowledgements

We thank Ken Russell for getting the AAO films. We are thankful to G. Williams for his help with additional linkings of objects and to Karri Muinonen for valuable comments on the manuscript.

References

Bowell, E., Lumme, K., 1989, Colorimetry and Magnitudes, in: *Asteroids II.* eds. R. P. Binzel, T. Gehrels, M. S. Matthews, The University of Arizona.

Bowell, E., Hapke, B., Domingue, D., Lumme, K., Peltoniemi, J. and Harris, A. W. 1989, Application of photometric models to asteroids, in: *Asteroids II.* eds. R. P. Binzel, T. Gehrels, M. S. Matthews, The University of Arizona.

Gradie J., Tedesco E., Compositional structure of the asteroid belt, Science, Vol. **216**, 25 June 1982.

van Houten, C. J., van Houten-Groenvald, I., Herget, P. and Gehrels, T., 1970: The Palomar-Leiden Survey of faint minor planets, Astron. Astrophys. Suppl. **2**, 339-448.

Ishida, K., Mikami, T. and Kosai, H., 1984. Size distribution of Asteroids, Astron. Soc. Japan **36**, 357-370.

Lagerkvist C.-I. and Magnusson P., 1990.Analysis of asteroid lightcurves. II. Phase curves in generalized HG-sytem. Astron. Astrophys. Suppl. **86**, 119-165.

Lagerkvist, C.-I., Hernius, O., Lindgren, M. and Tancredi, G., 1995, The Uppsala-ESO survey of asteroids and comets (UESAC). In proc. from the third international workshop on positional astronomy and celestial mechanics (ed. A. López Garcia), Valencia, subm.

Tancredi G., Lindgren M., 1994, Searching for comets encountering Jupiter: First campaign, Icarus **107**, 311-321.

Tedesco, E. F., Veeder, G. J., Fowler, J. W. and Chillemi, J. R., 1992, The IRAS Minor Planet Survey Data Base, National Space Science Data Center, Greenbelt, Maryland.

Williams, G. 1994. Private communication.

Zellner, B., and Bowell, E., 1977, Asteroid compositional types and their distributions, *Comets, Asteroids, Meteorites*, ed. A. H. Delsemme (University of Toledo Press, Toledo), p. 185.

DISCOVERY AND REDISCOVERY OF TROJAN ASTEROIDS

ERIC W. ELST

Royal Observatory at Uccle (Belgium)

Abstract. During 1987-1994, observational campaigns with different telescopes at several observatories have been initiated by the author in order to discover new Trojans. The importance of Trojan asteroids comes from celestial mechanics, where they represent the physical solution of the famous Lagrange triangular problem. Their importance lies also in the fact, that they may have some relation with comets. Furthermore, the Trojan belt may be as large as the belt of asteroids. Moreover, recently "families" have been discovered between the already well known Trojans. Enough reasons to continue to search for these interesting objects.

1. Introduction

From the Minor Planet Center (Marsden 1994) we have obtained four files that contain information on the already discovered Trojans. From these files we may learn that we have 120 numbered Trojans (L4 and L5 objects). The multiple-opposition and long-arc single opposition file contains about 50 objects, with 50 more from the short-arc non-survey file. A fourth file lists all objects (Van Houten-Gehrels Trojans) from their single-opposition surveys (1970, 1973, 1977). Our observations have been obtained at five different observatories, i.e. at the observatoires of Hoher List (Germany), Haute Provence (France), Caussols (France), Rozhen (Bulgaria) and ESO (Chile). We have used the GPO astrograph (ESO and Haute Provence), the Hoher List astrograph, the Rozhen Schmidt, the Caussols Schmidt and the ESO-Schmidt. To discover the slow moving Trojans on the photographical plates, we have applied several methods. In most cases the plates have been investigated with a Zeiss-blink-comparator (two plates per field). However with the larger Schmidts (ESO and Caussols) it was possible to discover the Trojans from a single plate by means of three separated exposures.

Table I gives a review of the plate material that we have obtained during March 1986 – September 1994.

2. Results

Table II gives a review of the discoveries that have been made from the investigation of the plate material in Table I. Of course not only Trojans have been discovered, since the plates were also extremely rich in main belt

Earth, Moon, and Planets **72**: 275-277, 1996.

TABLE I

The plate material (* means estimated)

Observatory	Instrument	Amount of plates	N	Reduction
Caussols	Schmidt	100*	< 100	80 %
ESO	GPO	1000*	< 10	80 %
ESO	Schmidt	450*	< 300	70 %
Haute Provence	GPO	60*	< 5	80 %
Haute Provence	Schmidt	550*	< 10	70 %
Hoher List	Astrograph	200*	< 15	30 %
Rozhen	Schmidt	200*	< 20	90 %

TABLE II

Review of discoveries (* means estimated)

Total number of discoveries	7500* (unnumbered+numbered+new designations)
New designations	6000*
Orbits	3000–4000* (improved $n > 3$; 3 pts-orbits (Gauss, Laplace, Väisälä) and 2 pts-Väisälä orbits).
Real discoveries	450* (objects that have been retraced, principal designation remains).
Numbered objects	58
Numbered Trojans	3 (Deipylos, Eurypylos, Ulysses)
Trojans waiting to be numbered	3 (1987 QN, 1987 YU1, 1990 VL6)
One-opposition Trojans	6 (1991 VZ5, 1992 GF3, 1993 BD4, 1994 CX13, 1994 CR18, 1994 ES6)

asteroids, from which a lot belong to the Hilda, Phocaea and Hungaria group.

During all these campaigns, we have rediscovered a lot of Trojans that had already been discovered by other observers. Table III gives a complete list of these rediscoveries.

New projects: As already has been mentioned, the Trojan belt (L4 and L5), may be as large as the main asteroid belt (up to 30,000 objects larger than 5 km). That means that we have discovered less than 1 % of the existing Trojans. There remains still a lot of (observational) work. But who is going to observe? The observation philosophy moves more and more to CCD-observation, which implies that the investigated field is small, and therefore the possibility of discovering a new Trojan extremely limited. A second question is: "Have all brighter Trojans already been discovered"?

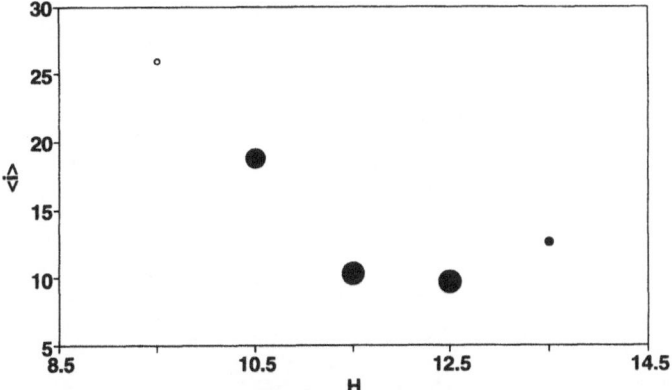

Fig. 1.   Mean inclination vs. absolute magnitude for short-arc, single-opposition non-survey Trojans. Symbol sizes indicate the number of objects included into the calculations.

TABLE III

Rediscovery of Trojans (MPC= page of the Minor Planet Center Bulletin).

Designation	Observatory	Designation	Observatory	MPC
1989 CT6	ESO	=5493 T-2	Palomar (1973)	16884
1989 ER2	"	=5187 T-2	" (1973)	16883
1989 SS4	"	=3104 T-3	" (1977)	16243
1989 VU3	"	=1988 TH1	" (1988)	18430
1989 TF6	"	=4101 T-3	" (1977)	15909
1990 EM	"	=5187 T-2	" (1973)	16883
1990 EP	"	=1973 SY	" (1973)	16693
1990 EE1	"	=5493 T-2	" (1973)	16884
1990 EY1	"	=1991 GX1	" (1991)	18440
1990 ED2	"	=1973 SM1	" (1973)	16693
1991 GD3	"	=9507 P-L	" (1960)	18445
1991 GA5	"	=1973 SD1	" (1973)	18412

This seems not to be the case, since from a plot (Figure 1) that combines the inclination with the absolute magnitude, from the short-arc single-opposition non-survey Trojans, there is a strong indication that there should exist still a lot of brighter high inclination Trojans, that are waiting to be discovered.

References

Marsden B.G.: 1994, *private communication.*

The Limitations of NEO-Uniformitarianism

D.I. Steel
Anglo-Australian Observatory, Private Bag, Coonabarabran, NSW 2357, Australia; and Department of Physics and Mathematical Physics, University of Adelaide, South Australia

Abstract.
The geological and biological sciences have gradually dispensed with the nineteenth-century concept of substantive uniformitarianism – or gradualism – whereby the physical and biological features of our planet are assumed to have been brought about by the long-term accumulation of small changes. The catastrophist alternative sees the changes as being wrought largely by discrete, exceptional events; one such type of event is an impact by a substantial asteroid or comet. It is argued here that scientists working on small solar system bodies generally still labour under a form of this gradualism, in that a conventional starting point is to presume a steady-state, and what is seen now is assumed to be diagnostic of the long-term average conditions. This is here termed *NEO-uniformitarianism*, the NEO referring to Near-Earth Objects. It is maintained herein that this area of science needs to revise its philosophical basis by allowing catastrophist principles to be entertained; that is, the presumption of a steady-state needs to be rejected until such time as evidence to support it is revealed. It is argued that the weight of evidence favours the contrary. For example, evidence is outlined for (a) Variations in the terrestrial cratering rate, disallowing any equating of the crater record with the presently-observed large impactor population; (b) The presence of significant NEO complexes which may be due to giant comet disintegrations within the last 20 kyr, hence solving the problem of the supply of short-period comets; (c) A misbalance between the present supply of meteoroids, there being too many to be supplied by presently-observed comets and also a surplus above the population needed to maintain the interplanetary dust complex; and (d) A substantial variation in the interplanetary dust flux in the past 20 kyr, as might be expected from (b and c).

Key words: asteroids, comets, cratering rate, meteoroids, interplanetary dust, inner solar system environment

1 Introduction

This is neither a research paper presenting new results, or a review; rather, it is a commentary upon the way in which solar system astronomers are currently approaching their science. An assumption made by many in investigating the characteristics of the small bodies in the solar system is that what is observed now (that is, in historical times, but especially in the past few centuries) is diagnostic of the long-term conditions. In this paper I argue that this assumption is incorrect and that, in the words of the Chinese proverb, 'we live in interesting [dangerous] times.' Here I maintain that, far from being in a steady-state, the ecology of small bodies in the solar system is characterized by large population fluctuations from time to time, and at present the conditions are dominated by the decay products of one or a few giant comet disintegrations. This concept, whereby the impact hazard to mankind is dominated by the occasional passage of the Earth through a gradually-dispersing stream of débris produced by such disintegrations, is termed

Earth, Moon, and Planets **72**: 279–292, 1996.
© 1996 *Kluwer Academic Publishers.*

'coherent catastrophism' (Steel, 1991; Asher *et al.*, 1994b).

I have entitled this paper 'The Limitations of NEO-Uniformitarianism' (the NEO referring to near-Earth objects) for the following reasons. Mainstream geology was founded upon the uniformitarian principles developed in the late eighteenth century and the first half of the nineteenth, predominantly by British geologists, at the expense of the French school of Georges Cuvier, the champion of catastrophism. The crux of the uniformitarian argument was that the face of the planet has been shaped solely by processes which are seen in action in the present epoch, and it is the long-term action of these that have resulted in the features that we witness in the present. We now realize that this is not the case, a vivid example being given by the location of the conference at which this paper was presented: the form and existence of the Åland Islands are largely due to the catastrophic impact which produced Lumparn many millions of years ago. Modern geological uniformitarianism entails rather different ideas, as described by Shea (1982). Alvarez *et al.* (1989) and Marvin (1990) discuss how the old concepts of uniformitarianism have limited the progress of geology; this is what Gould (1965) terms 'substantive uniformitarianism.' A biological corollary to that old concept of uniformitarianism or gradualism is Darwinian evolutionary theory, whereby gradual changes occurring in minute steps lead to new species. The catastrophist philosophy is to the contrary. In geology, catastrophism argues that extraordinary (indeed, often extraterrestrial) events are dominant, with the large impacts upon the Earth playing a major role in its history, although endogenic catastrophes are also significant. In biology, instead of evolutionary change occurring through gradual alterations over very long periods of time, many adaptation and speciation events are viewed as occurring rapidly: this is what has been called 'punctuated equilibrium' (Gould & Eldredge, 1993). Mass extinction events occurring due to major asteroid or comet impacts (or related events brought on by comets, such as dust veiling of the atmosphere causing rapid climatic deterioration) do not allow for gradual adaptational change in the flora or fauna, with previous evolutionary pressures being of no great consequence with regard to whether certain species or families survive or not, because the extremal conditions suddenly imposed have not previously been met. That is, a species succumbs in a mass extinction due to bad luck rather than bad genes (Raup, 1991), and a corollary of that idea is that survival is due to good luck in that a species has been equipped with the features necessary for survival by accident rather than evolutionary pressure.

These same principles can also be applied to astronomy. For example, it is often assumed that the comets that we see now are individual monolithic captures from long-period orbits (a uniformitarian view), rather than the decay products of one or a few giant comet disintegrations (a catastrophist view). By *NEO-Uniformitarianism* I imply an approach in which it is presumed that the phenomena which we see in action now (the last few centuries) are the same as those which have governed the inner solar system environment in the long term (millions to billions of years). I argue that there is evidence indicating this not to be the

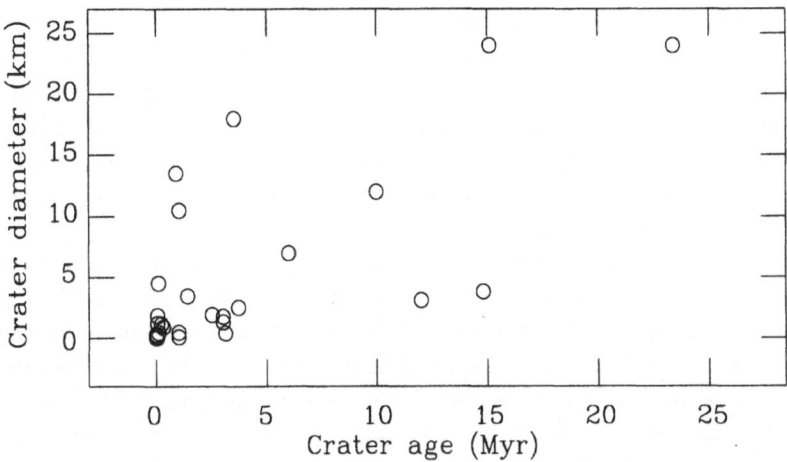

Fig. 1.   Diameter versus age for terrestrial impact craters less than 25 Myr old.

case, and that the present environment is dominated by a recent ($\sim 20\,\mathrm{kyr}$ ago) exceptional event, the repurcussions of which continue. *NEO-uniformitarianism* is intended to be a pejorative label, echoing the incorrect substantive uniformitarianism (gradualism) of the nineteenth century rather than the modern uniformitarian concepts as described by Gould (1965) and Shea (1982). I argue that not only are large fluctuations in the NEO population of great significance with regard to the impact rate upon the Earth for particles of all sizes, and therefore the extraterrestrial modulation of the terrestrial environment, but also that in the present epoch the inner solar system has been subject to just such an upwards deviation in the NEO population due to one or more disintegrating giant comets.

2   Terrestrial craters

Let us begin with the terrestrial cratering record. If the NEO-uniformitarians are correct then the craters should have been formed randomly in time. At the very simplest level, one can look at the youngest craters (ages $< 25\,\mathrm{Myr}$) to see whether their age distribution is random. This is clearly not the case (Fig. 1). 35 craters contribute, taken from the tabulation of Grieve (1982). Of those many occur in the bottom left corner of Fig. 1: there are 12 with diameters below 300 m and ages less than 50 kyr. Very likely this non-randomness is due to the size selectivity in erosion, identification, and so on, but the point is that the data cannot be used to argue in favour of randomness.

Because the Moon is subject to the same impactor flux as the Earth, and there is a more extensive record of lunar cratering (although with much poorer dating), non-randomness in the time of formation of the lunar craters would be of interest. It is well known that in the first $\sim$800 Myr of the Moon's history the influx was higher than at present (the late heavy bombardment). However, there is evidence for six major cratering episodes over the past 3800 Myr. Stothers (1992) has correlated these with the six major episodes of orogenic tectonism on Earth, which is of interest with regard to the Earth's history, but also may be interpreted as (i) Providing supportive evidence that the approximate dating of the lunar cratering is correct; and (ii) Indicating that on the longest time-scales (hundreds of Myr) the terrestrial influx is variable.

On time-scales of tens of Myr the influx is also variable. In the past two decades there have been several suggestions that the Earth's cratering record has a cyclic component, and it appears likely that >40% of terrestrial craters do indeed follow a $\sim$30 Myr periodicity (Yabushita, 1992). The idea that massive impacts are linked to geological boundary events is not a new one (Nininger, 1942; Urey, 1973), although the past decade or so has seen major advances. One might therefore ask whether geological upheavals and faunal mass extinctions also follow a similar periodicity as the craters, a question which has been answered in the affirmative (Rampino & Caldeira, 1993), indicating a general causative link. Thus on $\sim$10 Myr time-scales, the terrestrial large-body influx is variable, comet waves from the Oort cloud being the explanation in vogue.

Perhaps 50% or more of the near-Earth asteroids (NEAs) may be derived from comets (Wetherill, 1991), which would mean that the NEA population would be expected to fluctuate substantially at least on that $\sim$10 Myr time-scale. NEAs may also be derived from the main asteroid belt, however, and Menichella *et al.* (1995) have studied this source and shown that the supply can be highly variable on time-scales < 1 Myr, implying that the terrestrial cratering rate can also fluctuate markedly due to inputs from that source. Averaged over longer time-scales (> 10 Myr), however, Menichella *et al.* (1995) find that the NEA population derived from the main belt is unlikely to vary by more than a factor of two.

For NEAs both of cometary and main-belt origins, the results of Farinella *et al.* (1994) are important. They show that such bodies commonly evolve under planetary perturbations to enter orbits which fall into the Sun, such a fate being more likely than impacts upon the planets or ejection from the solar system. The appropriate time-scales are 0.1–1 Myr. This again warrants for the reservoir of potential Earth-impactors being variable on a short time-scale, whereas in the past it has been assumed that the limiting lifetime for NEAs is set by planetary collisions, so that the NEA population only changes over periods >10–100 Myr (e.g., see Bailey *et al.*, 1994).

To summarize this section, there is no support for the idea that the population of NEOs is constant in time, for any time-scale from 0.1–1000 Myr. Crater densities on the Moon and from certain areas of the Earth render average impact rates over periods towards the upper end of this range (Shoemaker, 1983). Therefore crater counts cannot tell us much about the NEO population, or the terrestrial impact danger, in any particular epoch; obviously the epoch of most interest to us is the present. It may be that the characteristic fluctuation time-scale is actually <0.1 Myr, but to investigate that we must look at other evidence.

3 Disintegrated giant comets

The results of Menichella *et al.* (1995) imply that the terrestrial impact rate due to main belt-derived NEAs can fluctuate by a large factor on time-scales <1 Myr. Singular events such as large comets fragmenting will also cause NEO/NEA population boosts. Clube & Napier (1984, 1986) have championed the idea of NEOs derived from disintegrating giant comets producing population spikes of fragmenting daughter comets/asteroids which last for 10–100 kyr. Thus we may expect a microstructure of terrestrial impacts quite unlike the normally-assumed constant population causing random impacts over time-scales of Myr.

For the purposes of this section the first thing to be determined is whether giant comets exist, and if so whether they are important. There is no known upper limit for the mass of a comet. The largest Apollo asteroid observed in the present epoch is about 8 km in size, whilst the periodic comets P/Halley and P/Swift-Tuttle are 10–20 km in dimension. Occasionally much larger long-period comets – of order 100 km or more – are seen, for example the great comet of 1577 or that of 1729 (Comet Sarabat). There are several outer solar system objects known which are thought to be cometary in nature (*e.g.*, see Stern, 1995). One example is the ∼150–200 km object known as (2060) Chiron; it is on an unstable orbit which may lead to an Earth-crossing path on a time-scale of ∼1 Myr (Hahn & Bailey, 1990). There are several similar objects known, such as (5145) Pholus, 1993 HA$_2$ and 1994 TA. One can argue that these all have significant probabilities of delivery into Earth-crossing orbits within ∼1 Myr (Asher & Steel, 1993). The Mars-grazer (5335) Damocles is ∼10–20 km in size and is likely to eventually attain an Earth-crossing orbit; this may be the first to be found of a large population of extinct Halley-type comets which are difficult to discover (Asher *et al.*, 1994a). At the time of writing 17 trans-Neptunian asteroids/comets have been identified which may be large (100–400 km) Kuiper belt objects on unstable orbits (Stern, 1995). The influence of such objects upon the terrestrial environment is discussed by Bailey *et al.* (1994).

Given the present state of our discoveries, the results of the orbital evolution studies mentioned in the previous paragraph, and the rate at which new outer solar system objects are being found, it would be unreasonable to expect the frequency of delivery of giant (> 100 km) comets into periodic Earth-crossing orbits to be less than one per 100 kyr. If this is the case, then such objects dominate the long-term average mass flux of Earth-crossing objects. They are therefore of considerable significance with regard to the past and future history of the NEO population.

Comets split frequently, as has been known for some time (Sekanina, 1982). Even in 1994, apart from witnessing the demise of a comet which split due to Jupiter's tidal force (P/Shoemaker-Levy 9), we have seen two comets with perihe-

lia in the inner solar system which have recently fragmented into multiple components whilst far from any planet: P/Machholz 2 (*IAUC 6090*) and P/Harrington
(*IAUC 6089*). Pittich & Rickman (1994) have studied cometary splitting from
the perspective of whether such events could supply the Jupiter family from giant
comet progenitors. They found that gravitational stirring by Jupiter would quickly
destroy the dynamical memory of the fragments (newly-produced comets), so that
the Jupiter family observed now could have been largely or entirely produced by
one or a few giant comets.

The daughter products most likely to retain their orbital similarity for long
enough to be recognized as being co-genetic would be those produced by a giant
comet (or a large fragment thereof) which before/during its disintegration attained
a sub-jovian orbit. With aphelia $Q < 4.5$ AU the fragments would not have their
orbits quickly randomized. Wetherill (1991) has investigated how such orbits may
be attained. What we are looking for is, in essence, a stream of large particles which
may or may not be active (in the sense of cometary activity). Meteor showers are
clear evidence that streams of smaller particles exist, the question being whether
meteoroid streams extend to ~km-sized particles. A suitable data set to inspect is
the growing catalogue of NEAs. Because P/Encke is an exception amongst comets,
having such a small orbit, and it is also associated with a well-known meteoroid
stream which constitutes ~50% of the broad sporadic meteor influx to the Earth
(see Štohl 1980, 1986a, 1986b, and Štohl & Porubčan, 1991, 1992), it is a suitable
initial test orbit to use in a search.

If P/Encke is one fragment from a broken-up giant comet, and there are other
fragments in a broad stream, there is no reason to expect any or all of these to
currently have a node at 1 AU. P/Encke itself does not, indicating that the associated Taurid meteor showers observed now result from the Earth's passage through
the periphery of the stream. Multiple branches of a meteoroid stream may be recognized through their longitudes of perihelion (ϖ) being close to the same value.
For example, a stream like the Taurids has a precession period for the argument
of perihelion (ω) of order 7 kyr, the value being critically dependent upon the particular meteoroid's orbit size (Table A1 of Asher & Clube, 1993). Within such a
period all four branches of the stream's intersections with the Earth – the four
meteor showers – will be produced (see Babadzhanov & Obrubov, 1992). However, ϖ has a precession period 5–10 times as long (Asher & Clube, 1993), meaning
that on each cycle of ω the showers will be produced at somewhat different values
of ϖ, and thus different values of Ω (different days of the year). For the Taurid
stream Babadzhanov *et al.* (1990) identify the Piscid and χ Orionid showers, as
distinct from the main Taurid showers, as being produced in this way, and Štohl
& Porubčan (1990) add the ρ Geminid showers. The existence of these distinct
showers allows the age of the stream since its formation – about 20 kyr – to be
estimated (see also Steel *et al.*, 1991), and the *differential* precession in ϖ (due to
the differing orbital sizes) means that a spread in ϖ is to be expected.

We (Asher *et al.*, 1994b) therefore searched through the NEA database for orbits similar in a, e and i to P/Encke or the Taurid meteors, and tabulated our results. Graphically, the results are shown in Fig. 2. Taking the 20 NEAs with (a, e, i) most similar to the Taurids we found that 14 had values of ϖ within the 100°–190° band delineated by the meteor data; P/Encke also, of course, appears within this band. The probability that 14 out of 20 might occur in that 90° band purely by chance is very small (actually 0.00003), given a null hypothesis that a random distribution in ϖ is to be expected.

Clearly the validity of that null hypothesis needs to be examined; for example, are there discovery selection effects which could lead to this concentration of values? Klačka (1995) argues that the null hypothesis (expected random ϖ) is incorrect and that selection effects would lead to a concentration near $\varpi = 155°$. Klačka's analysis, however, (i) Cannot explain the Taurid complex orbits with $\varpi = 160°–190°$; (ii) Does not explain the Hephaistos group (see below) at $\varpi = 220°–260°$; (iii) Does not explain why the exceptional Taurid meteors straddle the same range as the NEAs selected here (as shown in Fig. 2); and (iv) Does not explain the agreement with the present result by the distinct test applied by Napier (1993), in which the null hypothesis was that 'there is no significant concentration of NEAs with orbits similar to P/Encke.'

We therefore identified this group of 14 NEAs plus P/Encke as being a group of NEOs which have been formed by the disintegration of a much larger pregenitor about 20 kyr ago. If this grouping were due to some selection effects acting on the discovery probabilities of NEAs, then it would be remarkable that the group covers the same band in ϖ as the observed Taurid meteors. Looking beyond this group of 14, we find that five of the other six NEAs have $222° < \varpi < 251°$. Again, this has a very small probability of occurring by chance, given the null hypothesis. Of these five, (2212) Hephaistos is identified as the largest/the archetype, although Steel & Asher (1994) have noted that P/Helfenzrieder (a comet observed only in 1766, and which has the second-briefest period of all comets after P/Encke), with ϖ likely in the range 251°–255°, may also be associated with this complex. It therefore appears probable that there are two groups of NEOs in sub-jovian orbits; these may be genetically-related to each other, or may be the result of separate giant comet arrivals. The physical evidence from the two comets involved – P/Helfenzrieder being active only in 1766, whilst P/Encke suddenly became active later in 1786 – is also supportive of the concept that these groups are derived from disintegrated giant comets, some of the fragments now showing spasmodic cometary activity whilst others may by now be totally devolatilized.

Because these two groups would be orbitally dispersed and thus not recognizeable through the above analysis if the individual NEOs had been released more than ~50 kyr ago, there is evidence here for a recent disintegration and therefore a boost in the NEO population. The meteor evidence points to the arrival of the progenitor comet ~20 kyr ago (Steel et al., 1991). The relative activities/influxes of the different Taurid showers (the main Taurids, the Piscids and the χ Orionids) argue for later disintegration events within the last 10 kyr (Štohl, 1986b;

286 D.I. Steel

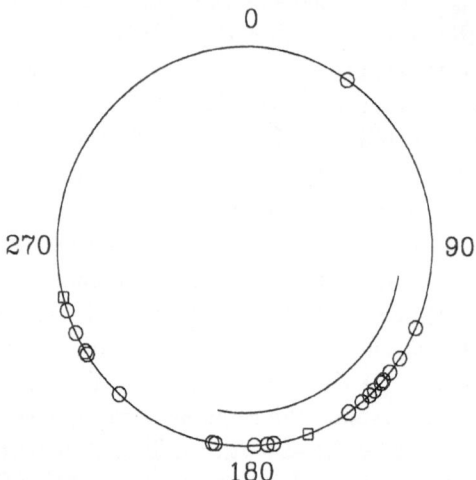

Fig. 2. The longitudes of perihelion (ϖ) of macroscopic objects with a,e,i similar to the Taurid meteors (see Asher *et al.*, 1994b). The arc interior to the ϖ-circle delineates the $\varpi = 100°\text{--}190°$ range of the Taurids. The circles represent near-Earth asteroids, 14 of the 20 being in the that range, as is P/Encke (square symbol), as expected. Another grouping occurs at $\varpi = 220°\text{--}260°$, (2212) Hephaistos being the best-known member. The second-shortest period comet after P/Encke (P/Helfenzrieder) may also be a member of that group (square at left).

Babadzhanov *et al.*, 1990), and many years ago Whipple & Hamid (1952) suggested that such episodes have occurred within the past 5 kyr. Although the NEOs plotted in Fig. 2 constitute only 10–15% of the total population known at present, it should be remembered that these are only the bodies with sufficiently small aphelia to have avoided orbit randomization (as discussed by Pittich & Rickman, 1994). Immediately after the hypothesized giant comet broke up in a Jupiter-family orbit the NEOs it produced would have dominated the Earth-crossing flux. Many of those objects would have entered the NEO population observed now with greatly-altered orbits which prohibit the recognition of a common origin.

4 Interplanetary dust flux

Whilst there seems to have been little attention paid to possible drastic variations in the population of asteroids and comets in the inner solar system on short time-scales, there has been some work done on the possibility that within recorded history there might be evidence for gross fluctuations in the interplanetary dust population (*e.g.*, Clube & Napier, 1984; Reach, 1992). Because most meteoroids, and hence dust, are produced by the largest comets (Hughes, 1990), the conclusion of the last section suggests that there could have been a gross enhancement in the interplanetary dust population in the past 20 kyr. One possibility might be that various civilizations recorded zodiacal light anomalies, the subject of the paper by Reach (1992). Here I will limit the discussion to the physical evidence available for laboratory investigation.

The first question is one of time-scale. For meteoroids (sizes $>100\,\mu$m) the limiting lifetimes are >100 kyr (Grün et al., 1985; Steel & Elford, 1986), whereas for the 10–$100\,\mu$m dust particles producing the zodiacal light, the Poynting-Robertson (P-R) effect limits the lifetimes to <10 kyr for the low densities thought to be appropriate for dust (Fig. 10 of Leinert et al., 1983). Thus the meteoroids and dust released directly by a giant comet, especially one which is undergoing spasmodic disintegration events, will be lost from the interplanetary environment at different rates. Meteoroids would be expected to persist for >100 kyr, whereas the dust will be lost within ~10 kyr (although there is a feedback effect in that the meteoroids will initially be lost more quickly due to the collisions with dust which limit the meteoroidal lifetimes: see Olsson-Steel, 1986). The observed space exposure ages of dust grains confirm their time-scale (Bradley et al., 1984). After this original comet-released dust has been lost, it is to be expected that there will be a surplus of meteoroids which gradually comminute so as to feed into the dust cloud, with the supply and loss of dust being out of balance; this is indeed what is observed at present (Grün et al., 1985). On the other hand, it is known that the presently-observed comets cannot supply meteoroids at the rate required to maintain the present population (Whipple, 1967; Kresák, 1980; Kresák & Štohl, 1990), mitigating for a recent massive (giant comet-derived) injection; both Whipple and Kresák invoke a massive P/Encke pregenitor as a plausible source.

In the first millennia after any giant comet break-up, then, one would anticipate a greatly-enhanced dust influx to the Earth-Moon system due to (i) Dust released directly from the cometary fragments; and (ii) Dust produced in the grinding up of the meteoroids by collisions with that temporarily-enhanced dust population (a positive-feed back effect which would continue only for so long as the dust spatial density is above the feedback threshold, or until the meteoroid supply is exhausted). The question is whether there is physical evidence for such an dust influx to the Earth and Moon.

Clube & Napier (1984) discussed evidence for an enhancement in the cosmic dust influx to the Earth from 20–14 kyr ago (the latter part of the last ice age), as derived from Greenland ice cores. This interpretation of the data has been attacked (La Violette, 1987) but defended (Clube & Napier, 1987). In view of this debate, however, I will invoke instead lunar microcratering data and their analysis as evidence of an enhanced dust flux ~20 kyr ago; the analysis (by Zook) is also pertinent because it provides a vivid example of NEO-uniformitarianism, the core subject of this publication.

Zook et al. (1977) discuss at length an interpretation of the solar flare tracks found in the melt-glass bases of dust impact pits in lunar rock. From the track density it is possible to deduce an exposure age, based upon an assumption that the solar cosmic ray intensity has been invariant over the ~100 kyr which is appropriate for the samples. However, the data of Hartung & Storzer (1974) indicated, subject to this assumption, that most of the pits were formed within the last 20 kyr, and thus that the dust flux has been higher by a factor of $\sim$10 within the past $\sim$10 kyr compared to the long-term average. Rather than accept that the dust flux has varied in such a way, Zook *et al.* (1977) and Zook (1978, 1980) argue that the root cause is that solar flare activity was much higher $\sim$10 kyr ago. One of the data sets underpinning this interpretation, concerning the ^{14}C in lunar rocks, has since been superseded (Goswami, 1991), and there seems to be a scarcity of support from solar physicists for the claimed solar cosmic ray flux variation. The line of reasoning behind the rejection of the hypothesis that the *dust* flux rather than the *solar flare* flux has varied was detailed by Zook *et al.* (1977) and summarized by Zook (1978). The line of reasoning was that:

(1) Meteoroids are derived predominantly from short-period comets;
(2) Either a large increase in the number of average short-period comets is required or there has been a recent capture of one or two massive comets;
(3) There are $\sim$100 short-period comets, and a factor of $\sim$10 increase in the meteoroid population would require a similar enhancement in the number of comet captures (which has a vanishingly-small probability) ;
(4) In the circumstance of massive comet capture, one or two large, dominant meteoroid streams would be expected, and these are not now observed;
(5) Such streams could not have appeared $\sim$10 kyr ago and then disappeared into the sporadic background;
(6) In any case small sporadic meteoroids have smaller orbits than large sporadics, and that would require >10 kyr for eccentricity reduction under the Poynting-Robertson effect;
(7) The same is true for meteoroids in several of the major streams;
(8) In view of the above an increase by $\sim$10 in the meteoroid population within the last $\sim$10 kyr appears highly improbable.

Given that one accepts point 3, and is thus forced towards one or more giant comet captures, one then has to argue against points 4–8. Zook's point 4 is incorrect, as indicated earlier in this publication, in particular with regard to the work of Štohl: the Taurid meteor complex does appear to be a massive stream, and at present the Earth passes only through its periphery rather than its core (see Steel *et al.*, 1991; and Asher *et al.*, 1994b), whilst Whipple (1967) and Kresák (1980), amongst others, have argued that this complex associated with P/Encke is the major source of the present overall population of meteoroids. In connection with point 5, Štohl (1986a, and several earlier papers) has shown that the sporadic population is largely *itself* a broad stream, which has been described earlier as requiring $\sim$10 kyr to spread to its present extent.

Points 6 and 7 are technical details that require an understanding of how meteor orbits are determined. The smaller meteoroids that Zook mentions were detected using radars, whereas the larger ones were observed photographically. There are important selection effects for each observation technique which are strongly speed-dependent (hence severely affecting the mean sizes of the orbits determined), and in particular the amount of ionization produced (affecting radar-detectibility) is not the same function of speed as is the luminosity (affecting the photographic-detectibility). The deceleration of the meteoroid in the atmosphere is also size-dependent, and in any case in radar programs some mean deceleration correction is normally applied. There are other factors which influence the veracity of any meteor speed (and orbit size) determination. Apart from this, and in line with Zook's note of the possibility that P-R drag is not the predominant factor in shrinking the orbits, smaller meteoroids have larger cross-section to mass ratios, making them more susceptible to orbit shrinkage due to, for example, sub-catastrophic decelerating collisions with small dust grains near perihelion.

Purely on the basis of the above discussion, it seems that Zook's conclusion (point 8) should be rejected. However, so far it has not been made clear that in any case Zook's points as listed above do not necessarily apply to the lunar microcraters in question. Those microcraters were mostly in the size range 20–100 μm, and thus formed by dust grains smaller than $\sim$10 μm in size. Meteors detected by radar techniques are typically due to meteoroids 100 μm–1 mm in size, and photographic meteors due to particles 1 cm and more across. The P-R lifetimes for $\sim$10 μm grains are very short, measured in millennia only, so that it would be expected that the particle population that actually caused the pits is no longer to be seen, although the dust population now would be enhanced over the long-term average because the elevated meteoroid population would, through comminution, be boosting the dust. The evidence which is still seen, because it has physical and dynamical lifetimes of >100 kyr, is the broad meteoroid stream which was produced in the giant comet disintegration The numerous NEOs liberated in that event – which was not necessarily a single episode, a hierarchical series of framentations being hypothesized) – also provide supportive evidence, as discussed in section 3.

I have discussed Zook's analysis of the lunar microcraters at length because I believe that it provides a good example of NEO-uniformitarianism: a disinclination to accept that an exceptional event may be responsible for phenomena observed in the inner solar system, despite (I believe) there being numerous pointers towards this being the simplest – and perhaps only tenable – explanation. Lest the reader believe that I am unfair on Zook, however, I would like to quote from his later paper, because the passage in question (concerning solar cosmic rays, SCRs) strengthens the major point of the present publication:

> "...because there is no accepted theoretical foundation for predicting the temporal behavior of SCR activity over these intermediate time intervals [1–100 kyr], no sounder philosophical reason exists for assuming an unvarying SCR activity averaged over these time intervals than exists for assuming a high degree of variability." (Zook, 1980).

My view is that the second part of that quote also applies to the asteroid, comet, meteoroid and dust population in the inner solar system, especially in view of the fact that there is both a theoretical and observational foundation for substantial temporal variations in NEO fluxes.

5 Interesting times

The position taken in this publication, and the concept of coherent catastrophism, may well be imagined to be beyond the pale by the reader working under the assumptions of NEO-uniformitarianism. That reader might baulk at these ideas in particular because, even given that the NEO populations might be expected to fluctuate substantially on time-scales of 0.1–1.0 Myr, it seems unlikely that we happen to be studying the sky in the epoch of such a large upward deviation in the NEO population as suggested herein; that is, it seems unlikely that a giant comet might have arrived so recently. However, the present hypothesis goes further, and circumvents that argument, because the hypothesis also suggests that climatic excursions on the Earth, such as the current interglacial, are largely controlled by the influx of extraterrestrial material (Asher & Clube, 1993). Thus not only do we live in interesting times, but also we would not be here to take an interest should these exceptional events not have occurred. The rise of civilizations over the past 5–10 millennia was made possible by, and perhaps even stimulated by, the arrival of the massive Encke pregenitor.

Acknowledgements

This work was supported by the Australian Research Council; travelling expenses were partly met by a grant from the Australian Academy of Sciences.

References

Alvarez, W., Hansen, T., Hut, P., Kauffman, E.G., & Shoemaker, E.M. (1989), "Uniformitarianism and the response of Earth scientists to the theory of impact crises." *Catastrophes and Evolution: Astronomical Foundations*, ed Clube, S.V.M., Cambridge University Press, Cambridge, U.K., pp. 13–24.

Asher, D.J., Bailey, M.E., Hahn, G., & Steel, D.I. (1994a). "Asteroid 5335 Damocles and its implication for cometary dynamics." *Mon. Not. Roy. Astron. Soc.*, **267**, 26–42.

Asher, D.J., & Clube, S.V.M. (1993). "An extraterrestrial influence during the current glacial-interglacial." *Ql. J. Roy. Astron. Soc.*, **34**, 481–511.

Asher, D.J., Clube, S.V.M., Napier, W.M., & Steel, D.I. (1994b). "Coherent catastrophism." *Vistas Astron.*, **38**, 1–27.

Asher, D.J., & Steel, D.I. (1993). "Orbital evolution of the large outer solar system object 5145 Pholus." *Mon. Not. Roy. Astron. Soc.*, **263**, 179–190.

Babadzhanov, P.B., & Obrubov, Yu.V. (1992). "Evolution of short-period meteoroid streams." *Cel. Mech. Dyn. Astron.*, **54**, 111–127.

Babadzhanov, P.B., Obrubov, Yu.V., & Makhmudov, N. (1990). "Meteor streams of Comet Encke." *Sol. Sys. Res.*, **24**, 12–19.

Bailey, M.E., Clube, S.V.M., Hahn, G., Napier, W.M., & Valsecchi, G.B. (1994). "Hazards due to giant comets: Climate and short-term catastrophism." *The Hazard due to Comets and Asteroids*, ed Gehrels, T., University of Arizona Press, Tucson, pp. 479–533.

Bradley, J.P., Brownlee, D.E., & Fraundorf, P. (1984). "Discovery of nuclear tracks in interplanetary dust." *Science*, **226**, 1432–1434.

Clube, S.V.M., & Napier, W.M. (1984). "The microstructure of terrestrial catastrophism." *Mon. Not. Roy. Astron. Soc.*, **211**, 953–968.

Clube, S.V.M., & Napier, W.M. (1986). "Giant comets and the galaxy: implications of the terrestrial record." *The Galaxy and the Solar System*, eds Smoluchowski, R., Bahcall, J.N., & Matthews, M.S., University of Arizona Press, Tucson, pp. 260–285.

Clube, S.V.M., & Napier, W.M. (1987). "The cometary breakup hypothesis re-examined: a reply." *Mon. Not. Roy. Astron. Soc.*, **225**, 55P–58P.

Farinella, P., Froeschlé, Ch., Froeschlé, Cl., Gonczi, R., Hahn, G., Morbidelli, A., & Valsecchi, G.B. (1994). "Asteroids falling into the Sun." *Nature*, **371**, 314–317.

Goswami, J.N. (1991). "Solar flare heavy-ion tracks in extraterrestrial objects." *The Sun in Time*, eds Sonett, C.P., Giampapa, M.S., & Matthews, M.S., University of Arizona Press, Tucson, pp. 426–444.

Gould, S.J. (1965). "Is uniformitarianism necessary?" *Amer. J. Sci.*, **263**, 223–228.

Gould, S.J., & Eldredge, N. (1993). "Punctuated equilibrium comes of age." *Nature*, **366**, 223–227.

Grieve, R.A.F. (1982). "The record of impact on Earth: Implications for a major Cretaceous/Tertiary impact event." *Geol. Soc. Amer., Spec. Pap.*, **190**, 25–37.

Grün, E., Zook, H.A., Fechtig, H., & Giese, R.H. (1985). "Collisional balance of the meteoritic complex." *Icarus*, **62**, 244–272.

Hahn, G., & Bailey, M.E. (1990). "Rapid dynamical evolution of giant comet Chiron." *Nature*, **348**, 132–136.

Hartung, J.B., & Storzer, D. (1974). "Lunar microcraters and their solar flare track record." *Proc. Fifth Lunar Sci. Conf.*, **Vol. 3**, 2527–2541.

Hughes, D.W. (1990). "The mass of meteor streams." *Mon. Not. Roy. Astron. Soc.*, **240**, 73–79.

Klačka, J. (1995). "The Taurid complex of asteroids." *Astron. Astrophys.*, in press.

Kresák, Ľ. (1980). "Sources of interplanetary dust." *IAU Symp. 90, Solid Particles in the Solar System*, eds Halliday, I., & McIntosh, B.A., Reidel, Dordrecht, pp. 211–222.

Kresák, Ľ., & Štohl, J. (1990). "Genetic relationships between comets, asteroids and meteors." *Asteroids, Comets, Meteors III*, eds Lagerkvist, C.-I., Rickman, H., Lindblad, B.A., & Lindgren, M., University of Uppsala, Sweden, pp. 379–388.

La Violette, P.A. (1987). "The cometary breakup hypothesis re-examined." *Mon. Not. Roy. Astron. Soc.*, **224**, 945–951.

Leinert, C., Röser, S., & Buitrago, J. (1983). "How to maintain the spatial distribution of interplanetary dust." *Astron. Astrophys.*, **118**, 345–357.

Marvin, U.B. (1990). "Impact and its revolutionary implications for geology." *Geol. Soc. Amer., Spec. Pap.*, **247**, 147–154.

Menichella, M., Paolicchi, P., & Farinella, P. (1995). "The main belt as a source of near-Earth asteroids." These proceedings.

Napier, W.M. (1993). "Earth-crossing asteroid groups." *Meteoroids and their parent bodies*, eds Štohl, J., & Williams, I.P., Slovak Academy of Sciences, Bratislava, pp. 123–126.

Nininger, H.H. (1942). "Cataclysm and Evolution." *Pop. Astron.*, **50**, 270–272.

Olsson-Steel, D. (1986). "The origin of the sporadic meteoroid component." *Mon. Not. Roy. Astron. Soc.*, **219**, 47–73.

Pittich, E.M., & Rickman, H. (1994). "Cometary splitting – a source for the Jupiter family?" *Astron. Astrophys.*, **281**, 579–587.

Rampino, M.R., & Caldeira, K. (1993). "Major episodes of geologic change: correlations, time structure and possible causes." *Earth Planet. Sci. Lett.*, **114**, 215–227.

Raup, D.M. (1991). "Extinction: Bad Genes or Bad Luck?" Norton, New York.

Reach, W.T. (1992). "On the origin of interplanetary dust within recorded history." *Meteoritics*, **27**, 353–360.

Sekanina, Z. (1982). "The problem of split comets in review." *Comets*, ed Wilkening, L.L., University of Arizona Press, Tucson, pp. 251–287.

Shea, J.H. (1982). "Twelve fallacies of uniformitarianism." *Geology*, **10**, 455–460.

Shoemaker, E.M. (1983). "Asteroid and Comet Bombardment of the Earth." *Ann. Rev. Earth*

Planet. Sci., **11**, 464–494.

Steel, D.I. (1991). "Our asteroid-pelted planet." *Nature,* **354**, 265–267.

Steel, D.I., & Asher, D.J. (1994). "P/Helfenzrieder (1766 II) and the Hephaistos group of Earth-crossing asteroids." *The Observatory,* **114**, 223–226.

Steel, D.I., Asher, D.J., & Clube, S.V.M. (1991). "The structure and evolution of the Taurid Complex." *Mon. Not. Roy. Astron. Soc.,* **251**, 632–648.

Steel, D.I., & Elford, W.G. (1986). "Collisions in the Solar System – III. Meteoroid survival times." *Mon. Not. Roy. Astron. Soc.,* **218**, 185–199.

Stern, S.A. (1995). "Chiron illuminated." *Nature,* **373**, 23–24.

Štohl, J. (1980). "On time-dependent models of the meteoric background complex." *Solid Particles in the Solar System, IAU Symp. 90,* eds Halliday, I., & McIntosh, B.A., Reidel, Dordrecht, pp. 141–144.

Štohl, J. (1986a). "The distribution of sporadic meteor radiants and orbits." *Asteroids, Comets, Meteors II,* eds Lagerkvist, C.-I., Lindblad, B.A., Lundstedt, H., & Rickman, H., University of Uppsala, Sweden, pp. 565–574.

Štohl, J. (1986b). "On meteor contribution by short-period comets." *Proc. 20th ESLAB Symp. on the Exploration of Halley's Comet, ESA SP-250, Vol. II,* 225–228.

Štohl, J., & Porubčan, V. (1990). "Structure of the Taurid meteor complex." *Asteroids, Comets, Meteors III,* eds Lagerkvist, C.-I., Rickman, H., Lindblad, B.A., & Lindgren, M., University of Uppsala, Sweden, pp. 571–574.

Štohl, J., & Porubčan, V. (1992). "Dynamical aspects of the Taurid meteor complex." *Chaos, Resonance and Collective Dynamical Phenomena in the Solar System, IAU Symp. 152,* ed Ferraz-Mello, S., Kluwer, Dordrecht, pp. 315–324.

Stothers, R.B. (1992). "Impacts and tectonism in Earth and Moon history of the past 3800 million years." *Earth, Moon & Planets,* **58**, 145–152.

Urey, H.C. (1973). "Cometary collisions and geological periods." *Nature,* **242**, 32–33.

Wetherill, G.W. (1991). "End products of cometary evolution: cometary origin of Earth-crossing bodies of asteroidal appearance." *Comets in the Post-Halley Era,* eds Newburn, R.L., Jr., Neugebauer, M., & Rahe, J., Kluwer, Dordrecht, pp. 537–556.

Whipple, F.L. (1967). "On maintaining the meteoritic complex." *The Zodiacal Light and the Interplanetary Medium, NASA SP-150,* ed Weinberg, J.L., NASA, Washington, D.C., pp. 409–426.

Whipple, F.L., & Hamid, S.E. (1952). "On the origin of the Taurid meteor streams." *Helwan Obs. Bull.,* **41**, 1–30.

Yabushita, S. (1992). "Periodicity in the crater formation rate and implications for astronomical modelling." *Cel. Mech. Dyn. Astron.,* **54**, 161–178.

Zook, H.A. (1978). "Temporal and spatial variations of the interplanetary dust flux." *Space Research,* **XVIII**, 411–422.

Zook, H.A. (1980). "On lunar evidence for a possible large increase in solar flare activity $\sim 2 \times 10^4$ years ago." *Proc. Conf. Ancient Sun,* eds Pepin, R.O., Eddy, J.A., & Merrill, R., Pergamon Press, Oxford, pp. 245–266.

Zook, H.A., Hartung, J.B., & Storzer, D. (1977). "Solar flare activity: evidence for large-scale changes in the past." *Icarus,* **32**, 106–126.

POSSIBLE COLLISIONS OF P/HALLEY AND
P/MACHHOLZ 1
WITH JUPITER AND THE TERRESTRIAL PLANETS

YU.V. OBRUBOV

Cybernetics department, Kaluga branch of MAA, Kaluga, 248007, Russia

Abstract. The perturbed motion of comet Halley and comet Machholz 1 1986 VIII was investigated within a time interval of about 20 millennia. The minimal distance of 0.043 AU between P/Halley and Venus may occur on April 4, 4868 AD. The distance of 0.036 AU between P/Halley and Jupiter will take place on April 1, 6616 AD.

The orbit of P/Machholz 1 crosses the orbits of Mercury and Venus eight times, that of the Earth six or eight times, and the orbit of Mars four times per a period of advance of the argument of perihelion. A distance of about 0.06 AU between P/Machholz 1 and the Earth may take place in August 2576 AD and 5751 AD and in February 4770 AD. The minimal comet-Earth distance of 0.035 AU occurs on September 14, 5971 AD. The closest encounter between P/Machholz 1 and Jupiter at the distance of 0.098 AU may be in May 4499 AD. These results may be considered as a forecast of possible collisions.

1. Introduction

It is known that the probability of collision between any Solar System bodies is different from zero only in the case of intersection of their paths (i.e., orbits). So, special attention must be paid to the orbital intersections and close encounters, which may occur in the near future.

2. Models and method

To investigate the orbital dynamics of comets Halley and Machholz 1 we have used the implicit single-sequence method for integrating orbits, developed by Everhart (1974). The calculations were done using a database of the orbits of the major planets within the time interval from 10,000 BC to 10,000 AD with the self-adjusting step length. The initial cometary orbits were taken from Marsden's catalogue (1989) in its computer version. The perturbations from 7 planets (Venus-Neptune) were taken into account.

It is necessary to note that we did not take into account the influence of nongravitational effects (which may be very strong for both comets) or the integration errors. However we can say that the integrator is a major error source after about 1,000 years of integration. But backward and forward integration shows that after 8,000 years at least four decimal numerals in eccentricity, inclination, argument of perihelion and longitude of ascending node remain trustworthy. So the results given below may present one of the

Earth, Moon, and Planets **72**: 293-300, 1996.

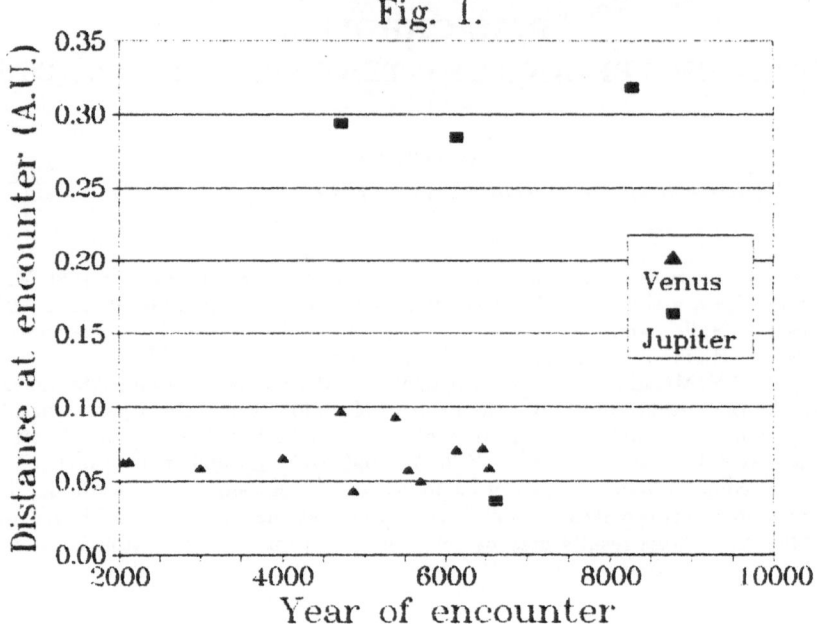

Fig. 1.   Close encounters of P/Halley with Venus, Earth and Jupiter

possible models of the orbital dynamics of P/Halley and P/Machholz 1 or the dynamics of objects in Halley-like and Machholz-like orbits.

The obtained results are in good accordance with those of Yeomans and Kiang (1981), Green et al. (1990), McIntosh (1990), Babadzhanov and Obrubov (1992), and Obrubov (1993), but show some contradiction with the results of Olsson-Steel (1987), because he did not find any encounter of comet Halley with the large planets. However, some encounters between P/Halley and Jupiter may take place in the future. It is possible that this discrepancy comes from integration by different methods. Here we shall consider the events which may take place in the nearest 8,000 yr.

3.   Close encounters and orbital dynamics

3.1. COMET HALLEY

Within the time interval under consideration comet Halley will undergo 4 close encounters with Jupiter at distances less than 0.3 AU and 12 encounters with Venus (< 0.07 AU) but no encounter with the Earth at a distance less than 0.1 AU. The times and distances at encounters are given by Fig. 1. It is obvious that the most significant variation in the orbital elements will occur after the possible encounter with Jupiter in 6616 AD at a distance of 0.036 AU. Fig. 2 shows the changes in the orbital elements of P/Halley due to planetary perturbations and the consequences of encounters. As a result of the close encounter in 6616 AD the semimajor axis jumps from 18 to 30 AU and one can see the jumps in the other orbital elements.

3.2. COMET MACHHOLZ 1

From the dynamical point of view P/Machholz 1 is a much more interesting object than P/Halley. During 8,000 yr there are about 30 close encounters with Jupiter (< 0.3 AU), 12 with the Earth (< 0.1 AU), and one encounter with Venus at a distance of about 0.006 AU (Fig. 3). The variation of the orbital elements are given in Fig. 4. Close encounters with the Earth may take place after the year 2250. After an increase in the semimajor axis close to the year 4250 due to jovian perturbations a quasi-resonant motion occurs. But a much greater influence on the dynamics of P/Machholz 1 will be caused by the encounters with Venus in 6975 AD and with Jupiter at the distance of about 0.1 AU in 6978 AD. These encounters will lead to capture into the very strong 2:1 resonance with Jupiter. The consequences of this resonance in terms of secular perturbations on Quadrantid-like meteoroid streams were investigated by Froeschlé and Scholl (1986). But the series of encounters with Jupiter in 8366–8591 AD will remove comet Machholz 1 from this type of motion. So the lifetime of the last resonance is equal to about 1,500 yr. However, the semimajor axis increases from 3 to 3.5 AU. It is interesting to note that the perihelion distance of P/Machholz 1 can be significantly less than 0.1 AU. This circumstance allows one to classify it as a sungrazing comet, and the question of the survival of the comet after hundreds of close encounters with the Sun arises as well.

4. Orbital intersections

It is known that the orbital elements determining a body's position in its Keplerian orbit, are not calculated perfectly. Thus, after integration over hundreds or even tens of orbital revolutions the body's position in its orbit becomes uncertain. For example the uncertainty in comet Halley's position after 50 revolutions becomes equal to its period (Sitarski and Ziołkowski 1986).

So due to uncertainty (biases) in cometary position, we must consider every orbital intersection as a moment of possible collision between the comet and a planet. Let us thus investigate the orbital intersections more carefully. Because the orbits of the major planets lie very close to the ecliptic plane we may consider the variations in cometary radii-vectors to ascending R_a

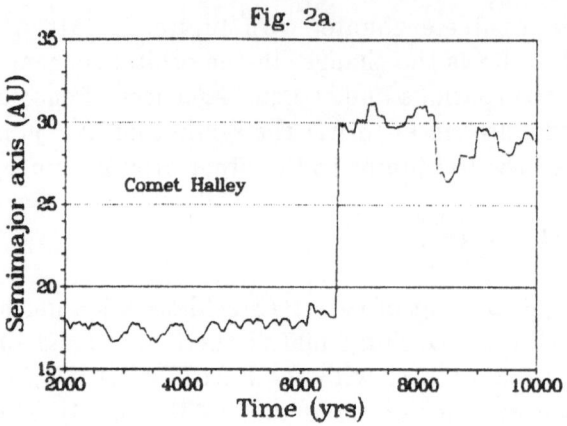

Fig. 2a.

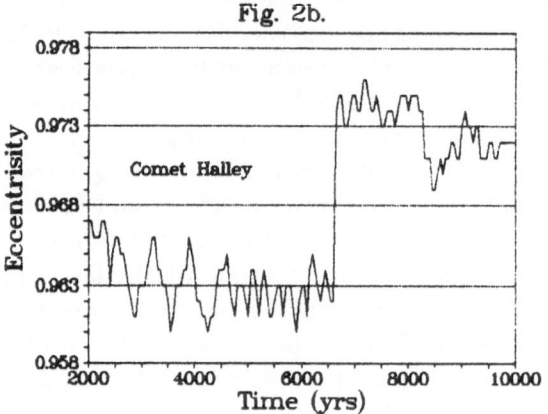

Fig. 2b.

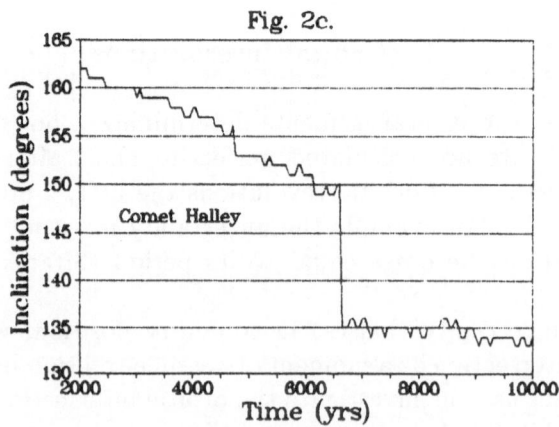

Fig. 2c.

Fig. 2.   Variations of orbital elements of comet Halley due to planetary perturbations

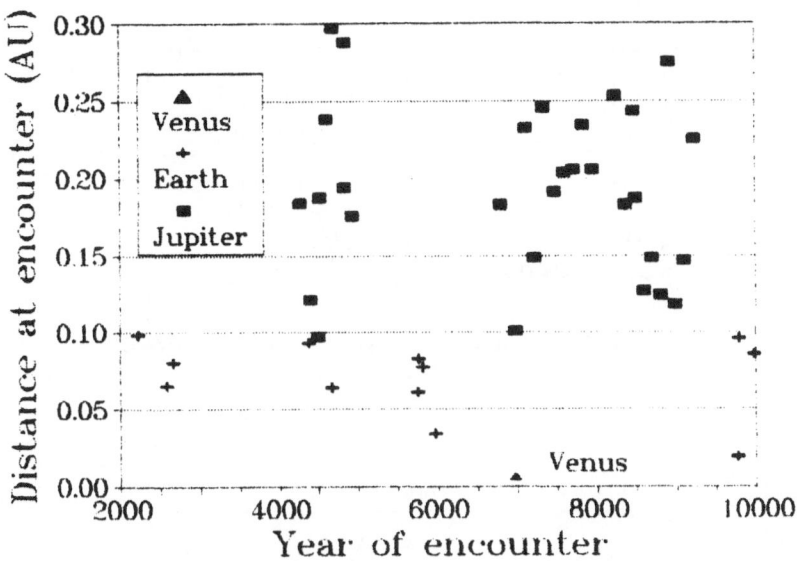

Fig. 3.   Close encounters of P/Machholz 1 with Venus, Earth and Jupiter

and descending R_d nodes. Fig. 5 shows the long-term variations of the nodal distances R_a and R_d of the P/Halley orbit. It is clear that during 8,000 yr this orbit does not intersect the orbit of any terrestrial planet but can touch the Venus orbit in 4650 AD. However the least distance between Venus and P/Halley is reached on March 4, 4868 AD. Intersections between the orbit of comet Halley and that of Jupiter may occur only after the closest encounter in 6616 AD. But there exists a possibility of a collision of comet Halley with Jupiter, as happened with comet Shoemaker-Levy 9 in July 1994. This possibility remains until the end of the time interval considered, but our calculation reveals only one close encounter at a distance of 0.319 AU on November 17, 8276 AD.

Fig. 6 gives the variations of R_a and R_d for comet Machholz 1. As seen this orbit intersects the orbits of the terrestrial planets very often: 10 intersections with Mercury's orbit, 9 with Venus' orbit, 13 with that of the Earth and 5 with Mars' orbit. Moreover, we have found 2 intersections with the orbit of Jupiter. But these intersections are possible only if comet Machholz 1 does not disintegrate after many passages near the Sun. During the gradual approach of the perihelion to the Sun (until the 25th century) there is only one intersection with the orbit of each terrestrial planet. However, we have found only one close encounter with the Earth on June 19, 2229 AD at a distance of 0.098 AU. It is necessary to note that in any case the route of the comet may be continued by large (and small) fragments of the cometary

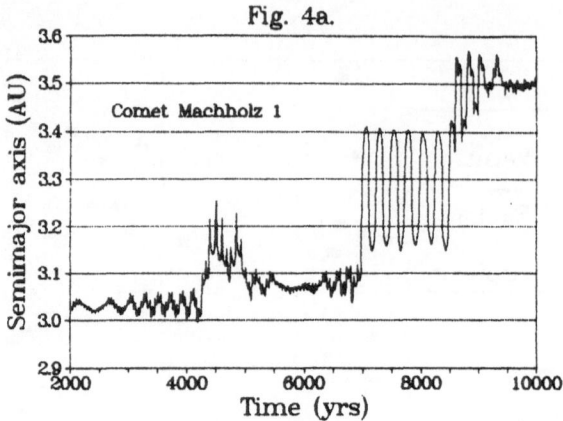

Fig. 4a.

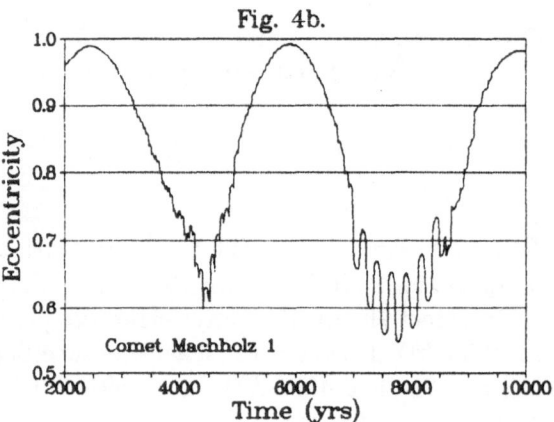

Fig. 4b.

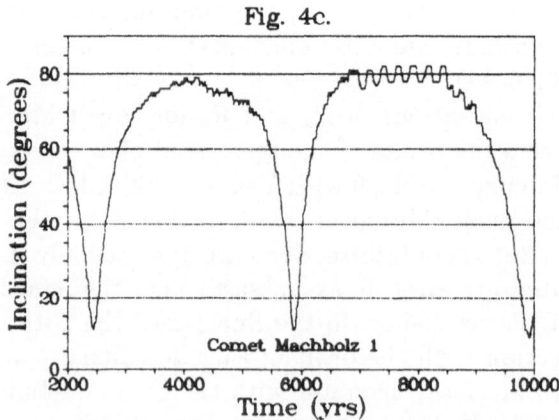

Fig. 4c.

Fig. 4. Variations of orbital elements of comet Machholz 1 due to planetary perturbations

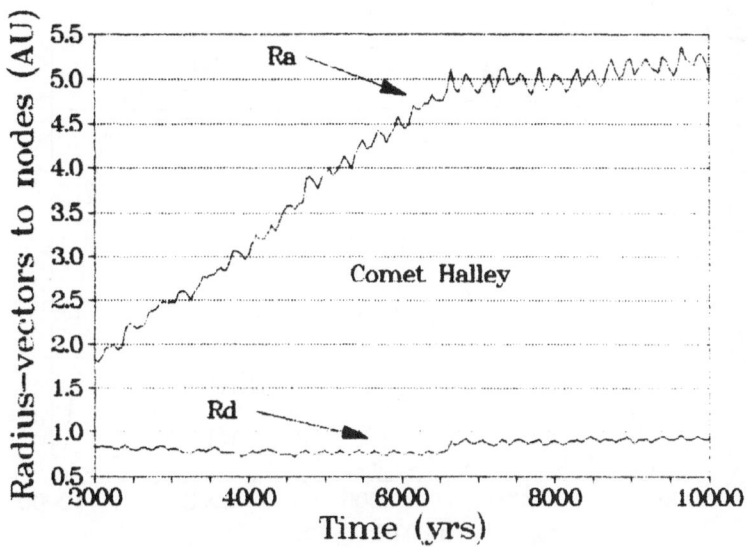

Fig. 5.   Variations of the radius-vectors to ascending R_a and descending R_d nodes of the P/Halley orbit

nucleus. Due to libration of the longitude of perihelion the comet (or its debris) can collide with the Earth only in the current periods of activity of the eight meteor showers, produced by the P/Machholz 1 meteoroid stream (Babadzhanov and Obrubov 1992, 1993).

5. Conclusions

During the nearest 8,000 yr comet Halley has a chance to collide with Venus or Jupiter. The orbit of comet Machholz 1 very often intersects the orbits of the terrestrial planets and the comet may collide with each planet including Jupiter. Close encounters play a significant role in the orbital dynamics of P/Machholz 1 and can change its type of motion and kind of resonance.

Acknowledgements

This work was supported in part by the grant of the American Astronomical Society and my participation at the Conference became possible due to a grant of the SOC.

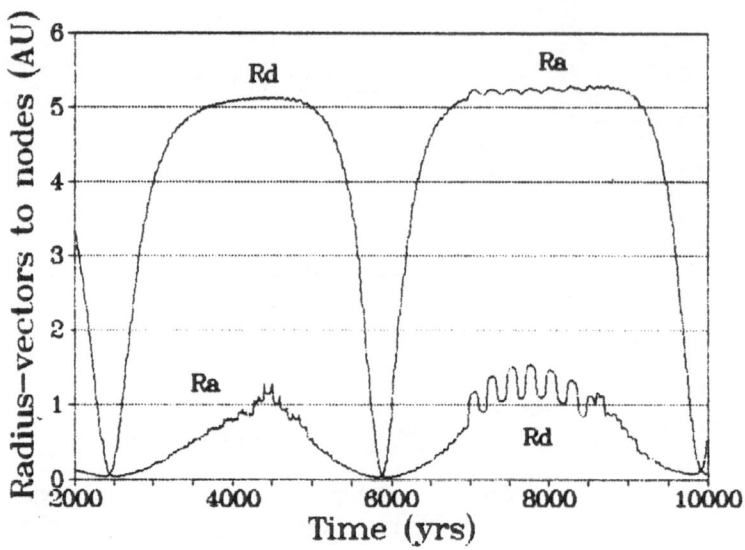

Fig. 6.   Variations of the radius-vectors to ascending R_a and descending R_d nodes of the P/Machholz 1 orbit

References

Babadzhanov, P.B., Obrubov, Yu.V.: 1992, P/Machholz 1986 III and Quadrantid meteoroid stream. Orbital evolution and relationship, in *Asteroids, Comets, Meteors 1991*, A.Harris, E.Bowell eds., Flagstaff, pp. 27-32.

Babadzhanov, P.B., Obrubov, Yu.V.: 1993, Unknown meteor showers of comet Machholz and asteroid Phaethon, *Astron. Vestnik* **27**, No 2, 110-118.

Everhart, E.: 1974, Implicit single-sequence method for integrating orbit, *Celest. Mech.* **10**, 35-55.

Froeschlé, C., Scholl, H.: 1986, Gravitational splitting of Quadrantid-like meteor streams in resonance with Jupiter, *Astron. Astrophys.* **158**, 259-265.

Green, D.W.E., Rickman, H., Porter, A.C., Meech, K.J.: 1990, The strange periodic comet Machholz, *Science* **247**, 1063-1067.

Marsden, B.G.: 1989, *Catalogue of cometary orbits*, 6th ed., IAU, 130 p.

McIntosh, B.A.: 1990, Comet Machholz and the Quadrantid meteor stream, *Icarus* **86**, 299-304.

Obrubov, Yu.V.: 1993, Long-period evolution and new meteor showers of comet P/Halley, in *Meteoroids and their parent bodies*, J. Štohl, I.P.Williams eds., Bratislava, pp. 67-72.

Olsson-Steel, D.: 1987, The dynamical life-time of P/Halley, *Astron. Astrophys.* **187**, 1-17.

Sitarski, G., Ziołkowski, K.: 1986, Investigation of the long-term motion of comet Halley: What is a cause of the discordance of results obtained by different authors?, in *Exploration of Halley's comet*, ESA SP-250, vol. III, pp. 299-302.

Yeomans, D.K., Kiang, T.: 1981, Long-term motion of Halley's comet, *Mon. Not. Roy. Astron. Soc.* **197**, 633-641.

STRUCTURES AND HAZARDS IN THE NEO SYSTEM

W.M. NAPIER

Royal Observatory, Blackford Hill, Edinburgh EH9 3HJ, U.K.

Abstract. Giant comets thrown into short–period orbits appear to be the dominant threat from space on timescales of human concern. A statistical examination of fine structure in the near–Earth object population reveals significant orbital groupings which may represent an intermediate stage in the disintegration of such bodies.

1. Introduction

Although ignored in recent risk assessments (*e.g.* The Spaceguard Survey: Morrison 1992; Chapman & Morrison 1994), the chief celestial hazard faced by mankind probably comes in the form of rare, giant comets (diameters $\gtrsim 100$ km) thrown from time to time into short–period, Earth–crossing orbits (Bailey *et al.* 1994). This is because they are the main carriers of mass into the inner solar system, and because, when such a comet disintegrates, the Earth is inevitably immersed within its material. The climatic effects of stratospheric dusting so induced may be prolonged and severe: the resulting 'cosmic winter' differs from a nuclear or impact one in that the duration of the stress may equal or exceed crucial time constants in the climatic system, such as the cooling time of the oceans (~ 10 yr) or the growth time of ice caps ($\sim 10\,000$ yr). Further, the target area presented by their disintegration products may be many powers of ten higher than that of a 1 km asteroid, yielding an erratic and virtually unpredictable impact hazard. It is not clear therefore that the impact hazard deduced from lunar cratering data (which is averaged over $\sim$3.9 Gyr) has much relevance to that expected over any given 0.1–1 Myr interval. A fuller discussion is given by Bailey *et al.* (1994), and only a few recent developments relevant to the 'giant comet' hypothesis are noted here: (i) recent surveys have revealed a significant population of large trans–Saturnian and trans–Neptunian bodies, presumed to be of cometary or quasi–cometary nature (Luu 1994, Levison & Duncan 1990); (ii) fast dynamical routes linking Chiron– and Halley–type orbits to short–period, Earth–crossing orbits have been discovered (see Bailey *et al.* 1994), the traversal times being $\sim 10^5$–10^6 yr; and (iii) cometary splitting appears now to be established as a common mode of disintegration (Chen & Jewitt 1994), with P/Comet Shoemaker–Levy 9 having convincingly demonstrated at least one mode of breakup potentially relevant to terrestrial phenomena (Wickramasinghe & Wallis 1994, and these proceedings).

Given these data, Bailey *et al.* (1994) estimate the cumulative flux of long–period comets, at the high mass end, to be $F \sim 1 \times (d/5)^2$ AU^{-1} yr^{-1}, d the diameter of the comet in km. A Chiron–sized body is therefore expected to cross the Earth's orbit once within the timescale of civilization, while the injection rate of giant comets from a chaotic, trans–Saturnian orbit into a *stable* Earth–crossing one is of order $10 \, Myr^{-1}$ (Hahn & Bailey 1990). If such a body has active lifetime 10^3–$10^4 \, yr$, the Earth acquires a stratospheric dust veil of optical depth $0.05 \lesssim \tau \lesssim 0.3$. All these numbers are uncertain by a factor of a few, but the mean interval $\sim 10^5 \, yr$ between arrival of giant short–period comets is comparable to that between ice ages in the present Cenozoic. Since the last ice age ended very recently ($\sim 11\,000$ BP), it is tempting to look for a smoking cometary gun, either in the ground or above it. I discuss here preliminary evidence for the latter.

2. Near–Earth asteroid groups

The data employed were the 315 NEO's known to July 1994. Given the frequency of comet splitting, one might search for recently extinct comets in the form of pairs of co–orbiting asteroids (*cf.* Steel *et al.* 1992). In Table 1 are listed 12 possible twin asteroids (six pairs) derived from the criterion that $\delta q \leq 0.02$, $\delta e \leq 0.02$, $\delta i \leq 3°$ and $\delta \varpi \leq 60°$ ($\varpi = \Omega + \omega$). Because of their rapid precession, (Ω, ω) were omitted. Forward and backwards integrations over $10^5 \, yr$ (by Dr David Asher) reveal that these twins are co–orbiting over this period.

pair	q	a	e	i	ϖ
1988 TA	0.803	1.54	0.479	2.5	300
1990 UQ	0.810	1.55	0.478	3.6	295
1993 HP1	0.973	1.98	0.509	8.0	189
1993 HC	0.980	1.99	0.507	9.4	148
1989 DA	0.986	2.16	0.544	6.4	129
1992 SY	0.994	2.21	0.550	8.0	121
2061 Anza	1.051	2.27	0.537	3.7	4
1987 SF3	1.052	2.25	0.533	3.3	320
1993 TQ2	1.153	1.99	0.419	6.0	91
1993 HO1	1.159	1.99	0.417	5.9	128
1986 NA	1.168	2.13	0.450	10.3	280
1992 SZ	1.176	2.18	0.460	9.3	319

Table 1. Probable asteroid twins.

A sample of $N = 315$ bodies can be paired in $\frac{1}{2} N^2 \sim 50\,000$ ways, and it is not immediately obvious that these apparent twinnings are significant. To assess this, comparison was made with synthetic near–Earth object (NEO) systems. The latter were constructed by adding, to the orbital elements of each asteroid, random numbers uniformly distributed in the range ± 0.2 AU in a, ± 0.05 in e, $\pm 4°$ in i and $\pm 45°$ in Ω and ω. This procedure preserves, in the synthetic NEO systems, all the major properties of the real one including the selection effects. The frequency distribution of chance pairs satisfying the above criterion was determined from these artificial NEO systems. In this way a chance expectation of 0.14 pairs was determined, as against 6 observed. The adopted criterion was more or less arbitrary, but such twinnings are easily detected even with large changes in the allowed ranges. Thus with $(\delta q, \delta e, \delta i, \delta \varpi)= (\pm 0.02$ AU, ± 0.02, $\pm 3°$, $\pm 60°)$ one obtains 15 pairs as against an expectation of 0.5 from 100 trials with synthetic systems (Fig. 1). The twinning phenomenon thus appears to be real at a high confidence level.

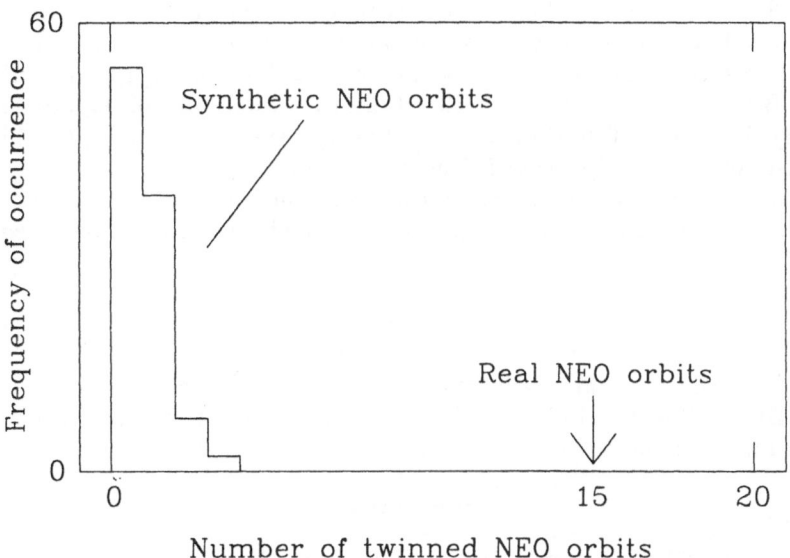

Fig. 1. Expected versus observed NEO twinnings as discussed in the text.

While the identification of such pairs as the degassed remnants of split comets remains conjectural, physical splitting of comets is the only mechanism currently known which could produce Earth–crossing twins in such abundance.

Similar trials were carried out on the candidate asteroids of the Taurid complex (Napier 1993, Asher *et al.* 1994). (a, e, i, ϖ) boxes of various sizes were constructed, centred on P/Encke's Comet, and comparisons made with synthetic NEO systems. These exercises confirm the reality of the system at a high confidence level. For example searching a box of width $(\delta a, \delta e, \delta i, \delta \varpi)= (\pm 0.2$ AU, ± 0.2, $\pm 8°$, $\pm 30°)$ yields 7 Taurid asteroids which are real at an overall confidence level $\sim 99.8\%$, with 1.3 interlopers. Larger boxes reveal more TC asteroids.

Valsecchi *et al.* (these proceedings) have discovered 'entry corridors' linking regions of the short–period comet system and the main asteroid belt respectively to the (a, e, i) phase space of the Taurid complex asteroids. They argue that the latter may be physically unconnected, and point out that the current D–criteria for this group would require the progenitor to have undergone a recent, drastic disruption, with fragment ejection velocities of several $km\,s^{-1}$. Such velocities are inconsistent with the results of hypervelocity experiments (a similar, unsolved problem exists for the main belt asteroid families: Farinella et al. 1993).

However the entry routes found by Valsecchi et al. do not predict the strong concentration of TC asteroids in the longitude range $140^0 \lesssim \varpi \lesssim 180^0$ straddling P/Encke's Comet (Fig. 1 of Asher et al. 1994). There are no obvious selection effects capable of yielding this concentration of ϖ, nor is it clear that the source regions are adequate to replenish the Taurid asteroid system on a one–to–one basis (Napier & Clube, in preparation). Now it has long been recognized that the progenitor of the Taurid meteors and the Stohl streams, within which the TC asteroids are immersed, must have been a comet of exceptional size. The evidence thus appears to be consistent with the hypothesis that a giant comet arrived through a Valsecchi 'gate' $\lesssim 10^5$ yr BP and underwent a hierarchy of disintegrations, one of its late fragments generating the current Taurid meteors 5 000 yr ago, the Piscids and χ Orionids having formed from it up to 18 000 yr BP, at the peak of the last ice age. Further work is required to assess whether rapid dynamical evolution and asymmetric outgassing of the cometary fragments around perihelion could yield the currently observed spread in orbital elements.

Acknowledgements

I am indebted to to Mark Bailey, Victor Clube and Giovanni Valsecchi for discussions, to Duncan Steel and a referee for critical reviews of this paper, and to David Asher for numerical integrations.

References

Asher, D.J., Clube, S.V.M., Napier, W.M. & Steel, D.I.: 1994, *Vistas in Astron.* **38**, 1.

Bailey, M.E., Clube, S.V.M., Hahn, G., Napier, W.M. & Valsecchi, G.B.: 1994, In *Hazards due to Comets and Asteroids.* (ed. T. Gehrels), University of Arizona Press, Tucson (in press).

Chapman, C.R. & Morrison, D.: 1994, *Nature* **367**, 33.

Chen, J. & Jewitt, D.: 1994, Icarus **108**, 265.

Farinella, P., Froeschlé, C. & Gonczi, R.: 1993, *IAU Symp. 160*, Asteroids, Comets, Meteors (eds. A. Milani et al.), 205, Kluwer, Dordrecht.

Hahn, G. & Bailey, M.E.: 1990, *Nature* **348**, 132.

Levison, H. & Duncan, M.: 1990, *Astron. J.* **100**, 1669.

Luu, J.: 1994, *IAU Symp. 160*, Asteroids, Comets, Meteors (eds. A. Milani et al.), 31, Kluwer, Dordrecht.

Morrison, D.: 1992, *The Spaceguard Survey: Report of the NASA International Near–Earth–Object Detection Workshop* (Pasadena: Jet Propulsion Laboratory).

Napier, W.M.: 1993, In *Meteoroids and their Parent Bodies* (eds. J. Stohl & I.P. Williams), 123, Slovak Acad. of Sciences, Bratislava.

Steel, D.I., McNaught, R.H. & Asher, D.J.: 1992, *Minor Planet Bull.* **19**, 9.

Valsecchi, G., Morbidelli, A., Gonczi, R. et al.: 1994, *Earth, Moon, Planets*, this issue.

Wickramasinghe, N.C. & Wallis, M.K.: 1994, *Mon. Not. R. astr. Soc.* **270**, 420.

IS 2329 ORTHOS A DEAD COMET?

P.B. BABADZHANOV

Institute of Astrophysics, Dushanbe, 734670 Tajikistan

Abstract. Probably most meteor showers have a cometary origin. Investigation of Near-Earth asteroids' orbital evolution to determine whether they have related meteor showers is necessary to determine which asteroids evolved from comets. The results of calculations show that asteroid Orthos' orbit is an octuple Earth-crosser. Therefore, if Orthos has an old meteoroid stream it may produce eight meteor showers observable on the Earth. The existence of four Orthos' Northern meteor showers is confirmed by our search in the published catalogues of meteor radiants and orbits or in the archives of the IAU Meteor Data Center (Lund, Sweden).

1. Introduction

Disintegration of cometary nuclei and as a result the formation of meteoroid streams are not in doubt. Thus the relationship between comets and meteor showers has been studied comparatively well and comets are considered to be the main source of meteoroids.

The problem of the possible connection between the asteroids and meteor matter is not new. The existence of meteor showers connected with asteroids Hermes (1937 UB), Apollo (1862) and Adonis (2101) was proposed by Hoffmeister in 1948. Lovell (1954) and Levin (1956) considered whether sporadic meteors have an asteroidal origin but meteor showers have a cometary origin. The discovery of 3200 Phaethon moving along the orbit of the Geminids meteor shower (Whipple 1983) was an important stimulus to investigations of the asteroid-meteor shower relationship (Fox *et al.* 1984; Babadzhanov and Obrubov 1987; Olsson-Steel 1988).

With the increased observational data on meteoroids and the discovery of an increased number of Earth-approaching asteroids the number of papers discussing the possible relationship between meteor showers and asteroids increased too (Sekanina 1973, 1976; Babadzhanov and Obrubov 1983; Drummond 1982; Olsson-Steel 1988; Kresák and Štohl 1989; Hasegawa *et al.* 1992; Hasegawa 1993; and others). Concerning this relationship, we first need to consider the Earth-approaching asteroids. According to one point of view a source of such asteroids may be the main belt of asteroids (Farinella *et al.* 1992). According to another hypothesis suggested by Öpik (1963), the majority of near Earth asteroids are extinct comets. Wetherill (1991) proposed Apollo objects were evenly divided between both sources.

Although both cometary and asteroidal origins have long been studied as the source of Near Earth Objects, which contribution is dominant still seems to be an unsettled problem. We propose that in order to distinguish

Earth, Moon, and Planets **72**: 305-310, 1996.

between the two origins it is important to determine which asteroids have associated meteor showers. If some asteroid is extinct comet, it may have relative meteoroid stream and possible meteor showers.

Usually, when determining the theoretical geocentric radiants of comets or asteroids, it is proposed that corresponding meteoroid streams are very flat or have relatively small thicknesses, narrow at perihelion and wide at aphelion. Such a notion corresponds to relatively young meteoroid streams and permit us to determine theoretical radiants of only one or two possible meteor showers. It is known that the meteoroids are ejected from the parent bodies with different velocities and, in the case of comets, at different points in the orbit and at different times. These processes lead to the initial dispersion of the meteoroid stream orbits. Since the ejection velocities are small (from few m/s to 1 km/s) in comparison with the parent body's orbital velocities $(10 - 100$ km/s), initially the meteoroids are spread basically along the parent body's orbit (due to differences in the orbital periods).

As previously shown (Babadzhanov and Obrubov 1987, 1989, 1992), the differences in the planetary perturbing action on the stream meteoroids cause meteoroid orbits to be at different evolutionary stages. As a result one meteoroid stream may produce from one to eight meteor showers depending on the Earth crossing class of the parent body's orbit. For example, the orbit of 3200 Phaethon is a quadruple crosser, and so it might be expected that Phaethon may have 4 meteor showers. The orbit of P/Machholz 1986 VIII is an octuple crosser and therefore it may produce 8 meteor showers (Babadzhanov and Obrubov 1983). In this paper we present the results of searching for asteroid Orthos' meteor showers.

2. Meteor showers of asteroid 2329 Orthos

The asteroid 2329 Orthos discovered in 1976 has following orbital elements (Batrakov 1988):

$$a = 2.402 \text{ AU}, \quad e = 0.659, \quad q = 0.820 \text{ AU},$$
$$\omega = 145°.78, \quad \Omega = 168°.87, \quad i = 24°.41 \quad\quad (1950.0)$$

Results of the calculations of secular perturbations show that Orthos' orbit is an octuple Earth-crosser. Therefore, if Orthos has an old meteoroid stream, then it may produce eight meteor showers, as does P/Machholz.

We calculated the secular variations of Orthos' orbital elements for the time interval of 12.5 millennia, which embraces one cycle of the argument of perihelion (Table I).

Fig. 1 shows secular variation in inclination i, eccentricity e and in radii-vectors of the ascending R_a and descending R_d nodes of Orthos' orbit versus the argument of perihelion ω. These dependencies show that the inclination and eccentricity of Orthos change within sufficiently wide range. Fig. 1 confirms that Orthos is an octuple Earth-crosser and hence it may produce 8 meteor showers observable on the Earth. Four of them have $i \simeq 17°$ and perihelion distances $q \simeq 0.75$ AU, as well as four other showers with $i \simeq 30°$ and $q \simeq 0.97$ AU.

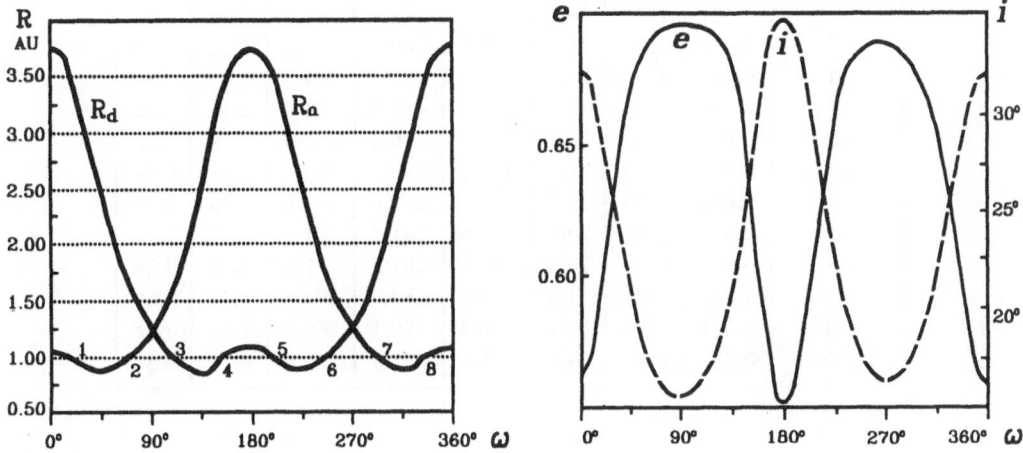

Fig. 1.   The dependence of eccentricity e, inclination i and radii-vectors to ascending R_a and descending R_d nodes of the asteroid 2329 Orthos on perihelion argument ω. The intersection (1) corresponds to the Mensaids, (2) μ-Lupids, (3) ζ-Bootids, (4) ι-Draconids, (5) γ-Draconids, (6) ω-Herculids, (7) Centaurids, (8) ι-Carinids.

Theoretical coordinates of geocentric radiants (α, δ), velocities V_G and solar longitudes L_0, at which 8 of Orthos' meteor showers may manifest their activity, are presented in Table II. They are named according to the position of their theoretical radiants.

In the known catalogues of meteor showers and associations there are some radiants close to 2 of 8 predicted Orthos' radiants. For example, the η-Draconids ($L_0 = 169°, \alpha = 249°, \delta = 63°, V_G = 20$ km/s) of Sekanina's (1973) catalogue correspond to theoretically predicted ι-Draconids. The value of Southworth and Hawkins' (1963) D-criterion between the observed η-Draconids and theoretical ι-Draconids results in $D = 0.07$ only, but their radiants differ significantly ($> 20°$) in right ascension.

The Orthos theoretical radiant ($L_0 = 168°.9, \alpha = 219°.0, \delta = 45°.8, V_G = 18.3$ km/s) calculated by Hasegawa et al. (1992) is very close to the ι-Draconids. The difference in these theoretical radiants can probably be understood in terms of planetary perturbing action.

TABLE I

Secular variations of the orbital elements of 2329 Orthos. $T = 0$ corresponds to 1950.0

$T \times 10^{-3}$ yrs	e	q AU	i°	Ω°	ω°	π°	R_d AU	R_a AU
0	0.659	0.82	24.4	168.8	145.8	314.6	0.88	2.98
-1	0.697	0.73	16.1	208.1	105.4	313.5	1.04	1.52
-2	0.676	0.78	19.6	263.7	49.4	313.2	2.33	0.90
-3	0.614	0.93	27.8	285.9	23.8	309.8	3.41	0.96
-4	0.565	1.05	31.7	295.8	5.4	301.2	3.74	1.05
-5	0.575	1.02	30.7	304.0	346.4	290.4	3.64	1.03
-6	0.633	0.88	25.0	317.4	326.2	283.6	3.04	0.94
-7	0.683	0.76	17.3	352.2	289.6	281.8	1.66	1.04
-8	0.678	0.77	20.0	47.6	234.2	281.8	0.93	2.15
-9	0.619	0.92	29.0	72.4	207.1	279.4	0.96	3.30
-10	0.561	1.05	34.0	82.7	188.9	271.6	1.06	3.69
-11	0.561	1.06	33.9	90.4	170.1	260.5	1.06	3.68
-12	0.617	0.92	28.5	101.5	151.3	252.8	0.97	3.24

TABLE II

Theoretical and observed geocentric radiants of 2329 Orthos' meteor showers

Shower	Theoretical				Observed			
	L_0°	α°	δ°	V_G km/s	L_0°	α°	δ°	V_G km/s
ω-Herculids	68	244	10	20	59	241	9	22
ζ-Draconids	114	270	51	20	128	269	51	20
ι-Draconids	158	228	61	20	154	229	59	21
ζ-Bootids	205	218	14	20	203	219	14	21
μ-Lupids	66	228	-47	20	Not observed			
Mensaids	117	102	-82	19	Not observed			
ι-Carinids	156	133	-60	19	Not observed			
Centaurids	199	191	-40	20	Not observed			

In Astapovich's (1973) list of photographic radiants of small meteor showers and in Lindblad's (1971) list we found radiant of ζ-Draconids also close to our theoretically predicted ζ-Draconids. Astapovich notes that this shower

TABLE III

Theoretical (T) and observed (O) orbital elements of 2329 Orthos' meteor showers

Shower		a AU	e	q AU	$i°$	$\Omega°$	$\omega°$	D
		Northern:						
ω-Herculids	T	2.40	0.687	0.754	17.9	67.9	246.7	0.08
	O	2.38	0.709	0.693	19.5	59.0	256.1	
ζ-Draconids	T	2.40	0.595	0.973	31.4	114.0	200.6	0.03
	O	2.52	0.606	0.993	31.4	128.4	200.2	
ι-Draconids	T	2.40	0.597	0.968	30.6	157.5	157.1	0.13
	O	3.54	0.722	0.983	31.0	154.0	159.3	
ζ-Bootids	T	2.40	0.696	0.730	16.4	204.7	109.9	0.06
	O	2.97	0.750	0.743	16.6	203.0	114.0	
		Southern:						
μ-Lupids	T	2.40	0.682	0.740	16.6	246.1	68.5	
Mensaids	T	2.40	0.592	0.980	29.7	297.3	17.3	
ι-Carinids	T	2.40	0.596	0.970	28.9	336.0	338.6	
Centaurids	T	2.40	0.680	0.768	17.9	19.2	295.4	

$(L_0 = 140°, \alpha = 271°, \delta = 54°, V_G = 22$ km/s$)$ is observed visually from XIX century and Lindblad found that it $(L_0 = 146°, \alpha = 269°, \delta = 59°, V_G = 24$ km/s$)$ is identical with Denning's shower No 198.

The observed ι-Draconids $(L_0 = 99°, \alpha = 280°, \delta = 64°, V_G = 22$ km/s$)$ by Sekanina (1973) is close to the theoretically predicted γ-Draconids $(D = 0.18)$, but their radiant differs by $10°$ in α and by $13°$ in δ.

In order to obtain confirmations of the relation of Orthos with meteor showers we carried out a search of individual meteors probably belonging to these showers in the catalogues of the IAU Meteor Data Center (Lindblad 1987). We assumed the dispersion of dates of activity of Orthos meteor showers within one month. The difference of observed radiants from theoretical ones was assumed to be within $\Delta\alpha = \Delta\delta = \pm5°$ and that of velocity within $\Delta v = \pm5$ km/s. We found 13 ω-Herculid meteors, 19 ι-Draconids, 17 ζ-Draconids and 6 ζ-Bootids. The average radiants and velocities of observed meteors are presented in Table II and orbital elements in Table III. The values of D-criterion given in the last column of Table III show a good accordance between theoretical and observed data.

The data of these tables show that the predicted and observed radiants, velocities and orbital elements of four northern showers of asteroid Orthos are in satisfactory agreement. For the remaining four southern predicted showers we could not find any observed shower or meteors. The existence of meteor showers connected with 2329 Orthos is an indication that this asteroid probably is a defunct comet.

Acknowledgements

This work was supported in part by the International Science Foundation grant MZ1000.

References

Astapovich, I.S..: 1973, *Astronomicheskij kalendar* (Perm. vol.), p. 579.
Babadzhanov, P.B. and Obrubov, Yu.V.: 1983, *Asteroids, Comets, Meteors*, ed. C.-I. Lagerkvist and H. Rickman, Reprocentralen HSC, Uppsala, p. 411.
Babadzhanov, P.B. and Obrubov Yu.V.: 1987, in *Interplanetary Matter*, ed. Z. Ceplecha and P. Pecina, Publ. Astron. Inst. Czecho-Sl. Acad.Sci., 2, Praha, p. 141.
Babadzhanov, P.B. and Obrubov, Yu.V.: 1989, in *Highlights of Astronomy*, ed. R. West, p. 287.
Babadzhanov, P.B. and Obrubov, Yu.V.: 1992, *Celest. Mech. Dyn. Astron.* 54, 111.
Babadzhanov, P.B. and Obrubov, Yu.V.: 1992, in *Asteroids, Comets, Meteors 1991*, ed. A. Harris and E. Bowell, LPI, Houston, p. 27.
Babadzhanov, P.B. and Obrubov, Yu.V.: 1993, in *Meteoroids and their parent bodies* (eds. J. Štohl and I.P. Williams), Astron. Inst. Slovak Acad. Sci., Bratislava, p. 49.
Batrakov, Yu.V. (editor): 1988, *Ephemerides of minor planets for 1989* (Institute of Theoretical Astronomy, Leningrad), 384 p.
Drummond, J.D.: 1982, *Icarus* 49, 143.
Farinella, P., Gonczi, R., Froeschlé, Ch. and Froeschlé, Cl.: 1992, in *Asteroids, Comets, Meteors 1991*, ed. A. Harris and E. Bowell, LPI, Houston, p. 167.
Fox, K., Williams, I.P. and Hughes, D.W.: 1984, *Mon. Not. Roy. Astron. Soc.* 208, 11.
Hasegawa, I.: 1993, *WGN, the Journal of the IMO* 21, 29.
Hasegawa, I., Ueyama, Y. and Ohtsuka, K.: 1992, *Publ. Astron. Soc. Japan* 44, 45.
Hoffmeister, C.: 1948, *Meteorströme*, Verlag Johann Ambrosius Barth, Leipzig, 286 p.
Kresák, L. and Štohl, J.: 1990, in *Asteroids, Comets, Meteors III*, ed. C.-I. Lagerkvist, H. Rickman, B.A. Lindblad and M. Lindgren, Reprocentralen HSC, Uppsala, p. 379.
Levin, B.Yu.: 1956, *Fiziçeskaja teoria meteorov i meteornoe veschestvo v solnecnoj sisteme*, Acad. Nauk., Moskva, 293 p.
Lindblad, B.A.: 1971, *Smiths. Contrib. to Astrophys.* 12, 1.
Lindblad, B.A.: 1987, in *Interplanetary Matter*, ed. Z. Ceplecha and P. Pecina, Publ. Astron. Inst. Czecho-Sl. Acad. Sci., 2, Praha, p. 201.
Lovell, A.C.B.: 1954, *Meteor Astronomy*, Clarendon Press, Oxford.
Olsson-Steel, D.: 1988, *Icarus*, 75, 64.
Öpik, E.J.: 1963, *Adv. Astron. Astrophys.* 2, 219.
Sekanina, Z.: 1973, *Icarus* 18, 253.
Sekanina, Z.: 1976, *Icarus* 27, 265.
Southworth, R.B. and Hawkins, G.S.: 1963, *Smiths. Contrib. Astrophys.* 7, 261.
Whipple, F.L.: 1983, *IAU Circ.* 3881.
Wetherill, G.W.: 1991, in *Near-Earth Asteroids*, (Abstracts), p. 37.

POSSIBLE ASSOCIATIONS OF DAYTIME FIREBALLS AND MINOR PLANETS

ICHIRO HASEGAWA

Otemae Junior College, Inano, Itami 664, Japan

Abstract. Chinese and Japanese historical records of daytime fireballs, and world-wide daytime meteorite falls in the catalogue have been investigated. Among them, there are 253 and 104 records of great daytime fireballs in China and in Japan respectively, and 506 meteorite falls in the daytime are recorded in the *Catalogue of Meteorites* (1985).

The same trends of seasonal and daily variations in the flux of daytime fireballs are clearly seen in both Chinese and Japanese records, and then the distributions of the daytime fireballs seem to suggest the association with meteorites and near-earth minor planets rather than with comets.

Possible relations with minor planets, such as (1566) Icarus, (3671) Dionysius, (4450) Pan, (4486) Mithra and others are suggested.

1. Introduction

Historical variations in the meteor flux as found in Chinese and Japanese records are reported elsewhere (Hasegawa 1992). The daytime β Taurids observed during June and July are considered to be members of the Taurid complex, and it is expected that they may be included in the historical records of daytime fireballs. On the other hand, a daytime fireball is thought to be a meteorite fall, and then the meteorites are considered to be associated with near-earth minor planets.

The records of daytime fireballs are easier to find than those of fireballs during the nighttime, and a catalogue of the daytime fireballs recorded in Chinese and Japanese histories was compiled (Hasegawa 1995). Records of daytime meteorite falls are extracted from the *Catalogue of Meteorites* (Graham et al. 1985).

Because the key data used in the identifications in this study are only the solar longitudes, we can not make sure the relations between daytime fireballs, meteorites and near-earth minor planets; however, we present here some possible associations among them.

2. Distributions and Associations

Both the distributions of local times of daytime fireball apparition and of meteorite fall show maxima at 16 and 18 o'clock, which are different from the case of cometary meteors.

Earth, Moon, and Planets **72**: 311-316, 1996.

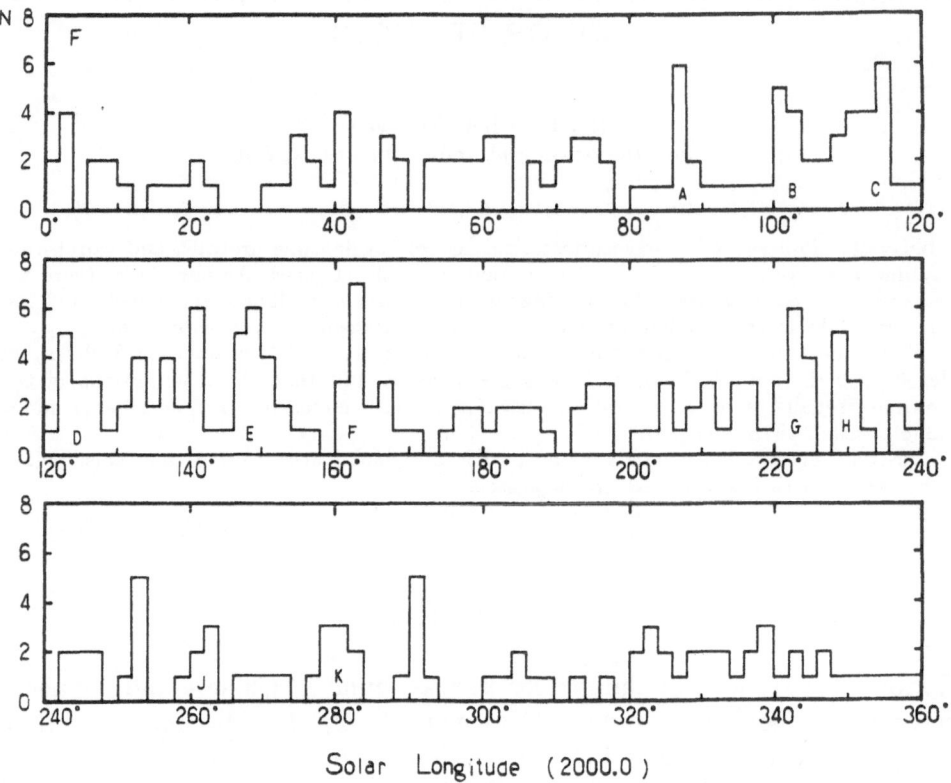

Fig. 1.   The number of daytime fireball records every two degrees in the solar longitude (2000.0 equinox).

The maxima in the seasonal variations in the flux of these two kinds of bodies occur during June and September, which are also different from the cometary meteors in nighttime.

In Figures 1 and 2, the distributions of the number of daytime fireballs (F) and daytime meteorite falls (M) in the solar longitude are shown respectively, and the mean numbers of records every two degrees in the solar longitude are:

Daytime fireballs: Mean = 1.8 ± 1.48,

Daytime meteorite falls: Mean = 2.8 ± 2.11.

Referring to these statistical parameters, possible associations or groups of daytime fireballs and meteorite falls are found and given in the attached tables. For the Association #6, it is noteworthy that the Kagarlyk meteorite (p. 184 in Graham et al. 1985) fell five hours later then the Tunguska event (p. 354) on the same day!

TABLE I
Associations of Daytime Fireballs, Daytime Meteorites
and Near Earth Minor Planets

	Association	Solar Longitude (2000.0), L	Range of L	Numbers
#1	Meteorite Falls	17.8, 18.0, 18.1, 18.2, 18.7, 18.8	1.0	6
#2	Meteorite Falls	50.3, 50.3, 50.7, 50.9, 51.0, 51.2, 51.6	1.3	7
#3	Meteorite Falls	63.2, 63.4, 63.5, 63.8, 64.2, 64.5	1.3	6
#4	Meteorite Falls	82.3, 82.5, 82.9, 83.5, 83.5, 83.7, 83.9	1.6	7

Prediction of Meteor Radiant Point associated with
(1566) Icarus: $L = 83.2$, $\alpha = 49°$, $\delta = +32°$, $\Delta = 0.037$ AU
1991 BA: $L = 84.3$, $\alpha = 94°$, $\delta = +26°$, $\Delta = 0.037$ AU
1991 OA: $L = 84.0$, $\alpha = 168°$, $\delta = -15°$, $\Delta = 0.065$ AU

#5	Meteorite Falls	86.6, 86.8, 87.0, 87.9, 88.5, 88.7, 89.7, 89.9, 89.9, 90.2, 90.4, 90.9, 90.9	4.3	13
	Daytime Firballs	86.2, 86.3, 86.8, 87.5, 87.9, 88.4	2.2	6

Prediction of Meteor Radiant Point associated with
(3671) Dionysius: $L = 90.0$, $\alpha = 90°$ $\delta = +35°$, $\Delta = 0.033$ AU
1989 UR: $L = 85.2$, $\alpha = 80°$, $\delta = +1°$, $\Delta = 0.094$ AU

#6	Meteorite Falls	98.7, 98.8, 99.1(Tunguska), 99.1, 99.2, 99.2, 99.3(Kagarlyk), 99.3, 100.2, 100.3, 100.3	2.0	11
	Daytime Fireballs	99.4, 100.2, 100.6, 101.2	1.8	4

Prediction of Meteor Radiant Point associated with
1991 JX: $L = 97.3$, $\alpha = 185°$, $\delta = -7°$, $\Delta = 0.036$ AU

#7	Meteorite Falls	112.6, 112.6, 112.6, 113.1, 113.8, 113.8, 113.8, 114.0, 114.1, 114.5, 114.8, 114.9, 115.1, 115.1	2.5	14
	Daytime fireballs	113.1, 113.8, 114.6, 114.7, 114.8, 115.1, 115.5, 115.7	2.6	8

Prediction of Meteor Radiant Point associated with
(3753) 1986 TO: $L = 113.6$, $\alpha = 95°$, $\delta = +51°$, $\Delta = 0.081$ AU

Table I (continued)

	Association	Solar Longitude (2000.0), L	Range of L	Numbers
#8	Meteorite falls	121.6, 122.6, 122.9	1.3	3
	Daytime Fireballs	121.9, 122.0, 122.2, 122.3, 122.6	0.7	5

Prediction of Meteor Radiant Point associated with
1992 LR: $L = 120.9$, $\alpha = 206°$, $\delta = -16°$, $\Delta = 0.033$ AU

| #9 | Meteorite Falls | 139.0, 139.1, 139.3, 139.5, 139.9, 140.2 | 1.2 | 6 |

Prediction of Meteor Radiant Point associated with
(4486) Mithra: $L = 140.7$, $\alpha = 154°$, $\delta = +14°$, $\Delta = 0.045$ AU
1987 SF3: $L = 141.4$, $\alpha = 229°$, $\delta = - 6°$, $\Delta = 0.053$ AU
1989 UQ : $L = 140.9$, $\alpha = 126°$, $\delta = +23°$, $\Delta = 0.014$ AU

| #10 | Meteorite Falls | 146.9, 147.5, 148.3, 149.2 | 2.3 | 4 |
| | Daytime Fireballs | 146.6, 147.0, 147.3, 147.7, 148.2, 148.6, 148.9 | 2.3 | 7 |

Prediction of Meteor Radiant Point associated with
(3362) Khuhu: $L = 147.9$, $\alpha = 140°$, $\delta = +33°$, $\Delta = 0.037$ AU
1972 RB: $L = 149.8$, $\alpha = 240°$, $\delta = + 4°$, $\Delta = 0.042$ AU

| #11 | Meteorite Falls | 157.7, 157.7, 158.1, 158.5, 158.5 | 0.8 | 5 |

Prediction of Meteor Radiant Point associated with
(1620) Geographos: $L = 157.4$, $\alpha = 145°$, $\delta = -23°$, $\Delta = 0.010$ AU
(4450) Pan: $L = 159.7$, $\alpha = 155°$, $\delta = + 2°$, $\Delta = 0.045$ AU
1986 JK: $L = 158.7$, $\alpha = 187°$, $\delta = - 4°$, $\Delta = 0.037$ AU
1991 EE: $L = 159.3$, $\alpha = 188°$, $\delta = +21°$, $\Delta = 0.030$ AU

| #12 | Meteorite Falls | 162.6, 163.1, 163.4, 163.9, 164.2, 164.4, 164.5, 164.8 | 2.2 | 8 |
| | Daytime Fireballs | 162.4, 163.0, 163.1, 163.7, 164.3, 164.9 | 2.5 | 6 |

Prediction of Meteor Radiant Point associated with
1978 CA: $L = 161.4$, $\alpha = 194°$, $\delta = +68°$, $\Delta = 0.018$ AU
1990 SM: $L = 164.3$, $\alpha = 164°$, $\delta = +19°$, $\Delta = 0.091$ AU

| #13 | Meteorite Falls | 201.7, 201.8, 202.0, 202.5, 202.7, 202.9, 203.0, 203.3, 203.3 | 1.6 | 9 |

Prediction of Meteor Radiant Point associated with
(1917)Cuyo : $L = 201.5$, $\alpha = 279°$, $\delta = +53°$, $\Delta = 0.100$ AU

TABLE I (continued)

Association	Solar Longitude (2000.0), L	Range of L	Numbers
#14 Meteorite Falls	222°.6, 224°.8, 224°.9, 225.1	2°.5	4

Prediction of Meteor Radiant Point associated with
1990 UN: L = 222.9, α = 29°, δ = + 3°, Δ = 0.036 AU

#15 Meteorite Falls	230.4, 230.9, 230.9, 231.3, 231.5, 231.5, 231.9, 231.9	1.5	8
Daytime Fireballs	228.5, 229.0, 229.1, 230.4, 230.6, 231.0	2.6	6

Prediction of Meteor Radiant Point associated with
(1862) Apollo: L =229.4, α = 224°, δ =-28°, Δ = 0.026 AU
1990 OS: L =231.6, α = 228°, δ =-25°, Δ = 0.070 AU

#16 Meteorite Falls	263.3, 263.6, 263.6, 263.6	0.3	4
Daytime Fireballs:	260.6, 260.9, 262.4, 262.9, 263.6	3.0	5

Prediction of Meteor Radiant Point associated with
(4179) Toutatis: L =259.6, α = 297°, δ =-22°, Δ = 0.006 AU

#17 Meteorite Falls	277.9, 278.2, 278.8	0.9	3
Daytime Fireballs	277.9, 278.4, 279.0, 279.9, 280.0, 281.3, 281.9	4.0	7

Prediction of Meteor Radiant Point associated with
1989 AZ: L =276.8, α = 293°, δ = +12°, Δ = 0.067 AU

#18 Meteorite Falls	313.0, 313.0, 313.1, 313.4, 313.5	0.5	5

#19 Meteorite Falls	328.7, 328.8, 328.9, 329.2, 329.4, 329.9, 330.3, 330.8, 330.9, 331.4, 331.8	3.1	11

Prediction of Meteor Radiant Point associated with
1991 VK: L = 328.3, α = 354°, δ = +15°, Δ = 0.051 AU

Possible associations of meteors with minor planets previously suggested in Table 2 of Hasegawa et al. (1992), such as (1566) Icarus, (3671) Dionysius, (4450) Pan, (4486) Mithra, 1984 KB and 1986 JK are pointed out here again.

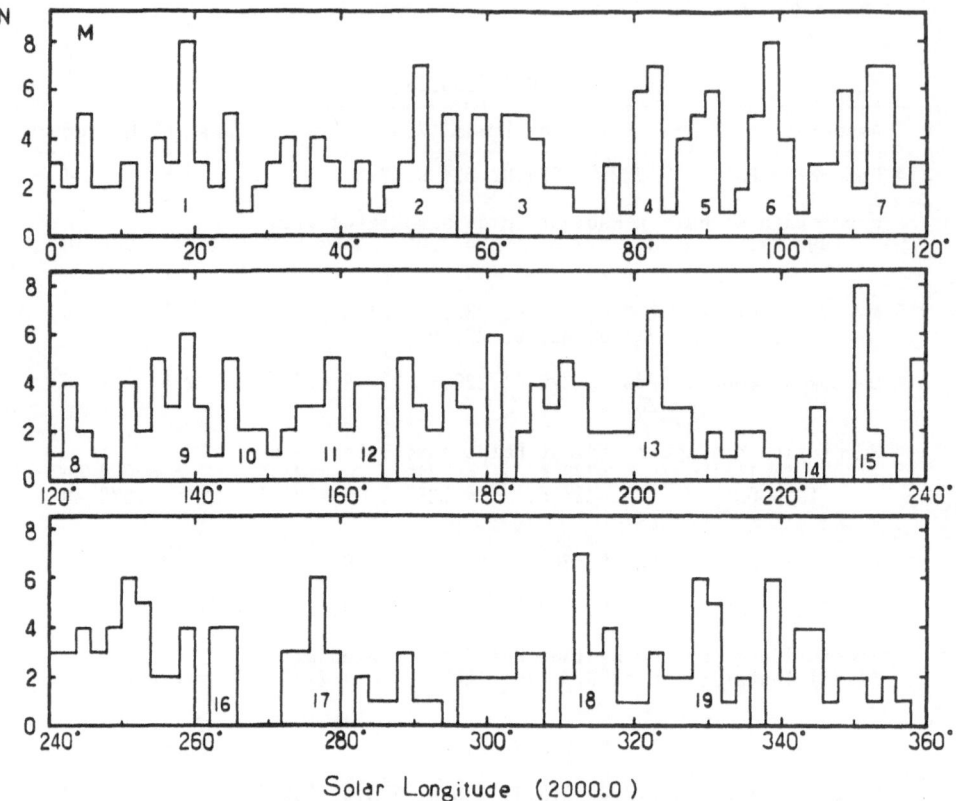

Fig. 2.   The number of daytime meteorite falls every two degrees in the solar longitude (2000.0 equinox). The numbers given in this figure correspond to the association # in Table I.

References

Graham, A.L., Bevan, A.W.R. and Hutchison, R.: 1985, *Catalogue of Meteorites*, 4th edition, The Univ. of Arizona Press, Tucson.

Hasegawa, I.: 1992, *Celes. Mech.* **54**, 129.

Hasegawa, I.: 1993, *WGN(IMO)* **21**, 29.

Hasegawa, I.: 1995, in preparation.

Hasegawa, I., Ueyama, Y. and Ohtsuka, K.: 1992, *Publ. Astron. Soc. Japan* **44**, 45.

ORBITAL EVOLUTION OF ASTEROIDS WITH THE EFFECTS OF MUTUAL GRAVITATIONAL ATTRACTION

M. YOSHIKAWA
Communications Research Laboratory
Hirai, Kashima, Ibaraki 314, Japan
e-mail : makoto@crl.go.jp

Abstract.

The effects of the mutual gravitational attraction between asteroids were analyzed by two N-body calculations, in which N=4,516 (the Sun, the nine planets, and 4,506 asteroids). In one calculation the gravity of the asteroids was taken into account, and in the other it was ignored. These calculations were carried out for a time period of about 100 years. The largest difference in the positions of the asteroids between these two calculations is about 10^{-3} AU. For the orbital elements of the semimajor axis, the eccentricity, and the inclination, the largest differences were 9×10^{-6} AU, 4×10^{-6}, and 5×10^{-4} degrees, respectively. It was found that the distribution of the differences of the semimajor axis between the two calculations is quite similar to the Cauchy distribution.

1. Introduction

So far, the orbits of more than 10^4 asteroids have been determined precisely or fairly precisely, and although this number seems to be very large, it is considered that we only know a few percent of the total number of asteroids (larger than 1km in diameter) that exist in the solar system.

Although there are many asteroids in our solar system, the size of an individual asteroid is quite small. The largest asteroid known is (1) Ceres, which has a diameter of about 910 km. The others are much smaller, so the masses of the asteroids are negligibly small in comparison with those of the planets. Therefore, we usually do not consider the masses of the asteroids when calculating their orbital motions.

Earth, Moon, and Planets **72**: 317-320, 1996.
© 1996 *Kluwer Academic Publishers.*

However the asteroids have masses, so we need to ask the following questions: Does the gravitational attraction between asteroids effect their orbital motions ? If so, by how much ? In order to get some answers to these questions, we carried out two N-body calculations with the gravity of asteroids and without it.

The aim of this work is not to make more precise ephemeris of asteroids but to estimate the effects of mutual gravitational attraction between asteroids. In other words, we want to know how the results of orbital calculations for asteroids are different between the dynamical model with the gravity of asteroids and that without it. Thus, this work is a simulation under the assumptions that initial values of the orbital elements of asteroids are exact and that we know the masses of all the asteroids considered here. If we want to make much precise ephemeris of certain asteroids, we should consider errors of their initial elements as well.

2. Methods

In this analysis, we considered a system of 4,516 bodies, ie. the sun, the nine planets, and 4,506 asteroids. The asteroids considered are the numbered asteroids from No.1 to No.4508, excluding No.719 and No.878. The motions of these bodies were calculated by two N-body calculations for a time period of 36,000 days (about 100 years) beginning Nov. 5, 1990.

In one N-body calculation, the gravitational attraction between the asteroids was taken into account. This is the calculation carried out by Yoshikawa and Nakamura (1994). In this calculation, the masses of most of the asteroids were estimated according to their brightness, because their masses have not been determined yet. For details, see Yoshikawa and Nakamura (1994). In this paper, we call this 'Calculation I'. In the other N-body calculation, the masses of the asteroids were neglected, so the asteroids were considered as test particles moving under the influence of the gravity of the Sun and the nine planets. We call this 'Calculation II'.

For both the N-body calculations, the 12th order Cowell method was used, with a step size of 0.2 days. After carrying out these calculations, the results were compared and the effects of the gravitational attraction between the asteroids were analyzed.

3. Results

When comparing the results of the two N-body calculations mentioned in the previous section, we found two types of orbital change caused by the gravity of the asteroids. We call one type 'stochastic changes', and the other type 'statistical changes'. The stochastic changes are caused by effective close encounters between asteroids, such as, close encounters with massive asteroids or close encounters at very short distances. The statistical changes are caused by a lot of close encounters that are not so effective and the gravity field of all the asteroids.

TABLE 1. Large differences in Positions after 100 years

Asteroid	Δr 10^{-4} AU	Number of encounters	Encountered asteroids (No.<1000)
1658	9.38	4	No.4, 620
2933	8.03	7	No.539, 621, 843
17	6.97	5	No.11, 779
3898	6.44	5	No.832
442	5.31	7	No.4, 313, 564

Some of examples of the stochastic changes are shown in Table 1. This table shows the largest five differences in positions (Δr) between the results of Calculation I and Calculation II. In this table, the numbers of close encounters and the asteroids that encountered with are presented. We can see that those differences are produced by close encounters with large asteroids such as No.4 and No.11, or by many close encounters. The largest difference in positions is about 10^{-3} AU after the 100 year period. This value is small, so usually we do not consider such differences. However, if we need to determine the positions of the asteroids precisely, we cannot neglect the gravity of the asteroids.

For the orbital elements, the largest differences caused by the gravity of the asteroids are $\Delta a = 9.3 \times 10^{-6}$ AU, $\Delta e = 3.6 \times 10^{-6}$, $\Delta i = 5.2 \times 10^{-4}$ degrees, $\Delta \omega = 4.1 \times 10^{-3}$ degrees, $\Delta \Omega = 3.2 \times 10^{-3}$ degrees, and $M = 2.0 \times 10^{-2}$ degrees (a : semimajor axis, e : eccentricity, i : inclination, ω : argument of perihelion, Ω : longitude of node, M : mean anomaly). We found that the differences are very small so normally we can ignore the effect of gravity of asteroids.

However, if we are considering chaotic motions, these small changes are important. As we know, chaos in the orbital motions of asteroids frequently occurs when, for example, they are in mean motion resonances. In such cases, small changes in orbital elements cause very large differences in relatively short intervals of time. The orbital changes by the gravity of asteroids can be one of the causes to make such large differences in the orbital evolution of asteroids. For example, asteroids just outside a chaotic region can be pushed into the chaotic region by the effect of mutual gravity.

The most important change is the change in the semimajor axis, because the mean value of semimajor axis does not change. Therefore, the change

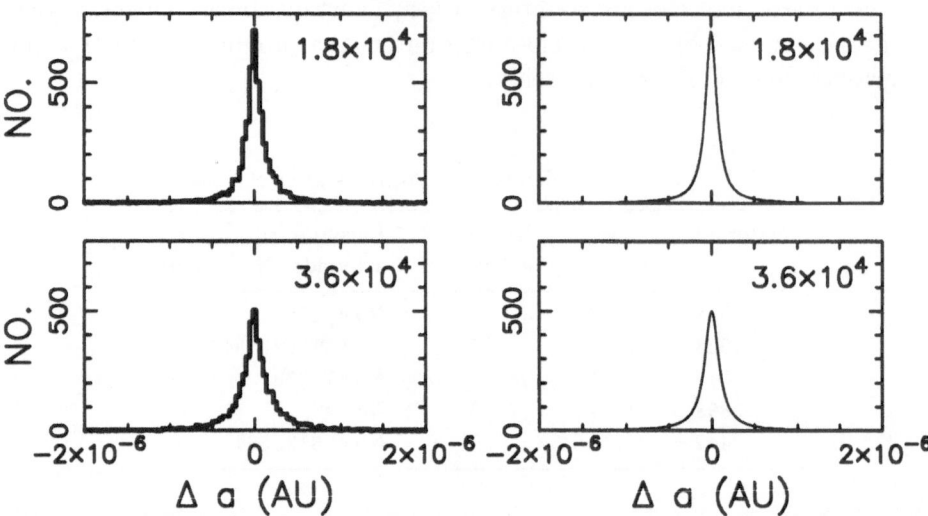

Figure 1. The distributions of the differences in the semimajor axis between Calculations I and II (left) and the Cauchy distributions that seem to represent these calculated distributions (right). The upper figures are for the results after 1.8×10^4 days and the lower figures are for the results after 3.6×10^4 days. For the Cauchy distributions, the values of Γ are 0.8×10^{-7} (upper) and 1.0×10^{-7} (lower), and the maximum values of the y-axis are the same as those of the histograms on the left.

of the semimajor axis is studied further. Fig.1 shows the distribution of the differences of the semimajor axes between Cal. I and Cal. II. This is what we call statistical change. The distribution is symmetric around 0, and this distribution fits the Cauchy distribution:

$$y = \frac{1}{\pi} \frac{\Gamma}{\Gamma^2 + x^2},$$

where Γ is the half width at half maximum. This result indicates that changes in the semimajor axis of the main belt asteroids can be understood by changes in the Cauchy distribution. This point needs a further analysis.

This study has shown the effects of mutual gravitational attraction between asteroids. Of course, the effects are not large, but they are important when we determine the orbits of asteroids precisely or when we consider asteroids in chaotic motion. In addition, we find that this analysis can be effective to study long-term evolution of the orbits of asteroids.

References

Yoshikawa, M. and Nakamura, T. (1994) Near-Misses in Orbital Motion of Asteroids, *Icarus*, **Vol. 108**, pp. 298–308

WHAT CAN METEOROID STREAMS TELL US ABOUT THE EJECTION VELOCITIES OF DUST FROM COMETS

I.P. WILLIAMS

Astronomy Unit, Queen Mary and Westfield College,
Mile End Rd, London E1 4NS, England UK

Abstract. The parent bodies of a number of major meteoroid streams are not in doubt and the orbits of these parents are also well determined. For these major streams individual orbits for a significant number of member meteoroids have also been determined. There is a significant spread in the determined values of the semi-major axis of individual meteoroids in a particular stream and this paper assumes that this spread is caused primarily by a variation in the ejection process and draws conclusions regarding the value of the ejection velocities from this.

1. Introduction

An association between comets and meteoroid streams was first suggested in the second half of the nineteenth century, probably by Schiaparelli (1867). Since that time, most investigations have re-enforced this conclusion and, when Whipple (1950) proposed his icy conglomerate model for a cometary nucleus, it became possible to understand this relationship in physical terms. As a cometary nucleus approaches the Sun, the increased radiation flux increases the temperature of the nucleus, causing the ices to sublime. The resulting outflow of gas can carry with it dust grains, or meteoroids, that were originally embedded in the icy matrix provided they are smaller than a few centimeters in diameter. For larger sizes, the gas drag is not capable of overcoming the gravitational attraction of the cometary nucleus. A mathematical formulation of this physical model was produced by Whipple (1951) and, according to this formulation, the ejection speed v in $cm\,s^{-1}$ of a meteoroid is given by

$$v^2 = 4.3 \times 10^5 R_c \left(\frac{1}{b\rho r^{2.25}} - 0.013 R_c \right), \tag{1}$$

where R_c is the radius of the nucleus in kilometers, r the heliocentric distance in AU, b the meteoroid radius and ρ its bulk density, both in cgs units. Numerous minor modifications of this formula have been proposed, but most give essentially the same numerical value for the ejection speed of meteoroids. This formula has been used by a number of authors in order to produce model meteoroid streams (see Williams 1993 for an account of some of these) and these models have been very successful in generating theoretical streams that generally match the main characteristics of actual streams. In most

Earth, Moon, and Planets **72**: 321-326, 1996.

models, the meteoroids are predominantly ejected when the comet is close to perihelion into a cone about the solar direction. The ejection velocity of each meteoroid is then added vectorially to the orbital velocity of the parent comet to give the heliocentric velocity of that meteoroid. From this velocity and the position of the comet at the time of ejection, the orbit of the meteoroid can be determined. Insertion of typical values for the parameters in Eq. (1) shows that the ejection speed of a meteoroid is generally in the range of 10 to 100 m s^{-1} and is thus significantly less than the orbital speed of the parent comet which will be typically 30 km s^{-1} or more. The heliocentric velocity of the meteoroid is thus almost the same as that of the parent comet which clearly implies very similar orbits. Consequently, when a number of meteoroids are ejected, they will form a meteoroid stream with a mean orbit that is broadly similar to the cometary orbit. The individual meteoroid orbits may then evolve because of gravitational perturbations from the planets, the effects of radiation pressure and Poynting-Robertson drag. However, each individual orbit will evolve in a broadly similar way so that the existence of a family of similar orbits as a meteoroid stream is maintained over a long period of time. As already mentioned many authors have been able to produce a good match between the theoretical and observed characteristics of meteor showers by this method. In this paper, we turn the problem around and discuss the possibility of starting with the observational data from actual showers, assume that this gives information about the initial orbits of the meteoroids and deduce the required values for the ejection speeds of the meteoroids.

With the availability of a large data set of meteoroid orbits in the IAU Meteor Orbit Catalogue at Lund (see Lindblad 1987) it is possible to obtain the orbital parameters of individual meteors belonging to specific streams and from this gain reliable information on the distribution of the orbital elements within most of the major streams. In particular, the mean value and the standard deviation for any of the orbital parameters can be obtained. This information relates to the current state within a meteoroid stream and, in order to obtain information about the initial state, we need to reverse the effects of gravitational perturbations, radiation pressure and drag. For individual meteoroids, this can be done, and indeed has been done (e.g. Wu and Williams 1992), though of course we do not know when the formation epoch was and so can only integrate the motion back in time in an open ended way. The aim of this paper is try to draw some general conclusions without involving long numerical integrations of specific meteoroids. Let us therefore consider the data available and the effects that are important. The generation of a visible meteor when a meteoroid interacts with the atmosphere of the Earth depends on the size, bulk density and speed of the meteoroid. For a given stream, both the impact velocity and density of meteoroids will be nearly constant and so the size range of meteoroids generating visible meteors in a given stream will be narrow, with a typical size of perhaps 1 mm. For such meteoroids, the effects of radiation pressure and the Poynting-Robertson drag is small and furthermore, since the size range is small, the effect will be almost the same on all meteoroids. These effects will not significantly increase the variation in the orbital parameters, though they may of course change the parameters themselves, the angular parameters changing by much more than the semi-major axis or perihelion distance. It has been known for a long time that certain orbital parameters evolve much faster than others under the influence of gravitational perturbations. In particular the semi-major axis, a, is a very slowly varying parameter unless very close encounters with a planet occur and many numerical integrations, for example Williams and Wu (1993), show this quite clearly. Consequently, if we select the semi-major axis as a parameter, then neither the mean value nor the range is likely to have changed significantly and in this paper we assume without further justification that the distribution of the semi-major axis as observed for meteoroids in a given stream contains the same general characteristics as the distribution did at the time of ejection of the meteoroids.

2. The Ejection Speed of Meteoroids

Standard orbit dynamics allows the evaluation of the speed v of a body at heliocentric distance r when moving on an orbit of semi-major axis a. For a family of different orbits, from the same standard equation we can obtain the difference between the values of v^2, in terms of the differences in $1/a$. At the instant of ejection, this orbital speed v is composed of two components, the orbital speed of the parent, v_C, and an ejection speed V relative to the parent. It is this ejection speed, V, that we are interested in. If the angle of ejection relative to the direction of instantaneous orbital motion of the comet is denoted by θ, then these speeds are related by

$$v^2 = v_C^2 + V^2 + 2v_C V \cos \theta. \tag{2}$$

As has already been mentioned, v_C will always be much greater than V so that V^2 can be ignored in comparison with v_C^2 while the difference between v^2 and v_C^2 will be related to the differences in the corresponding values of $1/a$. Combining these, and inserting numerical values for the known constants gives

$$2v_C V \cos \theta = 887 \left(1/a - 1/a_o \right), \tag{3}$$

where a_o is the semi-major axis of the comet measured in AU.

It is possible to obtain $1/a$ for each meteoroid for which an accurate orbit exists. Wu and Williams (1992) have extracted a set of 118 meteoroids belonging to the Quadrantid stream from the Lund Meteor Orbit Catalogue, while Williams and Wu (1993) similarly extracted 610 orbits for the Geminid stream. Data on many orbits also exists for the Perseids and Taurids (Williams et al. 1993) and the Eta Aquarids and Orionids (Wu and Williams 1993). However, there are two difficulties that have to be faced. First, we do not know the orbit of the parent comet at the time of ejection so that a_o and v_C are unknown. Second, we do not know the errors in the individual values of a obtained from the Lund Catalogue. Taking the second problem first, it is clear that some error exists in the data since the orbital data is heavily dependent on the estimated value for the speed of the meteor through the Earth's atmosphere. This problem is severe for highly eccentric orbits where the binding energy is close to zero and consequently $1/a$ is also small. We have eliminated from our data all orbits with positive binding energy, on the basis that these are likely to be errors, but the streams for which we have data are in general not of very high eccentricity and so the problem may be less severe. Individual orbits may have large errors associated with them, but we believe that the behaviour of the whole family contains significant information and so we assume that the standard deviation may be significant in giving an indication of the actual spread. In other words, we assume that the standard deviation in $1/a$ for a set of meteoroids gives a meaningful measure of the typical value of $1/a - 1/a_o$. If we assume that the mean value of $1/a$ gives the value of the semi-major axis of the parent, $1/a_o$, then the orbital speed of the comet at perihelion can be estimated. With these two assumptions, we obtain for each meteoroid stream a typical value for the range of values of $V\cos\theta$. In principle, this range of values could arise either from an intrinsic change in V, or from variations in the value of $\cos\theta$, or both. All models of grain ejection produce a single value for the ejection speed of a grain of given size, but that ejection can occur over a wide range of angles. If we assume that all angles are possible, then the maximum difference in the value of $V\cos\theta$ is given when $\cos\theta$ has its maximum value of 1, that is V.

3. Results and Discussion

From the Lund data we have obtained the implied value of V for the six major streams mentioned already and the resulting values are given in Table 1. For comparison, the ejection velocity as given by Whipple's formula for a grain of radius 1 mm and density 0.25 g cm^{-3} being ejected from a 5 km nucleus is also shown. A number of points are obvious. First, the ejection velocity, either implied or theoretical, is indeed much less than the orbital

TABLE I

Implied ejection velocity V, the Whipple value and the orbital velocity at perihelion, all in km s^{-1}

Stream	V	V(Whipple)	v(orbital)
Quadrantids	0.88	0.13	45
Geminids	0.20	0.80	109
Perseids	0.75	0.09	42
Taurids	0.34	0.28	65
Eta Aquarids	0.17	0.19	54
Orionids	0.16	0.19	54

velocity – the basic physics of the formation of a meteor stream is thus correct. Second, in the Taurid, Eta Aquarid and Orionid streams, the Whipple formula gives roughly the same ejection speed as is implied from the spread in observed energy. Third, for the Geminid stream the Whipple formula gives a larger value than our calculation. The Geminid parent may not be a normal comet; indeed the parent, asteroid 3200 Phaethon is known. Its orbit is certainly much smaller than those of most comets and this proximity to the sun may make the assumption of perihelion ejection only rather unrealistic. Modifying this would reduce the mean value given by Whipple (increasing r) while increasing the value obtained by us (decreasing mean v_C). This leaves two streams, the Quadrantids and the Perseids, where the implied velocity is considerably larger than the calculated one from the Whipple formula. There have been suggestions that the Perseid parent, comet Swift-Tuttle, is somewhat larger than normal, while the identity and the size of the Quadrantid parent may be in some doubt. However, an increase in radius alone will not explain the difference. It might also be argued that since the implied ejection speed for these two streams is somewhat larger than the average, it might be in error. However, the Perseids and the Quadrantids are the two streams where a very large sample of meteors were obtained so that the standard deviation should be a more reliable guide to the actual variations in ejection velocity. A higher ejection velocity would be obtained if the comet nucleus was only active over a fraction of its surface so that ejection would occur into a very narrow cone, thus significantly increasing the gas density and thus its effectiveness as an accelerator of grains.

References

Lindblad, B.A.: 1987, in *Interplanetary Matter* (Z. Ceplecha and P. Pecina, Eds.), Prague, p. 201.
Schiaparelli, G.V.: 1867, *Astronomische Nachrichten* **68**, 331.
Whipple, F.L.: 1950, *Astrophys. J.* **111**, 375.
Whipple, F.L.: 1951, *Astrophys. J.* **113**, 464.
Williams, I.P.: 1993, in *Meteoroids and their Parent Bodies* (J. Štohl and I.P. Williams, Eds.), Slovak Acad. Sci., Bratislava, p. 31.
Williams, I.P., Wu, Z.: 1993, *Mon. Not. R. astr. Soc.* **262**, 231.
Williams, I.P., Hughes, D.W., McBride, N., Wu, Z.: 1993, *Mon. Not. R. astr. Soc.* **260**, 43.
Wu, Z., Williams, I.P.: 1992, *Mon. Not. R. astr. Soc.* **259**, 617.
Wu, Z., Williams, I.P.: 1993, in *Meteors and their Parent Bodies* (J. Štohl and I.P. Williams, Eds.), Slovak Acad. Sci., Bratislava, p 77.

RADIATION PRESSURE CORRECTION
TO METEOR ORBITS

BO Å. S. GUSTAFSON and LARS G. ADOLFSSON

*Department of Astronomy, PO Box 112055, University of Florida,
Gainesville, Florida 32611-2055, USA*

Abstract. We present a method to calculate the radiation pressure force to gravity ratio on meteoroids from their atmospheric flight. Radiation pressure corrections to meteor orbits are negligible for fireballs; of the order of or less than the measurement errors ($\approx 1\%$) for photographic meteors; of the order of and in some cases substantially larger than the measurement errors ($\approx 10\%$) for radar meteors.

Key words: Radiation pressure, meteors

1. Introduction

Numerous studies have been devoted to the reduction of heliocentric meteoroid orbits from multistation photographic and radar data. These typically correct the observed velocity vector for diurnal aberration and zenith attraction due to the Earth's gravity (Whipple and Jacchia, 1957), but they neglect the effect of radiation pressure during translation of the state vector to heliocentric orbital elements. The radiation correction to orbits of massive meteoroids is negligible, but radiation pressure can shift the orbits deduced from faint meteors significantly.

Radiation pressure from sunlight partially cancels solar gravity. The resulting field equations on a stationary object have the same form as the unperturbed gravity field around a factor $1-\beta$ less massive star where β is the ratio of the radiation pressure force to gravity. It follows that because the potential energy of a meteoroid at 1 AU increases with β, the heliocentric orbit corresponding to a given state vector also increases in size. We must know β to estimate the true orbital energy.

Radiation pressure also distorts and amplifies existing distortions in the central force field. As radiation pressure increases, meteoroids bond less strongly to the Sun and the effect of planetary and other perturbations are amplified. This can be seen formally in the Gaussian perturbation equations where the perturbing force is expressed as a ratio to the central force. The distortion of the central force field has a negligible effect on the instantaneous orbit but affects the orbit's evolution over time. Orbital integrations to estimate a meteoroid's orbit at other epochs must account for this perturbation and for the Poynting-Robertson drag resulting from light absorption and scattering (Gustafson, 1994).

Earth, Moon, and Planets **72**: 327-332, 1996.

The Poynting-Robertson drag causes meteoroids to spiral toward the Sun at rates proportional to β. The spiraling motion results as the orbital energy and angular momentum gradually transfer to the scattered or reemitted light. The determination of β can therefore also be used to estimate the dispersion rate of meteoroids in a stream and give a measure of the age of meteor streams (Froeschlé et al., 1993). Such analysis and orbital integrations of meteoroid trajectories to past epochs (Gustafson, 1989b; 1990) have been plagued by the lack of direct knowledge of the value of β, which inspired this study. We show how β can be calculated from a meteoroid's atmospheric trajectory and how the immediately preceding heliocentric orbits of fireballs, photographic meteors, and radar meteors are affected by radiation pressure.

2. Calculation of β from meteor data

The ratio β is proportional to a meteoroid's effective cross-sectional area to mass ratio. The efficiency factor for radiation pressure, Q_{pr}, accounts for particle morphology and exposure geometry. For a perfect absorber we have $Q_{pr} \equiv 1$ independently of the shape and orientation. This is also a good approximation for any dark chondritic meteoroid producing photographic meteors and most detectable radio meteors. These meteoroids exceed a few times 10μm across (Bronshten, 1983) and are opaque to sunlight so that the large particle approximation and geometric optics apply (Gustafson, 1994). The optically large meteoroids have Q_{pr} values that are strictly independent of the particle shape when averages are made over random orientations Gustafson (1989a) as long as they are convex in shape. We can therefore choose any convex shape when we evaluate the efficiency factor Q_{pr}. This is easily done using geometric optics for a sphere of albedo w (van de Hulst, 1957) where $Q_{pr} = 1 - wg$. The gemoetrical factor g is inside the interval from unity (for totally forward scattering particles) to -1 (for total backscattering). We note that Q_{pr} is close to unity at the small albedos of a few percent that are typical for cometary and asteroidal chondritic material and we adopt $Q_{pr} = 1$. We can therefore evaluate β from the equation

$$\beta = C Q_{pr} A_0/m_0, \tag{1}$$

where the proportionality factor $C = 7.6 \times 10^{-5}\,\mathrm{g\,cm^{-2}}$ (Gustafson, 1994) and the subscripts denote that the cross-sectional area A_0 and the mass m_0 are values prior to atmospheric entry. We evaluate A_0/m_0 directly from meteor data.

The atmospheric flight of a meteoroid is a complex process. Severe problems emerge as we allow the meteoroid to be nonspherical, non-homogenous and fragmenting. Problems also arise in the modelling of transition between flow regimes. We expect to find a more reliable value of A_0/m_0 by confining our analysis to the uppermost part of the trajectory so that we can neglect fragmentation (unless the meteor has an abrupt beginning or has already started to fragment) and we can assume that the physical properties of the meteoroid remain constant. Further down along the trajectory this is not true.

To estimate A_0/m_0 we use classical single body meteor theory (e.g. Bronshten, 1983) where we treat only the cross-sectional area A and the mass m of the meteoroid as free parameters. It is then possible to integrate the system of classical equations to obtain two complementary estimates of A_0/m_0;

$$A_0/m_0 = \left(C_D \int_{h_1}^{h_2} \rho \, dh \right)^{-1} \cos z \, [\text{Ei}(\kappa \, v_1^2) - \text{Ei}(\kappa \, v_2^2)] \exp(-\kappa \, v_0^2) \quad (2)$$

and

$$A_0/m_0 = -2 \, \dot{v} / (C_D \rho \, v^2) \exp[\kappa(v^2 - v_0^2)], \quad (3)$$

where C_D is the drag coefficient, ρ the air density at height h, $\cos z$ the zenith distance, and v_0 the speed of the meteor at the onset of intensive evaporation. $\text{Ei}(x) = \int_{-\infty}^{x} e^t/t \, dt$ is the Exponential integral and $\kappa = (1 - \mu)\Lambda/(2C_D Q)$, where μ is the so called shape variation parameter, Λ the heat-transfer coefficient (fraction of energy that goes into meteoroid ablation) and Q the heat of ablation (the amount of energy needed to ablate a unit mass of meteoroid material). With introduction of the ablation coefficient $\sigma = \Lambda/(2C_D Q)$ and with the assumption of a self similar meteoroid, it is possible to write $\kappa = \sigma/6$.

We notice that Eq. (2) requires speeds, v_1 and v_2, at two heights, h_1 and h_2, in the atmosphere, and that Eq. (3) requires the speed, v, and the deceleration, $\dot{v}$, in one point. Equation (2) reduces to Eq. (3) if we let $h_2 \to h_1$ and substitute $dh/dt = -v \cos z$. It should also be noted that Eq. (2) is actually the integral of Eq. (3).

In Figure 1 we plot β, computed using the estimated A_0/m_0 (average of Eqs. (2) and (3)) in Eq. (1), using atmospheric flight data for twenty Harvard Meteor Program Geminid meteoroids (circles) and nine Perseids (crosses) from Jacchia *et al.* (1967). The β-values are plotted as a function of the photographic mass ($m_{\infty 2}$ adopted from Jacchia *et al.*, 1967). Based on their Knudsen numbers (c.f. Bronshten, 1983), all meteoroids were in the transition regime between the free-molecular flow regime and slip-flow in this part of their trajectory. From interpretation of photographic observations (Ceplecha, private communication) we use $\sigma = 0.012 \, \text{s}^2 \, \text{km}^{-2}$ for Geminids and $\sigma = 0.042 \, \text{s}^2 \, \text{km}^{-2}$ for Perseids with assumption of self similar meteoroids. We also adopt $C_D = 2$ (free-molecular flow) to estimate A_0/m_0 from Eq. (2) and Eq. (3). We notice that the Geminids seem to be

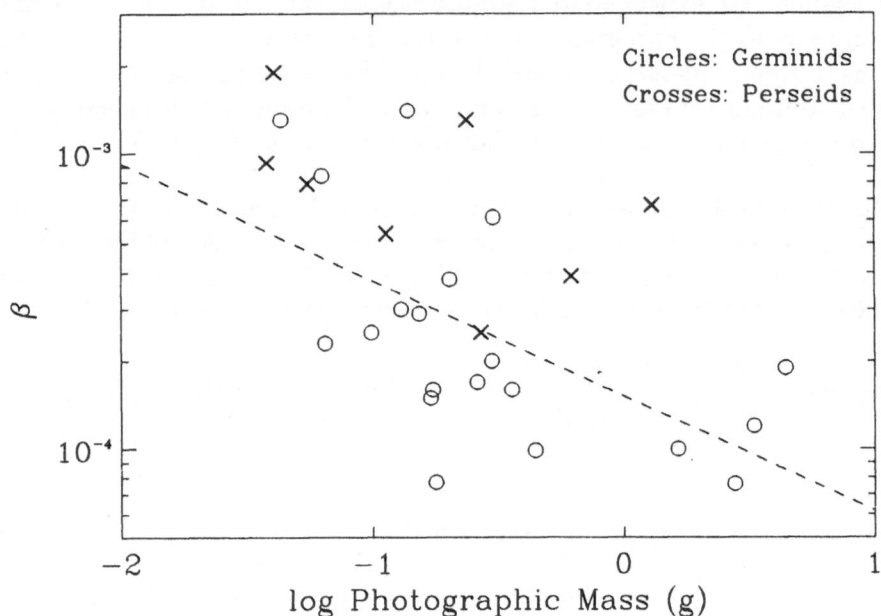

Fig. 1. Radiation pressure force to gravity ratio, β, as a function of photographic mass for twenty Geminid meteoroids (circles) and nine Perseids (crosses) photographed by Jacchia *et al.* (1967). The dashed line corresponds to the theoretical value, Eq. (1), for a sphere of bulk density $1\,\mathrm{g\,cm^{-3}}$.

denser than the Perseid meteoroids. This is consistent with previous investigations from which the Geminids are thought to be made of relatively tough, dense material. To provide a reference, the dashed line was generated using Eq. (1) and the surface area to mass ratio of a sphere of density $1\,\mathrm{g\,cm^{-3}}$. We notice that many points are located above this line indicating that the meteoroids are less dense or that they are aspherical. Another explanation might be that the photographic mass is overestimated leading to a left shift of all points in Figure 1. In any case, the value of β is not affected by these uncertainties and β is the only quantity used to generate new heliocentric orbits in the next section.

3. Radiation pressure correction to heliocentric orbits

Radiation pressure on a meteoroid vanishes as the meteoroid enters the Earth's shadow. Therefore, as we calculate a heliocentric orbit from meteor data, and integrate from the top of the Earth's atmosphere into interplanetary space, we need to account for radiation pressure ($\beta > 0$) as soon as the meteoroid leaves the Earth's shadow. The inclination and the longitude of the ascending node are not affected as radiation pressure only reduces the central force field. The effect on the argument of perihelion is automatically calculated in a full trajectory integration along with corrections to the semimajor axis a and the eccentricity e. Complete results will be presented in a separate article.

In this section, we use analytic formulae for the change in a and e from the *Vis Viva* integral and estimate the magnitude of change in a and e by comparing the cases $\beta = 0$ (denoted a_0 and e_0) and $\beta > 0$ (denoted a_β and e_β). Because the shift in the argument of perihelion is quite small it is not discussed here. To the first order in β we can then write

$$a_\beta - a_0 = a_0 \left(2a_0/r - 1\right)\beta \tag{4}$$

and

$$e_\beta - e_0 = (a_0/r - 1)(1 - e_0^2)\beta/e_0, \tag{5}$$

where r is the heliocentric distance. Realizing that $a_0 > r/2$, we see that the semimajor axis always increases with β. The eccentricity of a meteoroid's orbit decreases with β when the semimajor axis is in the interval $r/2 < a_0 < r$. Otherwise, the eccentricity increases and the orbit is hyperbolic whenever $\beta > e_0/[(1 + e_0)(a_0/r - 1)]$.

The correction for radiation pressure thus depends on the semimajor axis and β which in turn depends most strongly on the meteoroid's size. Let us consider the Geminids, $a = 1.36\,\text{AU}$ and $e = 0.896$, and the Perseids, $a = 28\,\text{AU}$ and $e = 0.965$ (Cook, 1973). We assume for illustrative purposes, that the meteoroids can be represented by spheres of bulk density $1\,\text{g cm}^{-3}$. We then compare the effect on meteoroids of masses $10^5\,\text{g}$ (fireballs), $10^0\,\text{g}$ (photographic meteors), and $10^{-5}\,\text{g}$ (radar meteors) using Eqs. (1), (4) and (5). We notice from Eqs. (4) and (5) that generally, the effect will be larger for the Perseids than for the Geminids since their semimajor axes are larger. We also notice that the normalized correction (%) in semimajor axis is approximately one order of magnitude larger than the correction in eccentricity.

We find that the correction for fireballs is less than $\approx 0.01\,\%$. This is negligible in comparison with the uncertainty in semimajor axis and eccentricity at the one standard deviation level as given by Ceplecha *et al.* (1983) for European Network fireballs. The semimajor axis increases by $\approx 0.5\,\%$ for a Perseid and $\approx 0.02\,\%$ for a Geminid photographic meteor. These corrections equal (Perseids) or are a magnitude less (Geminids) than the standard deviations given by Jacchia and Whipple (1961); we can not neglect the radiation pressure correction to photographic meteors with large semimajor axes. The correction to radar meteors is slightly less than $1\,\%$ for Geminids and $24\,\%$ for Perseids. The best radar observations made today allow us to determine a to about 10 % (Baggaley *et al.* 1992). We notice that the radiation pressure correction to radar meteors is actually substantially larger than the measurement errors. In addition, the correction is systematic.

In conclusion, we can not always neglect radiation pressure when computing a meteoroid's heliocentric orbit prior to atmospheric entry. Furthermore, it is important to determine the value of β even when it is numerically small so that we may investigate the long-term dynamical evolution of meteor streams, their life-span as a stream, and retrace the orbits to their origin following Gustafson (1989b).

4. Acknowledgements

This work was supported by NASA's Planetary Materials and Geochemistry Program through grant NAGW-2775. The authors also wish to thank Professor Zdenek Ceplecha for many helpful comments as a referee.

References

Baggaley, W. J., Steel, D. I., Taylor, A. D.: 1992 'A southern hemisphere radar meteor orbit survey', in *Asteroids Comets Meteors 1991*, Eds. A. W. Harris and E. Bowell, pp. 37–40, L.P.I., Houston.

Bronshten, V. A.: 1983, *Physics of meteoric phenomena*, Reidel, Dordrecht.

Ceplecha, Z., Boček, J., Novákova-Ježková, M., Porubčan, V., Kirsten, T., Kiko, J.: 1983 'European network fireballs photographed in 1977', *Bull. Astron. Inst. Czechosl.*, **34**, pp. 195–212.

Cook, A. F.: 1973 'A working list of meteor streams', in *Evolutionary and Physical Properties of Meteoroids*, Eds. C.L. Hemenway, P.M. Millman, A.F. Cook, pp. 183–191, NASA SP-319, Washington.

Froeschle, C., Gongzi, R., Rickman, H.: 1993 'New results on the connection between comet P/Machholz and the Quadrantid meteor streams: Poynting-Robertson drag and chaotic motion', in *Meteoroids and their parent bodies*, Eds. J. Štohl and I.P. Williams, pp. 169–172, Astron. Inst., Slovak Acad. Sci., Bratislava.

Jacchia, L. G., Verniani, F., Briggs, R. E.: 1967 'An analysis of the atmospheric trajectories of 413 precisely reduced photographic meteors', *Smithson. Contr. Astrophys.*, **10**, pp. 1–139.

Gustafson, B. Å. S.: 1989a 'Comet ejection and dynamics of nonspherical dust particles and meteoroids', *Ap. J.*, **337**, pp. 945–949.

Gustafson, B. Å. S.: 1989b 'Geminid meteoroids traced to cometary activity on Phaethon', *Astron. Astrophys.*, **225**, pp. 533–540.

Gustafson, B. Å. S.: 1990 'Are the Geminids high density porous flakes from a surface crust on Phaethon?', in *Asteroids Comets Meteors III*, Eds. C. - I. Lagerkvist, H. Rickman, B. A. Lindblad, and M. Lindgren, pp. 523–526, Uppsala Universitet, Reprocentralen HSC, Uppsala, Sweden.

Gustafson, B. Å. S.: 1994 'Physics of zodiacal dust', *Annu. Rev. Earth Planet. Sci.*, **22**, pp. 553–595.

van de Hulst, H. C.: 1957, *Light Scattering by Small Particles*, John Wiley & Sons, New York.

Whipple, F. L., and Jacchia, L. G.: 1957 'Reduction method for photographic trails', *Smithson. Contr. Astrophys.*, 1(2), pp. 183–206.

ON THE IMPORTANCE OF THE POYNTING-ROBERTSON EFFECT ON METEOROIDS

E.M. PITTICH

Astronomical Institute of the Slovak Academy of Sciences
Dúbravská cesta 9, 842 28 Bratislava, Slovak Republic

and

J. KLAČKA

Department of Astronomy and Astrophysics, Faculty for Mathematics and Physics
Comenius University, Mlynská dolina, 842 15 Bratislava, Slovak Republic

Abstract. A small generalization of the equation of motion for the Poynting-Robertson effect is tested in order to find the significance of new terms. The test is made for dust particles ejected at perihelia of the orbit of the comet Encke. The particles are released at the speed of 40 m s^{-1}. Gravitational perturbations of planets, Poynting-Robertson effect and solar corpuscular radiation (solar wind) are considered. Other nongravitational effects may be represented by new terms in the suggested form of the nongravitational force. Various values of normal and transversal components of the perturbing nongravitational force are used. The final results of numerical integrations are compared with those obtained on the basis of the Poynting-Robertson effect.

1. Introduction

Some previous papers studied the Poynting-Robertson effect and its influence on the long-term evolution of small dust particles in the solar system, e.g. Wyatt and Whipple (1950), Burns *et al.* (1979), Hughes *et al.* (1981), Jackson and Zook (1992), and, most thoroughly in Klačka (1992a, 1992b, 1993). Also comparison of the results of Pittich and Klačka (1993) and Klačka and Pittich (1994a) shows that consideration of nongravitational forces (solar electromagnetic radiation in the form of the Poynting-Robertson effect and solar corpuscular radiation – solar wind) in the evolution of meteoroid streams is very important on the scales of several thousands years (for $\beta = 4 \times 10^{-4}$, β is the ratio of the radiation force to the gravitation force). As to the real situation, the authors (Klačka and Pittich 1994c) found for the last 10,000 years' orbital evolution of comet Encke's meteoroid stream that nongravitational effects cause $a \approx 1.4 - 1.7$ AU (a – semimajor axis), $\pi \approx 120°$ (π – longitude of perihelion) and dispersions $\triangle\omega \approx 90°$ (ω – argument of perihelion) and $\triangle\Omega \approx 100°$ (Ω – longitude of ascending node), while gravitational perturbations of planets yielded $a \approx 2.2$ AU, $\pi \approx 160°$ and dispersions of ω and Ω were negligible.

Up to now we have considered only two nongravitational effects – Poynting-Robertson effect (P-R effect) and the effect of the solar wind (see also Pittich and Klačka 1994). However, there are some theoretical indications (Klačka, 1994a, 1994b, 1994d; Klačka and Kocifaj, 1994) that for a dust particle more complete equation of motion may be written in the form

$$\frac{d\mathbf{v}}{dt} = \beta \frac{\mu}{r^2} \left\{ \left(1 - \frac{\mathbf{v} \cdot \hat{\mathbf{S}}}{c}\right) \hat{\mathbf{S}} - \frac{\mathbf{v}}{c} + \alpha_T \, \hat{\mathbf{e}}_T + \alpha_N \, \hat{\mathbf{e}}_N \right\} + sw \,,$$

$$\hat{\mathbf{e}}_N = \frac{\hat{\mathbf{S}} \times \hat{\mathbf{v}}}{|\hat{\mathbf{S}} \times \hat{\mathbf{v}}|} \,, \qquad \hat{\mathbf{e}}_T = \hat{\mathbf{e}}_N \times \hat{\mathbf{S}} \,, \tag{1}$$

Earth, Moon, and Planets **72**: 333-338, 1996.

where sw represents the solar wind effect (Klačka, 1994a), α_T and α_N are dimensionless constants, **independent of radius vector r = $r\hat{\mathbf{S}}$ and velocity v** in the first approximation; they may also represent other nongravitational effects (besides the P-R effect and the effect of the solar wind). β is the ratio of the radiation force to the gravitation force, μ/r^2 represents the magnitude of the solar gravitational acceleration acting on the particle.

For more details, that such terms exist in Eq. (1) even for the process of interaction between dust particle and incident electromagnetic radiation, see (Klačka 1994a, Klačka and Kocifaj 1994). In principle, also a new radial term may be added. But its value should be much less than that for radiation pressure.

As seen from Eq. (1), the term containing α_N corresponds to the existence of normal acceleration (normal to the instantaneous orbital plane) and it **changes the inclination of the particle's orbit. The term containing** α_T causes additional **acceleration** ($\alpha_T > 0$) or **deceleration** ($\alpha_T < 0$) in comparison to the P-R effect (the term $-\mathbf{v} \cdot \hat{\mathbf{e}}_T/c = -v_T/c$) in particle's orbital motion. In the case of $\alpha_T = \alpha_N = 0$, Eq. (1) is reduced to the P-R effect.

The aim of this contribution is to test the importance of α_T- and α_N-terms in Eq. (1). The idea is based on the fact that the P-R effect and solar wind do not explain completely the observed Taurid meteor complex, as it is discussed in more detail in Klačka and Pittich (1994b). Thus, in the present contribution we have taken various values of α_T and α_N in order to find out whether better coincidence will be obtained between numerical simulations and observational data.

2. Theoretical Consideration

On the basis of Eq. (1) without the sw-term, one can calculate secular changes of orbital elements for a particle moving in gravitational and "nongravitational" fields of the Sun (see also Klačka 1994b).

If we consider $\alpha_T = 0$, one finally obtains

$$\frac{di}{dt} = \alpha_N \ X \ \cos\omega \ , \qquad \frac{d\Omega}{dt} = \alpha_N \ X \ \frac{\sin\omega}{\sin i} \ , \qquad \frac{d\omega}{dt} = - \ \alpha_N \ X \ \frac{\cos i}{\sin i} \ \sin\omega \ ,$$

$$X = \beta \ \frac{\sqrt{\mu}}{a^{3/2}} \ \frac{1}{e} \ \left\{ 1 - \frac{2}{\pi} \ \frac{1}{\sqrt{1-e^2}} \ \left(arctg\sqrt{\frac{1-e}{1+e}} + arctg\sqrt{\frac{1+e}{1-e}} \right) \right\} \ . \qquad (2)$$

The secular changes of semimajor axis and eccentricity are of the same form as those for the P-R effect:

$$\frac{da}{dt} = - \ \beta \ \frac{\mu}{c} \ \frac{2+3e^2}{a\,(1-e^2)^{3/2}} \ , \qquad \frac{de}{dt} = - \ \frac{5}{2} \ \beta \ \frac{\mu}{c} \ \frac{e}{a^2\sqrt{1-e^2}} \ , \qquad (3)$$

If we take into account that ω changes slowly in time due to gravitational perturbations of planets (see e. g. Klačka and Pittich 1994c), one can easily calculate on the basis of Eq. (2) that di/dt changes from $- \ 10^{-4}$ to $+ \ 10^{-4}$, for $\alpha_N = 5 \times 10^{-4}$, which is two orders of magnitude smaller than the change due to planetary perturbations. Thus, α_N may be neglected in Eq. (1) if its value is less than 10^{-3}.

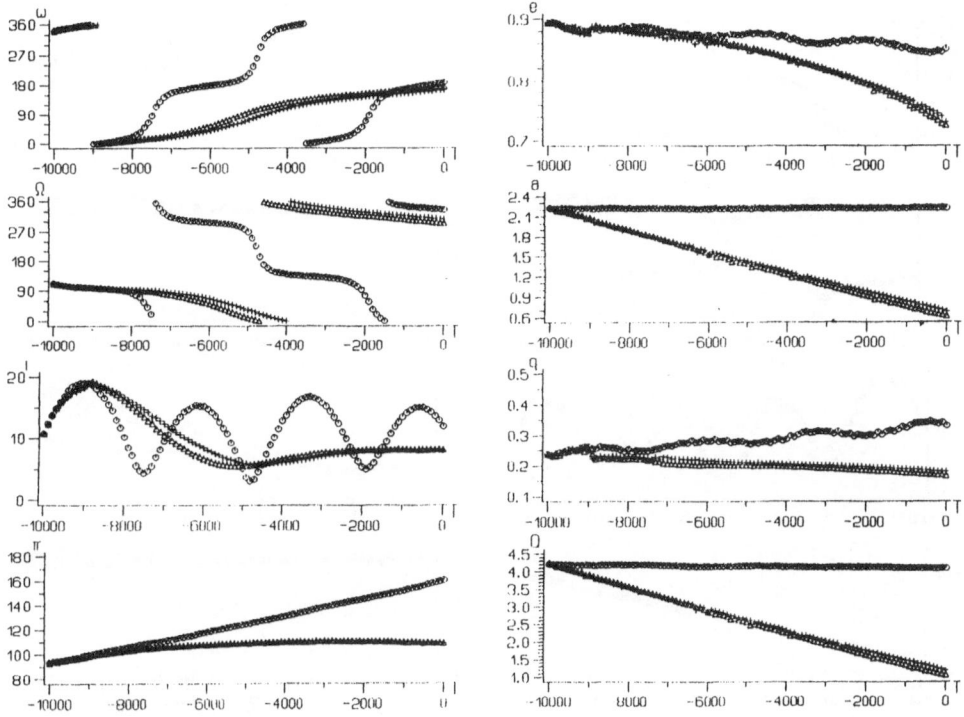

Fig. 1. Time evolution of orbital elements of comet Encke (⊙) and the model meteoroids within the period of past 10,000 years. The ejection velocity of the meteoroids in the direction of the cometary motion is: (△) +40 m s^{-1}, (+) −40 m s^{-1}. For both meteoroids $\beta = 4 \times 10^{-4}$ and $\alpha_T = -5 \times 10^{-4}$.

3. Computational Model

We traced the 10,000 years orbital evolutions of meteoroids, taking into account the gravitational perturbations of all planets, the effect of the solar wind, the effect of the solar electromagnetic radiation, and other nongravitational effects represented by Eq. (1). For the solution of the equations of motion we used the numerical integration program with the integrator RA15 (Everhart, 1985). The input data, ecliptical rectangular coordinates and velocities were calculated from the comet Encke orbital elements for the epoch 1987 July 24 (Marsden, 1989).

The model meteoroids were ejected from the comet Encke perihelion 10,000 years ago. For the model orbits the ejection velocity 40 $m\ s^{-1}$ was adopted. The particles were released in one direction – tangential to the cometary motion – and for two orientations in this direction (parallel and antiparallel to the motion).

As for the effect of the nongravitational forces (besides the effect of the solar wind), we used the equation of motion in the form of Eq. (1). Particles with the value of $\beta = 4 \times 10^{-4}$ were taken into account.

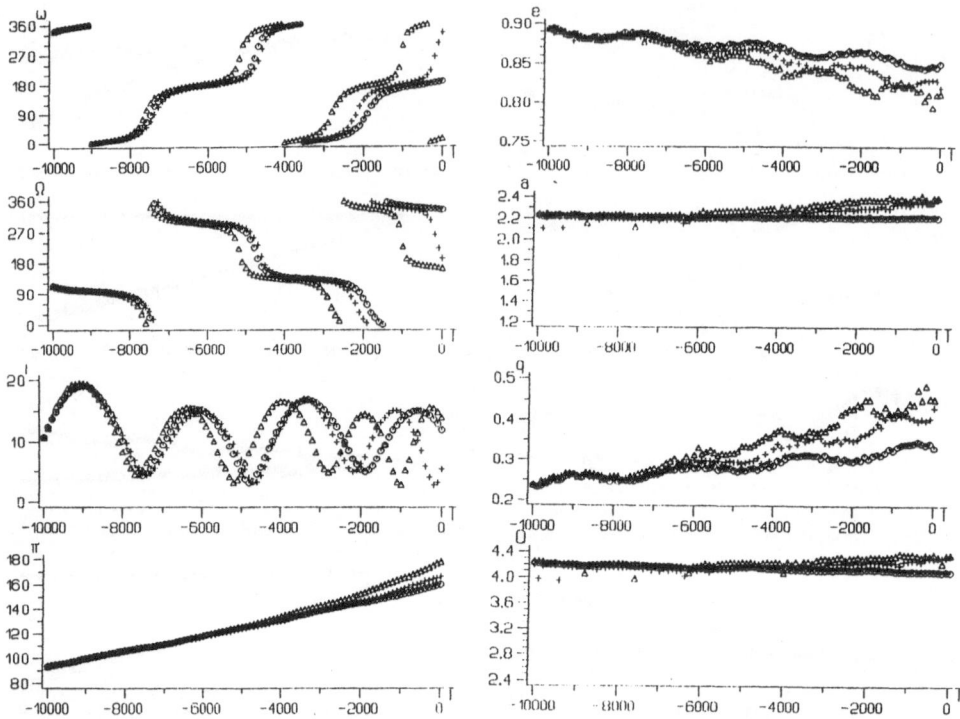

Fig. 2. Time evolution of orbital elements of comet Encke — ⊙ and the model meteoroids within the period of past 10,000 years. The ejection velocity of the meteoroids in the direction of the cometary motion is: (△) +40 m s^{-1}, (+) −40 m s^{-1}. For both meteoroids $\beta = 4 \times 10^{-4}$ and $\alpha_T = +5 \times 10^{-4}$.

We chose the following values: α_N: $\pm 5 \times 10^{-5}$, $\pm 5 \times 10^{-4}$, and α_T: $\pm 5 \times 10^{-4}$, $\pm 5 \times 10^{-3}$. The values of α_T and α_N around 10^{-4} are of the same order of magnitude as the term v/c in the P-R effect.

4. Results of Orbital Integrations

Results of orbital integrations are shown in Figures 1–3. The symbols of orbital elements are standard: e – eccentricity, q – perihelion distance, i – inclination, π – longitude of perihelion.

As for Figs. 1 and 2, we want to call the reader's attention mainly to the evolution of longitude of perihelion π. The current value of π for comet Encke is 160°, while for meteoroids we obtained values up to 180° for $\alpha_T = +5 \times 10^{-4}$ and only 105° for $\alpha_T = -5 \times 10^{-4}$ after 10,000 years evolution (the values $\alpha_T = \alpha_N = 0$ correspond to the Poynting-Robertson effect for which the value $\pi = 120°$ is obtained). An interesting and important result is that there has been practically no dispersion in π during the first 6,000 years of evolution for $\alpha_T = +5 \times 10^{-4}$; while for

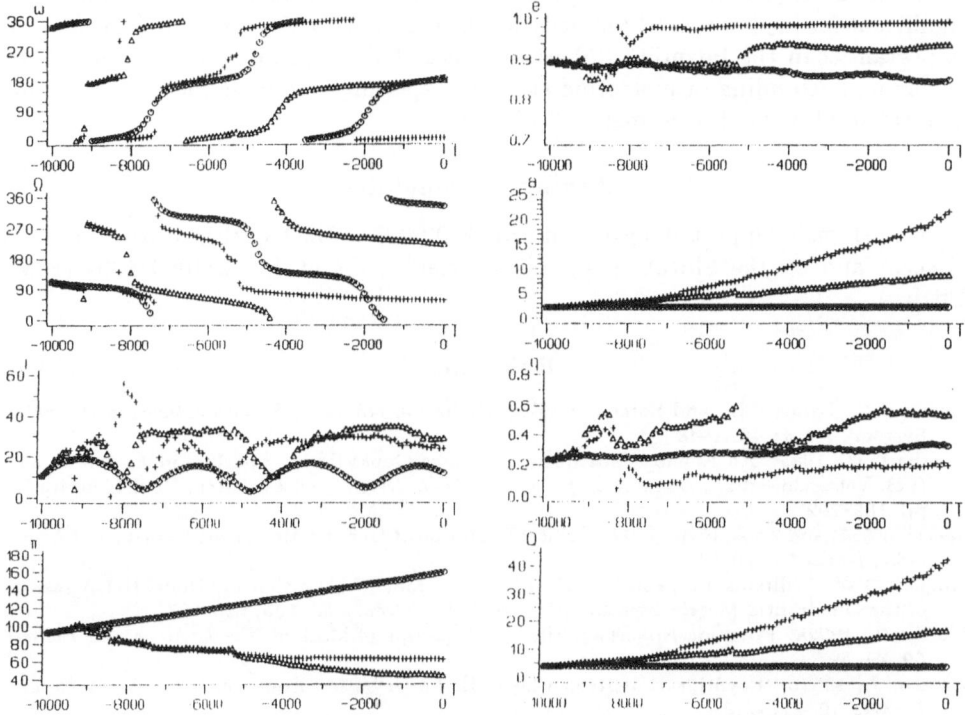

Fig. 3. Time evolution of orbital elements of comet Encke — ⊙ and the model meteoroids within the period of past 10,000 years. The ejection velocity of the meteoroids in the direction of the cometary motion is: (Δ) +40 m s^{-1}, (+) −40 m s^{-1}. For both meteoroids $\beta = 4 \times 10^{-4}$ and $\alpha_T = +5 \times 10^{-3}$.

$\alpha_T = -5 \times 10^{-4}$, the dispersion in π was negligible only for the first 2,000 years of evolution.

Figure 3 shows the rapid change of π for $\alpha_T = +5 \times 10^{-3}$. A very important and interesting feature is that π is a decreasing function of time during a greater part of orbital evolution of a particle. This is something completely different from smaller positive values of α_T, as one can see by comparison of Figs. 2 and 3 (and also Fig. 1 for negative α_T).

As for the value $\alpha_T = -5 \times 10^{-3}$, the change of all orbital elements is also very rapid and within 2,500 years the particle falls on the Sun (it is supposed that optical and mechanical properties of the particle are not changed during its motion).

As for $\alpha_N = \pm 5 \times 10^{-4}$, the obtained results of numerical integrations confirmed the theoretical consideration that even this value (not only $\pm 5 \times 10^{-5}$) is of negligible importance for the long-term orbital evolution of small interplanetary particles.

If Eq. (1) represents the general equation of motion of a dust particle under the action of nongravitational perturbations (besides the effect of the solar wind, which is separately considered in our numerical simulations), then the term containing α_N

for values of $|\alpha_N| < 10^{-3}$ is of negligible importance and the term containing α_T is significant for $|\alpha_T| \approx 5 \times 10^{-4}$ already on the scales of 10,000 years – it is even more important than the Poynting-Robertson effect. This value of α_T can possibly help in our understanding of meteoroid streams, especially the large meteoroid stream connected with the Taurid meteor complex.

Acknowledgements

This work was supported by Grant No. B-02-028 – ESO C&EE Programme (J. Klačka) and by the Slovak Academy of Sciences Grant No. 2/1050/1994 (E.M. Pittich).

References

Burns, J.A., Lamy, P.L., and Soter, S.: 1979, 'Radiation Forces on Small Particles in the Solar System', *Icarus* **40**, 1–48

Everhart, E.: 1985, 'An Efficient Integrator that uses Gauss-Radau Spacing' in A. Carusi and G.B. Valsecchi, ed(s)., *Dynamics of Comets: Their Origin and Evolution*, Reidel:Dordrecht, pp. 185–202

Jackson, A.A. and Zook, H.A.: 1992, 'Orbital Evolution of Dust Particles from Comets and Asteroids', *Icarus* **97**, 70–84

Hughes, D.W., Williams, I.P., and Fox, K.: 1981, 'The Mass Segregation and Nodal Retrogression of the Quadrantid Meteor Stream', *Mon. Not. R. astron. Soc.* **195**, 625–637

Klačka, J.: 1992a, 'Poynting-Robertson Effect. I. Equation of Motion', *Earth, Moon and Planets* **59**, 41–59

Klačka, J.: 1992b, 'Poynting-Robertson Effect. II. Perturbation Equations', *Earth, Moon and Planets* **59**, 211–218

Klačka, J.: 1993, 'Misunderstanding of the Poynting-Robertson Effect', *Earth, Moon and Planets* **63**, 255–258

Klačka, J.: 1994a, 'Interplanetary Dust Particles and Solar Radiation', *Earth, Moon and Planets* **64**, 125–132

Klačka, J.: 1994b, 'On the Stability of the Zodiacal Cloud', *Earth, Moon and Planets* **64**, 95–98

Klačka, J.: 1994c, 'Nongravitational Forces and Meteoroid Streams', *Earth, Moon and Planets* **65**, 191–196

Klačka, J. and Kocifaj, M.: 1994, 'Electromagnetic Radiation and Equation of Motion for Dust Particle' in K. Kurzyńska, F. Barlier, P. K. Seidelmann, and I. Wytrzyszczak, ed(s)., *Dynamics and Astrometry of Natural and Artificial Celestial Bodies*, Astronomical Observatory of A. Mickiewicz University, Poznań, pp. 187–190

Klačka, J. and Pittich, E.M.: 1994a, 'Long-term Orbital Evolution of Dust Particles Ejected from Comet Encke', *Planet. Space Sci.* **42**, 109–112

Klačka, J. and Pittich, E.M.: 1994b, 'Numerical Simulation of Comet Encke's Meteoroid Stream', presented at the *XXIInd Gen. Ass. of the IAU*, The Hague, Netherlands, August 1994.

Klačka, J. and Pittich, E.M.: 1994c, 'Orbital Dispersions of Comet Encke's Meteoroids', presented at the conference *Meteoroids*, Bratislava, Slovakia, August 1994.

Marsden, B.G.: 1989, *Catalogue of Cometary Orbits*, IAU, Central Bureau for Astronomical Telegrams, Cambridge, Massachusetts, 96 pp.

Pittich, E.M. and Klačka, J.: 1993, 'Orbital Model for Dust Particles of Comet Encke' in J. Štohl and I.P. Williams, ed(s)., *Meteoroids and Their Parent Bodies*, Astronomical Institute of the Slovak Academy of Sciences, Bratislava, pp. 395–398

Pittich, E.M. and Klačka, J.: 1994, 'Solar Wind and Motion of Dust Particles' in K. Kurzyńska, F. Barlier, P. K. Seidelmann, and I. Wytrzyszczak, ed(s)., *Dynamics and Astrometry of Natural and Artificial Celestial Bodies*, Astronomical Observatory of A. Mickiewicz University, Poznań, pp. 409–412

Wyatt, S.P. and Whipple, F.L.: 1950, 'The Poynting-Robertson Effect on Meteor Orbits', *Astrophys. J.* **111**, 134–141

LIGHT SCATTERING BY GAUSSIAN RANDOM PARTICLES

K. MUINONEN
Observatory, P.O. Box 14
FIN–00014 Univ. Helsinki, Finland

Abstract. We introduce multivariate lognormal statistics to describe the shapes of small particles, and compute scattering phase matrices in the ray optics approximation. The results help us understand light scattering by solar system dust particles, and thereby constrain the physical properties of, for example, asteroid regoliths and cometary comae. The present stochastic geometry could turn useful in modeling the shapes of asteroids.

1. Introduction

The ray optics approximation for light scattering by stochastically rough particles was studied by Peltoniemi *et al.* (1989). Here we provide a spherical harmonics method for generating three–dimensional lognormal—or simply Gaussian—particle shapes, and compute full 4×4 scattering phase matrices (Muinonen 1994). The results can be applied to explain recent laboratory measurements (Sasse 1993).

2. Stochastic Geometry

The lognormal probability density for the shape is

$$p_N(\mathbf{r}; a, \Sigma_r) = \frac{1}{(\sqrt{2\pi})^N \, r_1 \cdots r_N \, \sqrt{\det \Sigma_s}} \exp\left(-\frac{1}{2}\mathbf{s}^T \Sigma_s^{-1} \mathbf{s}\right),$$

$$\mathbf{r} = (r_1, \ldots, r_N)^T, \quad \mathbf{s} = (s_1, \ldots, s_N)^T, \tag{1}$$

where $\mathbf{r}$ and $\mathbf{s}$ contain the N radii and logarithmic radii—or the "logradii"— in the given directions (θ_j, ϕ_j), a and Σ_r are the mean and $N \times N$ covariance matrix for the radii, and Σ_s is the covariance matrix for the logradii (T is transpose). In detail, $s_j = \beta^2/2 + \log_e(r_j/a)$ with variance

Earth, Moon, and Planets **72**: 339–342, 1996.

$\beta^2 = \log_e(1 + \sigma^2)$, where $\sigma^2 \equiv \sigma_r^2/a^2$ and σ_r^2 are the normalized variance and variance for r_j's. It turns out that, when the radius is lognormally distributed, the lograradius is normally distributed. The covariance matrix elements are $\Sigma_{r,ij} = \sigma^2 C_r(\gamma_{ij})$ and $\Sigma_{s,ij} = \beta^2 C_s(\gamma_{ij})$ $(i, j = 1, \ldots, N)$, where C_r and C_s are the autocorrelation functions for the radius and lograradius, respectively, and γ_{ij} is the angular distance between the directions i and j. For isotropic shapes,

$$\sigma^2 C_r \;=\; \exp\left(\beta^2 C_s\right) - 1. \tag{2}$$

Three–dimensional particle shapes can be efficiently generated using a spherical harmonics expansion for the lograradius, and a Legendre expansion for its autocorrelation function:

$$r(\theta, \phi) \;=\; \frac{a}{\sqrt{1 + \sigma^2}} \exp\left(s(\theta, \phi)\right),$$

$$s(\theta, \phi) \;=\; \sum_{l=0}^{\infty} \sum_{m=0}^{l} P_l^m(\cos\theta)\left(a_{lm} \cos m\phi + b_{lm} \sin m\phi\right),$$

$$C_s(\gamma) \;=\; \sum_{l=0}^{\infty} c_l\, P_l(\cos\gamma), \tag{3}$$

where P_l^m's and P_l's are associated Legendre functions and Legendre polynomials, respectively. It can be shown that, if the coefficients a_{lm} and b_{lm} are independent Gaussian random variables with zero means and equal variances

$$\beta_{lm}^2 = (2 - \delta_{m0})\, \frac{(l - m)!}{(l + m)!}\, c_l\, \beta^2, \tag{4}$$

the lograradius will be normally distributed with zero mean and autocovariance function $\beta^2 C_s$.

A convenient choice for the autocorrelation function is

$$C_s(\gamma) \;=\; \exp\left(-\frac{1}{2} \frac{\sin^2\frac{1}{2}\gamma}{\sin^2\frac{1}{2}\Gamma}\right), \tag{5}$$

where Γ is the correlation angle—we obtain

$$c_l \;=\; (2l + 1)\, \exp(-\kappa)\, i_l(\kappa), \qquad l = 0, \ldots, \infty,$$

$$\kappa = (4\sin^2\tfrac{1}{2}\Gamma)^{-1}, \tag{6}$$

i_l being a modified spherical Bessel function. The stochastic shape is thus parametrized by σ and Γ, the relative standard deviation and the correlation angle.

3. Ray Optics Approximation

The 4×4 scattering phase matrix $\mathbf{P}$ relates the Stokes vectors of the incident and scattered light,

$$\mathbf{I}_s \propto \sigma_{sca} \, \mathbf{P} \cdot \mathbf{I}_i, \qquad \int_{4\pi} \frac{d\Omega}{4\pi} \, P_{11} = 1, \qquad (7)$$

where σ_{sca} is the scattering cross section. For particles large compared to the wavelength, the scattering phase matrix can be divided into the diffraction and geometric optics parts,

$$\mathbf{P} \approx \frac{1}{2\sigma_{sca}} \left(\sigma_{ext} \mathbf{P}^D + (2\sigma_{sca} - \sigma_{ext}) \mathbf{P}^G \right), \qquad (8)$$

where σ_{ext} is the extinction cross section, here twice the cross–sectional area of the particle. $\mathbf{P}^D$ can be approximated by Dirac's δ–function with proper normalization. $\mathbf{P}^G$ is computed using ray tracing methods refined from those in Muinonen *et al.* (1989).

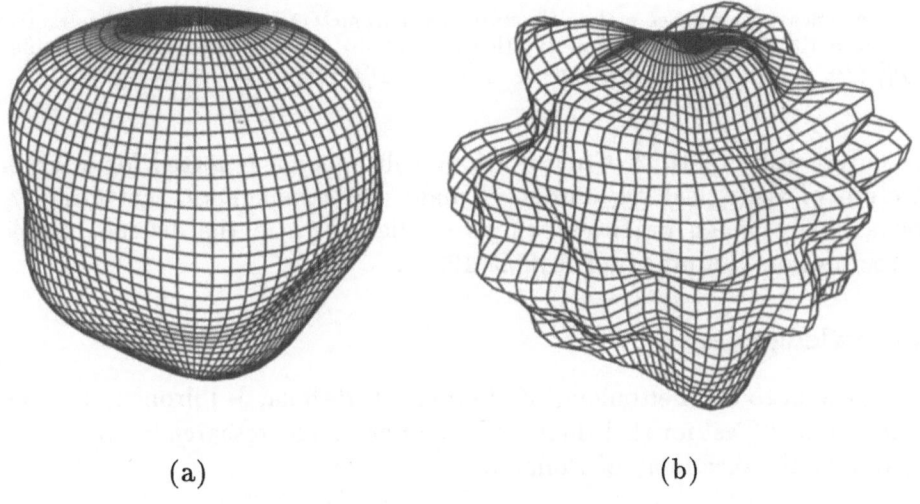

(a)   (b)

Figure 1. Three–dimensional shapes for sample Gaussian random particles with standard deviation $\sigma = 0.1$ and correlation angle (a) $\Gamma = 30°$ and (b) $\Gamma = 10°$.

4. First Results and Discussion

Figure 1 presents two Gaussian sample particles. The scattering phase matrix $\mathbf{P}^G$ (particle refractive index $m = 1.5$ and size much larger than wavelength) in Figure 2 was computed using 5×10^5 rays and sample stochastic shapes, maximum 11 internal reflections, and incorporating a relative

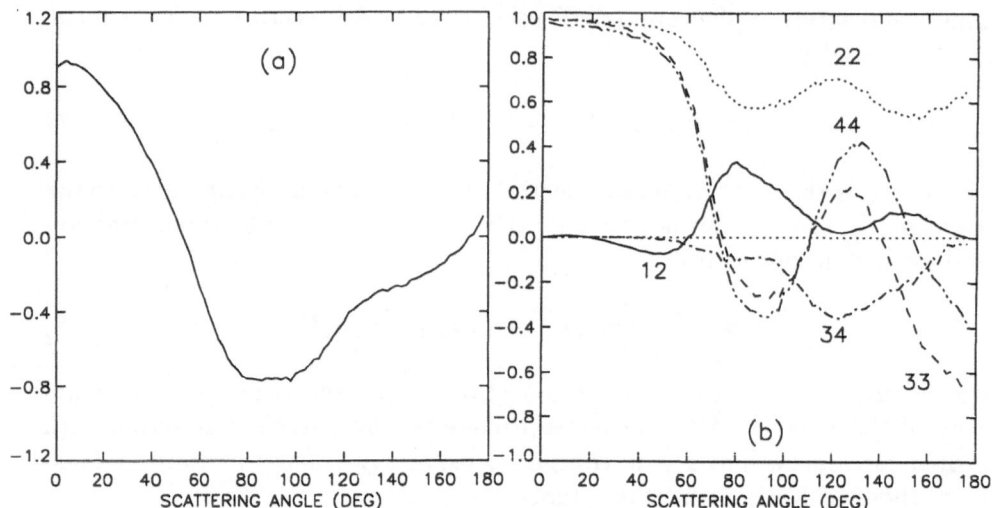

Figure 2. Scattering phase matrix for statistical parameters as in Fig. 1a and refractive index $m = 1.5$: (a) Phase function P_{11}^G ($\log_{10}$), and (b) other nonzero elements divided by P_{11}^G ($P_{21}^G = P_{12}^G$, $P_{43}^G = -P_{34}^G$, and '12' denotes $-P_{12}^G/P_{11}^G$).

flux cutoff value of 10^{-4}. Computations will be soon reported for a larger selection of statistical parameters. Finally, electromagnetic scattering by stochastically rough particles can be studied, e.g., in the discrete–dipole approximation (Lumme and Rahola 1994).

Acknowledgements

I am grateful to J.I. Peltoniemi, K. Lumme, J. Rahola, J. Piironen, T. Nousiainen, and P. Fast for their help and comments. The research is supported, in part, by the Academy of Finland.

References

Lumme, K., and Rahola, J. (1994). Light scattering by porous dust particles in the discrete–dipole approximation, *Astrophys. J.* **Vol. no. 425**, pp. 653–667.

Muinonen, K. (1994). Light scattering by stochastically rough particles, *Bull. Amer. Astron. Soc.* **Vol. no. 26**, p. 1173.

Muinonen, K., Lumme, K., Peltoniemi, J.I., and Irvine, W.M. (1989). Light scattering by randomly oriented crystals, *Appl. Opt.* **Vol. no. 28**, pp. 3051–3060.

Peltoniemi, J.I., Lumme, K., Muinonen, K., and Irvine, W.M. (1989). Scattering of light by stochastically rough particles, *Appl. Opt.* **Vol. no. 28**, pp. 4088–4095.

Sasse, C. (1993). Development of an experimental system for optical characterization of large arbitrarily shaped particles, *Rev. Sci. Instrum.* **Vol. no. 64**, pp. 864–869.

ARE CRATERING AND PROBABLY RELATED
GEOLOGICAL RECORDS PERIODIC?

S. YABUSHITA

Department of Applied Mathematics & Physics
Kyoto University, Kyoto 606, JAPAN

Abstract. The controversial topic of periodicity in geological records in relation to astronomical modeling is reviewed. Although impact cratering, frequency distribution of geomagnetic reversals, and mass extinction of fauna yield periods when certain tests are applied, none of them can be regarded significant in the sense of mathematical statistics. The first two records yield periods of 30 Myr, while the faunal-extinction record yields a period of $\sim 26 - 27$ Myr. It seems that although catastrophes in the form of large impacts trigger mass extinctions, certain geophysical or geological conditions need be satisfied for mass extinctions to be realized. One should not expect to find an indisputable periodicity in cratering record because random impacts by asteroids are dominant. Thus, the earth-crossing cometary flux modulated by the galactic tidal force appears consistent with the weak detected periodicity.

1. Introduction

The modern debate on the periodicity in catastrophies in the earth history was triggered when Raup & Sepkoski (1984) published a controversial paper in which they claimed that mass extinctions of marine fauna took place with a mean period of 27 Myr. This claim was shocking enough but it was followed by two papers, one by Alvarez & Muller (1984) and another by Rampino and Stothers (1984a), who argued that there is a similar period in the record of gigantic catastrophes in the form of large craters. Several astronomical models involving comet showers were proposed to account for the periodicity in crater formation rate (Rampino & Stothers 1984a, Davis et al. 1984, Whitmire and Jackson 1984, Whitmire and Matese 1985), and periodicities in geological records other than craters and mass extinctions were searched (Rampino & Stothers 1984b).

Here we review the current state of search for the periodicities, and discuss whether there are any astronomical models consistent with the periodicities.

2. Tools for testing the periodicity

A phenomenon is called periodic if it repeats itself with equal intervals of (astronomical) time. The counterpart of a periodic phenomenon is a random process. The border line between the two is, however, not very clear, unless

Earth, Moon, and Planets **72**: 343-356, 1996.

we are certain that the phenomenon under consideration is driven by a mechanism which is known to be periodical.

In the case under consideration, it is important first to establish periodicity (or periodicities), if any. The reason is that although astronomical (or geophysical) models can be presented, the relation between the cause and the effect is not very clearcut, except for impacts and impact cratering. In the case of geological records, one has an age (the time of formation) and diameter of a crater, a time when the direction of the geomagnetic dipole field reversed, a time when the rate of extinction of fauna reached a local maximum, and an interval when volcanic activity is high. These can all be represented by the time series $t_1, t_2, \cdots, t_N$. If there were an exact periodicity with a period P, there should be a relation of the form

$$t_i = \alpha + n_i P, \quad i = 1, 2, \cdots, N \tag{1}$$

where $n_i = 0, 1, 2, \cdots$, and α is the latest epoch.

In a situation where $t_1, t_2, \cdots, t_N$ are more or less uniformly distributed (no long term variation is present, which will be referred to later), deviation from an exact periodicity may be measured by the quantity

$$\frac{s^2}{P^2} = \frac{1}{N} \sum_{i=1}^{N} (t_i - \alpha - n_i P)^2 / P^2, \tag{2}$$

where the integer n's are so chosen that

$$|t_i - \alpha - n_i P| < P/2. \tag{3}$$

This is called linear measure of deviation from a strict periodicity. Clearly, the deviation s/P depends on the choice of the period, P, and the epoch, α. One may plot s/P against P for a given value of α. If s/P is sufficiently small, one may have detected a period, P; this process may by repeated by varying α, and when an absolute minimum of s/P is derived, that defines a period. The problem is whether the period arrived at may be regarded as significant. A criterion proposed by Broadbent (1955, 1956) is that the inequality

$$\sqrt{N}(\frac{1}{3} - \frac{s^2}{d^2}) > 1, \, d = \frac{P}{2}, \tag{4}$$

be satisfied. This inequality is satisfied only once in more than 1000 trials, if $t_1, t_2, \cdots, t_N$ are randomly distributed. The criterion may appear stringent, but since P and α are derived from given data, such a stringent requirement may be needed. It should be kept in mind that this criterion is valid provided that $t_1, t_2, \cdots, t_N$ do not exhibit a long-term trend.

Another method which is often adopted is the circular measure analysis. Following Lutz (1985), define

$$a_i = \sin \frac{2\pi}{P} t_i \, , \ b_i = \cos \frac{2\pi}{P} t_i$$

$$S = \frac{1}{N} \sum_{i=1}^{N} a_i \, , \ C = \frac{1}{N} \sum_{i=1}^{N} b_i \, , \tag{5}$$

$$R^2 = S^2 + C^2.$$

If $t_1, t_2, \cdots, t_N$ are randomly distributed, (a_i, b_i) would define a random walk and the sum R will be small, while if there were an exact periodicity, R would be large for that value of P. If R is plotted against P, then maximal values of R would correspond to periods in the series, $t_1, t_2, \cdots, t_N$.

This method appears simple enough, and at first sight there would be no complications. In reality however, there is what Lutz (1985) calls a wrapping effect. This arises from the situation when the assumed period is not an exact fraction of t_N. For instance, let $t_N = 165$ Myr and an assumed period $P = 30$ Myr. The vector (a_i, b_i) makes 5 wrappings but there is still the 6:th wrapping for t's between 150 and 165 Myr. Therefore, even if $t_1, t_2, \cdots, t_N$ were randomly distributed, this circular measure method gives a spurious period.

Another complication which arises is long-term variation in the series, $t_1, t_2, \cdots, t_N$. Of the geological records to be discussed in the following, crater formation rate and the frequency distribution of geomagnetic reversals apparently exhibit such long-term variations. Thus, the number of known craters decreases as one goes back in time, while there is an epoch when there is hardly any recorded magnetic reversals between 80 and 110 Myr BP. These long-term trends may give rise to apparent periods as the two methods mentioned here are applied. Thus, it is very important to bear in mind these complications before one arrives at a definite conclusion regarding the existence of periodicities in the geological records.

Although the two methods will be adopted in the following, we note a method which Stothers (1979) proposed, and which Rampino & Stothers (1984b) applied to discuss periodicities of geological records such as low sea levels, sea-floor spreading discontinuities and tectonic episodes. This method is a modification of the Broadbent criterion.

3. Periodicity in mass extinctions

Rampino & Caldeira (1992) derived peaks in mass extinctions based on a compilation of published extinction times of marine fauna and these are

TABLE I

Mass extinction peaks and probably related crater ages.
Data taken from Rampino & Caldeira (1992) and from Grieve (1993).

Mass extinction peak	Crater ages (Myr BP)	Diameter (km)	Name
1.6 Myr BP			
11.2			
36.6	36 ± 1	100.2	Popigai
	38 ± 4	28	Mistastin
66	65 ± 2	25	Kamensk
	64.98 ± 0.05	180	Chicxulub
	65.7 ± 1	35	Manson
91 ± 1	88 ± 3	24	Boltysh
	95 ± 7	25	Steen River
113 ± 3	115 ± 10	39	Carswell
144 ± 5	142 ± 0.5	22	Gosses Bluff
176 ± 3			
193 ± 3	186 ± 8	23	Rochechouart
216 ± 5	212 ± 2	100	Manicouagan
	219 ± 32	40	Saint Martin
	220 ± 10	80	Puchezh-Katunki
245 ± 5	247 ± 5	40	Araguainha Dome

reproduced in Table I. As one can see, there are 11 peaks. It is also seen that the distribution of $t_1, t_2, \cdots, t_{11}$ is uniform. In other words, there is no long-term variation contrary to crater ages or magnetic reversals.

A straightforward application of the Broadbent criterion gives rise to the result shown in Fig. 2 of Yabushita (1994). The linear measure of fitness shows a sharp drop at 26 Myr as may be seen. However, the criterion is not satisfied, because s/P does not drop below the level indicated by the criterion. However, the straightforward application of the Broadbent criterion is somewhat problematical, because the number of peak ages (11) is rather small, while inequality (4) is valid when N is large ($N \gtrsim 15$). In this case, however, one can perform a Monte Carlo simulation to see whether the derived period is statistically significant. First, s/P is plotted against P for the known peaks, $t_1, t_2, \cdots, t_{11}$, and the minimal value is obtained. In the case under discussion, $s/P = 0.142$. Secondly, uniformly random values $t_1, t_2, \cdots, t_{11}$ between 0 and 250 Myr are generated and the minimal value σ/P is found as calculated from Eq. (2). One then obtains a distribution of σ/P by repeating many trials. The distribution is presented in Fig. 1. One may then compare the s/P value with the distribution of σ/P to see at what

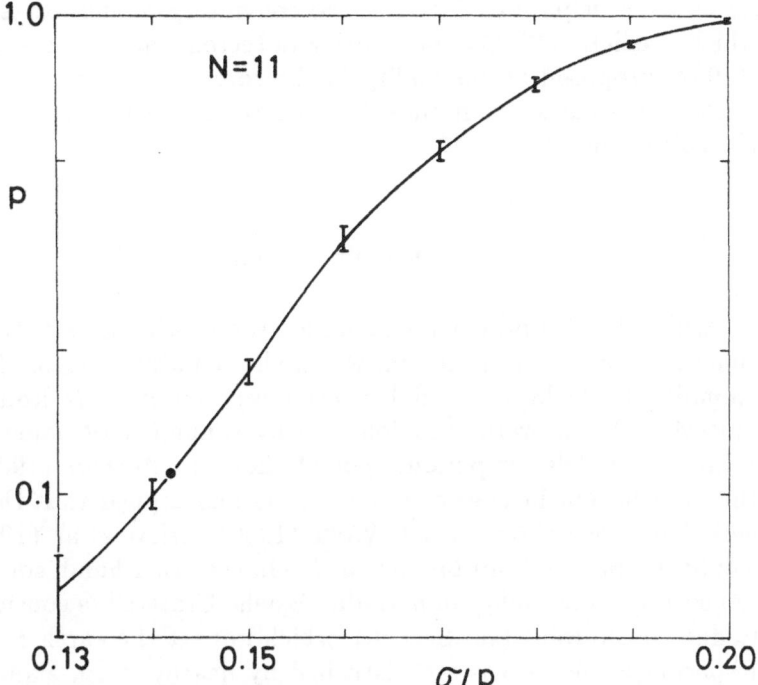

Fig. 1. Cumulative distribution of σ/P for $t_1, t_2, \cdots t_{11}$ where the t's are uniform random numbers between 0 and 250 Myr. The value of s/P calculated from mass-extinction peaks is denoted by a dot. The corresponding phase is 9.5 ± 0.5 Myr.

level the calculated s/P is significant. For the case of mass extinctions, the level of significance is 11%. In other words, even if $t_1, \cdots, t_{11}$ were uniformly randomly distributed, one gets the value of s/P as small as 0.142 in 11 cases out of 100 random samplings.

As will be shown later, cratering and geomagnetic reversal frequency distributions yield a period of 30 Myr, although at a level not satisfactory from the point of view of rigorous mathematical testing. The discrepancy between 30 and 26 Myr is significantly large. This difference naturally leads one to question whether mass extinctions are directly related to catastrophes in the form of large impacts. In Table I, we list the craters of known ages which agree with the times of peaks in extinction rate within probable errors. Apart from the latest extinction peak and the one at 176 Myr BP, there is reasonable correspondence. On the other hand, there are several large craters (with $D \geq 20$ km) that have no extinction counterpart (Yabushita 1994). In particular, those with ages 73 Myr ($D = 65$ km), 50 Myr (45 km) and 128 Myr (55 km) are sufficiently large to have caused mass extinctions. In other words, it appears that large impacts in the form of large cratering is a necessary but not a sufficient condition for mass extinction. Cratering may trigger mass extinction, but the geological or geophysical conditions ought to be such as to anticipate extinctions. No convincing model exists in geophysics which predicts a 27 Myr periodicity in tectonic activities. Courtillot & Besse (1987) proposed an instability in the core-mantle boundary where hot thermals are released from time to time which results in episodes of world-wide vulcanism.

4. Cratering record

Alvarez & Muller (1984) picked up 11 large craters with ages then known and obtained the frequency spectrum, which showed a peak at the frequency corresponding to 28 Myr. A similar result was obtained by Rampino & Stothers (1984a). These were then followed by a number of papers which proposed various models for periodic comet showers. Stothers (1988) then plotted the distribution in ages of large craters and argued that there is a 30 Myr periodicity. See Shoemaker & Wolfe (1986), Grieve et al. (1985) and Rampino & Stothers (1986) for further work. On the other hand, some workers questioned the detectability of periodic signals. Craters are considered to be largely due to asteroids (see Sect. 6), which impact the earth randomly. Only long-period comet orbits are disturbed by nearby stars, giant molecular clouds or more systematically, by the galactic tide, and their flux may vary periodically. Thus, if the asteroids occupy a large fraction of impacting bodies, it is natural to expect that no periodic signals will be detected.

We first investigate whether the ages of known craters exhibit a periodicity at a level that is beyond statistical argument. The number of craters available for such statistical testing is well over one hundred. According to Grieve (1993), the number of such craters is close to 150 at the fall of 1993. The present author made use of Grieve's compilation of 1991, which contains 139 craters. The 1991 list will be used in the following discussion. See also Grieve & Pesonen (1992).

Of the 139 craters, 35 are given only upper or lower limits to the ages. This reduces the number available for statistical testing to 104. Further, those with ages < 600 Myr are 99 in number. Of the 99, those with D (diameter) > 10 km are 40.

Were these crater ages more or less uniformly distributed in $(0, 600 \text{ Myr})$, it would have been a straightforward matter to test the periodicity hypothesis. Unfortunately, the situation is far from uniformity. In histogram distributions of crater ages (Weissman 1989, Yabushita (1994)), it is apparent that the number of recognized craters decreases with age. Assuming that the cratering rate remains constant over a time scale of 600 Myr, the decay constant may be calculated (Table II). In so doing, some care is needed in dealing with very young craters (those with ages < 25 Myr). There is a large excess of such craters. The excess may be genuine (Stothers 1988),

TABLE II

Decay constant of craters. Those with ages < 25 Myr are neglected. $\bar{D}$ denotes the average value.

	No. of craters	$\bar{D}$ (km)	Decay const.
Regardless of D	62	23.5	0.102/25Myr
$D \geq 2$ km	61	23.8	0.080/25
$D \geq 10$ km	34	38.4	0.164/25
$D < 10$ km	28	5.4	0.016/25

TABLE III

Periods detected from crater data.

Dataset	No. of samples	Period	Broadbent criterion satisfied ?
all	99	30 Myr	yes
		19.5	yes
		16.5	yes
$D \geq 2$ km	74	31	no
		16.5	no
$D \geq 5$ km	57	31	no
		16.5	no
$D \geq 10$ km	40	31	no
		16.5	no

indicating a recent comet shower which is also evidenced by an unrelaxed distribution of $1/a$ (a stands for the semi-major axes) of long-period comets (Yabushita 1983, Emel'yanenko & Bailey 1995), and departure from balance between supply and loss of comets by the galactic and planetary perturbations (Yabushita 1988). Thus, the decay constant should be determined from data which exclude those with ages < 25 Myr. Table II gives the decay constants so obtained.

4.1. RESULT NEGLECTING LONG-TERM TREND

Although it may seem unworthy of testing the periodicity hypothesis by neglecting the long-term trend, the periodicity was initially claimed by ignoring it. We therefore present the results of statistical testing ignoring the trend.

Figs. 2 and 3 of Yabushita (1994) give the plot of s/P against P. As before, the horizontal line represents the level below which the Broadbent criterion is satisfied. One easily notes that whether the criterion is satisfied or not depends much on which subset of the data one adopts. In Table III, we summarize the results obtained by the Broadbent criterion as reference, although we now know that it cannot be used without modification. It appears that those craters with $D > 10$ km least satisfy the criterion, while if the entire set is adopted, it seems satisfied. In other words, contrary to the claim that large ones are more or less periodic, the crater data adopted here apparently leads one to an opposite result. Finally, one finds that the entire set and the subsets yield the latest epoch between 1 and 3 Myr. Put it the other way, the solar system has just emerged or is about to emerge from an epoch of heavy bombardment.

4.2. LONG TERM TREND TAKEN INTO ACCOUNT

As referred to earlier, there is a loss of craters as one goes back in time. The Broadbent criterion is not appropriate when the time series $t_1, t_2, \cdots, t_N$ are not uniformly distributed. So one must test the hypothesis by a simulation, taking into account the long-term trend and comparing it with observational data. We now describe how such a simulation may be carried out.

Generate a series of random numbers $t_1, t_2, \cdots, t_N$ which obey an exponential distribution. This can be done by the formula

$$t_i = -\frac{1}{\lambda} \log \xi_i \quad i = 1, 2, \cdots N \tag{6}$$

where $\xi_1, \xi_2, \cdots \xi_N$ are a series of uniform random numbers in $[0, 1]$, and λ is the decay constant. If $t_i > 600$ Myr (or 240 Myr), it should be skipped. From the random numbers, calculate the sum

$$\frac{\sigma^2}{P^2} = \frac{1}{N} \sum_{i=1}^{N} [t_i - (\alpha + n_i P)]^2 / P^2 \tag{7}$$

and minimize σ/P by varying α and P. By carrying out a large number of samplings, the probability distribution of σ/P can be obtained. The observational data, on the other hand, gives the minimal value of s/P. If the s/P is sufficiently small, one may consider that there is a periodicity with a period which minimizes s/P.

Statistical testing of the sort described here has been carried out for various subsets of crater data (Yabushita 1995). Fig. 2 shows one of the distributions of σ/P, and the observational data is plotted as a dot on the curve. The result is summarized in Table IV.

One notes that although a period is detected by plotting s/P against P, the statistical testing shows that the minimal value of s/P is not sufficiently

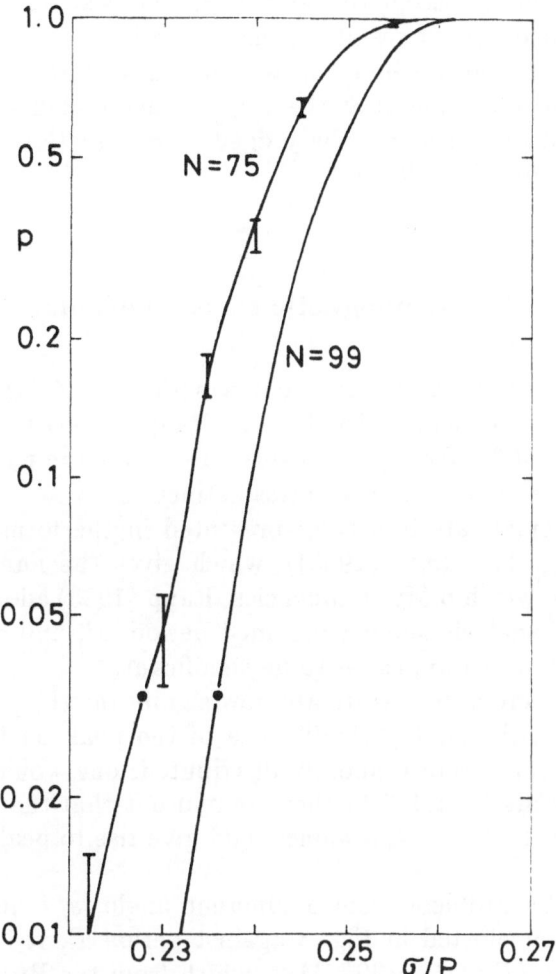

Fig. 2. Cumulative distribution of σ/P for $t_1, \cdots t_N$, where the t's are exponential random numbers with the decay constant derived from crater data. Dots indicate observed values of s/P.

TABLE IV

Dataset	detected period	s/P	probability
all ($N = 99$)	30	0.235	3.3%
	36.5	0.231	1%
$D \geq 2$ km	31	0.251	93%
$D \geq 10$ km	31.5	0.232	72%

small to compel one to accept the derived period as statistically significant. Other subsets yield similar results. Thus, one is led to conclude that crater data themselves should not be taken to yield an irrefutable period. Rather, the periodicity in cratering is at the verge of detectability and whether it may be accepted or not should be judged in conjunction with other data and with astronomical modeling.

5. Geomagnetic reversal events

Negi & Tiwari (1983) first argued that a period of 34 Myr is seen in the magnetic reversal frequency distribution. Raup (1985) then claimed that there a is period of 30 Myr in the reversal rate over the past 165 Myr, and associated the period with that of mass extinctions.

Magnetic reversal rate is usually presented in the form of a histogram distribution (Fig. 1 of Lutz (1985)), which gives the number of reversal events in a bin of width 5 Myr frequencies. Raup (1985) adopted the Fourier power spectrum analysis against assumed period, P, and obtained a peak at $P = 29.5$ Myr which appeared to be significant.

The problem here is to investigate how significant the peak at 29.5 Myr is. Lutz (1985) criticized the significance of the peak on the ground that even if $t_1, t_2, \cdots, t_{296}$ were randomly distributed, one would obtain similar peaks. Lutz & Watson (1988) further pointed out that the long-term trend in the time series $t_1, t_2, \cdots, t_{296}$ alone could give rise to peaks similar to the one obtained.

To look at the problem from a different angle, s/P for the magnetic reversals has been plotted in Fig. 3 against period P. It is seen here that there is a dip in s/P at $P = 30.5$ Myr, which from the Broadbent criterion is significant. Here, the solid curve gives s/P calculated from the real data, while the dotted curve gives s/P by taking into account the long-term trend only. One notes that the dip is deeper for the real data. There is, thus, more than the long-term trend alone in the dip and in this restricted sense, one may claim to have detected a period of 30.5 Myr. Finally, the phase (the latest epoch) of the assumed periodicity is 10 ± 0.5 Myr for the magnetic reversal frequencies.

Napier (1989) argued that both cratering and magnetic reversal records exhibit a period of 16.5 Myr. That a period is detected at $P = 16.5$ Myr in the cratering record has been confirmed (Yabushita 1991). This is not seen in Fig. 3 for the magnetic reversal record. Finally, the phase (latest epoch) of the assumed periodicity is 10 ± 0.5 Myr for the magnetic reversal frequencies. The detection of the $\sim 15 - 16$ Myr periodicity in the magnetic record requires a far more careful treatment of the data. Creer (1989) had used 5

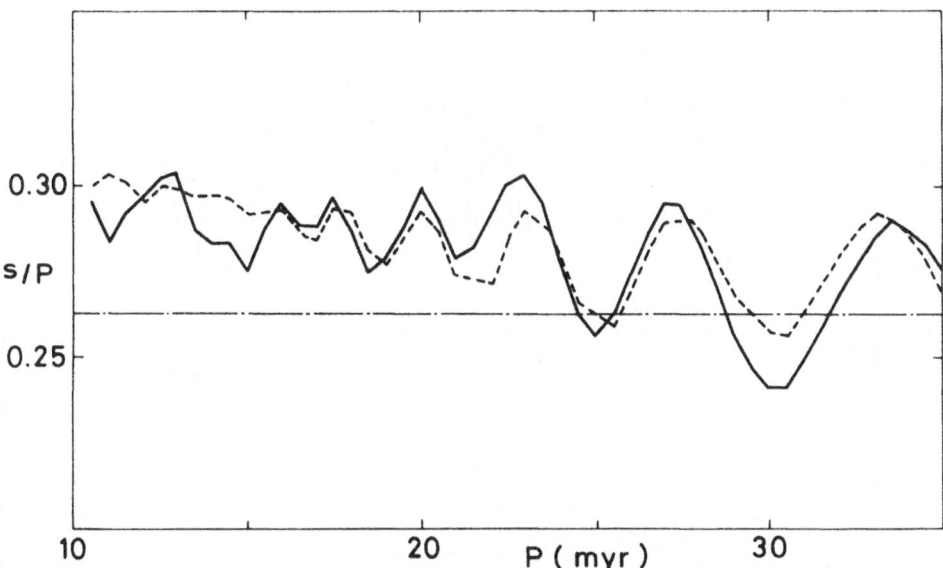

Fig. 3. s/P is plotted against P, where $t_1, t_2, \cdots t_{296}$ are observed times of magnetic reversal event. There appears to be a period at $P = 30$ Myr. The phase is 10 ± 0.5 Myr BP. The dotted curve corresponds to the long-term trend.

Myr moving windows to the reversal data to obtain frequency distribution and subtracted the long-term trend and extracted the 15 Myr periodicity.

6.   Discussion and Conclusions

Rampino & Stothers (1984a) first pointed out that the claimed periodicity in mass extinctions may be explained by periodic crossing of the Sun through the galactic mid-plane and proposed that intérstellar gas clouds could perturb the Oort cloud of comets, resulting in a so-called comet shower. Since cratering, magnetic reversals and mass extinction records yield periods close to 30 Myr, which is close to a period derived from the inferred density of matter in the galactic plane, the galactic crossing is an attractive idea. Since then, the z-component of the galactic gravitational field has been recognized (see references in Bailey et al. 1990) as a more efficient mechanism of supplying new comets from the Oort cloud than passing stars. Recently a detailed investigation has been made by Matese et al. (1995) on the cometary flux as the Oort cloud is perturbed by the gravitational field of the galaxy. It is assumed that the dark matter contributes one half of the matter distribution in the galactic plane, and the z-motion of the solar system has a period of 62 Myr. They then calculated how the flux of the Earth-crossing comets

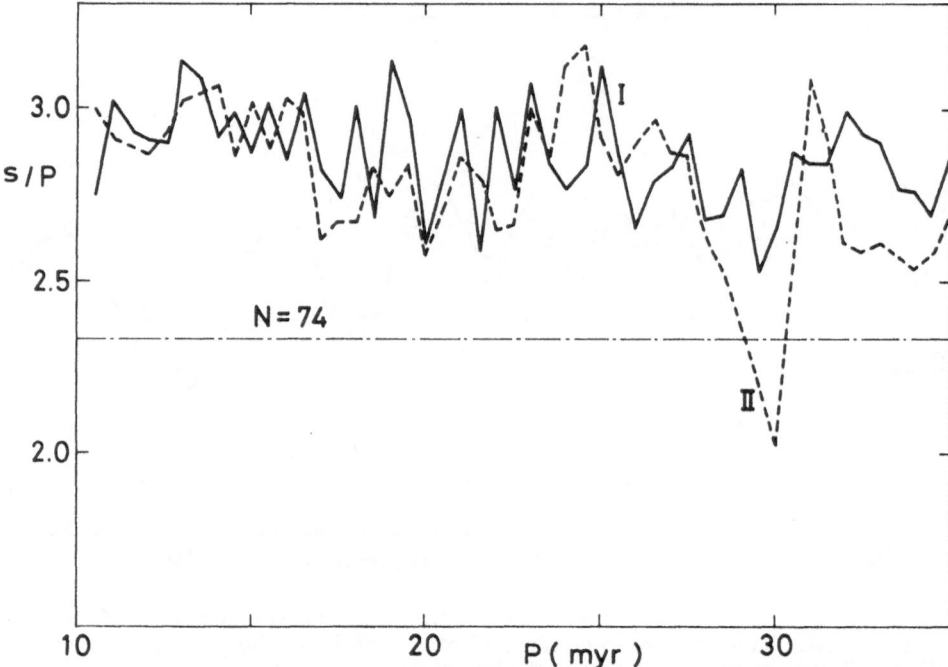

Fig. 4. s/P plotted against P for $t_1, t_2, \cdots t_{74}$ where the t's consist of periodic and random components. Here, the periodic component for curve I is 30% and 80% for curve II. Note that curve I resembles the s/P for known craters.

varies with time. According to Steel's (1987) calculation based on current observational data, the contribution to impactors by asteroids, short-period comets and long-period comets are 82, 12 and 6%, respectively. Weissman (1989) also gives similar figures. If S-P comets are captured from the flux of long-period comets (and not from the still unconfirmed Kuiper belt), the ratio of asteroids to comets will be 4 to 1.

In order to demonstrate how the s/P curve might look, a Monte Carlo simulation has been done by taking into account the effect of crater decay. $t_1, t_2, \cdots, t_N$ are made up of periodic as well as random components. The solid curve of Fig. 4 shows the s/P when the periodic component occupies 30%, while the dotted curve gives s/P when the periodic component is 80%.

The standard deviation from exact periodicity in cratering is somewhat arbitrarily taken at 5 Myr. One finds that for the former (30%), although there is a drop at $P = 30$ Myr, it is not very prominent and the entire behaviour of s/P resembles that of s/P calculated for ages of observed craters. Thus, there is no significant discrepancy between the astronomical modeling and observed impact records, although it is not possible to recognize a period in a statistically significant sense. However, this might after all be consistent with the cratering record. Although this record appears to indicate a period, it is not a compelling one.

We have discussed whether cratering records, magnetic reversal frequencies and mass extinctions exhibit any periodicity. Owing to long-term trends in the first two records, detected periods cannot be taken as significant when strict statistics is applied, while mass extinction peaks appear to yield a period which is significant at 10% level. The period derived is 26-27 Myr, compared with 30 Myr for cratering and magnetic reversals. One should keep in mind that the time-scale used in dating the mass extinction peaks may be stretched by 10% or so (Rampino, private communication). Stothers (1989) showed that extinction periods of 25-27 Myr, 25-30 Myr and 25-30 Myr are obtained on Harland, DNAG and Odin time scales, respectively. If the last two time-scales are valid, the discrepancies with the cratering may be removed. Because various subsets of crater data give rise to periods close to 30 Myr, it seems reasonable to take the results at face value. Astronomical modelling, on the other hand, predicts that only weak periodicity should be present in the cratering record, because random asteroidal impacts dominate cratering. In this sense, there is consistency between the galactic tidal force modulating the flux of Earth-crossing comets and the recorded cratering rate.

Acknowledgements

The author thanks a referee for making valuable comments which largely contributed for improving the presentation of the review.

References

Alvarez, W. & Muller, R.A.: 1984, *Nature* **308**, 718.
Bailey, M.E., Clube, S.V.M. & Napier, W.M.: 1990, *Origins of Comets*, Pergamon Press, Oxford.
Broadbent, S.R.: 1955, *Biometrica* **42**, 45.
Broadbent, S.R.: 1956, *Biometrica* **43**, 32.
Courtillot, V. & Besse, J.: 1987, *Science* **237**, 1140.
Creer, K.M.: 1989, in *Catastrophe and Evolution*, ed. Clube, S.V.M., Oxford Univ. Press.
Davis, M., Hut, P. & Muller, R.A.: 1984, *Nature* **308**, 715.
Emel'yanenko, V.V. & Bailey, M.E.: 1995, this volume.
Grieve, R.A.F.: 1993, Compilation of Crater Data obtainable upon request.
Grieve, R.A.F., Sharpton, V.L., Goodacre, A.K. and Carvin, J.B.: 1985, *Earth Planet. Sci. Lett.* **76**, 1.
Grieve, R.A,F. & Pesonen, L,J.: 1992, *Tectonophysics* **216**, 1.
Lutz, T.M.: 1985, *Nature* **317**, 404.
Lutz, T.M. & Watson, G.S.: 1988, *Nature* **334**, 240.
Matese, J.J., Whitman, P.G., Innanen, K.A. &Valtonen, M.J.: 1995, this volume.

Napier, W.M.: 1989, in *Catastrophe and Evolution*, ed. Clube, S.V.M., Oxford Univ. Press.
Negi, J.G. & Tiwari. R.V.: 1983, *Geophys. Res. Lett.* **10**, 713.
Rampino, M.R. & Caldeira, K.: 1992, in *Dynamics and Evolution of Minor Bodies with Galactic and Geological Implications*, eds. Clube, S.V.M., Yabushita, S. & Henrard, J., Kluwer.
Rampino, M.R. & Stothers, R.B.: 1984a, *Nature* **308**, 709.
Rampino, M.R. & Stothers, R.B.: 1984b, *Science* **226**, 1427.
Rampino, M.R. & Stothers, R.B.: 1986, in *The Galaxy and the Solar System*, eds. Smoluchowski, R., Bahcall, J.N. & Matthews, M.S., Univ. Arizona Press.
Raup, D.M.: 1985, *Nature* **314**, 341.
Raup, D.M. & Sepkoski, J.J.Jr.: 1984, *Proc. Natl. Acad. Sci.* **301**, 801.
Shoemaker, E.M. & Wolfe, R.F.: 1986, in *The Galayy and the Solar System*, eds. Smoluchowski, R., Bahcall, J.N. & Matthews, M.S., Univ. Arizona Press.
Steel, D.O.: 1987, *Mon. Not. R. astr. Soc.* **227**, 501.
Stothers, R.B.: 1979, *Astr. Astrophys.* **77**, 121.
Stothers, R.B.: 1988, *Observatory* **108**, 1.
Stothers, R.B.: 1989, *Geophys. Res. Lett.* **16**, 119.
Weissman, P.R.: 1989, *Proc. Conf. Global Catastrophes in Earth History*, Utah, U.S.A.
Whitmire, D. & Jackson,Jr., A.A.: 1984, *Nature* **308**, 713.
Yabushita, S.: 1983, *Mon. Not. R. astr. Soc.* **204**, 1185.
Yabushita, S.: 1988, *Mon. Not. R. astr. Soc.* **231**, 723.
Yabushita, S.: 1991, *Mon. Not. R. astr. Soc.* **250**, 481.
Yabushita, S.: 1994, *Earth, Moon and Planets* **64**, 207.
Yabushita, S,: 1995, *Mon. Not. R. astr. Soc.* (submitted).

TERRESTRIAL IMPACT CRATERS: THEIR SPATIAL AND TEMPORAL DISTRIBUTION AND IMPACTING BODIES

RICHARD A.F. GRIEVE

Geological Survey of Canada, Ottawa, Ont., Canada K1A 0Y3

and

LAURI J. PESONEN

Geological Survey of Finland Espoo, FIN-02150, Finland

Abstract. The terrestrial impact record contains currently ∼145 structures and includes the morphological crater types observed on the other terrestrial planets. It has, however, been severely modified by terrestrial geologic processes and is biased towards young (≤ 200 Ma) and large (≥ 20 km) impact structures on relatively well-studied cratonic areas. Nevertheless, the ground-truth data available from terrestrial impact structures have provided important constraints for the current understanding of cratering processes. If the known sample of impact structures is restricted to a subsample in which it is believed that all structures ≥ 20 km in diameter (D) have been discovered, the estimated terrestrial cratering rate is $5.5 \pm 2.7 \times 10^{-15} \mathrm{km}^{-2}\mathrm{a}^{-1}$ for $D \geq 20$ km. This rate estimate is equivalent to that based on astronomical observations of Earth-crossing bodies. These rates are a factor of two higher, however, than the estimated post-mare cratering rate on the moon but the large uncertainties preclude definitive conclusions as to the significance of this observation. Statements regarding a periodicity in the terrestrial cratering record based on time-series analyses of crater ages are considered unjustified, based on statistical arguments and the large uncertainties attached to many crater age estimates. Trace element and isotopic analyses of generally siderophile group elements in impact lithologies, particularly impact melt rocks, have provided the basis for the identification of impacting body compositions at a number of structures. These range from meteoritic class, e.g., C-1 chondrite, to tentative identifications, e.g., stone?, depending on the quality and quantity of analytical data. The majority of the identifications indicate chondritic impacting bodies, particularly with respect to the larger impact structures. This may indicate an increasing role for cometary impacts at larger diameters; although, the data base is limited and some identifications are equivocal. To realize the full potential of the terrestrial impact record to constrain the character of the impact flux, it will be necessary to undertake additional and systematic isotopic and trace element analyses of impact lithologies at well-characterized terrestrial impact structures.

1. Introduction

The highly active geologic environment of the Earth serves to remove, modify and mask the terrestrial impact record, making it harder to read than on the other terrestrial planets. Nevertheless, this disadvantage is more than offset by the availability of ground-truth information on impact structures in the terrestrial environment. Previously, we have summarized some of the general characteristics of the terrestrial impact cratering record, with particular emphasis on cratering mechanics, the geological and geophysical consequences of impact and the potential influence of impact on Earth evolution

Earth, Moon, and Planets **72**: 357-376, 1996.

(see Grieve and Pesonen, 1992 and references therein). Here, we reiterate some of the characteristics of the terrestrial impact record, to provide a context, but focus largely on what information the terrestrial record can or can not provide concerning the population of bodies that produced impacts on Earth. Approximately 145 terrestrial impact structures are currently known. The most recent listing of known terrestrial impact structures can be found in Grieve and Shoemaker (1994) but, with the present rate of new discoveries of 3-5 per year, it is already somewhat incomplete.

2. Morphology

Relatively uneroded structures display the basic morphologic progression from simple to complex crater forms with increasing crater diameter, observed on the other terrestrial planets. Simple impact structures have the form of a bowl-shaped depression with a structurally upraised rim. The rim area is overlain by ejecta deposits and the exposed crater floor represents the top of a sub-surface lens of allochthonous breccia. The canonical example of a near pristine terrestrial simple impact structure is Barringer or Meteor Crater, U.S.A. (Figure 1a). Simple impact structures occur up to diameters of $\sim$ 2-4 km on Earth. Above 4 km, all terrestrial impact structures generally have a complex form. When relatively uneroded, they are characterized by structurally complex and faulted rim areas, a flat annular trough and uplifted topographically high central structures, which contain rocks uplifted from deeper levels (Figure 1b). Various lines of observational and experimental evidence indicate that complex structures are a highly modified crater form with respect to simple structures (e.g., Melosh, 1989 and references therein).

Terrestrial complex impact structures also show the secondary forms observed on other planetary bodies, such as central peak craters, peak-ring craters and ring basins. It is not known categorically if there are examples of terrestrial multi-ring basins. Care must be exercised when comparing morphologic elements between individual terrestrial impact structures and, in particular, when comparing terrestrial and planetary craters (Pike, 1985). Original morphologic elements can be enhanced, modified or removed by various erosional processes on Earth. For example, Haughton, Canada a well-developed complex crater has an interior topographic ring, which has led to its description as a peak-ring crater (Robertson and Sweeney, 1983). More recent analyses indicate that this ring is not reflected in the geology or geophysical data (Bischoff and Oskierski, 1988; Scott and Hajnal, 1988). It is an erosional artifact rather than a primary feature. Morphometric relations are also affected by terrestrial processes to varying degrees. Some of

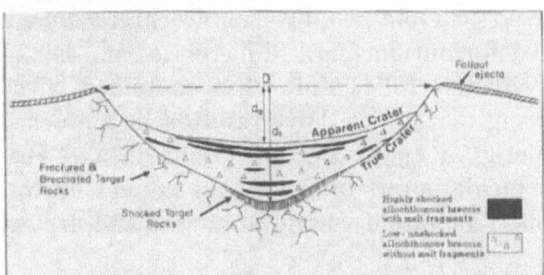

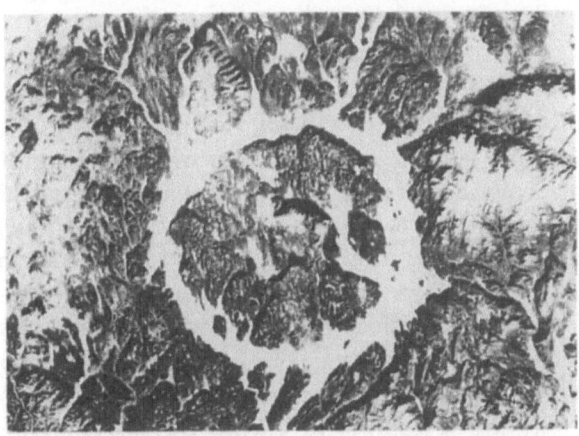

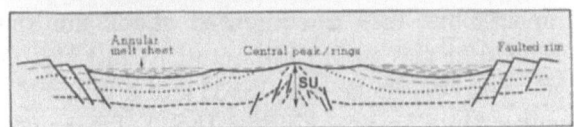

Fig. 1.   (a) Top: Oblique aerial photograph of the 50,000 year old, 1.2 km diameter
Barringer or Meteor Crater, U.S.A. This relatively fresh simple crater displays a typical
bowl shape with an upraised rim, overlain by hummocky fallout ejecta. Bottom: Schematic
cross-section of a simple impact structure. Note that apparent crater, as seen at the surface
(top), is underlain by a lens of allochthonous breccia containing zones of highly shocked
material. (b) Top: Vertical LANDSAT image of the 214 ± 1 Ma old, 100 km diameter Manicouagan impact structure, Canada. Annular lake (frozen) is 65 km in diameter and lies
interior to the estimated location of the original rim. Bottom: Schematic cross-section of a
complex impact structure in crystalline target rocks. Note its relatively shallow appearance
with a faulted rim area, down dropped annular trough and uplifted central structure. The
maximum amount of stratigraphic uplift (SU) in the center is indicated and the structure
is capped largely by an annular sheet of impact melt rocks.

the basic relations, such as depth/diameter (Figure 1), for relatively pristine terrestrial impact structures are given in Grieve and Pesonen (1992).

Any listing of the diameters of terrestrial impact structures is a mix of interpretations from topographical, geological and geophysical data. Individual diameter estimates can be different and controversial. As more and different types of data have become available for individual impact structures, estimates of their original diameter are revised. This is particularly evident when considering larger and older impact structures. For example, the Sudbury structure was initially listed as 55 km in diameter by Dietz (1964). Recent geological and geophysical interpretations, indicate considerable post-impact deformation (e.g., Milkereit *et al.*, 1992; Hirt *et al.*, 1993; Cowan and Schwerdtner, 1994) and, when combined with a greater understanding of cratering mechanics, have resulted in a much larger estimate of the original diameter of up to 280 km (Deutsch and Grieve, 1994). Data compilations of rim diameters of terrestrial impact craters should be used with some caution. They are dynamic in nature and are subject to constant revision.

3. Spatial distribution

The locations of known terrestrial impact structures are shown in Figure 2. These structures have evidence of an impact origin through the documented occurrence of meteoritic material and/or shock metamorphic features. The details of the character of shock metamorphic effects can be found in papers by French and Short (1968) and Roddy *et al.* (1977) and in Stöffler (1972, 1974) and Stöffler and Langenhorst (1994). The pressure-temperature range of shock metamorphism is illustrated in Figure 9 in Grieve and Pesonen (1992). To various degrees, terrestrial impact structures also have a number of other aspects in common, such as form, structure, geophysical characteristics, etc. There are a number of known terrestrial structures that have some of these aspects but lack documented shock metamorphic features. Recent examples include the 40 km diameter, ~ 140 Ma old Mjølnir structure in the Barents Sea (Gudlaugsson, 1993) and a possible 85 km diameter, ~ 35 Ma old structure in Chesapeake Bay, U.S.A. (Poag *et al.*, 1994). These structures are not included in Figure 2.

There are concentrations of known impact structures in N. America, Australia, and Europe, particularly Fennoscandia, through to the eastern part of the former USSR (Figure 2). This can be attributed to the fact that these are generally cratonic areas, where there have been programs to identify and study impact structures. We can not emphasize enough the importance of the influence of dedicated programs to identify impact structures on the local rate of discovery. For example, accelerated awareness of impact struc-

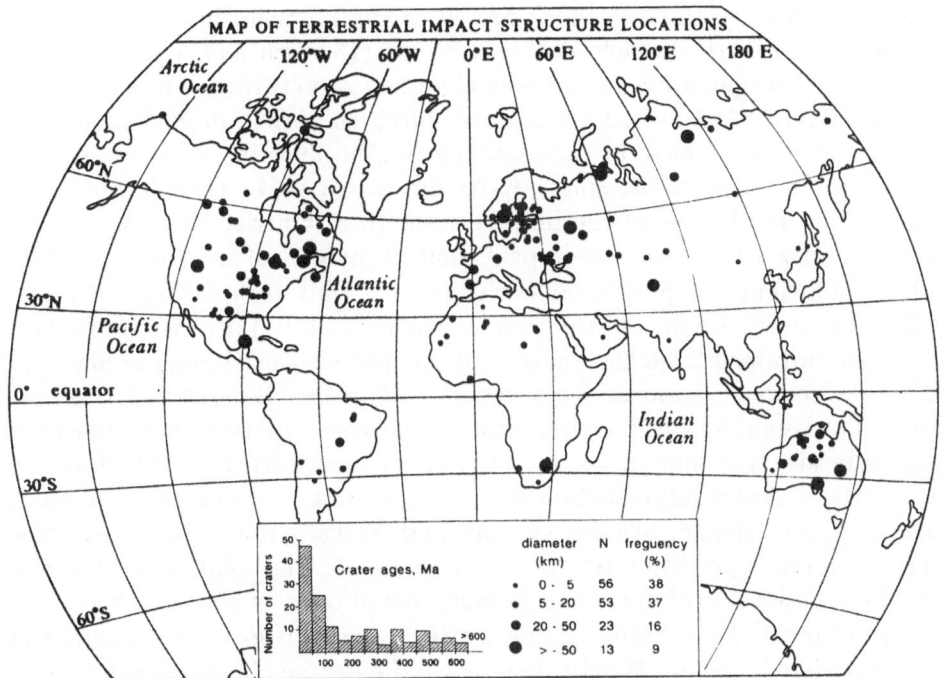

Fig. 2. Location of known terrestrial impact structures. Inset is histogram of age estimates and frequency percent of diameter ranges of known terrestrial impact structures. Note the majority are less than 200 Ma old, reflecting the removal of older structures from the record by terrestrial geologic processes.

tures and their characteristics in Fennoscandia has led to the recent confirmation of an impact origin for Gardnos, Norway (Dons and Naterstad, 1992), Lockne, Sweden (Lindström and Sturkell, 1992), Iso-Naakkima, Finland (Elo *et al.*, 1993), Lumparn, Finland (Svensson, 1994) and Suvasvesi, Finland (Pesonen, this vol.). Approximately 30% of the known terrestrial impact structures are completely buried by cover rocks. These were generally discovered through characteristic geophysical anomalies, which are associated with impact structures (Dabizha and Fel'dman, 1982; Henkel, 1992; Pilkington and Grieve, 1992) and subsequently explored through drilling for commercial and/or scientific purposes (e.g., Iso-Naakkima, Elo *et al.*, 1993).

4. Temporal distribution

Approximately 40% of the known terrestrial impact structures have been dated isotopically, generally from the analysis of impact melt rocks. Most of the materials (~ 90%) affected by impact in a cratering event, however, are subjected to insufficient shock pressures and post-shock temperatures to significantly disturb isotopic dating systems (Deutsch and Schärer, 1994). The bulk of isotopic dates for terrestrial impact structures are K-Ar or, more recently, ^{40}Ar–^{39}Ar dates. Fine-grained, often clast-rich, impact melt rocks are not particularly easy to date isotopically, because of inherited Ar from the clasts. In some cases, single K-Ar dates can be in considerable error with respect to the age of the impact event (e.g., Currie, 1971; Mak et al., 1976). In only a few cases are impact melt lithologies of sufficient grain size or have sufficient compositional variation to permit such dating techniques as Rb-Sr mineral isochrons (Deutsch and Schärer, 1994). Precise U-Pb dates have been obtained from the analysis of shocked zircons (Krogh et al., 1993) and new zircons crystallized from impact melt rocks (Hodych and Dunning, 1992). The remainder of known terrestrial impact structures have biostratigraphic or stratigraphic dates. In some cases, the biostratigraphic dates, on such units as crater-filling sediments, are as accurate and precise as isotopic dates, e.g., at Lockne, Sweden (Grahn and Nõlvak, 1993). Most are, however, minimum age estimates. In other cases, stratigraphic dates are only maximum estimates of age, the age being listed only as less than the age of the target rocks; for example, Eagle Butte, Canada, is formed in Cretaceous rocks and listed as < 65 Ma. In such cases, the degree of erosion can be used to provide a relatively crude constraint on the age.

Some broad trends, however, are clear. As with the spatial distribution, the temporal distribution of known terrestrial craters is not random. It is highly biased towards younger age estimates, with over 60% being younger than 200 Ma (Figure 2). As terrestrial impact craters are, at least initially, surface features in a highly active geologic environment, they can be removed or buried from observation relatively rapidly. The rate at which this occurs varies with the geologic history of the area.

5. Size-frequency distribution

There is an overall bias in the size-frequency distribution of terrestrial impact structures. In the Phanerozoic impact record, the cumulative size-frequency of terrestrial impact structures at large diameters is similar to that on other terrestrial planets (Figure 3). At diameters below ~ 20 km, however, the cumulative size-frequency distribution falls off and is generally taken to indicate the effects of erosion and, to a lesser extent, burial of smaller craters. The geological effects of impact at simple structures are visible to a depth of ≤ 1/3 the final rim diameter (Grieve and Pesonen, 1992). Thus, it is easy to appreciate why there is a deficit of small simple craters in the size-frequency distribution (Figure 3), with the geologic evidence for the largest terrestrial simple impact structures being removed by < 1.5 km of erosion. At larger complex structures, the depth/diameter relationship is relatively shallower (Figure 1b) but the absolute depths are often greater. In addition, the uplift of originally deeper lithologies in the center of complex structures provides an additional geologic manifestation of the impact event. The amount of stratigraphic uplift undergone by the deepest lithologies exposed in the central structures of complex impact structures is $\sim 1/10$ the final rim diameter (Figure 1b). Thus, even when the topography and interior impact products at a complex impact structure have been completely removed by erosion, it will still be recognizable as a roughly circular geologic anomaly. The shape of the cumulative size-frequency distribution in Fig. 3 appears to be an inherent property of the terrestrial record, as it has persisted as more impact structures have been added to the known sample over the years.

In principle, it is possible to use the size-frequency distribution of impact structures to derive the size-frequency distribution of impacting bodies; however, the size-frequency distribution is not well constrained because of small number statistics (Basaltic Volcanism Study Project, 1981). In addition, the conversion from impact structure diameters to impacting body sizes must be accomplished through energy-scaling relations, for an assumed impact velocity. Earlier scaling relations were based largely on the results of underground nuclear explosions and generally considered the kinetic energy of impact as a single term. More recent experimental data and dimensional analyses indicate that there is an additional dependence on impact velocity. Schmidt and Housen (1987) provide a summary of recent relations for crater dimension scaling. There is an other complication in that scaling equations refer generally to transient cavity dimensions, that is the cavity formed directly by ejection and displacement of material due to the cratering flow field (e.g., Melosh, 1989). It takes no account of cavity modification processes, which result in the final crater form and can be considerable in the case of complex structures. To relate transient cavity dimensions to final crater dimensions, an additional empirical scaling relation is required (Croft, 1985).

6. Cratering rate

The most complete record of impact cratering is recorded by relatively large (Figure 3), geologically young impact structures occurring in cratonic areas, which have been subjected to relatively thorough search programs. This results in the rate estimate being based on relatively small numbers of impact structures and, thus, having large uncertainties. Earlier estimates of the terrestrial cratering rate can be found in Shoemaker (1977) and Grieve and Dence (1979). From an examination of the degree of erosion at the specific impact structures, Grieve (1984) concluded that the original sample of impact structures of Grieve and Dence (1979) consisted of a younger group

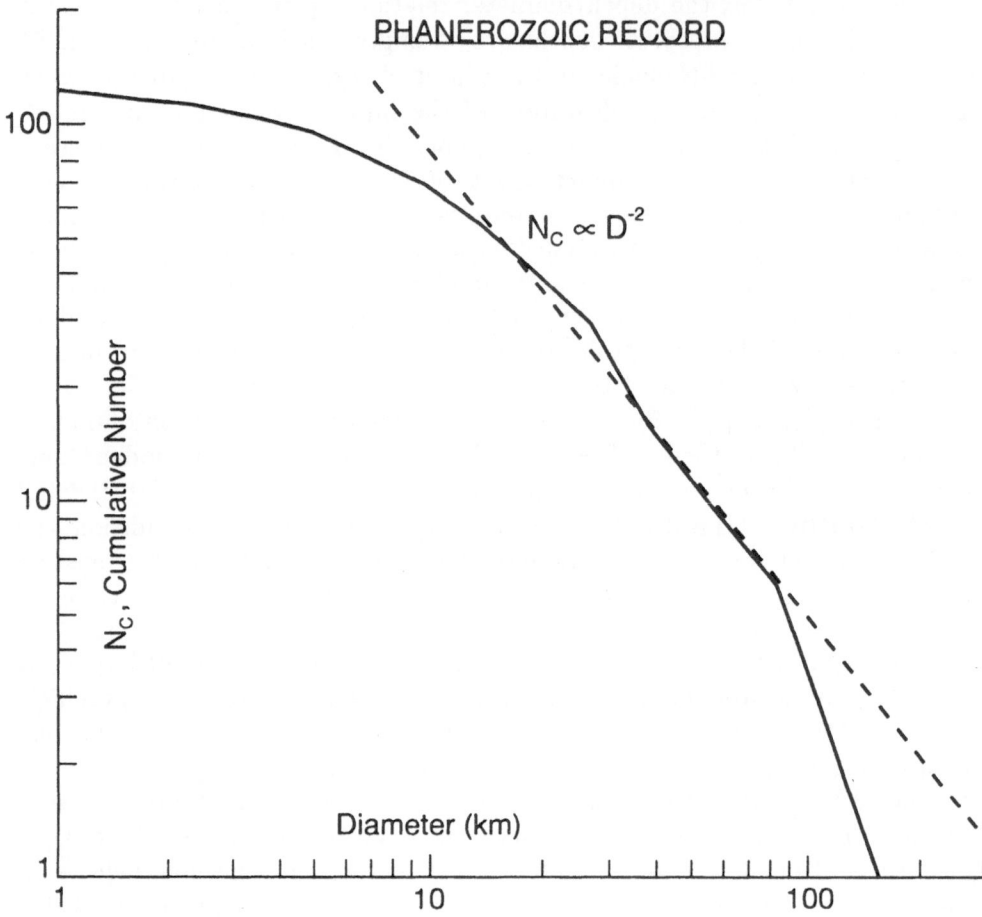

Fig. 3.   Logarithmic plot of cumulative number (N_c) – diameter of known terrestrial impact structures of Phanerozoic age. Note power law approximation, $N_c \propto D^{-2}$, of size-frequency distribution is no longer valid at $D < 20$ km, indicating a relative loss of smaller structures from the record.

(< 120 Ma), in which erosion was not a factor in their recognition, and an older group (> 120 Ma) in which impact structures may have been removed by erosion. From a reanalysis of the original data, the estimate of the cratering rate was revised upwards to $5.5 \pm 2.7 \times 10^{-15}$ km^{-2}a^{-1} for $D \geq 20$ km and impact structures ≤ 120 Ma in age. This rate estimate is comparable to the earlier estimate of Shoemaker (1977) and is very similar to an estimate based on astronomical observations of Earth-crossing asteroids and comets of $4.9 \pm 2.9 \times 10^{-15}$ km^{-2}a^{-1} for the last 100 Ma (Shoemaker *et al.*, 1990).

These estimates of the terrestrial cratering rate in relatively recent geologic time are approximately a factor of two higher than the average post-mare cratering rate on the moon (Basaltic Volcanism Study Project, 1981). With all the observational and model uncertainties associated with converting astronomical estimates of impacting body sizes to impact structure diameters and scaling between lunar and terrestrial impact conditions, it is unjustified to categorically state that the current cratering rate, based on terrestrial and astronomical data, is higher than that integrated over the last ~ 3.2 Ga, based on lunar cratering data. In addition, it is difficult to imagine a mechanism to increase the cratering rate with time, beyond the increasing involvement of cometary impacts (Shoemaker *et al.*, 1990).

It would be heartening to say that all that is required is to improve the statistics by the discovery and dating of new terrestrial impact structures. Unfortunately, this is not the case. The discovery of additional large, young impact structures throughout the world will not help, unless they are in an area where there is also a fair degree of confidence that the majority of the large impact structures have been found. Conversely, the discovery of additional large, young impact structures within known areas with relatively complete searches, such as those already used to derive cratering rate estimates, i.e., the N. American and Fennoscandia-European-Russian cratonic areas, would only acerbate the problem of an apparently higher rate in relatively recent geologic time. The direct dating of the age of a large number of individual post-mare lunar impact craters would be a solution but this is unlikely in the foreseeable future.

7. Periodic impacts

Considerable interest by the scientific community in the potential for impact to disrupt the biological balance on Earth followed the initial reports of evidence for the involvement of large scale impact at the Cretaceous-Tertiary (K/T) boundary. When Raup and Sepkoski (1984) reported evidence for a periodicity to the marine extinction record over the past 250 Ma, a number of works followed claiming a similar periodicity in the terrestrial cratering record (e.g., Alvarez and Muller, 1984; Davis *et al.*, 1984; Rampino and Haggerty, 1994). They suggested that this was the result of periodic cometary showers due to a variety of astronomical mechanisms. These periodic cometary showers were linked statistically to terrestrial mass extinction events and to a variety of other geological and geophysical phenomena (Rampino and Stothers, 1984). The claims were based on time-series analyses of subsets of the terrestrial cratering record. There were a variety of selection criteria but they conformed generally to large ($\geq$ 5 km), young ($\leq$ 250 Ma) and "well-dated" (± 20 Ma) impact structures.

Grieve *et al.* (1988) have argued against these conclusions, noting the biases in the terrestrial impact record and that, if the uncertainties in crater age estimates are taken into account, periodicities in the cratering record are questionable to weak. They also argued that, to have confidence in the reality of any period, the uncertainties attached to individual age estimates had to be $< 10\%$ of the period in question, which, in this case, was $\sim$ 30 Ma. This is not a general property of the terrestrial cratering record. Heisler and Tremaine (1989) reached a similar conclusion based on different statistical arguments and Baksi (1990) detected no periodicity, if the selected impact structures were restricted to those with age estimates of sufficient accuracy and precision. Weissman (1990) also found no evidence for periodic cometary showers and challenged the proposed mechanisms for producing periodic cometary showers. He did not, however, rule out random cometary showers.

Despite these arguments, periodic cometary showers, as defined by time-series analysis of the terrestrial cratering record, are still featured (e.g., Yabushita, 1992) and suggested as a causative agent for various geologic phenomena on Earth (e.g., Stothers and Rampino, 1990; Stothers, 1993; Rampino and Haggerty, 1994). The original listing used to define periodic cometary showers by most workers contained 103 known terrestrial impact structures. The most current list contains 145 known structures. In view of these additional data, as well as refinements to earlier age estimates, we have undertaken a time-series analysis. In doing this, we do not dismiss the severe problems due to the size (Figure 3) and temporal (Figure 2) biases in the terrestrial record. Nor do we exclude the problems of the relatively high age uncertainties in the record which, when taken into account, essentially preclude defining any statistically meaningful period. As in previous analyses, consideration was limited to craters with $D \geq 5$ km and ages ≤ 250 Ma with ≤ 20 Ma uncertainty (Table I). For comparison with previous work, we have continued to use an adaption of the method described by Broadbent (1955, 1956). Briefly, the procedure is to examine the hypothesis that a series of crater age estimates, y_i, may be expressed as:

$$y_i = A + r_i t \quad (i = 1, 2 \ldots) \tag{1}$$

where A (phase) and t (period) are constants, and r_i is zero or a positive integer. As a measure of the goodness-of-fit of the data to a period, we use the variable $\bar{Q}$ which is an r.m.s. measure of q_i,

$$\bar{Q} = \left\{ 1/n \sum_{i=1}^{n} (q_i)^2 \right\}^{1/2} \tag{2}$$

the departure from an integer of the individual observed ages divided by the period in question, such that:

TABLE I

Impact craters with $D \geq 5$ km and age estimates ≤ 250 Ma with ≤ 20 Ma uncertainty

Crater	Country	Diameter (km)	Age (Ma)	Method
Zhamanshin	Kazakhstan	13.5	0.90±0.10	Ar-Ar
Bosumtwi	Ghana	10.5	1.03±0.02	K-Ar
El'gygytgyn	Russia	18	3.5±0.5	K-Ar
Bigach	Kazakhstan	7	6±3	Strat.*
Karla	Russia	12	10±10	Strat.
Ries	Germany	24	15.1±1.0	Ar-Ar
Haughton	Canada	24	23.4±1.0	Ar-Ar
Logancha	Russia	20	25±20	Strat.
Popigai	Russia	100	35±5	Ar-Ar
Wanapitei	Canada	7.5	37±2	Ar-Ar
Mistastin	Canada	28	38±4	Ar-Ar
Logoisk	Belarus	17	40±5	Strat.
Montagnais	Canada	45	50.5±0.8	Ar-Ar
Ragozinka	Russia	9	55±5	Strat.
Marquez	U.S.A.	22	58±2	Strat.
Chicxulub	Mexico	180	64.98±0.05	Ar-Ar
Kamensk	Russia	20	65±2	Strat.
Kara, Ust-Kara	Russia	65	73±3	Ar-Ar
Manson	U.S.A.	35	73.8±0.3	Ar-Ar
Lappajärvi	Finland	23	77±0.4	Ar-Ar
Boltysh	Ukraine	24	88±3	K-Ar
Dellen	Sweden	15	89.0±2.7	Rb-Sr
Steen River	Canada	25	95±7	K-Ar
Avak	U.S.A.	12	100±5	Strat.
Carswell	Canada	39	115±10	Ar-Ar
Mien	Sweden	9	121.0±2.3	Ar-Ar
Tookoonooka	Australia	55	128±5	Strat.
Gosses Bluff	Australia	22	142.5±0.5	Ar-Ar
Rochechouart	France	23	186±8	Rb-Sr
Manicouagan	Canada	100	214±1	U-Pb
Puchezh-Katunki	Russia	80	220±10	Strat.
Araguainha	Brazil	40	247±5.5	Rb-Sr

* Strat. - Biostratigraphy

- For perfectly periodic data, $\bar{Q}$ is zero.

- For random data, $\bar{Q}$ is an approximately normally distributed variable with a mean of 0.29 and a standard deviation (s.d.) of $0.13/n^{1/2}$.

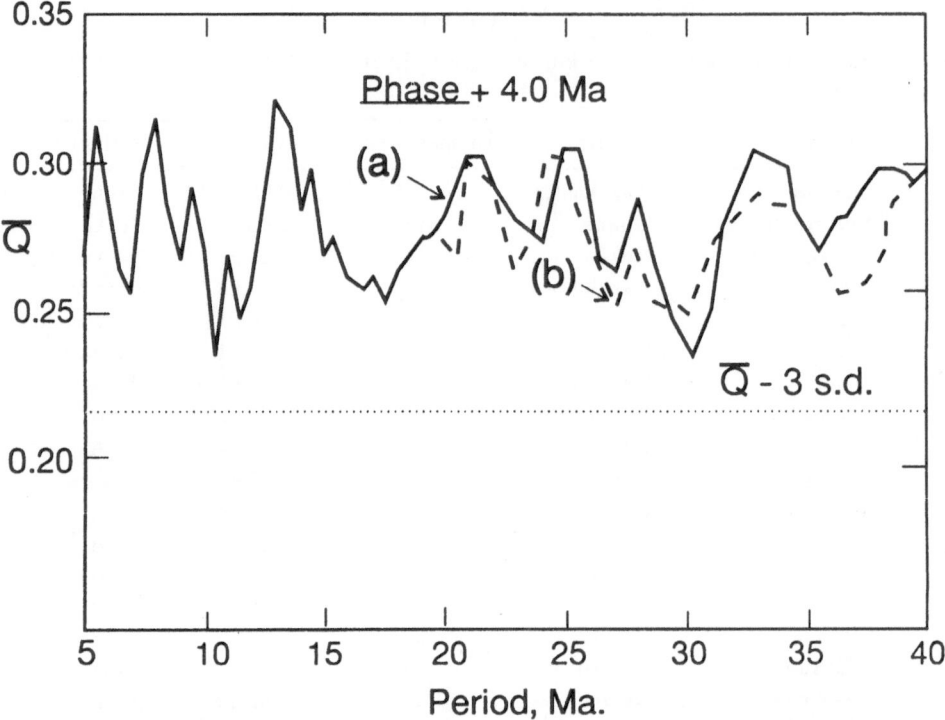

Fig. 4.   Plot of variable $\bar{Q}$ against period for a phase of +4 Ma of known terrestrial impact structures (solid line) with $D < 5$ km and ages < 250 Ma with uncertainties of $< \pm20$ Ma (Table II). Dashed line is for a subset of these structures with age uncertainties of $< \pm3$ Ma. Note neither data set indicates a significant period at the level of $0.29 - 3$ s.d. See text for details of $\bar{Q}$ and interpretation of the results.

Perfectly periodic ages combined with random ages yield $\bar{Q}$ values between 0.29 and zero, depending on the relative proportions of periodic and random data. We consider $\bar{Q}$ values less than $0.29 - 3$ s.d. as indicating the signal of a statistically significant period. This corresponds to less than 1 chance in 100 that the detection of a specific period is the result of a fortuitous combination of random data. A similar level of significance is quoted for periods from previous analyses (Alvarez and Muller, 1984; Rampino and Stothers, 1984). A 50:50 mixture of periodic and random ages, which is similar in proportion to the mix of random and periodic impact events, suggested by Shoemaker and Wolfe (1986) still yields a clear periodic signal (Grieve et al., 1988). The results on running the algorithm on the data in Table I produced no detectable period at the level of $\bar{Q}$ less than $0.29 - 3$ s.d. The run with the lowest $\bar{Q}$ value for any period-phase combination is for a period of ~ 30 Ma, and a phase of $+ 4$ Ma (Figure 4). This period is similar to some previous claims (e.g., Alvarez and Muller, 1984; Shoemaker and Wolfe, 1986) but the signal is even weaker than that in previous analyses.

If the terrestrial cratering record has a periodic signal it will not be perfect, due, in part, to orbital dynamics, which will smear out over several million years the time of impact of an injection of cometary bodies into the interior of the solar system. More importantly, real age estimates always have an attached uncertainty. To simulate age uncertainties, the algorithm determining $\bar{Q}$ for each possible phase and period was run 100 times with noise added to each date. The noise was added in such a manner that after 100 cycles it conformed to a normal distribution around each date, with a standard deviation equal to the uncertainty attached to the age estimate in Table II. Using this procedure, the $\sim$ 30 Ma signal was detected in $<$ 1% of the runs at the $0.29 - 3$ s.d. level. This level of detection is no higher than a number of other "periods" detected, when the age estimates were allowed to randomly fluctuate within their attached uncertainties.

The age estimates of some of the craters in Table I differ from those used previously in Grieve *et al.* (1988), because of more recent isotopic data. This illustrates a feature of the impact record that bears on the question of periodicity. Namely, not only is the precision of some age estimates for craters poor but so is the accuracy. Furthermore, there is a fundamental difference between obtaining a "date" for a particular event and its actual "age". A "date" is a numerical result produced from analytical procedures. An "age" is when that "date" can be specifically and unequivocally related to a geologic event. It is only in a few cases, where multiple dating methodologies have been applied and have produced consistent results, that enough confidence can be placed on the estimated age of an impact event for it to be used in time series analysis. We conclude that, despite continuing assertions, there is no compelling evidence for periodic impacts, due to cometary showers, in the terrestrial cratering record. Similarly, there are currently insufficient high quality data to comment upon whether or not the cratering record contains clusters of impacts of similar age that could reflect large-scale events in the asteroidal belt and relatively rapid resonant dynamics. There are, however, a number of cases (e.g., East and West Clearwater in Canada and Kamensk and Gusev in Russia), where the impact structures were formed by binary impacting bodies.

8. Impact body compositions

All small ($\leq$1 km) terrestrial impact structures are the result of the impact of iron or stony iron impacting bodies. This is due to the selective effect of the terrestrial atmosphere, which effectively destroys weaker stony and icy bodies as crater-forming bodies by atmospheric crushing, dispersal and retardation (Melosh, 1981). These small impact structures, which may occur as crater fields due to atmospheric break-up, are recent in age and it is often

TABLE II

Estimates of impacting body types at large terrestrial impact structures*

Crater	Country	Diameter (km)	Body Type
New Quebec	Canada	3.4	Chondrite
Brent	Canada	3.8	Chondrite
Gow	Canada	4	Iron?
Rio Cuarto	Argentina	4.5**	Chondrite
Ilyinets	Ukraine	4.5	Stone?
Sääksjärvi	Finland	5	Chondrite
Wanapitei	Canada	7.5	Chondrite
La Moinerie	Canada	8	***
Mien	Sweden	9	Stone?
Bosumtwi	Ghana	10.5	Iron
Ternovka	Ukraine	12	Chondrite
Nicholson Lake	Canada	12.5	Achondrite
Zhamanshin	Kazakhstan	13.5	Chondrite?
Dellen	Sweden	15	Stone?
Obolon	Ukraine	15	Iron
Lappajärvi	Finland	17	Chondrite
El'gygytgyn	Russia	18	Achondrite
Clearwater East	Canada	22	Chondrite
Rochechouart	France	23	Chondrite? or Iron?
Ries	Germany	24	Achondrite? or Chondrite?
Boltysh	Ukraine	25	Chondrite
Mistastin	Canada	28	Iron? or Achondrite?
Clearwater West	Canada	32	***
Kara	Russia	65	Chondrite
Manicouagan	Canada	100	***
Popigai	Russia	100	Chondrite
Chicxulub	Mexico	180	Chondrite

* All smaller impact structures appear to have been formed by iron or stony iron bodies, due to atmospheric effects (see text). With the exception of Rio Cuarto (Argentina) all listed estimates of impacting body types are based on geochemical anomalies at the impact structure.
** Longest dimension of largest crater.
*** Search for meteoritic signature unsuccessful, due either to a sampling problem or to an achondritic impacting body.

possible to recover pieces of the impacting body or pieces of material that spalled off the body prior to impact. Larger impacting bodies survive atmospheric passage and do so with impact velocities undiminished by atmospheric effects. In these cases, however, shock pressures and temperatures at the point of impact, which are a function of impact velocity, are sufficient to melt and/or vaporize the impacting body (Ahrens and O'Keefe, 1977). Estimates of the terrestrial r.m.s. impact velocity for cometary and asteroidal bodies are ~ 58 km s^{-1} and ~ 18 km s^{-1}, respectively (Shoemaker et al., 1990). Thus, at sizes where crater-forming stony impacting bodies are not crushed and dispersed by the atmosphere, their impact velocities are sufficient to guarantee that they do not survive as a physical entity on impact. There appears to be one exception. This is the $<$ 100,000 year old Rio Cuarto cratering field, Argentina, where a piece of chondritic meteorite has been recovered (Bunch and Schultz, 1992). The largest of the ten impact structures of Rio Cuarto is 4.5 km in its maximum dimension and it is believed that the preservation of a piece of the impacting body is due to the very oblique angle of the impact event, which resulted in projectile decapitation and highly elongated impact structures (Schultz and Lianza, 1992).

Although the impacting body is destroyed physically, some of the vaporized impactor material can be mixed into the impact lithologies; particularly, in impact melt rocks. Thus, through analyses for trace elements, such as siderophile and platinum group elements, which are rare in the terrestrial crust and relatively more abundant in certain meteorites, it is sometimes possible to detect a meteoritic signal. A number of factors, however, make the simple application of this procedure more difficult. Relative meteoritic contamination in, for example, impact melt rocks is generally at the 1% level or less, with absolute abundances of siderophile elements at the ppb level. Corrections must also be made for the terrestrial contribution to the trace element inventory in impact lithologies. Complications associated with this endogenic correction vary in proportion to the complexity of the lithologies in the target area. Fractionation of siderophile elements can occur, either directly through the vapor and melt phases produced in the impact event (Attrep et al., 1991; Mittlefehldt et al., 1992) or through terrestrial processes, such as weathering and alteration (Janssens et al., 1977; Lambert, 1982). These factors place limitations on the identification of impacting body composition. In some cases, however, careful studies have resulted in the identification of impacting body type, even, in a few cases, down to the level of known meteorite classes, e.g., the East Clearwater structure is believed to have been formed by a C1 chondritic body. Interestingly, no meteoritic signature has been detected in its twin structure, West Clearwater. One explanation is that the structures resulted from a pair of bodies which were in a gravitationally stable configuration but were also of different composition (Palme et al., 1981). In other cases, the identification of the composition of the impacting body is more controversial and, in yet, others no meteorite signal has been detected (Palme, 1982).

Recent interest has been revived in using Re/Os isotope systematics to detect a meteoritic component in impact melt rocks. They have been used to identify an extraterrestrial component in K/T boundary sediments

(Luck and Turekian, 1983) and with some limited success at East Clearwater impact structure (Fehn *et al.*, 1986). New, more sensitive, analytical techniques have extended the utility of Re/Os isotopes; particularly, where the target rocks are already relatively enriched in siderophile or platinum group elements (Koeberl *et al.*, 1994a). Low ^{187}Os/^{188}Os ratios are indicative of the admixture of meteoritic material.

The estimates of impacting body types in the literature are dominated by chondritic bodies (Table II). This is not unexpected, given the even greater dominance of chondrites as a percentage of meteorite falls but it may be somewhat misleading. For example, the reliability of impacting body identification at many of the impact structures in the former USSR is generally based on relatively few elements and inter-elemental ratios. Also, the geochemical signature of chondritic impacting bodies is the most easily identifiable, through their relatively high abundances of both siderophile elements and Cr. Identification of non-chondritic impacting bodies, e.g., at Mistastin (Table II), can be more of a negative result that the body was not a chondrite, rather than the firm indication that it was, as suggested, an iron body. Similar arguments apply to the identification of achondrites, which as differentiated bodies have compositions most similar to terrestrial rocks and tend to represent a negative result. It is possible that the increased sensitivity of Re-Os isotopic techniques may help detect additional cases of faint meteoritic signals in impact lithologies. For example, the technique has been used at Kalkkop (Table I), where an admixture of 0.05% meteoritic material was detected in one sample of suevitic breccia (Koeberl *et al.*, 1994a). Unfortunately, Re-Os systematics are, in themselves, not an effective discriminator between meteorite classes.

With the exception of Manicouagan (Table II), the largest known terrestrial impact structures appear to have been formed by chondritic bodies. While this does not confirm the suggestion of Shoemaker *et al.* (1990) that the large impact structures are more likely the result of cometary impacts, it is consistent with it, assuming that cometary bodies have chondritic abundances (Wetherill, 1974). Table II also assumes that the chondritic meteoritic signature at the K/T boundary is the result of the Chicxulub crater. Recent analytical data from Chicxulub support its K/T age and indicate elevated Ir values and meteoritic ^{187}Os/^{188}Os in one sample of the impact melt rock (Sharpton *et al.*, 1992; Koeberl *et al.*, 1994b). It is not known at present, however, if this Ir anomaly is complemented in other siderophile elements and whether or not they have chondritic relative abundances. In the case of Manicouagan, there is no apparent meteoritic signature in the melt rocks (Palme *et al.*, 1978, 1981). There is a small Cr enrichment, which may indicate an achondritic impacting body, but the situation is complicated by the occurrence of relatively Cr-rich mafic gneisses in the target rocks (Palme *et al.*, 1981).

9. Concluding remarks

Although limited by space, we have tried to give a flavor of the terrestrial impact record, particularly as it applies to constraining the flux with time and the nature of impacting bodies. The record contains biases due to terrestrial processes, which are particularly evident in the poor sample of smaller and older impact structures. Nevertheless, estimates of the terrestrial cratering rate can be derived, which are consistent with astronomical estimates. Given the nature of the current knowledge base of terrestrial impact structures, however, only the most general statements can be made concerning the character of the impact flux. Although the discovery rate of new impact structures continues at a reasonable pace, new discoveries do not necessarily improve constraints on the character of the impact flux. The addition of a newly discovered impact structure of uncertain age and geochemical character is, in itself, of no assistance. Systematic studies involving isotopic measurements for age estimates and trace element analyses are required, if there are to be major advances in the constraints provided by studies of the terrestrial impact record.

Acknowledgements

We would like to thank A. Therriault for critically reading an earlier version of the manuscript and M. Ford, J. Smith, M. Leino and S. Nässling for their assistance in manuscript preparation. Contribution from the Geological Survey of Canada 41994.

References

Ahrens T. J. and O'Keefe J. D.: 1977, Equations of state and impact-induced shock-wave attenuation on the moon. In *Impact and Explosion Cratering* (eds. D. J. Roddy, R. O. Pepin and R. B. Merrill), pp. 639-656. Pergamon Press.

Alvarez W. and Muller R.: 1984, Evidence from crater ages for periodic impact on the Earth. *Nature* **308**, 718-720.

Attrep M., Orth C. J., Quintana L. R., Shoemaker C. S., Shoemaker E. M. and Taylor E. S.: 1991, Chemical fractionation of siderophile elements in impactites from Australian meteorite craters (abstract). *Lunar Planet. Sci.* **XXII**, 39-40.

Baksi A. K.: 1990, Search for periodicity in global events in the geologic record: Quo vadimus? *Geology* **18**, 983-986.

Basaltic Volcanism on the Terrestrial Planets., 1981, Pergamon Press. 1286 pp.

Bischoff L. and Oskierski W.: 1988, The surface structure of the Haughton impact crater, Devon Island, Canada. *Meteoritics* **23**, 209-220.

Broadbent S.R.: 1955, Quantum hypotheses. *Biometrika* **42**, 45-57.

Broadbent S.R.: 1956, Examination of a quantum hypothesis based on a single set of data. *Biometrika* **43**, 32-44.

Bunch T. E. and Schultz P. H.: 1992, A study of the Rio Cuarto loess impactites and chondritic impactor (abstract). *Lunar Planet. Sci.* **XXIII**, 179-180.

Cowan E. J. and Schwerdtner W. M.: 1994, Fold origin of the Sudbury Basin. In *Ontario Geological Survey Special Volume 5, Proceedings of the Sudbury - Noril'sk Symposium* (eds. P.C. Lightfoot and A.J. Naldrett), pp. 45-55. Ont. Min. Northern Dev. and Mines.

Croft S. K.: 1985, The scaling of complex craters. *Proc. Lunar Planet. Sci. Conf.* 15th, 828-842.

Currie K. L.: 1971, Geology of the resurgent cryptoexplosion crater at Mistastin Lake, Labrador. *Canada Geol. Surv. Bull.* 207, 1-62.

Dabizha A. I. and Fel'dman V. I.: 1982, The geophysical properties of some astroblemes in the USSR (in Russian). *Meteoritika* 40, 91-101.

Davis M., Hut P. and Muller R. A.: 1984, Extinction of species by periodic comet showers. *Nature* 308, 715-717.

Deutsch A. and Grieve R. A. F.: 1994, The Sudbury Structure: Constraints on its genesis from Lithoprobe results. *Geophys. Res. Lett.* 21, 963-966.

Deutsch A. and Schärer U.: 1994, Dating terrestrial impact events. *Meteoritics* 29, 301-322.

Dietz R. S.: 1964, Sudbury structure as an astrobleme. *J. Geol.* 72, 412-434.

Dons J. A. and Naterstad J.: 1992, The Gardnos impact structure, Norway (abstract). *Meteoritics* 27, 215.

Elo S., Kuivasaari T., Lehtinen M., Sarapää O. and Uutela A.: 1993, Iso-Naakkima, a circular structure filled with NeoProterozoic sediments, Pieksomoki, Southeastern Finland. *Bull. Geol. Soc. Finland* 65, 3-30.

Fehn U., Teng R., Elmore D. and Kubik P. W.: 1986, Isotopic composition of osmium in terrestrial samples determined by accelerator mass spectrometry. *Nature* 323, 707-710.

French B.M. and Short N.M. (eds.): 1968, *Shock Metamorphism of Natural Materials.* Mono Book Corp. 644 pp.

Grahn Y. and Nõlvak J.: 1993, Chitinozoan dating of Ordovician impact events in Sweden and Estonia. A preliminary note. *Geol. Fören. Stockholm Förhand.* 115, 263-264.

Grieve R. A. F.: 1984, The impact cratering rate in recent time. *Proc. Lunar Planet. Sci. Conf.* 14th, *J. Geophys. Res., Supp.* 89, B403-B408.

Grieve R. A. F. and Dence M. R.: 1979, The terrestrial cratering record II. The crater production rate. *Icarus* 38, 230-242.

Grieve R. A. F. and Pesonen L. J.: 1992, The terrestrial impact cratering record. *Tectonophysics* 216, 1-30.

Grieve R.A.F. and Shoemaker E.M.: 1994, The record of past impacts on Earth. In *Hazards Due to Comets and Asteroids* (eds. T. Gehrels), (in press). Univ. Arizona Press.

Grieve R. A. F., Sharpton V. L., Rupert J. D. and Goodacre A. K.: 1988, Detecting a periodic signal in the terrestrial cratering record. *Proc. Lunar Planet. Sci. Conf.* 18th, 375-382.

Gudlaugsson S. T.: 1993, Large impact crater in the Barents Sea. *Geology* 21, 291-294.

Heisler J. and Tremaine S.: 1989, How dating uncertainties affect the detection of periodicity in extinctions and craters. *Icarus* 77, 213-219.

Henkel H.: 1992, Geophysical aspects of meteorite impact craters in eroded shield environment, with special emphasis on electric resistivity. *Tectonophysics* 216, 63-90.

Hirt A. M., Lowrie W., Clendenen W. S. and Kligfield R.: 1993, Correlation of strain and the anisotropy of magnetic susceptibility in the Onaping Formation: Evidence for a near-circular origin of the Sudbury Basin. *Tectonophysics* 225, 231-254.

Hodych J. P. and Dunning G. R.: 1992, Did the Manicouagan impact trigger end-of-Triassic mass extinction? *Geology* 20, 51-54.

Janssens M. J., Hertogen J., Takahashi H., Anders E. and Lambert P.: 1977, Rochechouart meteorite crater: Identification of projectile. *J. Geophys. Res.* 82, 750-758.

Koeberl C., Reimold W. U., Shirey S. B. and Roux F. G. L.: 1994a, Kalkkop crater, Cape Province, South Africa: Confirmation of impact origin using osmium isotope systematics. *Geochim. Cosmochim. Acta* 58, 1229-1234.

Koeberl C., Sharpton V. L., Schuraytz B. C., Shirey S. B., Blum J. D. and Marin L. E.: 1994b, Evidence for a meteoritic component in impact melt rock from the Chicxulub structure. *Geochim. Cosmochim. Acta* 58, 1679-1684.

Krogh T. E., Kamo S. L. and Bohor B. F.: 1993, Fingerprinting the K/T impact site and determining the time of impact by U- Pb dating of single shocked zircons from distal ejecta. *Earth Planet. Sci. Lett.* **119**, 425-429.

Lambert P.: 1982, Anomalies within the system: Rochechouart target rock meteorite. *Geol. Soc. Am. Sp. Pap.* **190**, 57-68.

Lindström M. and Sturkell E. F. F.: 1992, Geology of the early Palaeozoic Lockne impact structure, central Sweden. *Tectonophysics* **216**, 169-185.

Luck J. M. and Turekian K. K.: 1983, Osmium-187/Osmium-186 in manganese nodules and the Cretaceous-Tertiary boundary. *Science* **222**, 613-615.

Mak E. K., York D., Grieve R. A. F. and Dence M. R.: 1976, The age of the Mistastin Lake crater, Labrador, Canada. *Earth Planet. Sci. Lett.* **31**, 345-357.

Melosh H. J.: 1981, Atmospheric breakup of terrestrial impactors. In *Multi-Ring Basins* (eds. P. H. Schultz and P. B. Merrill), pp. 29-35. Pergamon Press.

Melosh H. J.: 1989, *Impact Cratering: A Geologic Process.* Oxford Univ. Press. 245 pp.

Milkereit B., Green A. and the Sudbury Working Group: 1992, Deep geometry of the Sudbury structure from seismic reflection profiling. *Geology* **20**, 807-811.

Mittlefehldt D. W., See T. H. and Hörz F.: 1992, Dissemination and fractionation of projectile materials in the impact melts from Wabar Crater, Saudi Arabia. *Meteoritics* **27**, 361-37.

Palme H.: 1982, Identification of projectiles of large terrestrial impact craters and some implications for the interpretation of Ir-rich Cretaceous/Tertiary boundary layers. *Geol. Soc. Am. Sp. Pap.* **190**, 223-233.

Palme H., Janssens M. J., Takahashi H., Anders E. and Hertogen J.: 1978, Meteoritic material at five large impact craters. *Geochim. Cosmochim. Acta* **42**, 313-323.

Palme H., Grieve R. A. F. and Wolf R.: 1981, Identification of the projectile at Brent crater, and further considerations of projectile types at terrestrial craters. *Geochim. Cosmochim. Acta* **45**, 2417-2424.

Pesonen, L. J.: 1995, Impact cratering record of Fennoscandia (this volume).

Pike R. J.: 1985, Some morphologic systematics of complex impact structures. *Meteoritics* **20**, 49-68.

Pilkington M. and Grieve R. A. F.: 1992, The geophysical signature of terrestrial impact craters. *Rev. Geophysics* **30**, 161-181.

Poag C. W., Powars D. S., Poppe L. J. and Mixon R. B.: 1994, Meteroid mayhem in Ole Virginny: Source of the North American tektite strewn field. *Geology* **22**, 691-694.

Rampino M. R. and Haggerty B. M.: 1994, Extra-terrestrial impacts and mass extinctions of life: In: *Hazards Due to Comets and Asteroids* (ed. T. Gehrels), Univ. of Arizona Press (in print).

Rampino M. R. and Stothers R. B.: 1984, Geological rhythms and cometary impacts. *Science* **226**, 1427-1431.

Raup D. M. and Sepkoski J. J.: 1984, Periodicity of extinctions in the geologic past. *Proc. Nat. Acad. Sci.* **81**, 801-805.

Robertson P. B. and Sweeney J. F.: 1983, Haughton impact structure: Structural and morphological aspects. *Can. Jour. Earth Sci.* **20**, 1134-1151.

Roddy D.J., Pepin R.O and Merrill R.B. (eds.): 1977, *Impact and Explosion Cratering.* Pergamon Press. 1301 pp.

Schmidt R. M. and Housen K. R.: 1987, Some recent advances in the scaling of impact and explosion cratering. *Internat. Jour. Impact Eng.* **5**, 543-560.

Schultz P. H. and Lianza R. E.: 1992, Recent grazing impacts on the Earth recorded in the Rio Cuarto crater field, Argentina. *Nature* **355**, 234-237.

Scott D. and Hajnal Z.: 1988, Seismic signature of the Haughton structure. *Meteoritics* **23**, 239-247.

Sharpton V. L., Dalrymple G. B., Marin L. E., Ryder G., Schuraytz B. C. and Urrutia-Fucugauchi J.: 1992, New links between the Chicxulub impact structure and the Cretaceous/Tertiary boundary. *Nature* **359**, 819-821.

Shoemaker E. M.: 1977, Astronomically observable crater-forming projectiles. In *Impact and Explosion Cratering* (eds. D. J. Roddy, R. O. Pepin and R. B. Merrill), pp. 617-628. Pergamon Press.
Shoemaker E. M. and Wolfe R. F.: 1986, Mass extinctions, crater ages and comet showers. In *The Galaxy and the Solar System* (eds. R. Smoluchowski, J. N. Bahcall and M. S. Matthews), pp. 338-386. Univ. Arizona Press.
Shoemaker E. M., Wolfe R. F. and Shoemaker C. S.: 1990, Asteroid and comet flux in the neighborhood of Earth. *Geol. Soc. Am. Sp. Pap.* **247**, 150-170.
Stothers R. B.: 1993, Impact cratering at geologic stage boundaries. *Geophys. Res. Lett.* **20**, 887-890.
Stothers R. B. and Rampino M. R.: 1990, Periodicity in flood basalts, mass extinctions and impacts; a statistical view and a model. *Geol. Soc. Am. Sp. Pap.* **247**, 9-18.
Stöffler D.: 1972, Deformation and transformation of rock-forming minerals by natural and experimental shock processes. I. Behavior of minerals under shock compression. *Fortschr. der Mineral.* **49**, 50-113.
Stöffler D.: 1974, Deformation and transformation of rock-forming minerals by natural and experimental shock processes. II. Physical properties of shocked minerals. *Fortschr. der Mineral.* **51**, 256-289.
Stöffler D. and Langehorst F.: 1994, Shock metamorphism of quartz in nature and experiment: 1. Basic observation and theory. *Meteoritics* **29**, 155-181.
Svensson N. B.: 1994, Lumparn - an impact crater on Åland, SW Finland and some effects of shock wave in the surrounding bedrock (abstract). *European Science Foundation. Second International Conference. Impact Cratering and Evolution of Planet Earth. The Identification and Characterization of Impacts.*
Weissman P.R.: 1990, The cometary impactor flux at the Earth. *Geol. Soc. Am. Sp. Pap.* **247**, 171-180.
Wetherill G. W.: 1974, Solar system sources of meteorites and large meteoroids. *Rev. Earth Planet. Sci.* **2**, 303-331.
Yabushita S.: 1992, Periodicity and decay of craters over the past 600 Myr. *Earth, Moon, and Planets* **58**, 57-63.

THE IMPACT CRATERING RECORD OF FENNOSCANDIA

L.J. PESONEN

Laboratory for Palaeomagnetism, Geophysics Department
Geological Survey of Finland, FIN-02150 Espoo, Finland

Abstract. The current database of craterform structures in Fennoscandia contains 22 structures of impact origin and about fifty other structures which lack sufficient evidence for impact. The discovery rate of new structures has been one or two per year during the past ten years. The proven impact structures are located in southern Fennoscandia and the majority have been found in Proterozoic target rocks. The age of the structures varies from prehistoric to $\leq 1\,000$ Ma and their diameters (D) from 0.04 km to 55 km. Nine of the structures contain impact melt. A characteristic feature of the Fennoscandian impact record is a relatively large number of small (≤ 5 km) but old (> 200 Ma) structures: this is a result of success of geophysical methods to discover small but old impact structures in an eroded shield covered with relatively thin overburden. Some of the large circular structures in satellite images and/or in geophysical maps may represent deeply eroded scars of very old impacts, but due to the lack of shock metamorphic features, impact-generated rocks or identified ejecta layers, they cannot yet be classified as impact sites. Two huge structures are proposed here as possible impact sites on the basis of circular satellite images and distinct geophysical anomalies: the Lycksele structure in northern Sweden ($D \sim 120$ km, see also Witschard, 1984) and the Valga structure in Latvia/Estonia ($D \sim 180$ km). However, endogeneous explanations, like buried granites, basement domings, or fault-bounded blocks are also possible for these structures. Hints, such as distal ejecta layers or impact produced breccia dykes, of an Archaean or Early Proterozoic impact structure have not been found in Fennoscandia so far. New ways of searching for these structures are proposed with particular emphasis on high-resolution integrated geophysical methods. The impact cratering rate in Fennoscandia is $\sim 2.0 \cdot 10^{-14}$ km^{-2} a^{-1} (for craters with $D > 3$ km) corresponding to about two events per every 100 Ma for the last 700 Ma. Due to erosion, this is a minimal estimate but is higher than the global rate probably due to strong research activity for finding impact structures in Fennoscandia.

1. Introduction

Impact cratering is a ubiquitous process in our Solar System. The study of terrestrial impact structures has recently gained increased attention in the earth science community for three reasons. First, impact structures have great potential for containing economically important resources, such as diamonds (Popigai, Russia), metallic ores (Sudbury, Canada), building materials (Ries, Germany) and water reservoirs (Lappajärvi, Finland) (e.g. Masaitis, 1992; Lehtimäki, 1994). Second, scientists have become aware of the geological and biological consequences of impact on the evolution of the Earth, for example, the Cretaceous-Tertiary impact event, and related mass extinctions (Grieve and Pesonen, 1992; 1995). Third, impact craters have preserved down-dropped sediments of both pre- and post-impact origin which serve as unique rocks (not otherwise available due to erosion) for geol-

Earth, Moon, and Planets **72**: 377-393, 1996.

ogists to obtain complete stratigraphic records of eroded shields (Lindström, 1993).

The Fennoscandian (Baltic) Shield has a long geologic history and thus the cumulative effects of impact-cratering have played a major role in the creation of tectonic patterns and in the redistribution of elements in the ancient crust, atmosphere and biosphere. It is in a unique position for impact cratering research for several reasons: (i) the cratonic part of the shield has peneplane topography and has been stable since the last major orogeny (Svecofennian) at 1.9–1.8 Ga ago and is well-exposed, (ii) there are presently more than 72 craterlike structures in Fennoscandia (Henkel and Pesonen, 1992), of which 22 are now identified as being extraterrestrial in origin, and (iii) high-quality satellite (LANDSAT, NOAA, SPOT) images for the entire Fennoscandia region are available, as well as high-resolution airborne and ground geophysical and geological data, providing a remarkable database for impact cratering research (Pesonen, 1991).

The purpose of this paper is to give a new look at the database of Fennoscandian impact structures by first adding seven new impact structures in the earlier database of Henkel and Pesonen (1992). The possibility that some of the large structures seen in satellite images (e.g. Witschard, 1984; this work) and in geophysical maps (Korhonen, 1990, 1994; this work) are impact structures will be discussed with particular emphasis on two huge structures (Lycksele in Sweden and Valga in Latvia/Estonia).

2. Criteria indicative of impact

The criteria used for accepting a craterlike structure as being an impact structure are outlined by Henkel and Pesonen (1992). Thus, the structure must have either (i) historical record of an impact event, (ii) meteorite fragments or contaminated meteorite material in the rocks, e.g. enriched siderophile element abundances, or (iii) shock metamorphic features, e.g., shatter cones or planar deformation features (PDF's), high-pressure mineral pseudomorphs (coesite, stishovite) or solid state or fusion glasses (Stöffler, 1972; Grieve and Robertson, 1979; Grieve and Pesonen, 1992; 1995). These criteria are not, however, generally met when very old and eroded structures are to be identified as meteorite impact craters since they often lack shock effects in rocks due to their deep erosional level, or post-impact deformations are masking the impact signatures. Other criteria involving specific geophysical and petrophysical characteristics (e.g. circular gravity and magnetic anomalies, radial dependence of physical properties of rocks from the impact centre), discoveries of impact produced igneous melts or breccia dykes, or successful linkage of distal ejecta layers to a certain structure, may eventually become useful criteria for identifying very old structures as impact sites (Pesonen, 1993; Österman et al., 1995). Here we rely on a combination of the previous criteria in describing the Fennoscandian impact record (Table I). Two new structures with associated circular geophysical anomalies are presented in this paper (the Lycksele structure in Sweden and the Valga structure in Latvia/Estonia) although their impact origin is not yet proven (see Chapter 5). This also applies to some other large circular structures in Fennoscandia (e.g. Marras, Nunjes, Uppland) as previously discussed (Henkel and Pesonen, 1992; Korhonen, 1990; 1994). All these structures may be classified as impact structures in the future provided that shock metamorphic features or impact-generated rocks will be discovered in the deeply eroded target rocks of the structures, or distal ejecta layers linked to these structures will be found.

3. Fennoscandian impact craters

Table I summarizes the morphometric and other data of the 22 impact structures in Fennoscandia. Included are data for the locations of the structures, ages, diameters (D), presence of impact melt, target lithologies and environments, and types of structures (see also Henkel and Pesonen, 1992; Puura et. al., 1994). The locations are plotted in Fig. 1, where the center of each structure is indicated with a solid dot, the size of which is proportional to the present diameter of the structure. The numbers refer to the list in Henkel and Pesonen (1992), except those from Lithuania (63, Misarai; 64, Vepriai), Latvia (65, Dobele) and Russia (66, Mishinogorskaya), which were not included in the previous catalogue (Puura et al., 1994). Structures 35a (Suvasvesi N, Finland) and 37 (Lumparn, Finland), which were previously listed as *possible* impact structures (Henkel and Pesonen, 1992) are now *proven* impact structures based on studies of drill core materials and on petrological and geophysical data (see Chapter 4). The impact structures are located in Sweden (6), Finland (6), Norway (1), Russia (2), Estonia (4), Lithuania (2) and Latvia (1). All are on present continental land although some (e.g., Lockne, Tvären and Kärdla; Puura et al., 1994) have been formed in shallow epicontinental seas during the Ordovician (Table I). It is noticeable that these three marine structures are of nearly similar age ($\sim 450-460$ Ma) and therefore the possibility that they are caused by a fragmented asteroid or comet shower cannot be missed (Lindström et al., 1992; Puura et al., 1994). Some of the impact craters listed in Table I refer to a crater field (e.g. 38, Kaali; Pirrus and Tiirmaa, 1990); here only the one with largest diameter is listed. One of the structures (35a, Suvasvesi N, Finland) is part of a probable double impact structure of which the northern member has been so far proven to be an impact structure (Sect. 4).

TABLE I

Data of impact structures in Fennoscandia

No	Structure	Country	Lat., °N	Long. °E	Age [Ma]	s.d. [Ma]	meth.	D [km]	melt.	env.	type
5	Dellen	S	61.8	16.8	89	2.7	Ar-Ar	19	+	cont	C
6	Siljan	S	61.1	15.0	368	1.1	K-Ar	55	−	cont	M
12	Tvären	S	58.8	17.4	~ 455	10	bstr	2	−	mar	S
20	Mien	S	56.4	14.9	121	2.3	Ar-Ar	9	+	cont	C
25	Granby	S	58.4	14.9	~ 470	30	bstr	3	−	mar	S
31	Lappajrvi	F	63.2	23.7	77	0.4	Ar-Ar	23	+	cont	C
32	Söderfjärden	F	62.9	21.7	~ 550		K-Ar, str	6	−	cont	C
33	Sääksjärvi	F	61.4	22.4	~ 560	12	Ar-Ar	5	+	cont	C
35a	Suvasvesi N	F	62.7	28.0	~ 250		str	4	+	cont	S
36	Jänisjärvi	R	62.3	31.4	698	22	K-Ar	14	+	cont	C
37	Lumparn	F	60.2	20.1	> 500, < 1200		str	10	−	cont	C
38	Kaali[a]	E	58.4	22.4	0.00395	0.001	h, ^{14}C	0.110	−	cont	S
42	Gardnos	N	60.7	9.2	~ 500	10	str	5	+	mar	C
48	Lockne	S	63.0	14.7	~ 455	10	bstr	7	+	mar	M
52	Ilumetsä[a]	E	57.9	27.4	0.002		^{14}C	0.08	−	cont	S
53	Tsöörikmäe[a]	E	58.1	27.5	0.0095		^{14}C	0.04	−	cont	S
54	Kärdla	E	59.0	22.7	~ 455	5	str	4	−	mar	C
61	Iso-Naakkima	F	62.2	27.1	> 500		str	3	−	cont	S
63	Misarai	Li	54.0	24.6	~ 595	145	str	5	+	mar	C
64	Vepriai	Li	55.2	24.6	~ 165	30	biostr	7.5	−	mar	C
65	Dobele	La	56.6	23.3	~ 205	35	biostr	4.5	−	cont	C
66	Mishinogorskaya	R	58.7	28.0	< 1000		str	4.5	−	cont	C

No	refers to catalogues of Henkel and Pesonen (1992) and Puura et al. (1994) and references therein
Structure	name of proven impact structure, [a] very small crater
Country	S=Sweden, F=Finland, R=Russia, E=Estonia, N=Norway, Li=Lithuania, La=Latvia
Lat (°N), Long (°E)	coordinates (of the centre) of the structure
Age, s.d.	estimated age, and standard deviation
meth.	method used in dating: K-Ar, Ar-Ar (=^{40}Ar-^{39}Ar), str (=stratigraphic), bstr (=biostratigraphic), ^{14}C (carbon 14) and H (=historic estimate).
D	diameter of structure
melt.	impact melt is (+), is not (−) present
env.	environment of target at impact time, cont=continental, mar=marine (shallow sea)
type	type of structure: S=simple, C=complex, M=multiring (see text)

Fig. 1 shows also that all impact structures are so far found in south-central part of Fennoscandia. This bias may reflect the lack of activity in searching impact structures in the north or difference in the glacial cover (moraines) between northern and southern Fennoscandia.

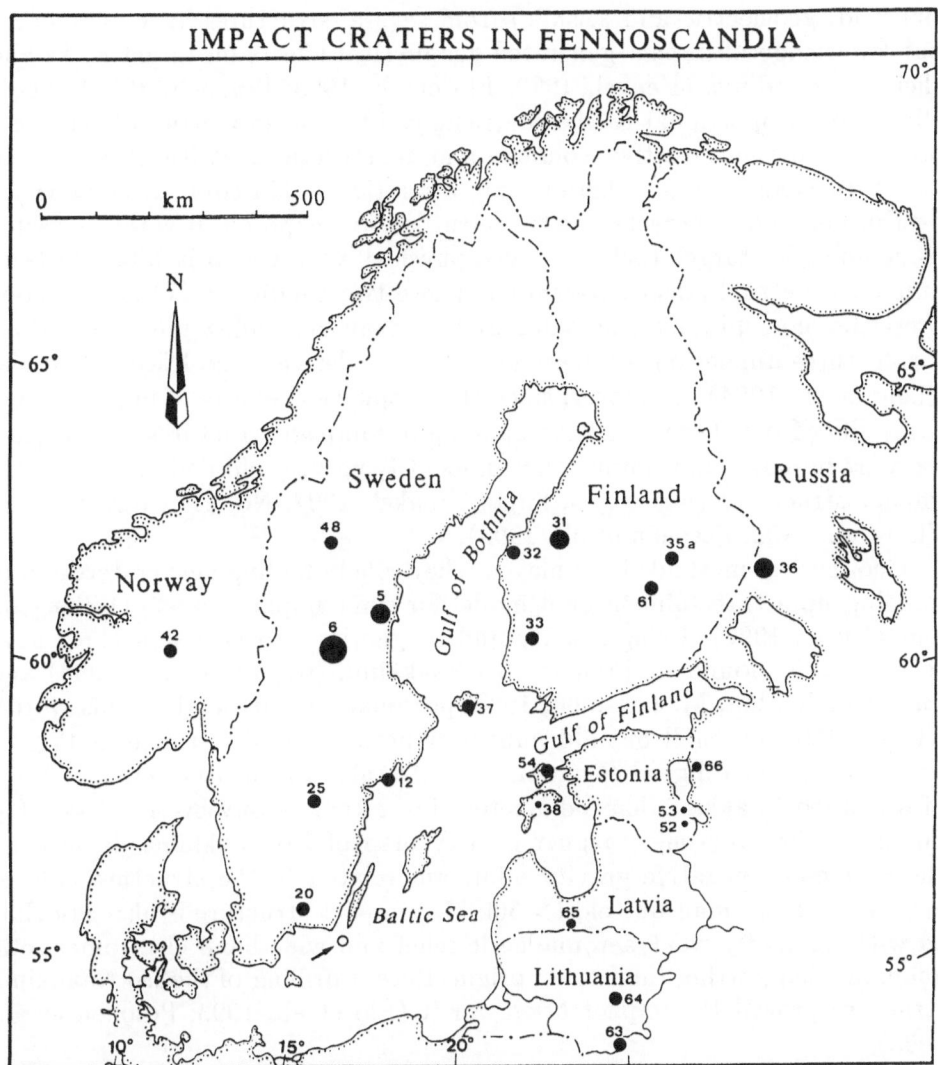

Fig. 1.   The twenty two proven impact structures in Fennoscandia. Dots show centre point of impact and its size is proportional to the diameter (D). Data are listed in Table I. Numbers (up to 62) refer to the catalogue of Henkel and Pesonen (1992) and 63-66 to present work (see text).

4. Recognition of old impact craters: the role of geophysics

The increased coverage of high-resolution airborne geophysical data (flight altitude 30 m, line spacing 200 m) and detailed digital elevation data in the low-relief regions of Fennoscandia has increased the prospects to dis-

cover new impact structures. Integrated studies of geophysical data (potential field, geoelectric and seismic) from several structures have augmented
our knowledge of impact-generated geophysical effects on target rocks and
their surroundings (Henkel, 1992; Elo et al., 1993; Pesonen, 1993). These
effects are proportional to the contrasts in physical properties of the rocks
involved, to their relative volumes and to structural attitudes (Pesonen,
1993). In Fennoscandia, there is only a relatively thin cover of Quaternary
overburden and the crystalline basement is often exposed. It is thus possible
to sample the target rocks for petrophysical analyses to isolate effects of
impact in petrophysical parameters. These two conditions aid in the interpretation of geophysical anomalies and in many cases allow unique solutions
for the three-dimensions of the structures to be derived (e.g. Kärdla, Estonia;
Plado et al., 1994). Moreover, since the geophysical effects of impact structures differ from those of the surrounding pre-impact structures, geophysical
data are of paramount importance in identifying very old and deeply eroded
impact structures (e.g. No. 6, Siljan; Henkel, 1992; No. 61, Iso-Naakkima;
Elo et al., 1993; Pesonen et al., 1995).

Geophysical methods have played a key role in finding new craters and in
building up a three-dimensional model for some impact craters (Pilkington
and Grieve, 1992). Using gravity and magnetics two new structures have
recently been found in Finland, Iso-Naakkima (61, Elo et al., 1993) and
Suvasvesi N (35a, M. Lehtinen, 1994, personal communication; this work).
These relatively small but old impact structures are discernible in Fig. 2a
on a shaded-relief digital elevation map of Finland as distinct circular lakes
of which Iso-Naakkima has a diameter of ~ 2 km and Suvasvesi ~ 4 km. Fig
2b shows the greytone Bouguer gravity map of Iso-Naakkima delineating
the ~ 4 mGal negative gravity minimum related to the structure (Elo et
al., 1993). This small but old (> 500 Ma) impact structure is also associated with distinctly weak aeromagnetic relief and wide-band electromagnetic
anomalies supporting the impact origin. Recent drilling of the Iso-Naakkima
structure proved the impact origin for it (Elo et al., 1993; Pesonen et al.,
1995).

Fig. 2c shows the total field aeromagnetic anomaly map (shaded relief)
of the Suvasvesi N structure in central-east Finland showing a ~ 4 km wide
circular weak magnetic relief of this complex impact structure, at the centre
of which there is a ~ 0.8 km wide distinct circular negative (200 nT) anomaly. A recent drilling in the peak anomaly penetrated an 80 meter thick layer
consisting of impact melt and suevite breccias (M. Tyni, personal communication, 1994). Microscopic investigation of the melt rocks (M. Lehtinen, 1994,
personal communication; K. Kinnunen, 1995, personal communication) show
planar deformation features (PDF's) in quartz indicative of moderate shock
(> 10 GPa) thus proving the impact origin for Suvasvesi N. Paleomagnetic
properties of the (unoriented) melt samples show the presence of very stable

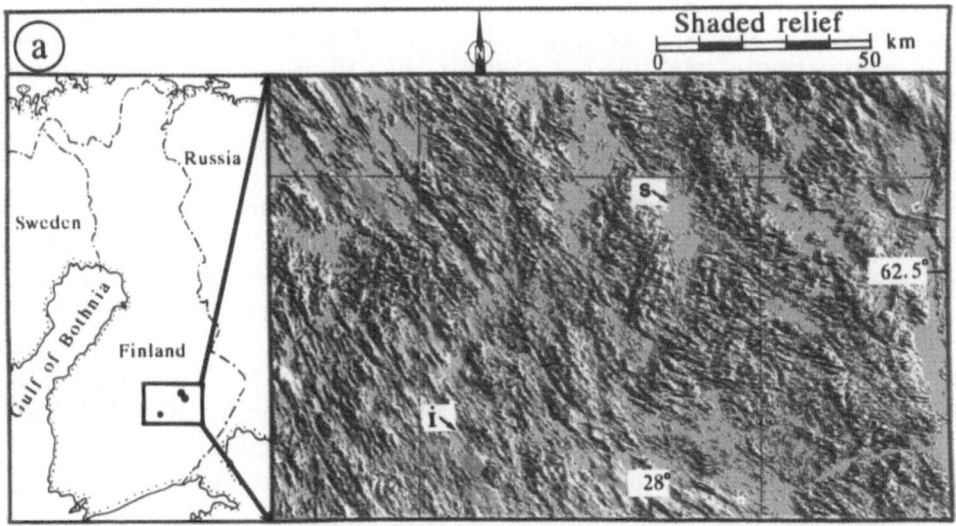

Fig. 2.   Two examples of recently discovered new impact structures in Finland with high-resolution geophysical methods. (a) **Left**: index map of Finland showing the study area (=rectangular box) with Iso-Naakkima (lower left) and Suvasvesi N (upper right). Note that latter is part of a (probable) double impact structure. **Right**: a shaded relief topographic image of study area showing the two structures as morphologically distinct circular lakes, where **i** (**s**) denote Iso-Naakkima (Suvasvesi N) structures, respectively. (b) Shaded-relief Bouguer gravity anomaly map of Iso-Naakkima (i) structure. Amplitude of minimum anomaly corresponds to ~ -4 mGal and D is ~ 2 km (Elo et al., 1993). (c) Total field-aeromagnetic map (shaded relief) of the nearly 4 km wide Suvasvesi N impact structure delineating a weak circular aeromagnetic relief, at centre of which occurs a distinct negative peak anomaly of ~ -200 nT, which was drilled in 1993 (see text). Data kindly provided by Seppo Elo, Matti Tyni and Martti Lehtinen (personal communications, 1994).

and strong (Q-values > 5) natural remanent magnetization (NRM), after first eliminating a small viscous overprint due to present Earth's magnetic field. Preliminary interpretation of the negative magnetic anomaly (Fig. 2c), coupled with paleomagnetic results of the melt samples, favour a Permian or younger age for Suvasvesi N event (Pesonen, 1995).

These two examples demonstrate that small (< 5 km) and relatively old impact structures, which appear as a rare class in the Fennoscandian impact database (Table I), can be found with high-resolution geophysical methods. They owe their preservation against erosion to cover sediments capping the down-dropped structure. Many recently found impact structures also have distinct geophysical signatures; e.g. the Lumparn structure in Åland, Finland ($D \sim 10$ km, age < 1200 Ma, > 500 Ma; Svensson, 1994; Lehtinen, 1994), has a nearly circular, weak aeromagnetic relief indicating shock modification of the magnetic properties of rocks of the structure (Pesonen, 1995, unpublished data).

5. Huge multiring structures in Fennoscandia

Henkel and Pesonen (1992) and Pesonen (1993) have pointed out that there should be evidence in Fennoscandia of old huge impact structures, which might be discernible in satellite images, in digital elevation maps, or in geophysical maps. Previously (Henkel and Pesonen, 1992), three such structures (Uppland (Sweden), Nunjes (Norway) and Marras (Finland); Fig.3) have been proposed as possible impact structures, based on geophysical signatures (Korhonen, 1990; 1994). A restudy of LANDSAT and NOAA satellite images of the Fennoscandia by present author confirms the observations of Witschard (1984) of the presence of several huge multiring structures in Fennoscandia as shown in Fig. 3. Two examples (Lycksele in Sweden and Valga in Estonia/Latvia) of huge multiring structures in Fennoscandia are shown in Figures 4 and 5.

Lycksele can be seen as a circular multiring structure with a diameter of ~ 120 km in LANDSAT and NOAA satellite images of Northern Europe (Fig.4a; see also Witschard, 1984). The topographic map (Fig.4b; Eriksson and Henkel, 1994) also shows a circular ring with a diameter of ~ 70 km in the center of the satellite image. The free-air gravity anomaly map of Fennoscandia (Granar et al., 1993; Fig. 4c) shows an NNE-SSW trending belt of gravity highs with a circular peak anomaly of ~ 20 mGal slightly to south from the satellite image of Lycksele structure (compare Figs. 4a and 4c). Fig. 4d shows the total magnetic field anomaly map of northern Sweden (Eriksson and Henkel, 1994). The Lycksele structure is associated with nearly circular weak magnetic relief with a diameter of ~ 100 km, at the center of which there is a positive magnetic ring anomaly of ~ 20 km

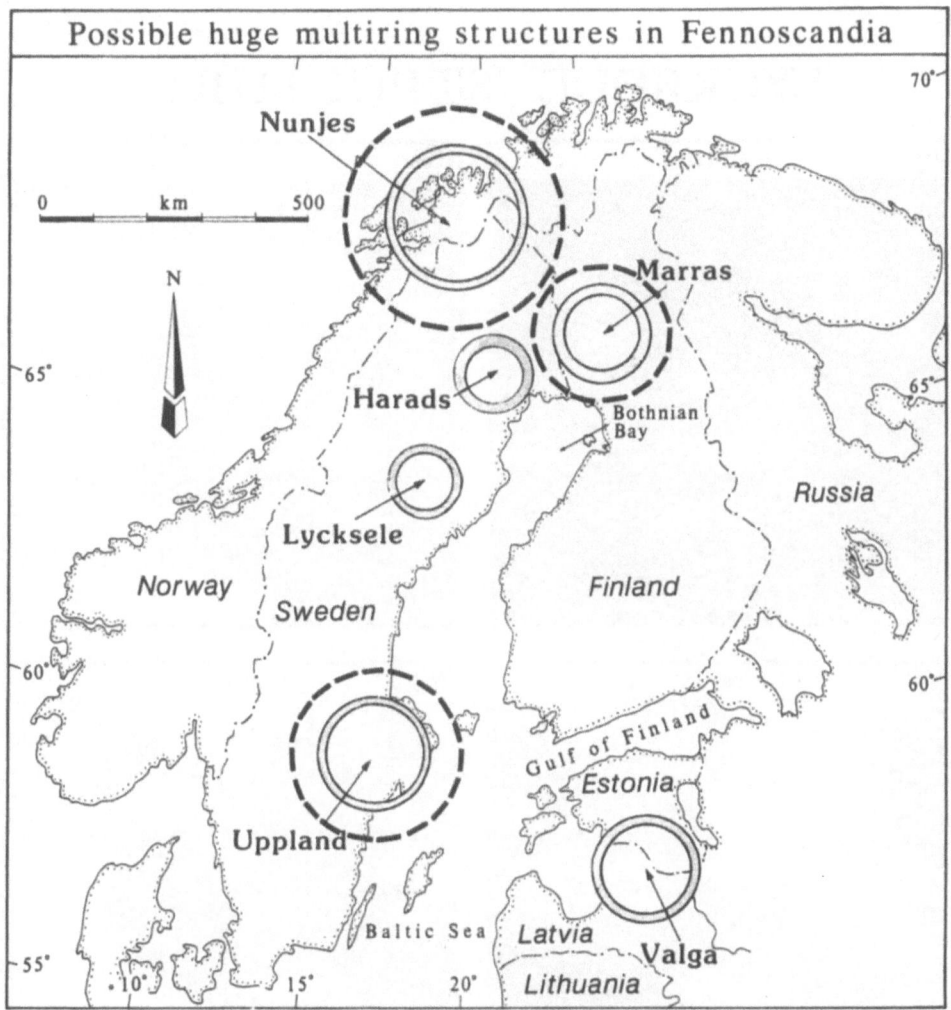

Fig. 3.   Possible large multiring structures in Fennoscandia drawn from satellite images and from geophysical maps (Witschard, 1984; Henkel and Pesonen, 1992; Korhonen 1990, 1994). Structures discussed in present paper are Lycksele (central-north Sweden; Witschard, 1984) and Valga (Latvia/Estonia; this work). For other structures (Nunjes, Marras, Uppland) see Witschard (1984), Henkel and Pesonen (1992), Korhonen (1993) and references therein.

in size (Fig.4d). Recent Mid-Norden residual gravity anomaly map (Aaro et al., 1994) reveal a pattern of circumring negative gravity anomalies (up to −20 mGal) close to, but slightly to the west of the Lycksele structure, with

LYCKSELE STRUCTURE

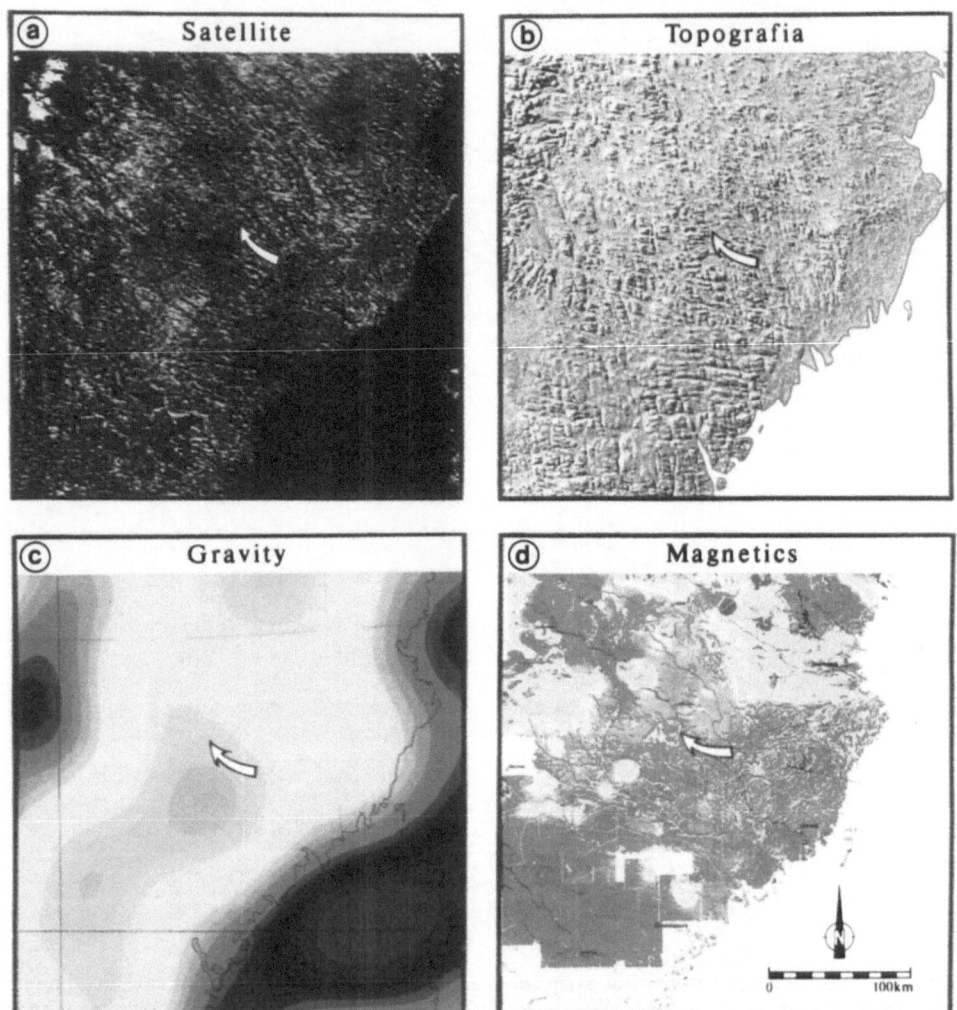

Fig. 4.   (a) LANDSAT–satellite image of northern Sweden showing the proposed Lycksele structure with $D \sim 120$ km. (b) The topographic relief illuminated from NW. (c) Free-air gravity anomaly map (Granar et al., 1993). (d) Total field aeromagnetic anomaly map (Eriksson and Henkel, 1994). Note: Figs. 4b, 4c and 4d have been reduced into the (roughly) same scale as Fig. 4a.

VALGA STRUCTURE

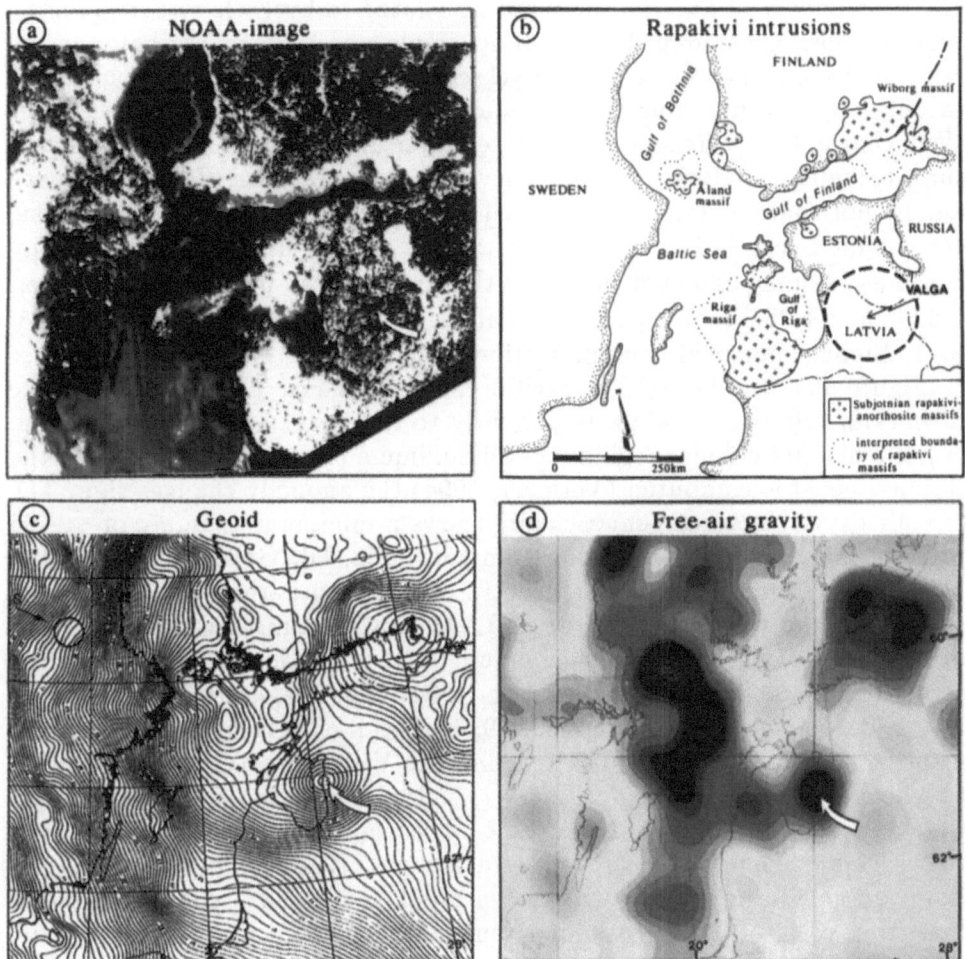

Fig. 5.   (a) NOAA satellite image of Fennoscandia region showing the proposed Valga structure in Latvia/Estonia (for scale, see (b)). (b) Valga structure is shown as dashed line (circle) together with major rapakivi massifs in southern Fennoscandia (Puura and Huhma, 1993). (c) Nordic geoid map (Kakkuri, 1992) showing a pronounced geoid (minimum) anomaly slightly west of Valga (see also Vermeer, 1994). (d) Free-air gravity map of Fennoscandia showing the negative (∼ −65 mGal) gravity minimum coinciding with geoid minimum (Granar et al., 1993; see text).

circular pattern of positive anomalies at its center. The magnetic data of the Mid-Norden do not yield any circular anomalies for Lycksele, however (Aaro et al., 1994), but the aeromagnetic anomaly map of Scandinavia (Eriksson and Henkel, 1983) shows an elongated negative (−400 nT) magnetic anomaly some 50 km south of Lycksele; this anomaly (which matches with the free-air gravity anomaly of Fig. 4c), may not be related to Lycksele structure, however. The geological map shows that no simple correlation exists between the regional geologic pattern and the Lycksele multiring structure although contacts between granites and supracrustals sometimes follow arcuate structures (Witschard, 1984). Recent detailed geophysical investigations of the structure, including analysis of digital elevation data, residual Bouguer gravity and its horizontal component, coupled with density determinations of rocks, verify that Lycksele is a multiring structure caused either by impact or by basement doming and that the age of the structure can be restricted between 1.8 Ga and 1.25 Ga based on cross-cutting relationships (Nisca et al., 1995; S.-Å. Elming, personal communication, 1995).

Figure 5 shows the NOAA satellite image of the circular Valga structure in Latvia/Estonia. Close to, but slightly to the west of this structure (Fig. 5c), is a distinct circular regional geoid minimum (Kakkuri, 1992). When the residual geoid is computed (Vermeer, 1994) the anomaly changes sign. The free-air gravity map of Fennoscandia shows a minimum anomaly of $\sim$ −65 mGal coinciding with the geoid minimum (Fig. 5d; Granar et al., 1993). The gravity anomalies do not exactly coincide with the structure in the satellite image (compare Fig. 5c and 5d), but the most recent GPS-geoid of Estonia/Latvia shows this minimum further to the east, closer to the structure in satellite image (Vermeer, 1994). The magnetic anomaly maps of the Baltic countries (Korhonen, 1993) reveal a NNW-SSE trending belt of positive magnetic anomalies crossing the Valga structure: their relation to Valga is unclear. Geologically the structure lies in the Proterozoic granulite terrane covered by Phanerozoic rocks: no evidences of shock in the rocks have been reported (Puura and Huhma, 1993). Since the structure, however, also lies close to some known rapakivi massifs (e.g. the large Riga massif, see Fig. 5b) the possibility remains that it reflects a buried rapakivi body or a basement dome. In this respect it is noteworthy that both the Wiborg and Åland rapakivi massifs in southern Finland (Fig. 5b) are not associated with a single gravity minimum but with double minima (Fig. 5d), which are slightly offset from the exposed rapakivi bodies. This may also explain why the Valga gravity anomalies are offset from the structure seen in satellite image. Whatever is the explanation for Valga sructure, it must be investigated in more detail to show whether it is a buried impact structure (see also Vermeer, 1994) in the granulite basement and now covered with younger rocks, or whether it is an endogeneous feature .

6. Crater ages

Table I lists the ages of Fennoscandian impact craters. Ten craters (Lappajärvi, Dellen, Mien, Sääksjärvi, Söderfjärden, Siljan, Jänisjärvi, Ilumetsä, Tsöörikmae and Kaali) have been dated with isotopic methods; the rest are dated with stratigraphic or biostratigraphic methods. Isotopic dating methods occasionally give ages which cannot be interpreted as the age of the impact due to various reasons like only partial resetting by the impact or contamination (Deutsch and Schärer, 1994). An example of the large range of ages which can be obtained with radiometric dating is the Dellen structure (No. 5), for which radiometric ages vary 90-250 Ma (Müller et al., 1990; Deutsch et al., 1992). The ^{87}Rb-^{87}Sr age of 89 ± 2.7 Ma is currently favoured (Deutsch et al., 1992). Several craters have been dated stratigraphically (e.g. Lumparn, Kärdla, Gardnos, Misarai; Puura et al., 1994) or biostratigraphically (e.g. Tvären, Lockne, Iso-Naakkima, Vepriai; Puura et al., 1994). In these cases the age is a minimum (biostratigraphy) or a maximum (stratigraphy) age. The palaeomagnetic apparent polar wander (APW) dating has also been applied to a few Fennoscandian impact craters (e.g. Lappajärvi, Dellen and Siljan) with variable success (e.g. Bylund, 1974; Pesonen et. al., 1990; Elming and Bylund, 1991; Pesonen, 1994; Pesonen et al., 1995).

The general pattern of the ages of Fennoscandian impact craters does not follow the global pattern, which reveals a clear bias towards young ($<$ 120 Ma) craters, due to erosion (Grieve and Pesonen, 1995). The bias in Fennoscandia is towards both young ($<$ 100 Ma) and old ($>$ 500 Ma) craters with a deficiency of impact craters with ages of $\sim 100-500$ Ma. This may be a coincidence since a new, probable Mesozoic ($\leq$ 250 Ma) impact crater, has recently been found in the Baltic Sea by seismic methods (the Ivar structure in Fig. 1; Flodén and Bjerkeus, 1994). It is noteworthy that some of the old ($>$ 200 Ma) craters have very small diameters ($\leq$ 5 km; e.g. Iso-Naakkima and Suvasvesi N, Finland). These craters owe their preservation against erosion, since they have been down-dropped and capped with preimpact or postimpact sediments (Elo et al., 1993).

7. Impact cratering rate in Fennoscandia

Impact cratering rate is here defined as the number of structures per unit area for the time interval covered by the ages of impacts. These parameters are not easily assessed (Grieve, 1984; Henkel and Pesonen, 1992). The area ($\sim 1.1 \cdot 10^6$ km^2) and the age interval used here refer to the area in Fennoscandia covered by presently known impact structures with $D > 3$ km, age $\leq$ 700 Ma and with age uncertainty $<$ 100 Ma. When the 14 structures fullfilling these criteria (Table I) are considered, about two impacts per

every 100 Ma have occurred in Fennoscandia during the last 700 Ma, yield-
ing a cratering rate of $\sim 2 \cdot 10^{-14}$ km^{-2} a^{-1}. This is about four times higher
than the mean terrestrial cratering rate of $0.5 \cdot 10^{-14}$ km^{-2} a^{-1} calculated
for proven impact craters in North America-Central Europe with $D \geq 3$ km
and age < 700 Ma. The discrepancy most likely reflects the high (particularly since 1990) success rate to find new impact structures in Fennoscandia.
Since these cratering rates are based on presently identified impact structures, they are probably too low due to processes removing, deforming and
masking impact structures on the Earth.

8. Discussion

The current database of impact structures in Fennoscandia reveals 22 proven
impact structures with a discovery rate of one or two new structures per
year. The ages of impact structures range from 0.0035 Ma to ~ 1 Ga and
diameters from 0.04 to 55 km. All typical morphologies of terrestrial impact
structures, such as simple, complex and multiring structures, are present in
Fennoscandia. The largest impact structure so far identified in Fennoscandia
(and in Western Europe) is the Siljan multiring structure in Sweden, with
$D \sim 55$ km and age of ~ 360 Ma. The youngest, on the other hand, is the
Kaali crater in Estonia with $D \sim 0.11$ km and age $\sim 3\,950$ a.

Nine of the proven impact structures (Mien, Dellen, Lockne, Lappajärvi,
Sääksjärvi, Jänisjärvi, Suvasvesi N, Gardnos and Misarai) contain variable
amounts of impact melt. For the last 700 Ma the cratering rate for Fennoscandia is $2.0 \cdot 10^{-14}$ km^{-2} a^{-1} (for craters with $D > 3$ km), which is higher than
the corresponding mean terrestrial cratering rate estimated for North Amer-
ica and Central Europe. The discrepancy probably reflects the high search
activity of impact craters in Fennoscandia. Recently, new impact structures
have been discovered in Fennoscandia on the basis of high-resolution geophysical methods (gravity, magnetics) coupled with petrological observations
of shock metamorphic effects in samples drilled through the structures (e.g.
Iso-Naakkima, Suvasvesi N and Lumparn). A notable feature in these newly
found impact craters is their small size ($D < 5$ km) but relatively old age
(> 200 Ma) making them a rare class in the global data base of terrestrial impact structures. Although we have not yet found any Archaean/Early
Proterozoic impact structure in Fennoscandia it is quite probable that during the Early Precambrian (from -4.5 to -3.5 Ga ago) impact cratering was
a major crust-shaping process comparable to later orogenies. The remnants
of large impact structures, although deformed by folding and faulting and
obliterated by erosion, may still be discernible as relict anomalies in high-
resolution geophysical maps, digital elevation data and in satellite images as
shown with examples in the present work (Lycksele and Valga structures).

However, to prove them to be of impact origin rather than endogenous (e.g. ring intrusions, basement domes or fault bounded crustal blocks) requires discoveries of shock effects in the target rocks, which unfortunately decay in the course of progressive erosion, or discoveries of impact produced rocks like igneous melts or pseudotachylite and breccia dykes, or distal ejecta layers. The role of isotope geochemistry in proving the impact origin of such rocks will likely increase in the near future.

Acknowledgements

Matti Leino, Salme Nässling and Sisko Sulkanen helped in preparing the figures. Seppo Elo, Maija Kurimo and Mauri Terho assisted in geophysical analysis. Paolo Farinella and Martti Lehtinen gave fruitful comments on the manuscript. The samples from the Suvasvesi impact crater were kindly provided by Matti Tyni and Martti Lehtinen. Kari K. Kinnunen helped with microscopic study of the Suvasvesi samples. Harri Järvinen and Heikki Pietarinen helped in word processing and James S. Thompson and David Miller corrected the English of the manuscript. I thank all these persons for their help.

References

Aaro, S., Elo, S., Gustavsson, N., Hult, K., Kihle, O., Ruotoistenmäki, T., Sindre, A., Skilbrei, J., Tervo, T. and Thorning, L.: 1994, 'Mid-Norden: Gravity and Magnetic Maps', in: *Abstracts, 21:a Nordiska Geologiska Vintermötet*, J.-A. Perdahl (ed.), Tekniska Högskolan i Luleå and Geologiska Föreningen of Sverige, p. 2.

Bylund, G.: 1974, 'Paleomagnetism of the probable meteorite impact, the Dellen structure', *Geol. Fören. Stockholm Förh.* **96**, 275-278.

Deutsch, A., Buhl, D. and Langenhorst, F.: 1992, 'Rb-Sr and Sm-Nd dating of impact melts: Dellen (Sweden) and Araguainha (Brazil)', in: *Terrestrial Impact Craters and Craterform Structures with a Special Focus on Fennoscandia*, L.J. Pesonen and H. Henkel, (eds.), *Tectonophysics* **216**, 205-218.

Deutsch, A. and Schärer, U.: 1994, 'Dating terrestrial impact events', *Meteoritics* **29**, 301-322.

Elming, S.-Å. and Bylund, G.: 1991, 'Palaeomagnetism and the Siljan impact structure, central Sweden', *Geophys. J. Int.* **105**, 757-770.

Elo, S., Kuivasaari, T., Lehtinen, M., Sarapää, O., and Uutela, A.: 1993, 'Iso-Naakkima, a circular structure filled with Neoproterozoic sediments, Pieksämäki, south eastern Finland', *Bull. Geol. Soc. Finland* **65**, part I, 3-30.

Eriksson, L. and Henkel, H.: 1983, 'Deep structures in the Precambrian interpreted from magnetic and gravity maps of Scandinavia', *Int. Basement Tectonics Ass. Publ.* **4**, 351-358.

Eriksson, L. and Henkel, H.: 1994, 'Geofysik', in: C. Fredén (ed.), *Berg och jord*, Sveriges Nationalatlas, 76-101 (in Swedish).

Granar, L., Eriksson, L., Larkin, S., Fan, H. and Sjöberg, L.E.: 1993, 'Free-air gravity anomaly map of Fennoscandia', SGU Serie Ba 49, MO Print, Uppsala, Sweden.

Flodén, T. and Bjerkeus, M.: 1994, 'The proposed Ivar impact structure in the southern Baltic', in: *The Identification and Characterization of Impacts*, Törnberg, R. (ed.), The 2nd International Workshop, Lockne -94, Impact Cratering and Evolution of Planet Earth (Abstract).

Grieve, R.A.F. and Robertson, P.B.: 1979, 'The terrestrial cratering record I. Current status and observations', *Icarus* **38**, 212-229.

Grieve, R.A.F.: 1984, 'The impact cratering rate in recent time', *Geophys. Res.* **89**, Suppl., B403-408.

Grieve, R.A.F. and Pesonen, L.J.: 1992, 'The terrestrial cratering record', in: *Terrestrial Impact Craters and Craterform Structures with a Special Focus on Fennoscandia*, L.J. Pesonen and H. Henkel (eds.), *Tectonophysics* **216**, 1-30.

Grieve, R.A.F. and Pesonen, L.J.: 1995, 'Terrestrial impact craters: their spatial and temporal distribution and impacting bodies', this volume.

Henkel, H.: 1992, 'Geophysical aspects of impact craters in eroded shield environment – with special emphasis on electric resistivity', in: *Terrestrial Impact Craters and Craterform Structures with a Special Focus on Fennoscandia*, L.J. Pesonen and H. Henkel (eds.), *Tectonophysics* **216**, 63-90.

Henkel, H. and Pesonen, L.J.: 1992, 'Impact craters and craterform structures in Fennoscandia', in: *Terrestrial Impact Craters and Craterform Structures, with a Special Focus on Fennoscandia*, L.J. Pesonen and H. Henkel (eds.), *Tectonophysics* **216**, 32-40.

Kakkuri, J.: 1992, 'Traces of impact crater in the geoid', in: *Terrestrial Impact Craters and Craterform Structures with a Special Focus on Fennoscandia*, L.J. Pesonen and H. Henkel (eds.), *Tectonophysics* **216**, 41-44.

Korhonen, J.V.: 1990, 'The concentric structure of central Finnish Lapland, an astrobleme?', in: *Symposium on Fennoscandian Impact Structures, Espoo and Lappajärvi, Finland*, Programme and Abstracts, L.J. Pesonen and H. Niemisara (eds.), Geological Survey of Finland, p. 32.

Korhonen, J.V.: 1993, 'The magnetic maps in 1 : 1 Million of Finland and nearby regions: a look at the projects', A poster presentation in the IX Geophysics Negotiation Days, Rovaniemi (Finland), Nov. 10-11, 1993 (in Finnish).

Korhonen, J.V.: 1994, 'Geophysical characteristics of two major circular structures in Finland', in: *The Identification and Characterization of Impacts*, Törnberg, R. (ed.), The 2nd International Workshop, Lockne -94, Impact Cratering and Evolution of Planet Earth (Abstract).

Lehtimäki, J.: 1992, 'Seismic soundings for locating wells in Lappajärvi', An Open File Report (Dec. 18, 1992), Regional Office of Southern Finland, Geological Survey of Finland, 4pp. (in Finnish)

Lehtinen, M.: 1994, 'Mikä tappoi dinosaurit – romahtiko taivas vai pettikö maa?', *Luonnon Tutkija* **98**, No1, 16-20 (in Finnish).

Lindström, M., Flodén, T., Puura, V. and Suuroja, K.: 1992, 'The Kärdla, Tvären and Lockne craters – possible evidences of an Ordovician asteroid swarm', *Proc. Est. Acad. Sci Geol.* **41**, No 2, 45-53.

Lindström, M.: 1993, 'Terrestrial craters: impact and past impact formations', in: *Impact Cratering and Evolution of Planet Earth*, Post-Nördlingen Newsletter, A. Montanari and J. Smit (eds.), European Science Foundation, p. 11.

Masaitis, V.: 1992, 'Impact craters: are they useful', *Meteoritics* **27**, 21-27.

Müller, N., Hartung, J.B., Jessberger, E.K. and Reimold, W.U.: 1990, '^{40}Ar – ^{39}Ar ages of Dellen, Jänisjärvi, and Sääksjärvi impact craters', *Meteoritics* **25**, 1-10.

Nisca, D.H., Thunehed, H. and Elming, S-Å.: 1995, 'The Lycksele ring structure in northern Sweden: an impact or a dome?', *Annales Geophysicae*, Part III, Space and Planetary Sciences, Supplement III to Vol.13, p. C 740.

Österman, M., Deutsch, A. and Agrinier, P.: 1995, 'Geochemical variation in the Foy Offset Dyke, Sudbury impact structure', *Annales Geophysicae*, Part III, Space and Planetary Sciences, Supplement III to Vol.13, p. C 741.

Pesonen, L.J., Marcos, N. and Pipping, F.: 1990, 'Paleomagnetic, rock magnetic and other geophysical results from the Lake Lappajärvi impact crater, central-western Finland', Open File Report Q29.1/90/2 – Laboratory for Paleomagnetism, Geological Survey of Finland, Espoo, 53 pp.

Pesonen, L.J.: 1991, 'Probing impacts in Fennoscandia', *EOS, Trans. Act. Am. Geoph. Un.* **72** (No. 2), 11-16

Pesonen, L.J.: 1993, 'Terrestrial craters: geophysics. Report on topic 4', in: Post-Nördlingen Newsletter, A. Montanari and J. Smit (eds.), ESF, Impact Cratering and Evolution of Planet Earth, 8-13.

Pesonen, L.J.: 1994, 'Magnetism and palaeomagnetism of impact craters – examples from Fennoscandia', *EOS, Transact. Am. Geoph. Un.* **75** (No 16), p. 121.

Pesonen, L.J.: 1995, 'The Suvasvesi N structure – a new impact crater in Central-East Finland', *Annales Geophysicae*, Part III, Space and Planetary Sciences, Supplement III to Vol. 13, p. C 741.

Pesonen, L.J., Järvelä, J. and Pietarinen, H.: 1995', 'Palaeomagnetism and petrophysics of the Iso-Naakkima impact structure, central-east Finland', Open File Report, Laboratory for Paleomagnetism, Geophysics Department, Geological Survey of Finland (in prep.).

Pilkington, M. and Grieve, R.A.F.: 1992, 'The geophysical signature of terrestrial impact craters', *Rev. Geophysics* **30**, 161-181.

irrus, E. and Tiirmaa, R.: 1990, 'The meteorite craters in Estonia', in: *Symposium on Fennoscandian Impact Structures, Espoo and Lappajärvi, Finland*, Programme and Abstracts, L.J. Pesonen and H. Niemisara (eds.), Geological Survey of Finland, 51-52.

Plado, J., Pesonen, L.J., Elo, S., Puura, V., and Suuroja, K.: 1994, 'Geophysical research into the Kärdla meteorite crater, Hiiumaa, Estonia', in: *The Identification and Characterization of Impacts*, Törnberg, R. (ed.), The 2nd International Workshop, Lockne -94, Impact Cratering and Evolution of Planet Earth (Abstract).

Puura, V. and Huhma, H.: 1993, 'Palaeoproterozoic age of the East Baltic granulite crust', *Prec. Res.* **64**, 289-294.

Puura, V., Lindström, M., Flodén, T., Pipping, F., Motuza, G., Lehtinen, M., Suuroja, K. and Murnieks, A.: 1994, 'Structure and stratigraphy of meteorite craters in Fennoscandia and the Baltic region: a first outlook', *Proc. Est. Acad. Sci. Geol.* **43**, No 2, 93-108.

Stöffler, D.: 1972, 'Deformation and transformation of rock-forming minerals by natural and experimental shock processes. I . Behaviour of minerals under shock compression', *Fortschr. der Mineral.* **49**, 50-113.

Svensson, N.-B.: 1994, 'Lumparn - an impact crater on Åland, SW Finland and some effects of shock wave in the surrounding bedrock', in: *The Identification and Characterization of Impacts*, Törnberg, R. (ed.), The 2nd International Workshop, Lockne -94, Impact Cratering and Evolution of Planet Earth (Abstract).

Vermeer, M.: 1994, 'A fast delivery GPS-gravimetric geoid for Estonia', *Reports of the Finnish Geodetic Institute* **94**, 1, 7 pp.

Witschard, F.: 1984, 'Large magnitude ring structures in the Baltic shield – metallogenetic significance' *Econ. Geol.* **79**, 1400-1456.

Video observations, atmospheric path, orbit and fragmentation record of the fall of the Peekskill meteorite

Z. Ceplecha
Academy of Sciences, 251 65 Ondřejov Observatory, Czech Republic

P. Brown
Department of Physics, University of Western Ontario, London, Ontario N6A 3K7, Canada

R. L. Hawkes
Department of Physics, Engineering & Geology, Mount Allison University, Sackville, New Brunswick E0A 3C0, Canada

G. Wetherill
Carnegie Institution of Washington, Department of Terrestrial Magnetism, 5421 Broad Branch Road NW, Washington DC 20015, USA

M. Beech
Department of Astronomy, University of Western Ontario, London, Ontario N6A 3K7, Canada

and

K. Mossman
Department of Physics, Engineering & Geology, Mount Allison University, Sackville, New Brunswick E0A 3C0, Canada

Abstract. Large Near-Earth-Asteroids have played a role in modifying the character of the surface geology of the Earth over long time scales through impacts. Recent modeling of the disruption of large meteoroids during atmospheric flight has emphasized the dramatic effects that smaller objects may also have on the Earth's surface. However, comparison of these models with observations has not been possible until now. Peekskill is only the fourth meteorite to have been recovered for which detailed and precise data exist on the meteoroid atmospheric trajectory and orbit. Consequently, there are few constraints on the position of meteorites in the solar system before impact on Earth. In this paper, the preliminary analysis based on 4 from all 15 video recordings of the fireball of October 9, 1992 which resulted in the fall of a 12.4 kg ordinary chondrite (H6 monomict breccia) in Peekskill, New York, will be given. Preliminary computations revealed that the Peekskill fireball was an Earth-grazing event, the third such case with precise data available. The body with an initial mass of the order of 10^4 kg was in a pre-collision orbit with $a = 1.5$ AU, an aphelion of slightly over 2 AU and an inclination of 5°. The no-atmosphere geocentric trajectory would have lead to a perigee of 22 km above the Earth's surface, but the body never reached this point due to tremendous fragmentation and other forms of ablation. The dark flight of the recovered meteorite started from a height of 30 km, when the velocity dropped below 3 km/s, and the body continued 50 km more without ablation, until it hit a parked car in Peekskill, New York with a velocity of about 80 m/s. Our observations are the first video records of a bright fireball and the first motion pictures of a fireball with an associated meteorite fall.

Key words: meteorite fall, fireball, Earth-grazing, video-records, trajectory, orbit, Peekskill

1 Introduction

In 1992 on October 9 at 23^h48^m UT, many eyewitnesses on the east-cost of the U.S. saw a fireball brighter than the full moon. It lasted more than 40 seconds.

Earth, Moon, and Planets **72**: 395–404, 1996.

This happened to be early evening local time when many people, particularly fans of football games, were outdoors. At least 15 video recordings of this event were made from different locations. As well, two high resolution photographs showing extreme fragmentation of the body were made. At least 70 individual fragments separated by more than 20 km distance are visible on these photographs. Almost 4 minutes after the last light from the fireball extinguished, one of the fragments weighing 12.4 kg struck a parked car at Peekskill, New York (Wlotzka 1993). It turned out to be an ordinary chondrite, H6 monomict breccia. This is only the fourth meteorite to have been recovered for which detailed and precise data exist on the meteoroid atmospheric trajectory and thus also on orbit (Brown et al., 1994). We present preliminary results based on video recordings of this fireball from 5 stations.

2   Reductional procedures

The toughest part of computing the atmospheric trajectory for this fireball was the first step; namely computing azimuths and elevation angles to the body from individual video frames. Relative timing is, of course, perfect in this case, but the position of any object depends on the direction and focal distance of the camera. Moving and zooming cameras following the fireball flight are not only a good recording device, but also quite a problem for absolute calibration of the direction of any vision line. Photometry is an even worse problem, because the automatic focus and gain controls on camcorders inhibit collection of precise photometric data. Calibration of measured directions was mostly done by using terrestrial objects recorded together with the fireball. Some of the terrestrial objects were related to star images taken from exactly the same location later on. For each station, the individual video frames were digitized, measured and calibrated. One of the 5 stations showed too much systematic deviations with respect to the other 4 stations: thus we were forced to exclude this station from the preliminary analysis. The following preliminary data are based on measurements of 254 individual directions from 4 stations.

3   Atmospheric trajectory and orbit

The classical method of computation of a fireball trajectory uses the intersections of planes, where each plane is defined by the position of the station and all directions from this station to the fireball trajectory. But this method failed completely in the case of the Peekskill fireball. It was obvious that a curved fireball trajectory upon an even more curved Earth's surface were the reason for this failure of the classical approach usually used for much shorter and steeper trajectories. It soon became clear that the Peekskill fireball was an Earth-grazing event, actually the third ever such event with precise data available. We used a method and the software developed by J. Borovička (1990) for locating a curved trajectory as

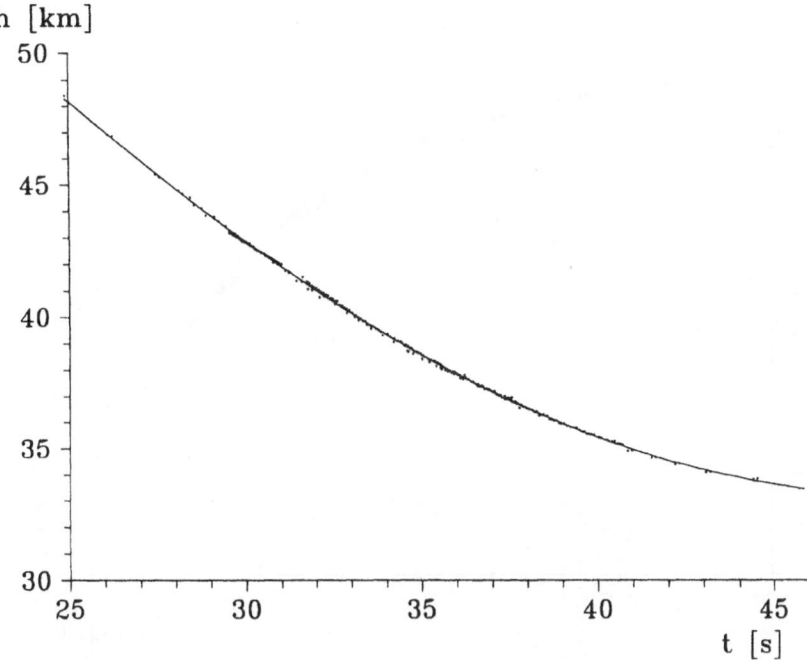

Fig. 1. Height as function of time: points correspond to individual observed values, the curve is the average change of height with time.

close as possible to all 254 lines of vision. After application of this procedure, the standard deviation for one line of vision turned to be ±0.9 km. This is about 20× to 50× less precise than is typical for multistation photographs of fireballs, but it is compensated by the extremely long trajectory. From the total luminous trajectory of almost 800 km, 250 km were video recorded.

Fig. 1 presents the change of height with time for the observed part of the Peekskill trajectory. The individual points belong to the individual values as measured from the video recordings, while the smooth curve is the model average. Our model was a circle of 12250 km over the curved Earth's surface, which corresponds to height changes due to gravity during a known time interval. It can also be viewed as an osculating circle of the hyperbolic Earth-bound orbit close to the perigee. Very good precision is evident from Fig.1.

Because of the precise relative timing of individual frames, we were also able to compute the velocity with very high precision. In Fig. 2 the velocity is plotted against time. Especially the early part of the trajectory which showed very little change of velocity (due to the large size of the initial body) resulted in an accurate initial velocity with a high precision of $v_\infty = 14.72 \pm 0.05$ km/s. Thus, also the orbit (Table II) is quite well determined. The velocity was computed using a model of ablation and fragmentation along a curved trajectory (Ceplecha et al., 1993).

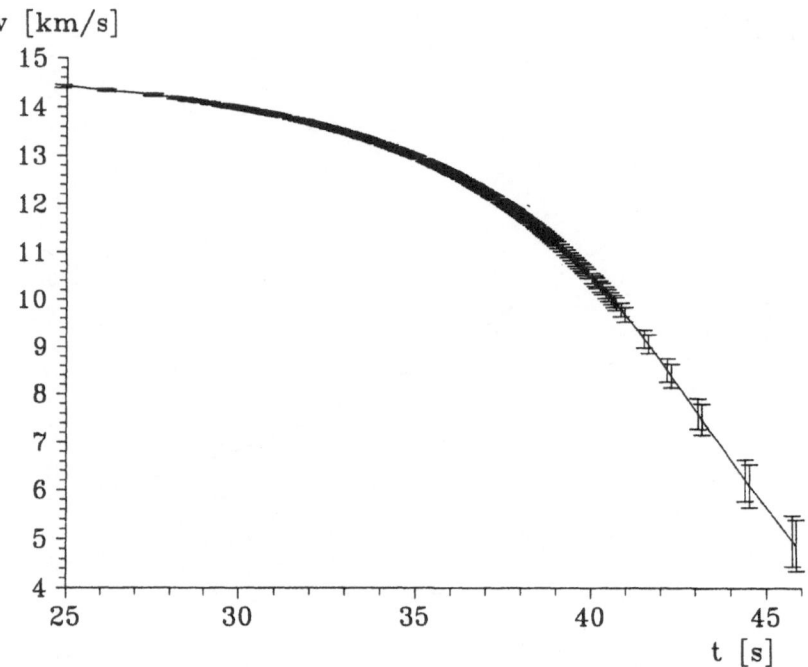

Fig. 2.   Velocity with its standard deviations as function of time.

The orbit was computed by a method usual for precise meteor data as described
in details by Ceplecha (1987, p. 229) including also the nomenclature customary
in meteor branch. Especially, v_∞ is the initial (or no-atmosphere) velocity, i.e. the
velocity the body would have had at the point, where it is measured, but without
the atmosphere. v_G is then the geocentric velocity, i.e. the initial velocity corrected
for the Earth's rotation and gravity. In our case we corrected the observed velocity
$v = 14.45 \pm 0.03$ km/s and the observed right ascension and declination of the
radiant, $\alpha_R = 226.7° \pm 0.4°$ and $\delta_R = -16.2° \pm 0.2°$, at a point of longitude
$\lambda = 281.794°$ E, latitude $\varphi = 39.806°$ N , height $h = 46.41$ km; $(v \to v_\infty \to v_G)$.

The standard deviation for one distance measured along the fireball trajectory
resulted in ± 1.0 km, almost the same value as in the direction perpendicular to
the trajectory. The computed terminal velocity is in general agreement with the
position of the meteorite fall at Peekskill. The vertical projection of the trajectory is
represented in Fig. 3 using the usual frame of geographic coordinates. The Peekskill
meteorite position lies only 9 kilometers away from the prolongated trajectory.

The Earth-grazing trajectory is evident on a schematic representation of all
three Earth-grazing fireballs which have precise data available (Fig. 4). If there
had been no atmosphere, the Peekskill body would have continued to a perigee of
22 km above the Earth's surface and then again skipped back into space. But due to
atmospheric ablation, fragmentation and deceleration, the Peekskill meteoroid did

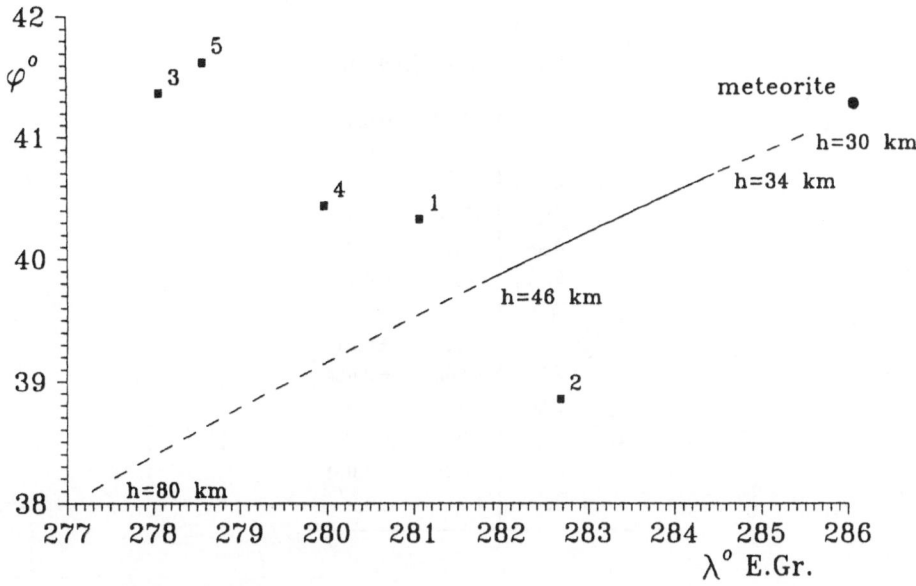

Fig. 3. Vertical projection of the Peekskill fireball trajectory in the frame of geographic longitude λ and latitude φ. Trajectory computed from video records is given by the full line, while the dashed parts correspond to the extrapolated trajectory combined with the visual sightings of the fireball. Heights, h, are given at four points along the trajectory. Location of the meteorite is denoted by a small full circle. Small full squares with numbers 1 to 5 denote positions of the stations we used in our preliminary analysis.

TABLE I

Summary of data on trajectory.

number of measured directions	254
standard deviation for one case	
perpendicular to the trajectory	± 0.9 km
along the trajectory	± 1.0 km
length of the measured trajectory	253 km
height of perigee	22 km
terminal height	30 km
initial mass	1.3×10^4 kg
initial size	$\approx 1.7 \times 1.7 \times 1$ m
ablation coefficient	0.090 ± 0.003 s^2/km^2
terminal (meteorite) mass	12.4 kg
impact velocity	80 m/s

TABLE II

Radiant and orbit (2000.0)

α_R	deg	226.7	±0.4
δ_R	deg	−16.2	±0.2
v_∞	km/s	14.72	±0.05
α_G	deg	209.0	±0.6
δ_G	deg	−29.3	±0.2
v_G	km/s	10.1	±0.1
a	AU	1.49	±0.03
e		0.41	±0.01
q	AU	0.886	±0.004
Q	AU	2.10	±0.05
ω	deg	308	±1
Ω	deg	17.030	±0.001
i	deg	4.9	±0.2

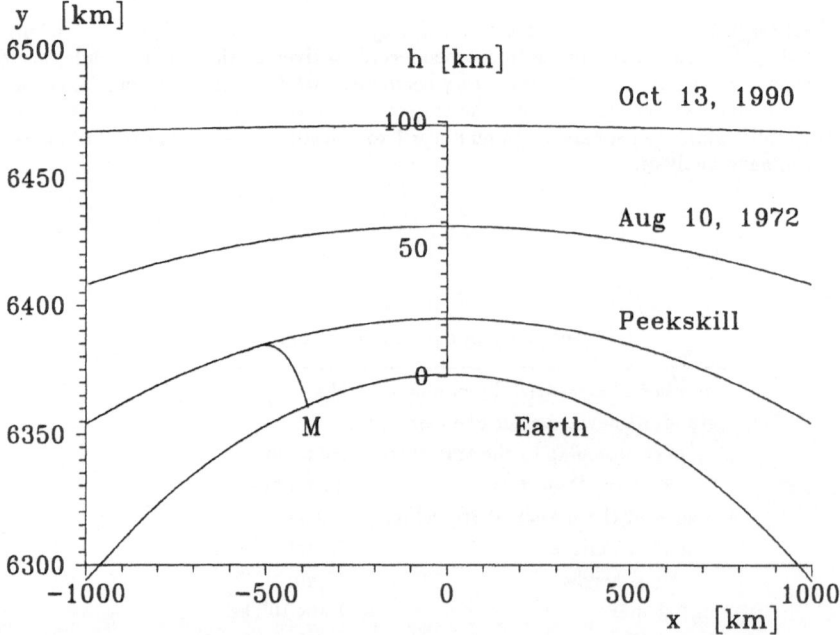

Fig. 4.   Schematic representation of the Peekskill Earth-grazing trajectory in comparison to the other two Earth grazing fireballs which have precise data on their trajectories available. All trajectories are put to the same apsidal line. x is the distance from the Earth center parallel to the flight at perigee, y is the distance from the Earth center in the direction to perigee (very much enlarged relative to the x distances), h is the height at perigee, M is the location of the Peekskill meteorite.

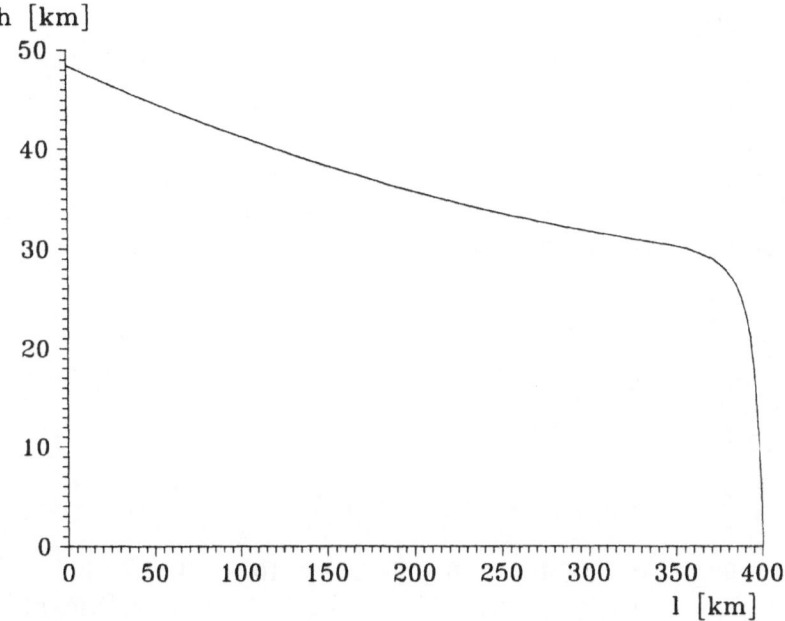

Fig. 5.   Height h as function of horizontal distance l including the dark flight.

not reach the perigee point, started its dark flight trajectory at a height of 30 km and landed 460 km before the perigee. The other two Earth grazing fireballs in Fig. 4 both went through perigee and continued again on changed orbit around the Sun. The upper one of Oct 13, 1990 lost only 1% of its mass. The famous daylight fireball of Aug 10, 1972 – also over the U.S. – lost 90% of its mass (Borovička and Ceplecha 1992, Ceplecha 1979, 1994). Thus meteoroids with fusion crust produced through interaction with the Earth's atmosphere exist also in interplanetary space.

Fig. 5 represents heights as function of horizontal distance determined for the whole trajectory, i.e. including also the dark-flight computed using the actual wind field below 30 km as given by meteorological measurements. Fig. 6 demonstrates a narrow interval of heights close to 30 km, where most of the slow-down took place. Two points of gross fragmentation are also included in these model computations.

Dynamic pressure as function of height is given in Fig. 7. The Peekskill body during its flight through the atmosphere underwent severe fragmentation under dynamic pressures between 7 and 10 Mdyn/cm^2 at heights of 41 km and lower, while the meteorite has an estimated strength somewhere close to 300 Mdyn/cm^2, more than one order of magnitude higher. A mechanism other than only the dynamical pressure acting on the recovered meteorite material must be responsible for this. (*Referee's note: The parent meteoroid was likely partially fragmented from previous collisions so that the stress necessary to break it into smaller pieces*

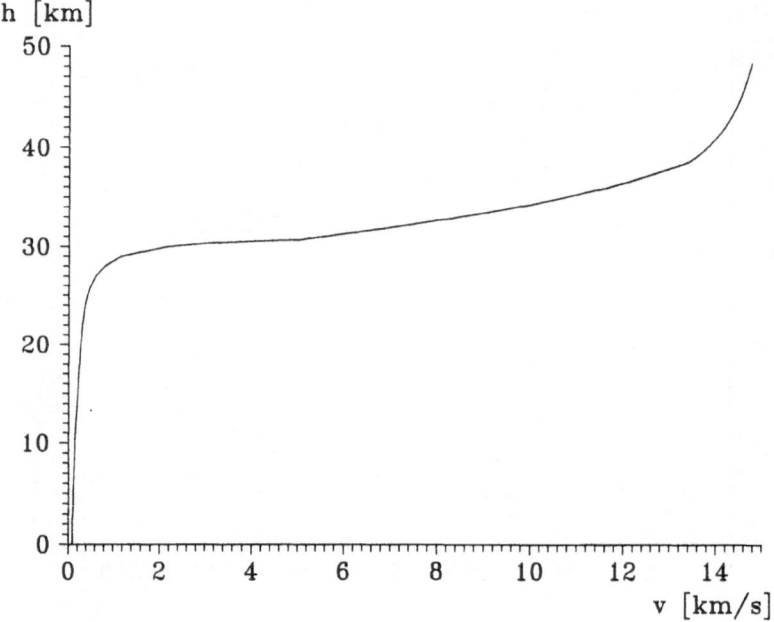

Fig. 6.   Height h as function of velocity v including the dark flight.

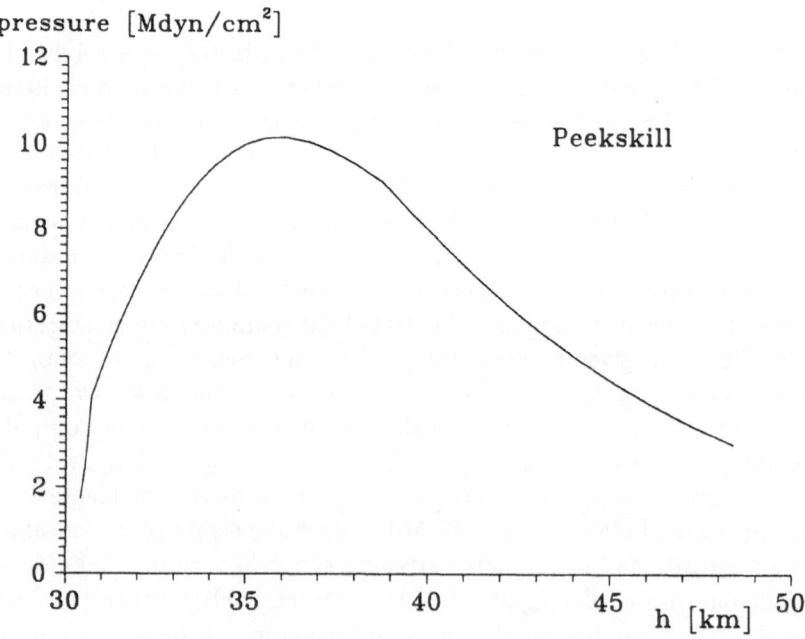

Fig. 7.   Dynamic pressure as function of height h.

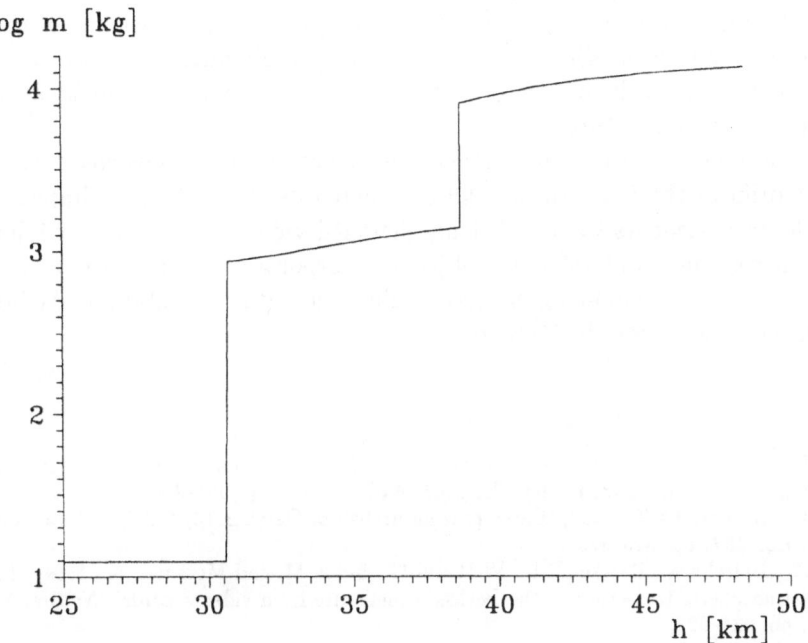

Fig. 8. Mass m as function of height h. This is just a model computation based on the preliminary computations and demonstrating two large fragmentation events important for the Peekskill
meteorite fall.

was much less than the strength of the surviving, very well consolidated, fragments.)
The dark-flight of the recovered meteorite started from a height of 30 km, when
the velocity dropped below 3 km/s, and the body continued an additional horizontal distance of 50 km without ablation, until it hit a parked car with a vertical
velocity of about 80 m/s.

Fig. 8 represents the change of mass of the precursor body of the Peekskill
meteorite with height. Computation of mass is very sensitive to the measured
deceleration. The low deceleration values at the beginning of the trajectory are
the most model dependent from all the data we determined. In any case, to get
the body from the recorded trajectory to the location, where the meteorite was
recovered, at least 10^4 kg of its initial mass is needed. The best model of the
initial Peekskill body is represented by a meteoroid with a 40% flattened shape
with dimensions of $1.7 \times 1.7 \times 1$ m. This is in good agreement with independent
computations of the initial size of the parent meteoroid from measurements of
radionuclides, noble gases and cosmic-ray tracks (Graf et al., 1994).

Summary of the principal data on trajectory and orbit as derived from the
video recordings are given in Table I and Table II. The high value of the ablation coefficient 0.090 s^2/km^2, compared to average value for stony meteorites of
0.014 s^2/km^2, is due to severe continuous fragmentation. We used the dynamic

model to follow only the main precursor of the Peekskill meteorite and we cannot thus distinguish different types of ablation: fragmentation is just one of the ablation forms. Quite a detailed study of motions of individual fragments is possible using the video records and we intend to perform and publish more sophisticated analyses in the near future.

We continue our work on further refinement of the atmospheric trajectory and the orbit of the Peekskill fireball and meteorite. We anticipate improvements through measurements from additional digitized video frames, more reliable positional measurements of reference objects, incorporation of data from additional stations, and better modeling of motion, ablation and fragmentation. We intent to publish our final results in Meteoritics.

References

Borovička J.: 1990, 'The comparison of two methods of determining meteor trajectories from photographs', *Bull. Astron. Inst. Czechosl.* **Vol. no. 41**, pp. 391–396

Borovička J., Ceplecha Z.: 1992, 'Earth-grazing fireball of October 13, 1990', *Astron. Astrophys.* **Vol. no. 257**, pp. 323–328

Brown P., Ceplecha Z., Hawkes R.L., Wetherill G., Beech M. and Mossman K.: 1994, 'The orbit and atmospheric trajectory of the Peekskill meteorite from video records', *Nature* **Vol. no. 367**, pp. 624–626

Ceplecha Z.: 1972, 'Earth-grazing fireballs (The daylight fireball of Aug 10, 1972)', *Bull. Astron. Inst. Czechosl* **Vol. no. 30**, pp. 349–356

Ceplecha Z.: 1987, 'Geometric, dynamic, orbital and photometric data on meteoroids from photographic fireball networks', *Bull. Astron. Inst. Czechosl* **Vol. no. 38**, pp. 222–234

Ceplecha Z.: 1994, 'Earth-grazing daylight fireball of August 10, 1972', *Astron. Astrophys.* **Vol. no. 283**, pp. 287–288

Ceplecha Z., Spurný P., Borovička J. and Keclíková J.: 1993, 'Atmospheric fragmentation of meteoroids', *Astron. Astrophys.* **Vol. no. 279**, pp. 615–626

Graf Th., Marti K., Xue S., Herzog G.F., Klein J., Koslowsky V.T., Andrews H.R., Cornett R.J.J., Davies W.G., Greiner B.F., Imahori Y., McKay J.W., Milton G.M., Milton J.C.D., Metzler K., Jull A.J.T., Wacker J.F., Herd R. and Brown P.: 1994, 'Size and exposure history of the Peekskill meteoroid', *Meteoritics* **Vol. no. 29**, pp. 469–470

Wlotzka F.: 1993, *Meteoritical Bull. No. 75*, *Meteoritics* **Vol. no. 28(5)**, pp. 692–703

IMPACTS INTO OCEANS AND SEAS

I.V. NEMTCHINOV and T.V. LOSEVA

Institute for Dynamics of Geospheres, 38 Leninsky prospect (build 6),
Moscow 117979, Russia. Email: idg@glas.apc.org

and

A.V. TETEREV

Belorussian State University, 4 Prospekt F. Skorina 314, Minsk, 220080, Belarus

Abstract. Impacts of cosmic bodies into oceans and seas lead to the formation of very high waves. Numerical simulations of 3-km and 1-km comets impacting into a 4 km depth ocean with a velocity of 20 km/sec have been conducted. For a 1-km body, depth of the interim crater in the sea bed is about 8 km below ocean level, and the height of the water wave is 10 m at a distance of 2000 km from the impact point. As the water wave runs into shallows, a huge tsunami hits the coast. The height of the wave strongly depends on the coastal and sea bed topography.

If the impact occurred near the shore, the huge mass of water strikes the cliffs and the near shore mountain ridges and can cause displacement of the rocks, initiate landslides, and change the relief. Thus, impact into oceans and seas is an important geological factor.

Cosmic bodies of small sizes are disrupted by aerodynamic forces. Fragments of a 100-m radius comet striking the water surface create an unstable cavity in the water of about 1 km radius. Its collapse also creates tsunami.

A simple estimate has been made using the light curves from recent atmosphere explosions detected by satellites. The results of our assessment of the characteristics of meteoroids which caused these intense light flashes suggests that fragments of a 25-m stony body with initial impact velocity 15 to 20 km/sec will hit the surface. For a 75-m iron body striking the sea with a depth of 600 m, the height of the wave is 10 m at 200-300 km distance from the impact.

1. Introduction

Impacts of a cosmic body (a comet or an asteroid) into an ocean or sea is more frequent than the impact on continents as the major part of the Earth's surface is covered by water. Such an impact leads to formation of a large unstable cavity in water which collapses. The compressive pulse and collapse of the crater produce a surge and huge waves that propagate to large distances. Numerical simulations (Ahrens and O'Keefe, 1987; Roddy et al., 1987) have given a detailed picture of the hydrodynamical processes caused by an impact of a large asteroid (5 km radius, modeled as a stony sphere) into the ocean with a depth of 5 km. At 5 sec after the impact, the water rim is located at 10-17 km from the impact point. At 30 sec, the water rim is at 35 km (the average velocity is 1 km/sec). The ratio of the crater rim diameter to the diameter of the impacting body is about 7. The depth of the crater is 39 km, or 8 times the body's radius. At 120 sec, the water rim height is 4 km and the velocity is still 0.2 km/sec. The interim crater radius

Earth, Moon, and Planets **72**: 405-418, 1996.
© 1996 *Kluwer Academic Publishers.*

in the Earth was 40 km. The water rim moves ahead of the ground rim. The huge tsunami may lead to disastrous consequences all over the world.

2. Impacts of objects a few km in diameter

Impacts of such asteroids and comets are rare events (Morrison et al., 1994). We have conducted similar numerical simulations for smaller bodies for which frequencies of the impacts are much higher. We describe the result of the vertical impact of a 3-km radius comet (modeled as a water sphere) into an ocean with a depth of 4 km (Nemtchinov et al., 1994b, 1994c). In this case the ratio of the impactor radius to the ocean depth is 0.75, whereas in the previous case this ratio was unity. In Fig. 1, the lines of equal density (isohores) are given at 10 sec after a vertical impact with a velocity of 20 km/sec. We used the same equation of state for the comet as for oceanic water. The semi empirical Equation of State (EOS) has been calculated and given to us by G.S. Romanov's group (Heat and Mass Transfer Institute, Minsk, Belarus). We suppose that rocks of the oceanic bed are similar to granite, and used Tillotsen EOS (see Melosh, 1989). The presence of air above the water surface, and of the wake formed after the body's flight through the atmosphere has been taken into account. The Eulerean numerical scheme of Belotserkovsky and Davidov (1982) was used to describe the axially-symmetric motion of water and granite. It was combined with Hirt and Nickols (1981) Volume of Fluid (VOF) method to track the interfaces between water and air, and water and granite.

Density contours labeled 1-9 correspond to densities of 0.01, 0.13, 0.26, 0.38, 0.50, 0.63, 0.75, 0.88 and 1.0 g/cm^3. The upper part of the water lip is a rarefied mixture of steam and water droplets. Dots correspond to the oceanic bed. The ejected material reaches the water wave at 18 km. The height of the rim of the ground is 5 km, its radius is 11-12 km, the depth of the interim crater is 12 km. In this case, the ratio of the radius to the ocean depth is 0.75, whereas in previous case this ratio was unity.

At 25 sec after the impact, the height of the wave is 4 km, and the velocity of the water rim is about 1 km/sec. The depth of the transient crater is 12 km, its radius is 12 km or 4 times the radius of the impacting body. The water rim breaks away from the ground rim, which is located at approximately 15 km radius. Thus, a decrease in the diameter of the impacting body changes the mechanism of water-wave generation.

For a 1-km radius comet, the ratio of the body's radius to the ocean depth is 0.25, much less than unity. Therefore, the shock wave created by the body's intrusion into water moves for a sufficiently long time in the water with a sufficiently high velocity, and all this volume of water (with mass of about 30 times larger than the mass of the comet) is vaporized before the

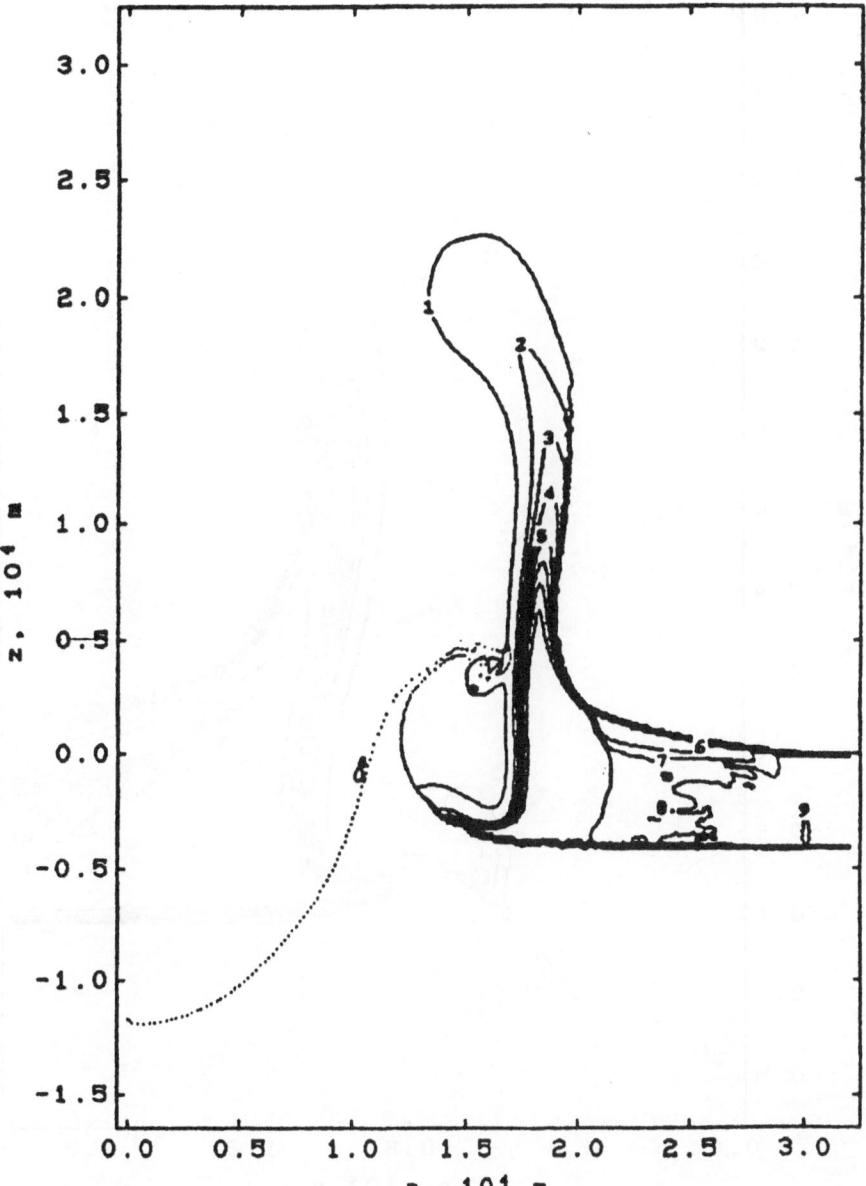

Fig. 1. Isohores at 10 sec after a vertical impact of a 3-km radius comet with a velocity of 20 km/sec into an ocean with a depth of 4 km.

body strikes the ocean bed. The results are presented in Fig. 2 at 20 sec after the impact. The height of the water wave is 4 km above the initial water level, and the water rim is located at 13 km from the impact. At 42 sec

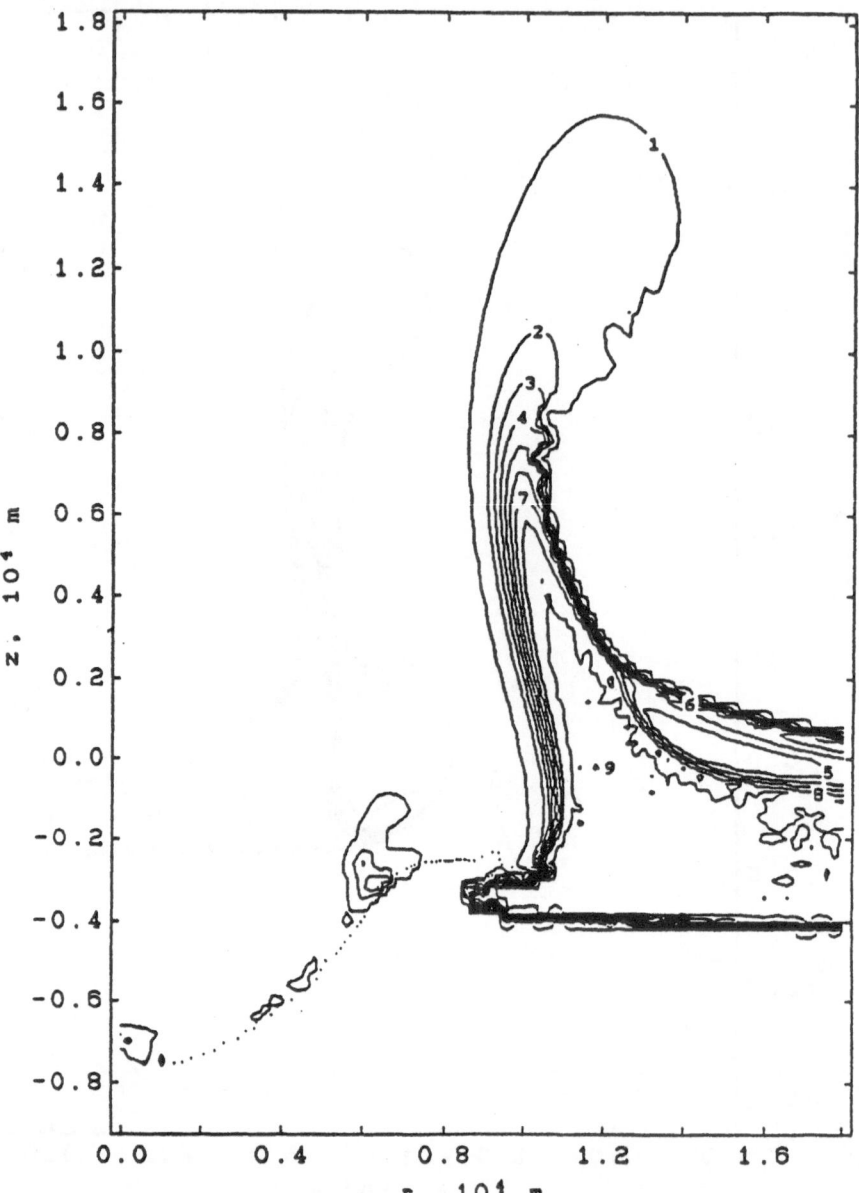

Fig. 2.   Isohores at 20 sec after a vertical impact of a 1-km radius comet with a velocity of 20 km/sec into an ocean with a depth of 4 km.

(Fig. 3), the height of the wave is 1 km at 18 km from the impact point, and the average velocity of the rim propagation is still 0.4 km/sec.

The height of the ground rim is −2.5 km, thus it is below the initial water level. The height of the water wave decreased 4 times in the interval between

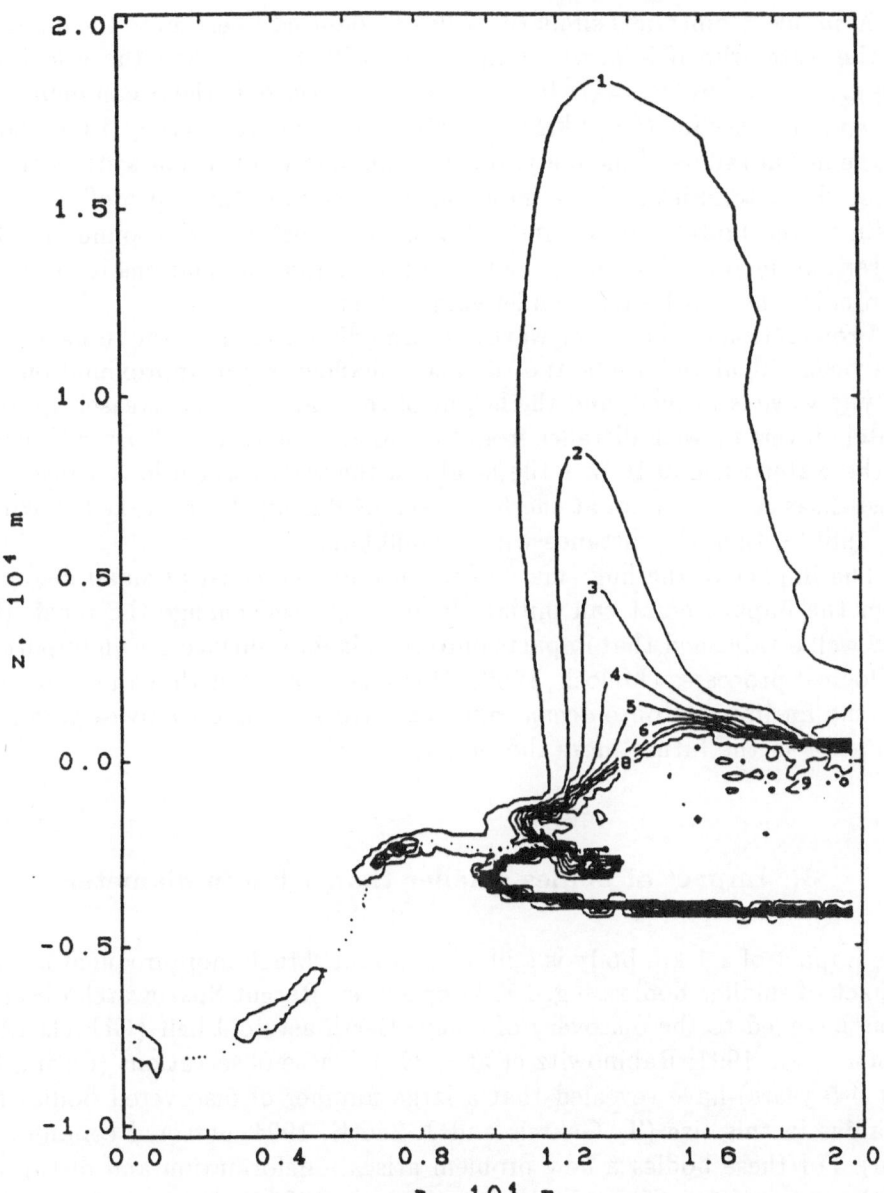

Fig. 3.   Isohores at 42 sec after a vertical impact of a 1-km radius comet with a velocity of 20 km/sec into an ocean with a depth of 4 km.

20 sec and 42 sec, due to the break away of the water wave from the ground rim and filling the water crater. The depth of the interim crater in the sea bed is about 8 km, and its radius is also 8 km, which is 8 times the diameter of the body.

A simple geometrical similarity rule can be used to estimate the diameter of the water rim if it moves with the velocity higher than the velocity of the gravity waves ($V_g = \sqrt{gH}$, where g is gravity, H is the ocean depth, i.e. $V_g = 0.2$ km/sec for $H = 4$ km). The diameter is approximately proportional to the initial radius of the body. But this similarity rule is not a strict one, as other characteristics curl are important: the ratio of the depth of the water basin to the body's radius, and ratio of the velocity to the sound speed in water and in rock. So a large number of numerical simulations is necessary to predict the depth of the underwater crater.

Propagation of the water waves at large distances from the impact point has been calculated using the so-called shallow water approximation. An N-type wave is formed, and the height of the water wave decreases approximately inversely with distance from the impact point. If the "critical" height of the water wave ≈ 10 m – the height of the waves which have caused the most disastrous tsunami at the Kuril islands during this century (Shokin et al., 1988) – then the distance will be 2000 km.

The impact of the huge mass of water onto the coast at small distances from the impact point can initiate land slides, and change the relief. It is now well established that impacts onto the planet's surface are an important geological processes (Melosh, 1989). Our results suggest that the same may be true for impacts onto oceans and seas. We are going to investigate such processes in the future more thoroughly.

3. Impact of bodies smaller than 1 km in diameter

The impact of a 1 km body is still a rare event. Much more frequent are the impact of smaller bodies, e.g. 100 m or 200 m. Recent Spacewatch observations have led to the discovery of a near-Earth asteroid belt (Gehrels 1991, Scotti et al., 1991; Rabinowitz et al., 1993). These observations (during the last 4-5 years) have revealed that a large number of discovered bodies has a radius in this size (T. Gehrels and J. Scotti, 1994, personal communication). For these bodies a new problem arises – deformation and disruption by the atmosphere. Simple estimates (Melosh, 1981, 1989) have determined the following critical radius: 290 m for a comet, 160 m for a stony body, 100 m for an iron body. A meteoroid of the critical diameter, upon striking vertically, is spread to attain the aspect ratio ≈ 8. According to this model, the meteoroid's shape becomes that of the pancake.

We have conducted numerical simulations of the deformation and disruption of a comet with a radius of 100 m, 3 times less than the critical radius by Melosh (1989), and approximately equal to the critical radius by Chyba et al. (1993).

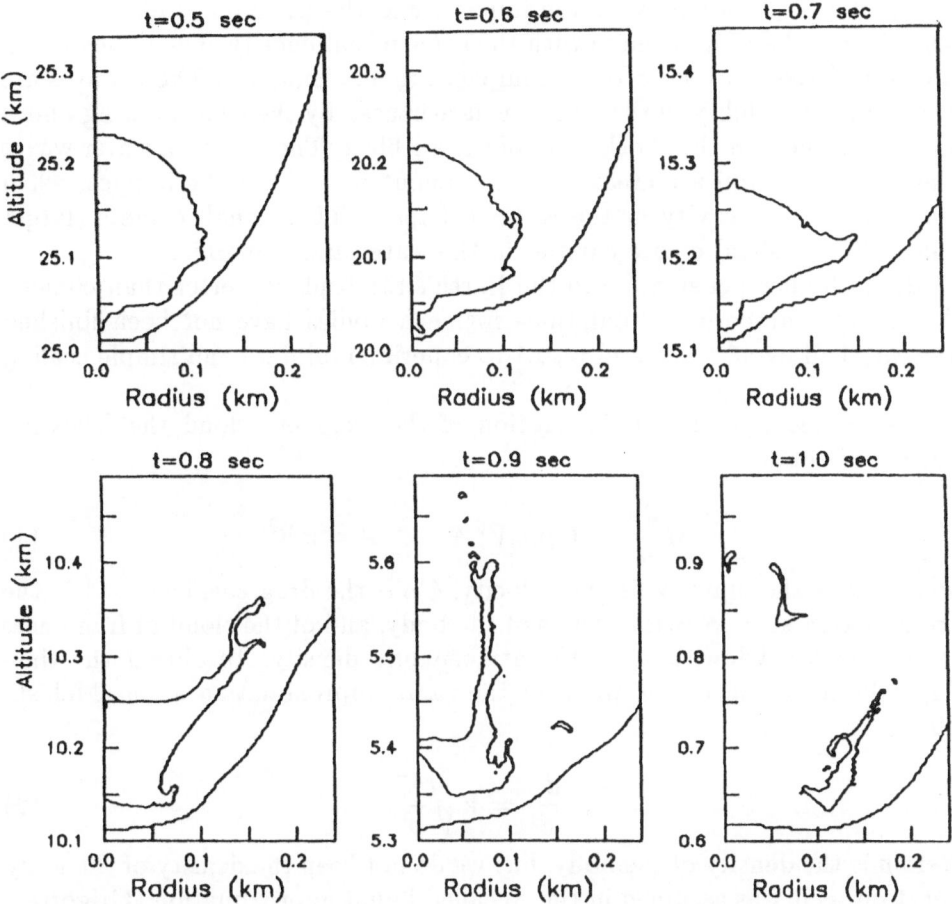

Fig. 4. Density contours at different moments of time of a 100-m radius comet striking the atmosphere with an initial velocity of 50 km/sec (at t=0 the comet was at the altitude of 50 km).

The results of the numerical simulation are presented in Fig. 4 (initial position of the body was 50 km, initial velocity 50 km/s). The body is becoming flat, but its radius is increasing only 1.5 times. Later, a cavern is formed in the blunt nose, the body becomes a hollow shell (at a height of 5 km above the surface), and its fragments impact on the surface. Such a collection of fragments do not produce the same kind of craters as a sphere (O'Keefe and Ahrens, 1982).

In Fig. 5, the result of the impact on a water surface is given (here time is after the impact, in sec). The impact of a hollow shell creates a hollow cavity in water, but a large water wave is formed. Later a large cavity in water, 1 km in radius, is formed (10 times the initial radius of the body). This is due to the high compressibility of water. The disruption of the 100 m radius body did not prevent the impact, and the mechanical effect is only slightly less than for a sphere with the same initial density. The waves which are clearly seen at 1 sec after the impact are not tsunamis. These waves are created by instability of the water surface caused by the wind blowing above the surface behind the shock wave of the air blast. The real high water waves leading to tsunami are created later – about 5 sec after the impact, when the 1 km radius cavity in the water collapses. Thus, small comets (larger than 100 m radius) create cavities in the water and tsunami.

Stony bodies penetrate into the Earth's atmosphere better than comets. As intricate numerical simulations for such bodies have not been finished yet, we shall avoid them and resort to experimental data and simple scaling laws.

We use as a model of the motion of the fragment cloud the following system of equations:

$$M\frac{dV}{dt} = C_D\rho_a V^2 A, \qquad A = \pi R^2 \tag{1}$$

where M is the mass, V is the velocity, C_D – the drag coefficient, A is the cross section area, R is the radius of the body, and of the cloud of fragments (after the breakdown), ρ_a is the atmospheric density. To obtain the time dependence of the radius we used the same approximation as in (Melosh, 1989):

$$\frac{dR}{dt} = V\sqrt{\frac{\rho_a}{\rho}} \tag{2}$$

Here ρ is the density of the body, but we do not keep the density of the body constant as it was assumed in the so called liquid approximation (Grigorian, 1979; Melosh, 1981; Chyba et al., 1993). Instead we suppose that the density of the "sand-bank" (the cloud of fragments filling the volume and creating a common bow shock, see e.g. Teterev and Nemtchinov, 1993) is determined by the conservation of mass

$$\rho R^3 = \rho_0 R_0^3 \tag{3}$$

where R_0 is the initial radius, and ρ_0 is the initial density. If we assume that the atmosphere is exponential:

$$\rho_a = \rho_E \exp(-h/H) \tag{4}$$

where h is the height above the Earth's surface, ρ_E is atmosphere density at the height $h = 0, H$ is the characteristic scale of the atmosphere. The system $(1) - (4)$ has an analytical solution:

$$\left(\frac{R}{R_0}\right)^{\frac{5}{2}} - 1 = \left(\frac{\rho_E}{\rho_0}\right)^{\frac{1}{2}} \frac{5H}{R_0 \sin\theta} \left[\exp\left(-\frac{h}{H}\right) - \exp\left(-\frac{h_b}{H}\right)\right] \tag{5}$$

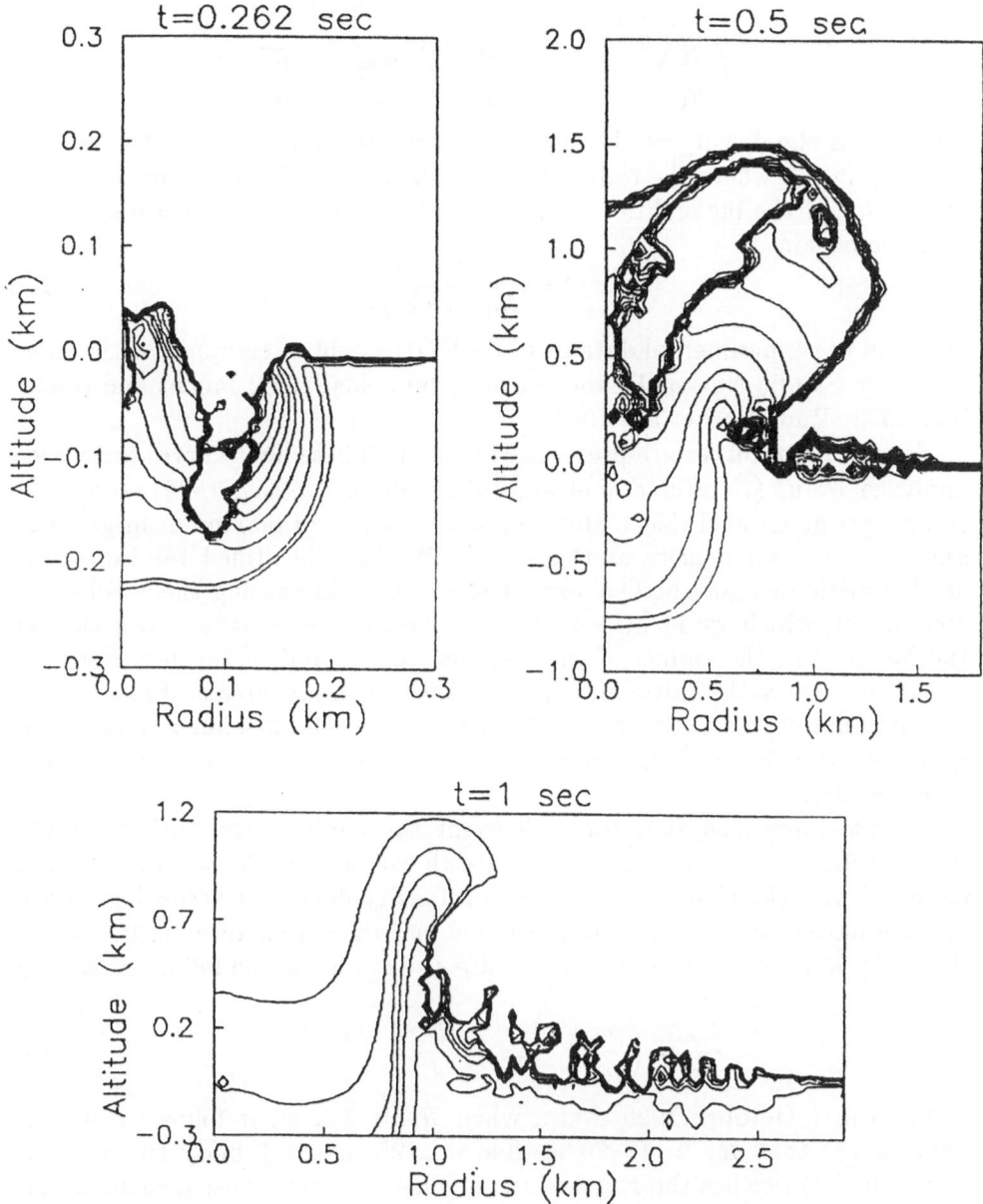

Fig. 5. The result of the impact on a water surface of a 100-m radius comet with an initial velocity 50 km/sec.

where θ is the angle of the trajectory inclination to the horizon and h_b is the height of the breakup. Using formula (4) once more we rearrange Eq.

(5):

$$\left(\frac{R}{R_0}\right)^{\frac{5}{2}} - 1 = \frac{5H}{R_0 \sin\theta}\left(\sqrt{\frac{\rho_\omega}{\rho_0}} - \sqrt{\frac{\rho_b}{\rho_0}}\right) \qquad (6)$$

where ρ_b is the density at the altitude of the breakup, and ρ_ω is the density at altitude h_ω, where the radius R is by a factor of ω higher than the initial radius R_0. If the inequalities $(\rho_\omega/\rho_b)^{1/2} \gg 1$ and $\omega^{5/2} \gg 1$ are satisfied, then we obtain

$$\omega^{5/2} = \frac{5H}{R_0 \sin\theta}\sqrt{\frac{\rho_\omega}{\rho_0}} \qquad (7)$$

We shall use experimental data from US DOD satellites equipped with detectors from Sandia National Laboratories (Reynolds, 1992; Jacobs and Spalding, 1993; Tagliaferri et al., 1994).

Several light curves are presented in Fig. 6. These light curves have been analyzed by us (Nemtchinov et al., 1994a, 1995; Svetsov, 1994), and some results are given at Table I. Here h_m is the height of maximum brightness, and ρ_m is the air density at this height. We have identified the impactors as chondritic or stony bodies, except the meteoroid causing the 1 February 1994 event, which we identify as an iron. Our analyses have shown that at the height h_m, the radius of the fragment cloud and of the bow shock is about 4-7 times the initial radius R_0, and the body breaks down due to aerodynamic forces at the height h_b, which is 10-20 km higher than h_m. If $h_m = 0$ the cloud of fragments reaches the Earth's surface with a rather high velocity.

We have assumed that for each event the angle of the trajectory, the initial velocity, density and factor ω are the same, but the size of the body increases and the altitude of the peak intensity decreases according to Eq. (7). We have extrapolated data presented in Table I to determine the size of the body R_*, for which h_ω is zero, and $\rho\omega = \rho_E$, using the following scaling laws.

$$\frac{R_*}{R_0} = \left(\frac{\rho_m}{\rho_E}\right)^{-\frac{1}{2}}, \qquad \frac{E_*}{E} = \left(\frac{\rho_m}{\rho_E}\right)^{-\frac{3}{2}} \qquad (8)$$

For the 1 October 1990 event, when $R_b = 2$–3 m, it follows from the relation (8) that $R_* = 17$–26 m. Thus, a 25 m stony body (its mass is 120×10^6 kg) reaches the Earth surface. There are two consequences. First, the Tunguska meteoroid, which has this size or even larger, cannot have been a stony body, as suggested by Chyba et al. (1993), but a comet, as it did not reach the Earth surface, but exploded at the height of about 8 km. Second, such a body striking the surface will create a crater in the water and resulting tsunami. Radius of the iron body for which height of the peak intensity $h_m = 0$ is 13 to 15 m. A stream of fragments and vapor impacting the ocean surface, though with the average density much less than the density of water, will still create water waves and tsunamis. Thus

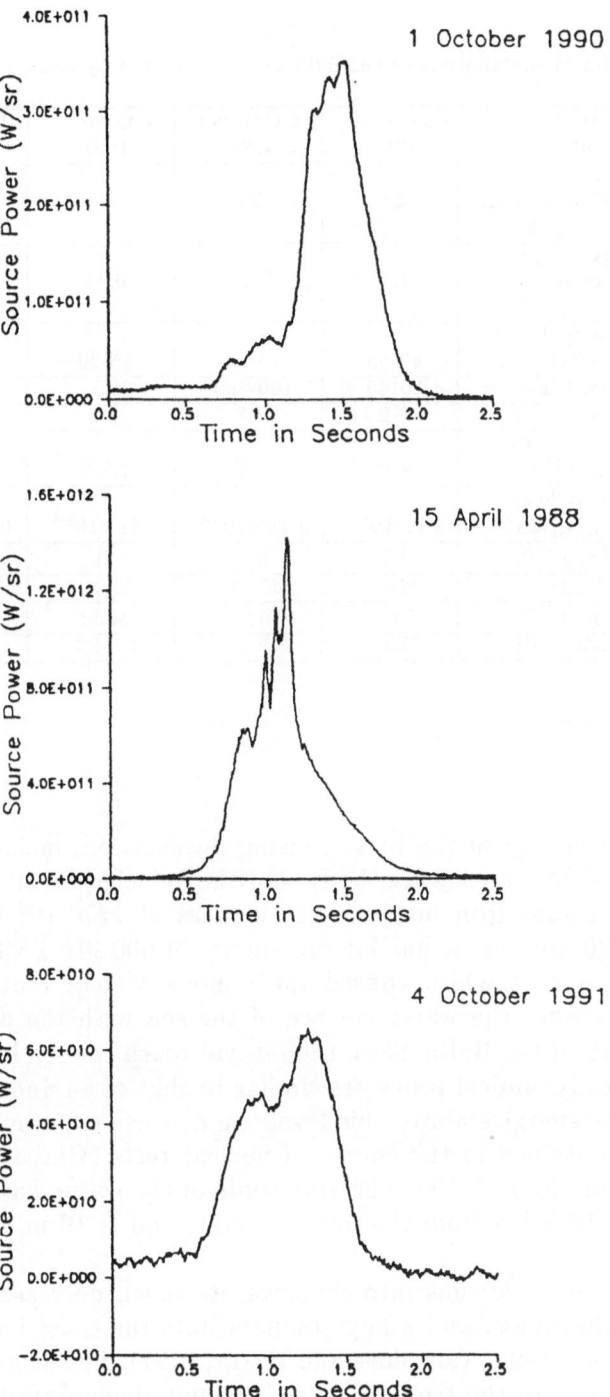

Fig. 6.   The light curves detected by Sandia's detectors.

TABLE I

The results of the analysis of the light curves detected by Sandia's detectors

Event, Date	15 April 1988	1 October 1990	4 October 1991	1 February 1994 *)
h_m, height of maximum brightness, km	43	30	33	21
Energy of light impulse, E_s, kt TNT	1.7	0.57	0.14	4.4
Velocity V, km/sec	40-50	15-20	15-20	15-20
Mass, tons	25-45	100-300	25-75	1,200-2,500
Radius R, m	1-2	2-3	1-2	3.4-4
Kinetic energy E, kt TNT	8-9	5-8	1.2-2	40-70
Density at height h_m, ρ_m, g/cm^3	2.7×10^{-6}	0.18×10^{-4}	0.11×10^{-4}	0.89×10^{-4}
$(\rho_m/\rho_E)^{-1/2}$	22	8.5	11	3.8
R_*, m	22-44	17-26	11-22	13-15
$(\rho_m/\rho_E)^{-3/2}$	10,000	600	1300	55
E_*, Mt TNT	83-94	3.0-4.8	1.5-2.5	2.2-3.8

*) preliminary data.

the minimum energy of the blast causing devastation, including tsunamis, is about 2 to 4 Mt (for an iron body and the velocity 15–20 km/sec).

If a 75-m radius iron body having a mass of 12×10^9 kg moves with the velocity 20 km/sec, it has kinetic energy of 600 Mt TNT (such a body is greater than that which caused the famous Meteor crater, see Melosh (1989). If it strikes the water surface of the sea with the depth $\approx$ 600 m (average depth of the Baltic Sea), then it will reach the sea bed and explode creating hydrodynamical processes similar to that of an underwater nuclear explosion. For energies above this level, we can use experimental data and scaling laws obtained in the course of nuclear tests (Glasstone and Dolan, 1977; Hills and Goda, 1993). The amplitude of the water wave is 300–800 m at a distance of 3 km from the impact point, and is 10 m at a distance of 200–300 km.

As the water wave runs into shallows, its speed decreases and its front increases in sharpness, and a huge tsunami hits the coast (Hills and Goda, 1983; Hills et al., 1994; Yabushita and Hatta, 1994). The heights of the deep-water waves, and of the tsunami wave, strongly depend on the coastal and sea bed topography, including the shape of the coast line, and the existence of underwater ridges and canyons (Nemtchinov et al., 1994).

Acknowledgements

This work is supported in part by Contract AI-3118 with Sandia National Laboratories.

References

Ahrens, T.J., and O'Keefe, J. 1987. Impact on the Earth, ocean and atmosphere. *J. Impact Eng.* **5**, 13-32.

Belotserkovsky, O.M. and Davidov, Yu.M. 1982. *Method of large particles in gas dynamics.* Moscow, Science Publ. Co, 392 p.

Chyba, C.F., Thomas, P., and Zahnle, K. 1993. The 1908 Tunguska explosion: atmospheric disruption of a stony asteroid. *Nature*, **361**, 40-44.

Gehrels T. 1991. Scanning with charge coupled devices. *Space Sci. Rev.* **58**, 347-375.

Glasstone, S., and Dolan, P. 1977. *The effects of nuclear weapons.* US Dep. of Defense and US Dept. of Energy. US Government printing office.

Grigoryan, S.S. 1979. Motion and disintegration of meteorites in planetary atmospheres. *Cosmic Research* **17**, 724-740.

Hills, J.G., and Goda, N. 1993. The fragmentation of small asteroids in the atmosphere. *Astron.J.* **105**, 1114-1144.

Hills, J.G., Nemtchinov, I.V., Popov, S.P., Teterev, A.V. 1994. Tsunami generated by small asteroid impacts. *Hazards due to Comets and Asteroids* (Ed. T. Gehrels). Univ. Arizona Press, Tucson and London, pp. 779-789.

Hirt, C.V., Nickols, B.T. 1981. Volume of fluid (VOF) method for the dynamics of free boundaries. *J. Comut. Phys.* **39**, 211-228.

Jacobs, C., and Spalding, R. 1993. Fireball observation by satellite-based Earth-monitoring optical sensors. *Hazards due to Comets and Asteroids* (Ed. T. Gehrels). Univ. Arizona Press, Tucson, p. 45 (abstract).

Melosh, H.J. 1981. Atmospheric breakup of terrestial impactors. In *Multi-ring Basins* (Eds. P.H. Schultz and R.B. Merril), pp. 29-35.

Melosh, H.J. 1989. *Impact cratering: a geological process.* Clarendon press, Oxford.

Morrison, D., Chapman, C.R., Slovic, P. 1994. The impact hazard. *Hazards due to Comets and Asteroids* (Ed. T. Gehrels). Univ. Arizona Press, Tucson and London, pp. 59-91.

Nemtchinov, I.V., Popova, O.P., Shuvalov, V.V., Svetsov, V.V. 1994a. Radiation emitted during the flight of asteroids and comets through the atmosphere. *Planet. Space Sci.*, **42**(6), 491-506.

Nemtchinov, I.V., Teterev, A.V., Popov, S.P. 1994b. Waves created by comet impact into ocean. *Lunar Planet. Sci.* XXV, 989-990.

Nemtchinov, I.V., Teterev, A.V., Popov, S.P. 1994c. Estimates of the characteristics of waves and tsunami produced by asteroids and comets falling into oceans and seas. *Solar System Research*, **28**, 260-274.

Nemtchinov, I.V., Popova, O.P., Shuvalov, V.V., Svetsov, V.V. 1995a. On the photometric mass and radiation size of large meteoroids. *Solar System Research* (submitted).

Nemtchinov, I.V., Svetsov, V.V., Golub', A.P., Kosarev, I.B., Popova, O.P., Shuvalov, V.V., 1995b. Sandia Network bolides and the assessment of the meteoroid's characteristics. *Planet. Space Sci.* (submitted).

O'Keefe, J.D., and Ahrens, T. 1982. Cometary and meteorite swarm impact on planetary surfaces. *J. Geophys. Res.* **87**, 6668-6680.

Rabinowitz, D.L., Gehrels, T., Scotti, J.V., McMillan, R., Perry, M., Wisniewski, W., Larson, S., Howell, E. and Mueller, B. 1993. Evidence of a near-Earth asteroid belt. *Nature* **363**, 492-493.

Reynolds, D.A. 1992. Fireball observation via satellite. In *Proc. Near-Earth-object inter-ception workshop* (Eds. G.H. Canavan, J.C. Solem, J.D.G. Rather). Los Alamos National Lab., Los Alamos, pp. 221-226.

Roddy, D.J., Shuster, S., Rosenblatt, M., Grant, L., Hassig, P., and Kreyenhagen, K. 1987. Computer simulations of large asteroid impacts into oceanic and continental sites- preliminary results on atmospheric, cratering and ejecta dynamics. *Int. J. Impact Eng.* **5**, 123-135.

Scotti, J.V., Rabinowitz, D., and Marsden, B. 1991. Near miss of the Earth by a small asteroid. *Nature* **354**, 287-289.

Shokin, Yu.I., Chubarov, L.P., Marchuk, A.G., Simonov, K.V. 1989. *The numerical exper-iment in the problem of tsunami.* Novosibirsk, Nauka Publ. Co, 168 p.

Svetsov, V.V., Nemtchinov, I.V., Teterev, A.V. 1995. Disintegration of large meteoroids in the Earth's atmosphere: theoretical models. *Icarus* (in press).

Tagliaferri, E., Spalding, R., Jacobs, C., Worden, S.P., Erlich, A. 1994. *Hazards due to Comets and Asteroids* (Ed. T.Gehrels). The University of Arizona press, Tucson and London, 199-220.

Teterev, A.V., and Nemtchinov, I.V. 1993. The sand bag model of the dispersion of the cosmic body in the atmosphere. *LPSC XXIV*, Houston, pp. 1415-1416 (abstract).

Yabushita, S., Hatta, N. 1994. On the possible hazard of the major cities caused by asteroid impact in the Pacific Ocean. *Earth, Moon and Planets* **65**, 7-13.

VELOCITY DISTRIBUTION OF METEOROIDS IN THE VICINITY OF PLANETS AND SATELLITES

K.V. KHOLSHEVNIKOV

St.Petersburg University, email: kvk@aispbu.spb.su

and

V.A. SHOR

Institute of Theoretical Astronomy, St.Petersburg, Russia; email: shor@ita.spb.su

Abstract. Collisions in the Solar System play an important role in its history. Impact processes depend essentially on the velocity distribution of meteoroids colliding with a chosen planet. According to Carleman's theorem it is sufficient to find the set of $M_k =$ mathematical expectation of v^k, v being the collisional velocity. We suppose that M_k for meteoroids of asteroidal nature differs slightly from that for asteroids themselves. So among all numbered minor planets we select those which may potentially collide with the chosen major planet. Then we calculate v at intersection points and count the average over all such points and all selected asteroids. The gravitation of a body-target may be taken into account or not. Numerical results are collected in four Tables.

Key words: Velocity distribution, Meteoroids

1. Introduction

Collisions in the Solar System play an important role in its history. Consequences of such events depend essentially on the relative velocity v of the impactors. Treating v as a random variable, its complete description may be given by the corresponding distribution function. According to Carleman's theorem [see e.g. (Prokhorov and Rosanov, 1973, §4.3)] the distribution function is uniquely determined by its momenta $M_k =$ mathematical expectation of v^k. The main idea of calculating M_k is as follows. Let us fix a body-target, say the s-th major planet Q_s, and a potential projectile, say a minor planet Q. Choose among all the numbered minor planets those which have the semi-major axis a and eccentricity e satisfying the inequalities

$$a(1-e) < a_s(1+e_s), \qquad a(1+e) > a_s(1-e_s), \tag{1}$$

where the elements with index s refer to Q_s.

As the lines of nodes and apsides rotate, the orbits of Q and Q_s intersect each other from time to time. The relative velocities at intersection points can be calculated analytically without difficulties. When averaging over all possible intersection points and all selected asteroids we obtain a set of momenta $\{M_k\}$. We expect that M_k for meteoroids of asteroidal nature differ very little from those for the asteroids themselves (Steel, 1985). The case of the body-target being one of the natural satellites is analogous though more

Earth, Moon, and Planets **72**: 419-423, 1996.

complicated. The gravity of the Q_s can be taken into account in a rather simple way.

A more detailed discussion and the description of algorithms will appear soon in *Astronomical and Astrophysical Transactions*. The closed formulae for v and numerical process for averaging over all intersection points are also given there. For M_2 and M_4 the averaging can be fulfilled in analytical form, too. The results coincide up to four decimals at least.

2. Mercury

For $s = 1$ (Mercury) there are 6 asteroids catalogued in (Batrakov, 1992) satisfying the inequalities (1). In the first row of Table I five momenta M_k are given, the planetocentric velocity v of a collider being calculated without taking into account the gravity of Mercury. In the second and the third rows $V_k = (M_k)^{1/k}$ and $\lambda_k = V_k/V_1$ are given, correspondingly. The last row represents λ_k for Maxwell's velocity distribution designated by $\tilde{\lambda}_k$.

In this paper distances are measured in km, and velocities in km/s. So the first line contains quantities of different dimensions $(km/s)^k$; the third and fourth line quantities are dimensionless.

TABLE I

Velocity distribution characteristics when colliding with Mercury; gravitation is not taken into account; the last row represents λ_k for Maxwell's distribution

k	-1	1	2	3	4
M_k	0.03588	33.08	1281	56090	2684000
V_k	27.87	33.08	35.79	38.28	40.48
λ_k	0.8426	1	1.082	1.157	1.224
$\tilde{\lambda}_k$	0.7854	1	1.085	1.162	1.233

$$N = 6, \quad \sigma = 13.66 \ km/s$$

Below Table I we give N = number of minor planets as potential colliders and $\sigma = \sqrt{M_2 - M_1^2}$ = the mean squared deviation of the velocity.

We described the collision velocity distribution providing the planet's gravity is negligible. In reality this is the planetocentric velocity distribution of the meteoroid swarm on the border of the planet's sphere of action. Table II contains the same quantities as Table I (except Maxwell's momenta) calculated with due regard to the planet's attraction. More precisely, Table II contains the characteristics of the planetocentric velocity field at distance R (R = radius of the planet) from the centre of Mercury. The corresponding

TABLE II

Velocity distribution characteristics when colliding with Mercury; gravity is taken into account

k	-1	1	2	3	4
M_k	0.03527	33.40	1299	56990	2731000
V_k	28.35	33.40	36.04	38.48	40.65
λ_k	0.8489	1	1.079	1.152	1.217

$$N = 6, \quad R = 2439.7 \ km, \quad v_0 = 4.249 \ km/s, \quad \sigma = 13.54 \ km/s$$

correction depends on the parabolic velocity $v_0(R)$ only. The rotation of the planet was not taken into account. The correction depends on the velocity direction and place of impact, but it is rather small even for the Earth and Mars.

In Table III for each minor planet number N^*, satisfying condition (1) for Mercury, we give the minimal and the maximal velocities of its collision with the planet over all possible intersection points. The values in the second and third columns correspond to the case when gravity is not taken into account; those in the fourth and fifth columns are obtained in consideration of the planet's gravity.

3. Other terrestrial planets and their satellites

We have no place to describe the velocity field near other body–targets so minutely as in the previous case. Let us give a table only. For each planet and the Moon the first row in Table IV contains the main quantities calculated without taking into account its gravity; the second row contains the same with due regard of its gravity. The attraction of Phobos and Deimos does not change velocities within the accuracy accepted.

4. Discussion

It should be noted that the results of our calculations of minimal and maximal collisional velocities of minor planets with Mars (to save space we do not give them in a special Table) differ systematically from those published in (Steel, 1985). According to our calculations the lower boundary of collisional velocities is as a rule several hundred meters per second greater and the upper boundary several hundred meters per second smaller. The variation of this difference is between 0 and 1.5 km/s. Minor planet (1727) Mette

TABLE III

Minimal and maximal velocities for Mercury-crossers; the second and third columns correspond to the case when gravity is not taken into account; the fourth and fifth ones - when gravity is taken into account

N^*	v_{min}	v_{max}	v_{min}^*	v_{max}^*
1566	34.99	54.31	35.25	54.48
2101	19.03	23.25	19.50	23.63
2212	17.31	34.19	17.82	34.45
2340	16.86	17.21	17.38	17.73
3200	43.65	64.12	43.85	64.26
3838	34.35	35.96	34.62	36.21

$$\Delta v = (v_{max} - v_{min})_{mean} = 10.48 \ km/s$$
$$\Delta v^* = (v_{max}^* - v_{min}^*)_{mean} = 10.39 \ km/s$$

is a typical example. Its perihelion distance is equal to 1.665 AU, that is' very close to the aphelion distance of Mars (1.666 AU). The collision of a

TABLE IV

The main characteristics of velocity distribution; gravity is taken into account in the second row for each body–target

Planet or Satellite	N	V_{-1}	V_1	V_2	V_3	V_4	σ
Mercury	6	27.87	33.08	35.79	38.28	40.48	13.66
		28.35	33.40	36.04	38.48	40.65	13.54
Venus	25	18.36	21.02	22.39	23.75	25.04	7.73
		21.90	23.68	24.67	25.71	26.75	6.95
Earth	45	15.41	18.15	19.48	20.70	21.78	7.07
		20.15	21.65	22.46	23.28	24.07	5.99
Mars	165	9.98	11.87	12.82	13.74	14.61	4.85
		11.69	13.03	13.77	14.53	15.28	4.45
Moon	45	15.54	18.25	19.56	20.77	21.84	7.04
		15.83	18.43	19.70	20.89	21.95	6.98
Phobos	165	10.74	12.43	13.30	14.17	15.01	4.73
Deimos	165	10.25	12.06	12.99	13.90	14.77	4.82

body having the same orbital elements a, e, i as Mette has, with Mars is possible only within a small vicinity of perihelion. Moreover, the collision velocity with Mars in all points of this vicinity is practically the same and equal to 9.38 km/s. In Steel's paper for this planet v_{min} and v_{max} are given as 8.7 km/s and 10.2 km/s, respectively. Such a distinction appears as a consequence of different setting up the problem. Steel's results correspond to the assumption, that the distribution of nodes and perihelia of both Mars and minor planet is uniform. So our results refer to the present epoch, and Steel's ones to a period of perihelion precession of Mars, i.e. about one hundred thousand years.

When examining the Tables above, we can draw the following conclusions.

1) The velocity distributions are similar to Maxwell's one, but definitely do not coincide with it. We shall try to find the analytical form of distribution functions in the nearest future.

2) The quantities V_k, σ decrease with the distance of the target body from the Sun. The reason is evident: the further from the Sun – the larger the potential and the less the kinetic energies of projectiles are.

3) The planetary or lunar gravity augments V_k (this is obvious without calculations) and makes λ_k approach 1. In other words, the gravity of the target body makes the random variable v "less random". This is reflected also in the diminishing of the mean squared deviation σ.

Acknowledgements

This work is partially supported by a grant of RFFI.

References

Batrakov, Yu.V., editor (1992): *Ephemerides of Minor Planets for 1993*, ITA RAN, St.–Petersburg.

Prokhorov, Yu.V., Rosanov, Yu.A. (1973): *Probability Theory*, Nauka, Moscow.

Steel, D.I. (1985): *Mon. Not. R. Astr. Soc.* **215**, 369.

COMETS, METEORITES AND ATMOSPHERES

T. OWEN

Institute for Astronomy, University of Hawaii, Honolulu, HI 96822, USA

and

A. BAR-NUN

Department of Geophysics & Planetary Sciences,
Tel-Aviv University, Ramat-Aviv, Tel-Aviv, Israel

Abstract. The relatively low value of Xe/Kr in the atmospheres of Earth and Mars seems to rule out meteorites as the major carriers of noble gases to the inner planets. Laboratory experiments on the trapping of gases in ice forming at low temperatures suggest that comets may be a better choice. It is then possible to develop a model for the origin of inner planet atmospheres based on volatiles delivered by comets added to volatiles originally trapped in planetary rocks. The model will be tested by results from the Galileo Entry Probe.

1. Introduction

This is a summary of the talk given at Mariehamn, which was based in part on a paper currently in press (Owen and Bar-Nun 1995). The central hypothesis of this paper is that impacts by icy planetesimals (comets) have been largely responsible for the inventories of volatile elements that we find on the inner planets today. An obvious corollary of this model is that impacts can also remove volatiles, especially if a planet is small, like Mars. It is the balance between delivery and removal that ultimately determines how much of an atmosphere a given body will possess.

2. The Difficulty with Meteorites

The traditional approach to the problem of atmospheric origin has been to propose one or more classes of chondritic meteorites as the volatile carriers that augmented whatever gases were trapped in the rocks composing the bulk of these planets (Turekian and Clark 1975, Anders and Owen 1977, Dreibus and Wänke 1987, 1989). These meteorites are thought to have supplied a late-accreting veneer of volatile-rich material that degassed upon impact and through subsequent processing on the planets, ultimately producing the atmospheres we observe today. On Earth alone, this simple picture has been strongly modified by the existence of liquid water, plate tectonics, and life, but a reconstruction of the Earth's volatile inventory reveals nearly the same abundances of carbon and nitrogen (per gram of rock) that we find today in the atmosphere of Venus (Donahue and Pollack 1983) and

Earth, Moon, and Planets **72**: 425-432, 1996.

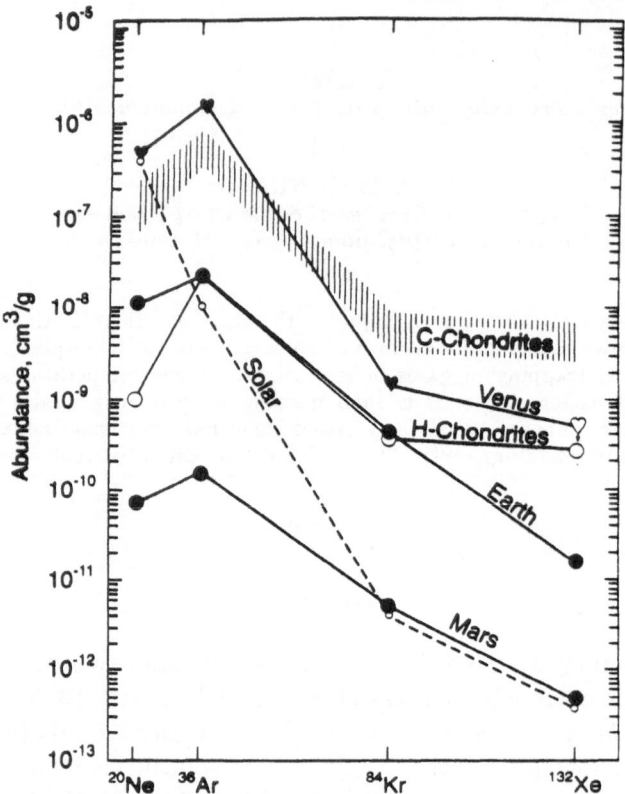

Fig. 1.   Abundances of the noble gases per gram of rock for the atmospheres of Venus, Earth, Mars and the ordinary and carbonaceous chondrites. Solar relative abundances are shown for reference.

about the same proportion of C/N that we see in the current atmospheres of both Venus and Mars (Owen et al. 1977).

These pleasing similarities come to an end when we consider the noble gases. Taking Earth as our standard, we find that Venus has far more neon and argon than Earth possesses, more even than exists in type I carbonaceous chondrites, per gram of rock. Furthermore, the relative abundances of krypton and argon on Venus are dramatically different from those on Earth, Mars, or in the meteorites. The ratio of Ar/Kr on Venus is closer to the solar value than the ratio we find in these other sources (Figure 1).

In contrast, Mars shows nearly the same relative abundance pattern as the Earth, but the absolute abundances of noble gases per gram of rock are over two orders of magnitude *lower* than those in Earth's atmosphere or the ordinary chondrites (Figure 1). Finally, on both Mars and Earth, Kr/Xe ∼ 10, whereas in the meteorites, this ratio is near unity.

On Earth, this last deficiency is often referred to as the "missing xenon problem". For many years, people assumed that this missing xenon must be trapped in shales, ice, or clathrate hydrates. It now seems clear that it simply isn't there (e.g. Wacker and Anders 1984).

3. A Cometary Solution

This was the background for our attempt to see if comets might be a suitable substitute for the meteorites. While this idea has certainly come up before (e.g. Sill and Wilkening 1978), it has suffered from the absence of data. We now have much better information about abundances of elements and compounds in the interstellar medium, in comets, and in planetary atmospheres. There are still no observations of noble gases in comets, which is not surprising in view of the fact that the ground-state transitions of these elements produce emission lines only in the far ultraviolet region of the spectrum. The mass spectrometers on the Giotto spacecraft also failed to detect any noble gases (Geiss 1988).

We have therefore relied on laboratory investigations to simulate the formation of cometary ices in the outer solar nebula (Bar-Nun et al. 1985, 1988; Owen et al. 1991). These experiments have shown that amorphous ice forming at temperatures in the range 16–190 K will trap gases in amounts that are a strong function of the temperature at which the trapping occurs. Hydrogen and helium are only trapped at very low temperatures; above 25 K even neon is not retained by the ice. A solar mixture of argon, krypton and xenon is trapped in its original proportions at 30 K, but is severely fractionated at 50 K.

This result led us to suspect that comets might indeed serve as the volatile carriers we were seeking. We tried the simple hypothesis of assuming that both Mars and Earth obtain their heavy noble gases (argon, krypton and xenon) from two reservoirs – one internal: the rocks that make up the bulk of the planet, and one external: impacting icy planetesimals (comets). We constructed a three-element plot on which the proportions of these three noble gases in the atmospheres of Mars and Earth would correspond to two points (Figure 2). The ratios of Ar/Xe and Kr/Xe found in various chondritic meteorites and the sun are also illustrated as is the domain of possible values for the atmosphere of Venus, where only an upper limit for xenon exists at the present time.

In our simple model, a straight line connecting Mars and Earth in such a plot will extend into the domain of the external reservoir at one end and into the internal reservoir at the other. In the logarithmic plot illustrated in Figure 2, this "mixing line" appears curved, since it is a simple linear function. Here we see that the point from the laboratory experiments that

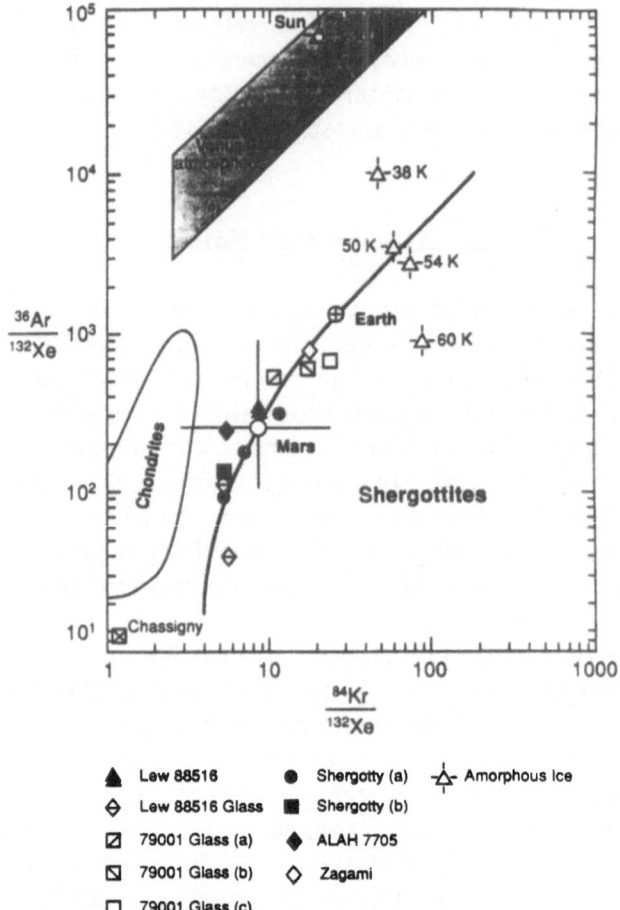

$\frac{^{36}Ar}{^{132}Xe}$

$\frac{^{84}Kr}{^{132}Xe}$

Fig. 2.   Ratios of the heavy noble gases in planetary atmospheres, chondritic meteorites and the SNC meteorites. The mixing line described in the text is the heavy line through the points for the atmospheres of Earth and Mars. Only an upper limit exists for Xe on Venus, hence the trapezoidal uncertainty.

corresponds to the mixture of noble gases trapped by ice at 50 K falls right on an extrapolation of this mixing line. Icy planetesimals that formed at about 50 K thus constitute a plausible external reservoir for the heavy noble gases on these two planets. This is a significant result, because 50 K is the canonical temperature for the Uranus-Neptune region of the solar nebula where most of the Oort cloud comets are thought to have formed (Boss et al. 1989, Oort 1990).

The rocky reservoir is then located at the other end of the mixing line, somewhere below the point for Mars. Why are Mars and Earth so far apart on this diagram? Evidently Mars is missing much of its cometary complement of gases. This conclusion is consistent with the present thinness of the

Martian atmosphere and the low abundances of the noble gases on Mars that we see in Figure 1. Melosh and Vickery (1989) have shown that Mars must have lost an amount of atmosphere equivalent to at least 100 times the atmosphere we see today as a result of impact erosion during the early heavy bombardment. Impact erosion also appears to be the only process that can account for the high values of $^{129}Xe/^{130}Xe$ and $^{40}Ar/^{36}Ar$ that we find on Mars, respectively 2.5 and 100 times the terrestrial ratios (Owen et al. 1977, Owen 1992). Evidently many of the cometary impacts that on Earth would have delivered volatiles, failed to do so on Mars. Both impacting comets and asteroids would contribute to an erosion of the Martian atmosphere that would have been more severe than that which occurred on the more massive Earth.

Figure 2 also contains points for noble gases measured in a number of Shergottite meteorites (see Owen et al. 1992 and Owen and Bar-Nun 1993 for references). These meteorites are now generally acknowledged to have originated on Mars, as demonstrated in part by the gases they contain (Bogard and Johnson 1983, Becker and Pepin 1984). What we see in Figure 2 is that samples of the noble gases extracted from these meteorites (sometimes different samples of the same stone or even different temperature releases from the same sample) exhibit proportions of argon, krypton and xenon that lie along the mixing line. Some samples fall above the Mars atmosphere point (based on the Viking Lander measurements of Owen et al. 1977) and some fall below. Our interpretation of these data is that the meteorites represented by the lowest points contain a mixture of Martian atmosphere with samples of the internal reservoir. The recent measurements of Becker and Pepin (1993) in the samples of glass from LEW 88516 tend to strengthen the mixing line formulation. As we move up the mixing line toward the Earth, it appears that the meteorites are adding cometary gas to the mixture they contain. This suggests that the impactor that blew them off the surface of Mars was a comet. On the other hand, the large error bars on the Viking Mars atmosphere measurement allow the possibility that the highest points on this curve, for Zagami and EETA 79001, may represent the Martian atmosphere, and the lower points simply correspond to increasingly greater proportions of gas from the internal, rocky reservoir. Additional analysis of the gases from the other SNC meteorites, especially the Nakhlites, will be required to settle this question.

With this model, we can explain the unusual noble gas pattern found in the atmosphere of Venus by invoking an impact by one or more comets from the Kuiper Disk (Kuiper 1951, Duncan et al. 1988, Jewitt and Luu 1993). These objects will have formed at temperatures near 30 K, thereby trapping the noble gases in nearly solar proportions. Obviously comets from both the Kuiper Disk and from the Uranus-Neptune region must have struck all these inner planets. What we are therefore suggesting is that Venus received a dominating share of its heavy noble gases from the colder source.

Because neon is only trapped in ice at temperatures below 25 K, we suggest that the neon in these atmospheres is a relic of gas trapped almost exclusively in the rocks that make up most of the mass of these planets. The severe fractionation $^{20}Ne/^{22}Ne$ that we observe today may then reflect early escape processes whose effects on nitrogen and the other noble gases have been masked by subsequent cometary delivery of these volatiles.

4. Conclusions, Implications, and Tests

The laboratory experiments on the trapping of argon, krypton and xenon in ice at temperatures characteristic of the Uranus-Neptune region of the solar nebula demonstrate a fractionation pattern of noble gas abundances found in the atmospheres of Mars and Earth. This suggests that these gases were delivered to the inner planets by icy planetesimals, a.k.a. comets. If that is true, other volatiles would be brought in as well, most notably water, carbon and nitrogen. The hydrogen that is bound up in carbon and nitrogen compounds in the comet nuclei would become available on the planets to make compounds, offering the potential for early reducing atmospheres. On Mars, because of its small size, impacts apparently carried off more atmosphere than they produced, leaving the thin envelope we find today.

This model can be tested in a number of ways, as described in our complete paper (Owen and Bar-Nun 1995). The first test is simply to see if argon rather than neon is present in dynamically new comets. This will require either a dedicated rocket launch for observations of the UV spectrum of a bright comet, or a comet mission that includes the option of examining volatiles trapped in the interior of a comet nucleus.

Future missions to Mars and Venus can determine more details about atmospheric noble gases. For example, we would predict that the krypton/xenon ratio on Venus is very close to the solar value. Accurate values for the krypton and xenon isotopes on Mars would allow more meaningful interpretation of the SNC data, including the possible role of a comet in expelling the Shergottites from Mars.

Perhaps the first test of the general model of icy planetesimal delivery of volatiles will come from the Galileo spacecraft encounter with Jupiter in December 1995. The Galileo Probe will carry a mass spectrometer capable of measuring abundances and isotopic ratios to a depth of about 10 bars. We already know that $C/H \approx 3\times$ solar in Jupiter's atmosphere, but the value of N/H is somewhat uncertain. The enrichment of carbon is commonly attributed to the release of gases from the condensed matter (icy planetesimals) that formed the original core of the planet. The model therefore predicts that N/H will be approximately solar, rather than enhanced to the same degree as C/H. The reason is that ice formed in Jupiter's region of the solar nebula would not have been able to trap N_2, only nitrogen compounds. While most of the carbon in the original solar nebula was presumably in the form of compounds with relatively low-volatilities, most of the nitrogen was present as N_2. Hence N/C in the icy material should have been sub-solar, as it was found to be in Halley's Comet (Geiss 1988).

On the other hand, the envelope of nebula gas that has provided most of Jupiter's atmosphere would contain that N_2. Therefore, we expect N/H $\approx$ solar and N/C < solar in the atmosphere. Because this nitrogen came from N_2, rather than from N-compounds that the model suggests were the source of our atmospheric nitrogen, the isotope ratio $^{14}N/^{15}N$ may be different on Jupiter as well.

The mass spectrometer on the probe will have an opportunity to measure both N/C/H and $^{14}N/^{15}N$, although the former measurement will be particularly difficult owing to the tendency for NH_3 to absorb on the metallic walls of the instrument and its inlet system.

5. Acknowledgements

This research was supported in part by NASA grants NAGW 2631 and NAGW 2650. T.O. thanks the Organizing Committee for financial support.

References

Anders, E. and Owen, T.: 1977, Mars and Earth: Origin and abundances of volatiles. *Science* **198**, 453–465.

Bar-Nun, A., Herman, G., Laufer, D., and Rappaport, M.L.: 1985, Trapping and release of gases by water ice and implications for icy bodies. *Icarus* **63**, 317–332.

Bar-Nun, A., Kleinfield, I. and Kochavi, F.: 1988, Trapping of gas mixtures by amorphous water ice. *Phys. Rev. B* **38**, 7749–7754.

Becker, R.H. and Pepin, R.O.: 1984, The case for a Martian origin of the Shergottites: Nitrogen and noble gases in EETA 79001. *Earth and Planet. Sci. Lett.* **69**, 225–242.

Bogard, D.D. and Johnson, P.: 1983, Martian gases in an Antarctic meteorite? *Science* **221**, 651–654.

Boss, A.P., Morfill, G.E. and Tscharnuter, W.M.: 1989, Models for the formation and evolution of the solar nebula. In *Origin and Evolution of Planetary and Satellite Atmospheres*, ed. S.K. Atreya, J.B. Pollack, and M.S. Matthews, U. of Ariz. Press, Tucson, pp. 487–512.

Donahue, T. and Pollack, J.B.: 1983, Origin and evolution of the atmosphere of Venus. In *Venus*, ed. D.M. Hunten, L. Colin, T.M. Donahue, and V.I. Moroz, U. of Ariz. Press, Tucson, pp. 1003–1036.

Dreibus, G. and Wänke, H.: 1987, Volatiles on Earth and Mars: A comparison. *Icarus* **71**, 225–240.

Dreibus, G. and Wänke, H.: 1989, Supply and loss of volatile constituents during the accretion of terrestrial planets. In *Origin and Evolution of Planetary and Satellite Atmospheres*, ed. S.K. Atreya, J.B. Pollack and M.S. Matthews, U. of Ariz. Press, Tucson, pp. 268–288.

Duncan, M., Quinn, T. and Tremaine, S.: 1988, The origin of short-period comets. *Astrophys. J.* **328**, L69–L73.

Geiss, J.: 1988, Composition in Halley's Comet: Clues to origin and history of cometary matter. *Rev. Mod. Astron.* **1**, 1–27.

Jewitt, D.C. and Luu, J.: 1993, Discovery of the candidate Kuiper belt object 1992 QB1. *Nature* **362**, 730–732.

Kuiper, G.P.: 1951, Origin of the solar system. In *Astrophysics*, ed. J.A. Hynek, McGraw-Hill, New York, pp. 357–424.

Melosh, J. and Vickery, A.M.: 1989, Impact erosion of the primordial atmosphere of Mars. *Nature* **338**, 487–489.

Oort, J.H.: 1990, Orbital distribution of comets. In *Physics and Chemistry of Comets*, ed. W.F. Huebner, Springer-Verlag, Berlin, pp. 235–245.

Owen, T.: 1992, The composition and early history of the atmosphere of Mars. In *Mars*, ed. H.H. Kieffer, B.M. Jakosky, C.W. Snyder, and M.S. Matthews, Univ. of Ariz. Press, Tucson, pp. 818–834.

Owen, T., Biemann, K., Rushneck, D.R., Biller, J.E., Howarth, D.W., and Lafleur, A.L.: 1977, The composition of the atmosphere at the surface of Mars. *J. Geophys. Res.* **82**, 4635–4639.

Owen, T., Bar-Nun, A., and Kleinfield, I.: 1991, Noble gases in terrestrial planets: Evidence for cometary impacts? In *Comets in the Post-Halley Era*, ed. R.L. Newburn, Jr., M. Neugebauer, J. Rahe, Kluwer, Dordrecht, pp. 429–438.

Owen, T., Bar-Nun, A. and Kleinfield, I.: 1992, Possible cometary origin of heavy noble gases in the atmospheres of Venus, Earth and Mars. *Nature* **358**, 43–46.

Owen, T. and Bar-Nun, A.: 1993, Noble gases in atmospheres. *Nature* **361**, 693–694.

Owen, T., and Bar-Nun, A.: 1995, Comets, impacts and atmospheres. *Icarus*, in press.

Sill, G.T. and Wilkening, L.L.: 1978, Ice clathrate as a possible source of the atmospheres of the terrestrial planets. *Icarus* **33**, 13–27.

Turekian, K.K. and Clark, S.P., Jr.: 1975, The non-homogeneous accumulation model for terrestrial planet formation and the consequences for the atmosphere of Venus. *J. Atmos. Sci.* **32**, 1257–1261.

Wacker, J.F. and Anders, E.: 1984, Trapping of xenon in ice and implications for the origin of the Earth's noble gases. *Geochim. et Cosmochim. Acta* **48**, 2372–2380.

EVOLUTION, PUNCTUATIONAL CRISES AND THE
THREAT TO CIVILIZATION

S.V.M. CLUBE

Department of Physics, University of Oxford, Oxford, U.K.

Abstract. The relationship between "punctuated equilibrium" and "impact crises" is critically examined in the light of our present knowledge of asteroids and comets. It turns out that the emphasis on relatively narrow epochs associated with occasional "NEO" impacts is probably misplaced. Rather priority should be given to the wider and more frequent epochs associated with multiple "NEO" debris impacts which result in so-called "punctuational crises" afflicting the planets. These comprise the global coolings, super-Tunguska events and generally enhanced fireball flux produced by the larger orbital debris whenever an active, dormant or dead comet fragments and produces a trail. Taken as a whole and in conjunction with the target, the response function is inevitably complex. Nevertheless we broadly expect that the strength of a punctuational crisis will vary as the progenitor comet mass, the inverse dispersion of its debris and the inverse delay since fragmentation. The encounter of P/SL-9 with Jupiter may be taken as representing an extreme punctuational crisis where the dispersion and delay were exceptionally small. The more familiar crises affecting the Earth with less extreme values of dispersion and delay, which have resulted in civilization being disturbed a good many times during recent millennia, are no less important however. Indeed, the next such threat to civilization ostensibly has a roughly 1 in 4 lifetime chance. Any support for the Spaceguard programme which detracts from consideration of these punctuational crises, whatever their strength, would seem now to be peculiarly wide of the mark.

1. Punctuated equilibrium

The idea that evolution on Earth proceeds at a uniform pace towards some undefined state of perfection in the remote future has given way in recent years to one involving successive states of "punctuated equilibrium". Thus, in keeping with evolution's supposedly progressive nature, it is assumed that both the environment and the distribution of living species remain in successive uniformitarian states for characteristically long periods of time and that these also begin and end with much briefer periods when both the environment and the distribution of species undergo very rapid upheaval. There is a presumption then that the upheavals are progressive. However, inasmuch as the uniformitarian states essentially correspond to equilibrium complexes comprising both physico-chemical and bio-chemical species, and these complexes are subject to selective subtraction and addition during cosmic inputs, the punctuations can also be regarded as random non-equilibrium states. If these are rapidly imposed (*eg* through catastrophes) and rapidly removed (*eg* through natural selection) then it is likely that the subsequent

Earth, Moon, and Planets **72**: 433-440, 1996.

uniformitarian states are broadly regressive (*ie* towards the mean) while the punctuations are progressive only so far as the immediately successful species are concerned. It follows that biological evolution then has a general cosmic pseudomorphic character, as first proposed by Spengler (1991), and that this must also extend to the timescale of the evolution of civilizations.

With improvements also during recent years in our knowledge of the astronomical environment, there has been a tendency to suppose these punctuations might reasonably be associated with isolated "impact crises" due to encounters with bodies in Earth-crossing orbits. While, as a matter of definition, it might be considered arguable exactly what level of crisis qualifies as a punctuation in the terrestrial record, one can perhaps assume, as a matter of principle, that only those impinging bodies capable of significantly influencing *the whole globe* for a brief period should be considered. It is on just such a basis apparently that the near-Earth objects (NEOs) greater than a kilometre or so in size have come to be the new focus of attention so far as terrestrial evolution is concerned. Indeed, through this rather simplistic perception, recognizing also that the human species is a global phenomenon, the notion that km-plus NEOs are the most serious threat to civilization has recently gained some impetus (*eg* Chapman & Morrison 1994). Civilizations, treated as a significant evolutionary characteristic of the human species on centennial to millennial timescales, tend however to be regarded as a local rather than a global phenomenon; in which case, the theory of punctuated equilibrium would seem to require that single or multiple sub-km NEOs, capable of depositing massive dust veils or inducing super-Tunguska events, represent the most likely serious threat to civilization.

2. Cumulative catastrophic record

Our knowledge of the impactor flux reflecting the general state of the inner Solar System environment is largely based on the cumulative counts of impact craters formed on lunar mares since the end of the heavy bombardment phase. As a result, the diameter-flux relationship for the largest impactors arriving at the Earth is commonly represented by a simple uniformitarian power law:

$$\dot{\Phi}(D) = kD^{-\alpha}, \qquad 1 \leq D \leq 10^{2.5} \text{km} \qquad (1)$$

Continuity considerations require that this relationship is applicable to the impactors which are currently observed in potential Earth-crossing orbits. Among these we must include the Earth-crossers having intermediate and long period orbits which reach out beyond Jupiter but their influence, in comparison with that of the Earth-crossers in sub-Jovian space deriving principally from short-period asteroidal (mainbelt) and cometary (Jupiter family) reservoirs, is so small that they can reasonably be neglected for the purpose of the present discussion. It follows that the bodies of interest, comprising mostly sub-asteroidal and sub-cometary fragmentation products at the Earth (meteorites, meteoroids) are recognized as surviving in one or other of two sub-Jovian orbital regimes for "dynamical lifetimes" of $\sim 10^8$ yr or $\sim 10^6$ yr respectively, depending on the eccentricity of their progenitor injection orbits - below or above 0.5, say. Thus, for these short-period reservoirs separately dominating the terrestrial influx, it is essentially orbital eccentricity which determines the likelihood of a grazing encounter with a major celestial body such as would tend to remove these smaller bodies from inner Solar System space, and it is recognized that the terrestrial planets are most likely to play this role when $e < 0.5$ (Wetherill 1988, 1991) while the Sun and Jupiter are most likely to play this role when $e > 0.5$ (Farinella 1995).

It does not follow of course that transitions between these orbital regimes are excluded but the top-heavy mass distribution, fragmentation spectrum and physical lifetime of the bodies under consideration, in relation to their dynamical lifetime, are such that the cometary-meteoroidal impactors within the considered range of mass, unlike their asteroidal-meteoritic counterparts, cannot be expected to achieve a completely relaxed distribution in solar ecliptic longitude and latitude. In other words, the minor body flux of cometary origin normally takes a period of time to become fully sporadic which is significantly in excess of its physical survival time. This is in accordance with the observed steepening of the uniformitarian law at lower levels of mass since smaller bodies have usually experienced additional hierarchical fragmentation and are thus more likely to be present in sufficient numbers at the appropriate longitudes to be fully sampled during terrestrial passages, whence

$$\dot{\Phi}(D) = kD^{-\beta}, \qquad 10^{-3} \leq D \leq 1 \text{ km} \tag{2}$$

where $\beta > \alpha$ (Shoemaker 1983). The relationships (1) and (2) are based on the lunar cratering record but we can also determine the diameter-flux relationship for smaller impactors currently arriving at the Earth, as derived from the bodies in space which are observed either *in situ* or penetrating the atmosphere (Rabinowitz *et al* 1993, Ceplecha 1992, Tagliaferri *et al* 1994), whence it turns out that

$$\dot{\Phi}_0(D) \simeq 10 - 100 \times \dot{\Phi}(D), \quad D \leq 10^{-2} \text{km} \tag{3}$$

This may evidently be understood as a temporary condition, also in accordance with the observational steepening, and has for some while been attributed to a still disintegrating, very large comet (Kresák 1981) of the kind now believed to be present in the inner Solar System from time to time ($\Delta t \sim 10^5$ yr; Bailey *et al* 1994). The steepening thus straightforwardly implies that the commonest "evolutionary events" on Earth relating to the low mass end of the Solar System minor body population (*ie* $10^{-1} \leq D \leq 1$ km) are due to correlated encounters with the hierarchically disintegrated products of successive giant comets. The appeal to a contemporary giant comet does of course represent a general departure from uniformitarianism on timescales $\sim 10^6$ yr, the average interval between giant comets settling in sub-Jovian space. This leaves open the question whether there are additional modulations of the terrestrial record on timescales < and > 10^6 yr which would also be indicative of a predominantly cometary influence on terrestrial evolution.

3. Punctuational crises

To many investigators, the idea that isolated impact crises rising above some global threshold are the only astronomical influence we need consider when dealing with evolutionary processes in geology and biology is simply not compatible with the evident complexity of the terrestrial record. Thus it is widely recognized that long-term climatic and other factors must also be involved and it has been known for seventy years that the terrestrial record is marked by periodic and stochastic modulations on timescales between 10^6 and 10^9 yr indicative of a Galactic driving force (*eg* Holmes 1927). Indeed it is for these reasons that many investigators in recent years have given greater credence to a cometary (Oort cloud) rather than a (Mainbelt) asteroidal source of "punctuational crises".

The role model for punctuations then is not the ostensibly narrow epoch associated with a random (km-plus) asteroid; rather it is the considerably broader epoch associated with the relatively short-lived, orbitally correlated, disintegration products of a not-so-random (km-plus) comet. There are two points to be noted here. First, these considerably broader epochs may be characterized by one or more global cooling and/or super-Tunguska events occurring as a prelude to or in association with an enhanced fireball flux: the epochs of the Justinian cold period (Baillie 1994) and the 17th century mini-ice age can be considered possible exemplars. Thus, events of this kind are to be expected as a consequence of high-level dust insertions and low-level multimegaton explosions such as may be produced, depending on their cohesive strength, by sub-cometary masses of about 0.1 - 1 km in size. Secondly, both the cometary mass function (which is top-heavy) and the tendency of comets to undergo rapid disintegration determine that a high degree of coherence may be present in the incidence of sub-km and km-plus comets on Earth at any one time.

Cometary material in general is capable of being active, dormant or dead and for inner Solar System material of this kind whose distribution does not evolve and which has broadly unchanging orbital and constitutional characteristics, it would certainly be expected that the frequency of punctuational crises (as now defined) at any epoch would also be broadly unchanging and reflective of the integrated "minor body" mass in residence. With a variable mass content however, such as arises with the top-heavy mass distribution of comets settling randomly in inner Solar System space, the pattern of punctuational crises may be expected to take on the general character of a glacial-interglacial with both periodic and random groups of events on timescales $< 10^5 - 10^6$ yr reflecting the orbital and fragmentation history of a particular giant comet (*eg* Asher & Clube 1993). In other words, as a consequence of the cometary mass distribution, we envisage punctuational crises which are themselves hierarchically nested in the overall manner of glacial-interglacials, each lasting in effect for $\sim 10^4 - 10$ orbits in accordance with the size of the parent comet within the nested hierarchy *ie* from a few hundred to a few kilometres in size.

It is thus in the general nature of the intermittently top-heavy population of comets deposited in inner Solar System space that our planet is bound to experience glacials and interglacials, the latter being themselves interspersed with global coolings of shorter duration which are in association with super-Tunguska events and sustained enhancements of the fireball flux. Several facts then come together to inform us as to the likely nature of the current environment:

(a)   the disintegration products of comets (meteoroids) currently incident upon the Earth outweigh the disintegration products of asteroids (meteorites) by one or two orders of magnitude (*eg* Tagliaferri *et al, loc cit*)

(b)   the sporadic flux of meteoroids in inner Solar System space is dominated by a single elliptical torus (*eg* Štohl 1983) suggestive of a recently disintegrated, very large, Taurid comet (*eg* Steel *et al* 1994).

(c)   the current interglacial follows on a recent glacial at $\sim 20,000$ $\pm 10,000$BP broadly suggestive of a recently disintegrated, very large comet and is itself interspersed with sustained enhancements of the fireball flux known on several occasions to be correlated with severe global coolings (Asher & Clube 1993, Clube 1994, Baillie 1994).

(d)   dynamical studies (*eg* Asher *et al* 1993, Valsecchi *et al* 1995) are consistent (within a factor ~ 2)with a $\sim 20,000$ yr hierarchically disintegrated Taurid comet characterised by fragmentation speeds ~ 1 kms^{-1} such as are plausibly associated with the expected close (terrestrial) planetary passages within the Roche limit.

We may conclude that it is the cometary punctuational crises which are dominant on timescales $\leq 10^6$ yr.

4. The Holmes cycle

Unfortunately the cratering record is not yet well enough resolved to describe with certainty any of its possible modulations on timescales $\sim 10^6 - 10^9$ yr (Grieve 1989). Nevertheless to the extent that geological and biological signatures may be understood as proxy-signatures for punctuational crises, there is good evidence for a late Phanerozoic cycle of 26.3 Myr (Rampino & Caldeira 1992) broadly confirming previous determinations based on the extinction cycle alone (Raup & Sepkoski 1984 cf Holmes 1927). It also appears that the most recent maximum phase of this cycle coincides with a mid-Miocene peak from which the early Pleistocene peak preceding the Sun's latest Galactic plane passage is clearly distinguished. On the assumption that geomagnetic reversal events in particular provide a reasonably undistorted record of the major glacials due to large (inner Solar System) comets, the 26.3 Myr cycle is interleaved with another cycle of the same period but lower amplitude; the two together being then uniquely associated with variations in the Galactic "dark matter" gravitational field acting upon the Oort cometary cloud, due to the changes in space density and encounter speed expected as a consequence of the Sun's vertical oscillation about the Galactic plane (Clube & Napier 1995 cf Matese et al 1995). Quite apart from the not-unimportant implications for dark matter and its nature, these cycles are in fact strong *prima facie* evidence of a very persistent influence on terrestrial affairs due to very large comets and we may conclude that cometary punctuational crises are dominant on timescales $\geq$ as well as $\leq 10^6$ yr.

It is perhaps an interesting aside on the theory of punctuational crises that the latest two peaks of the compound Holmes cycle correlating with two broad spasms of terrestrial activity seem to be associated with the first appearance of hominids (15–7 Myr BP) and with the rapid evolution of *homo sapiens* (6–0 Myr BP) respectively, while the particular activity that goes with the latest, very large, comet seems remarkably well correlated with civilized man's low point at the end-Pleistocene during a global climatic recession and the rise of civilization during the Holocene. Civilization in other words is merely the latest random facet of galactoterrestrial control expressed through the action of comets on the resident gene pool!

5. Commentary and conclusion

The punctuational crises discussed here are identical, in principle, to the recent P/SL-9 encounter with Jupiter. The essential differences are the target, the source of fragmentation and the degree of orbital correlation. But the aspect of punctuational crises which gives them their special distinction over isolated impact crises is their predictability following fragmentation and their capacity therefore to induce social destabilization. In the case of P/SL-9, since another planet was involved, mankind was able to take a detached view of the subsequent proceedings. In the case of an Earth encounter however, since the enhanced fireball flux is indicative of its more massive correlates, the view can never be detached (Clube & Asher 1995). Such enhancements have in fact occurred frequently in the past - ostensibly in association with the Taurid stream (Clube 1994) - and the next occurrence has a roughly 1 in 4 lifetime chance. Historically, inasmuch as these enhancements have frequently been interpreted as indicating the imminence of "last times", predisposing even the most advanced societies to break up and lose control, the corrective response has recently been either (pre-Newton) to censor any deviations from the perceived celestial paradigm, whatever form these took, or (post-Newton) to trivialize any threat posed by comets (Clube 1995). Neither censorship nor trivialization is likely to be effective in future and it is not clear therefore that civilization is currently well placed to handle the next punctuational crisis. Indeed, any benefit accruing to civilization through the uncritical endorsement of Spaceguard (Chapman & Morrison 1994; Harris 1995) does little at present to alleviate the pressures due to the next punctuational crisis.

References

Asher D.J. & Clube S.V.M., 1993, Quart.J.Roy.Astron.Soc., **34**, 481.
Bailey M.E., Clube, S.V.M., Hahn G., Napier, W.M. & Valsecchi G.B., 1994, *"Hazards due to comets and asteroids"* (ed. T. Gehrels *et al*, Univ. of Arizona Press), 479.
Baillie, M.G.L., 1994, The Holocene, **4**, 212.
Ceplecha Z., 1992, Astron.Astrophys., **263**, 361.
Chapman C.R. & Morrison D., 1994, Nature **367**, 33.
Clube, S.V.M., 1994, *"How science works in a crisis: the mass extinction debate"* (ed. W.Glen, Stanford Univ. Press), 152.
Clube, S.V.M., 1995, *"The inspiration of astronomical phenomena"* (ed. R.White *et al*), in press.
Clube, S.V.M. & Asher, D.J., 1995, Southern Sky 9, 24.
Clube, S.V.M. & Napier W.M., 1995, Mon.Not.Roy.Astron.Soc., submitted
Farinella P., 1995, These proceedings.
Grieve R.A.F., 1989, *"Catastrophes and Evolution"* (ed. S.V.M. Clube, C.U.P.), 57.
Harris A.W., 1995, Conference Summary (These proceedings).
Holmes A., 1927, *"The Age of the Earth: an Introduction to Geological Ideas"* (Benn,

London).

Kresák L., 1981, Bull.Astron.Inst.Czechosl. **32**, 19.

Matese J.J., Whitman P.G., Innanen K.A. & Valtonen M.J., 1995, These proceedings.

Rabinowitz D.L., Gehrels, T., Scotti J.V., McMillan R.S., Perry M.L., Wisnewski W., Larson S.M., Howell, E.S. & Mueller B.E.A., 1993, Nature **363**, 704.

Rampino M.R. & Caldeira K., 1992, Celest.Mech.Dyn.Astron. **54**, 143.

Shoemaker E.M., 1983, Ann.Rev.Earth.Planet.Sci. **11**, 461.

Spengler, O., 1991, *"The Decline of the West"* (Oxford Univ. Press, abridged). Steel D.I., Asher, D.J., Napier, W.M. & Clube S.V.M., 1994, *"Hazards due to comets and asteroids"* (ed. T.Gehrels *et al* Univ. of Arizona Press), 463.

Štohl J., 1986, Asteroids, Comets, Meteors II (ed. C-I Lagerkvist *et al*, Univ. of Uppsala Press), 565.

Tagliaferri E., Spalding, R., Jacobs, C., Woerden, S.P. & Erlich, A., 1994, *"Hazards due to comets and asteroids"* (ed. T.Gehrels *et al* Univ. of Arizona Press), 199.

Wetherill G.W., 1988, Icarus **76**, 1.

Wetherill G.W., 1991, *"Comets in the Post-Halley Era, I"*, (ed. J.R. Newburn *et al*, Kluwer), 537.

THE "SHIVA HYPOTHESIS": IMPACTS, MASS EXTINCTIONS, AND THE GALAXY

MICHAEL R. RAMPINO* and BRUCE M. HAGGERTY

Earth Systems Group, New York University, New York, New York 10003

Abstract. The "Shiva Hypothesis", in which recurrent, cyclical mass extinctions of life on Earth result from impacts of comets or asteroids, provides a possible unification of important processes in astrophysics, planetary geology, and the history of life. Collisions with Earth-crossing asteroids and comets $\geq$ a few km in diameter are calculated to produce widespread environmental disasters (dust clouds, wildfires), and occur with the proper frequency to account for the record of five major mass extinctions (from $\geq 10^8$ Mt TNT impacts) and $\sim$ 20 minor mass extinctions (from 10^7-10^8 Mt impacts) recorded in the past 540 million years. Recent studies of a number of extinctions show evidence of severe environmental disturbances and mass mortality consistent with the expected after-effects (dust clouds, wildfires) of catastrophic impacts. At least six cases of features generally considered diagnostic of large impacts (e.g., large impact craters, layers with high platinum-group elements, shock-related minerals, and/or microtektites) are known at or close to extinction-event boundaries. Six additional cases of elevated iridium levels at or near extinction boundaries are of the amplitude that might be expected from collision of relatively low-Ir objects such as comets.

The records of cratering and mass extinction show a correlation, and might be explained by a combination of periodic and stochastic impactors. The mass extinction record shows evidence for a periodic component of about 26 to 30 Myr, and an $\sim$ 30 Myr periodic component has been detected in impact craters by some workers, with recent pulses of impacts in the last 2-3 million years, and at $\sim$ 35, 65, and 95 million years ago. A cyclical astronomical pacemaker for such pulses of impacts may involve the motions of the Earth through the Milky Way Galaxy. As the Solar System revolves around the galactic center, it also oscillates up and down through the plane of the disk-shaped galaxy with a half-cycle $\sim$ 30$\pm$3 Myr. This cycle should lead to quasi-periodic encounters with interstellar clouds, and periodic variations in the galactic tidal force with maxima at times of plane crossing. This "galactic carrousel" effect may provide a viable perturber of the Oort Cloud comets, producing periodic showers of comets in the inner Solar System. These impact pulses, along with stochastic impactors, may represent the major punctuations in earth history.

1. Introduction

Mass extinctions are geologically brief episodes when large numbers of existing species ($\sim$ 25% to > 90%) disappeared (Raup and Sepkoski, 1984). Paleontologists recognize five major mass extinctions of marine organisms, and about 20 other identifiable peaks of extinction above the background during the Phanerozoic Eon, the past $\sim$ 540 Myr (Sepkoski, 1982, 1992, 1994) (Fig. 1). These extinctions of marine life coincide with extinctions of non-marine organisms (e.g., vertebrates, insects, land plants) (e.g., Rampino,

* also at NASA, Goddard Institute for Space Studies, 2880 Broadway, New York, New York 10025

Earth, Moon, and Planets **72**: 441-460, 1996.
© 1996 *Kluwer Academic Publishers.*

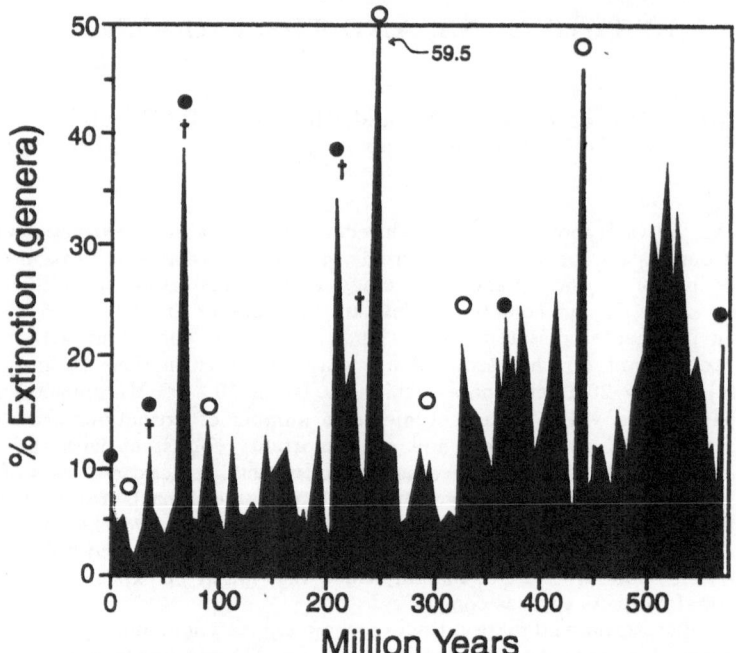

Fig. 1.   Percent extinction of marine genera per geologic stage (or substage) during the Phanerozoic (data from Sepkoski, 1992, 1994, and pers. comm.). The following local maxima are recognizable in Sepkoski's data (end of stage, as defined, may not be exactly coincident with the extinctions) age in Myr (approx.), as dated by most recent geologic time scales, : 1. Pliocene (1.6), 2. Mid-Miocene (14), 3. Upper Eocene (35.4), 4. Maastrichtian (65.0), 5. Cenomanian (92), 6. Aptian (112), 7. Tithonian (144), 8. Callovian (163), 9. Pliensbachian (193), 10. Norian (205), 11. Carnian (225), 12. Tatarian (245), 13. Guadelupian (253), 14. Stephanian (286), 15. Serpukhovian (320), 16. Famennian (362), 17. Frasnian (367), 18. Eifelian (381), 19. Ludlovian (411), 20. Ashgillian (438), 20a. Caradocian? (448), 21. lower Llanvirnian (478), 22. Trempeleauan (505*), 23. Dresbachian (515*), 24. Botomian (520*), 25. Proterozoic/Cambrian? (540*). Data were culled by removing rare genera known from single localities of exceptional preservation. Note that Cambrian extinction peaks may be anomalously high as a result of the relatively poor record of diversity, and have recently been redated (*). Diagnostic stratigraphic evidence of impact (solid dots), possible stratigraphic evidence of impacts (open circles), and large, dated impact craters (crosses) (see text and Tables).

1988; LaBandiera and Sepkoski, 1993; Benton, 1995), suggesting global environmental perturbations as a cause.

The possible causes of mass extinctions of life remains a subject of intense debate. With the discovery of considerable evidence for the impact of a comet or asteroid precisely at the Cretaceous/Tertiary (K/T) mass extinction boundary (65 Myr) by L.W. Alvarez et al. (1980; for a recent review, see Glen, 1994), much attention has focused on large-body impacts as an agent of mass extinctions. A periodic component of 26 to 30 Myr in the record of mass extinctions was detected (Raup and Sepkoski, 1984, 1986), and a similar periodic component was soon reported in terrestrial impact craters (Rampino and Stothers, 1984a,b; Alvarez and Muller, 1984). Although controversial (see Weissman, 1986), these results led to several astrophysical hypotheses involving generation of periodic comet showers from the Oort comet cloud (Rampino and Stothers, 1984a; Whitmire and Jackson, 1984; Davis et al., 1984; Whitmire and Matese, 1985).

The name "Shiva (or Siva) Hypothesis", after the Hindu deity of destruction and renewal, has been suggested for this hypothesis relating recurrent mass extinctions to cyclical astrophysical causes (Gould, 1984; Goldsmith, 1986; Rampino, 1990). The name seems particularly apt. Shiva is perhaps the most ancient deity worshipped in the world today (Campbell, 1987), and in his role as a cosmic dancer, as Gould (1984) writes, "he holds in one hand the flame of destruction, in another (he has four in all) the *damaru*, a drum that regulates the rhythm of the dance and symbolizes creation. He moves within a ring of fire – the cosmic cycle – maintained by an interaction of destruction and creation, beating out a rhythm as regular as any clockwork of cometary collisions."

The idea that mass extinctions on the Earth might be paced by astrophysical cycles is far-reaching, and Raup (1989) suggested that "the subject involves so many separate scientific disciplines – from paleontology to astrophysics – that no one individual is competent to judge the merits of all the arguments and counterarguments." However, a general theory of mass extinction by impact catastrophe would represent a powerful predictive generalization in the geological and paleobiological sciences (Alvarez, 1986), and links to astrophysical dynamics make for a intuitively pleasing scientific paradigm. We believe that enough information now exists to develop a coherent and, most importantly, a testable hypothesis.

2. Impacts and Environmental Catastrophes

Most studies have come to the conclusion that the impact of an asteroid or comet ≥ 10 km in diameter (releasing $\geq 10^{24}$ J, 10^8 Mt TNT) would cause a global catastrophe of enormous proportions (e.g., Chapman and Morrison, 1994; Toon et al., 1994), and the severe end-Cretaceous crisis seems to have involved an ~ 10 km diameter impactor (Alvarez et al., 1980). In looking at a general connection between impacts and extinctions, it is important to determine the threshold impactor size predicted to cause a global environmental crisis that could result in an identifiable peak in extinction of life.

Chapman and Morrison (1994) suggested that a > 5 km object ($> 10^7$ Mt) would cause a global disaster, and calculations by Raup (1990) support the idea that the impact of a ~ 5 km diameter asteroid would be sufficient to cause extinction pulses clearly recognizable above the background rate. Large impacts of these magnitudes are predicted to result in major global environmental disasters primarily related to production of dense clouds of fine ejecta, and surface fires from intense heating caused by re-entry of ejecta into the earth's atmosphere (see Morrison, 1992; Chapman and Morrison, 1994).

In a more detailed analysis, Toon et al. (1994) recently calculated that an impact releasing between 10^7 and 10^8 Mt would be sufficient to produce a dust cloud of very large optical depth covering the entire planet. A global cloud of fine debris could reduce global atmospheric transmission below the limit of photosynthesis for several months (Gerstl and Zardecki 1982). Model calculations predict that land-surface temperatures could decrease by $\sim 15°C$ in less than a week under such conditions (Toon et al., 1982; Pollack et al., 1983; Covey et al., 1990).

Toon et al. (1994) also find that the threshold impact energy required to cause global-scale wildfires through intense heating from ballistic ejecta re-entering the atmosphere was $\geq 10^8$ Mt. The heat emitted globally by such ejecta is capable of igniting combustible material (Melosh et al., 1990), and large amounts of soot have been discovered at the K/T boundary suggesting combustion of a significant fraction of the terrestrial biomass (Anders et al. 1986, Wolbach et al., 1988).

Other disastrous environmental effects of large impacts that have been proposed include enhanced greenhouse effect from water vapor injected into the atmosphere by an oceanic impact (Croft, 1982; O'Keefe and Ahrens, 1982), or from CO_2 released by impact into a carbonate-rich terrane (O'Keefe and Ahrens, 1989), and the possible creation of large amounts of H_2SO_4 aerosols, and thus a marked cooling, from sulfur in the impactor, and if the target rocks contained deposits of $CaSO_4$ (Sigurdsson et al., 1992). Atmospheric shock waves from a large impact are calculated to create large amounts of NO, possibly producing nitric acid rain with a pH of ~ 1, and NO_x in the stratosphere that could rapidly remove the ozone layer.

Astronomical observations of earth-crossing asteroids and comets, and the cratering records of the inner planets, are consistent with a waiting time of $\geq 10^8$ yr for asteroids and comet impacts of 10^8 Mt, and perhaps 2 to 3×10^7 yr for objects producing explosions releasing a few times 10^7 Mt of energy (Fig. 2) (e.g., Shoemaker et al., 1990; Chapman and Morrison, 1994). The observations would thus predict 5 or 6 major mass extinctions (related to $\geq 10^8$ Mt impacts), and about 20 to 30 less severe events (related to $\sim 10^7 - 10^8$ Mt events) during the Phanerozoic Eon (the last 540 Myr), as simulated by Raup (1990). This agrees with the Phanerozoic record of extinctions that can be interpreted as showing 5 major and ~ 20 minor extinction pulses (Fig. 1). The results suggest that the record of extinction events could be explained by

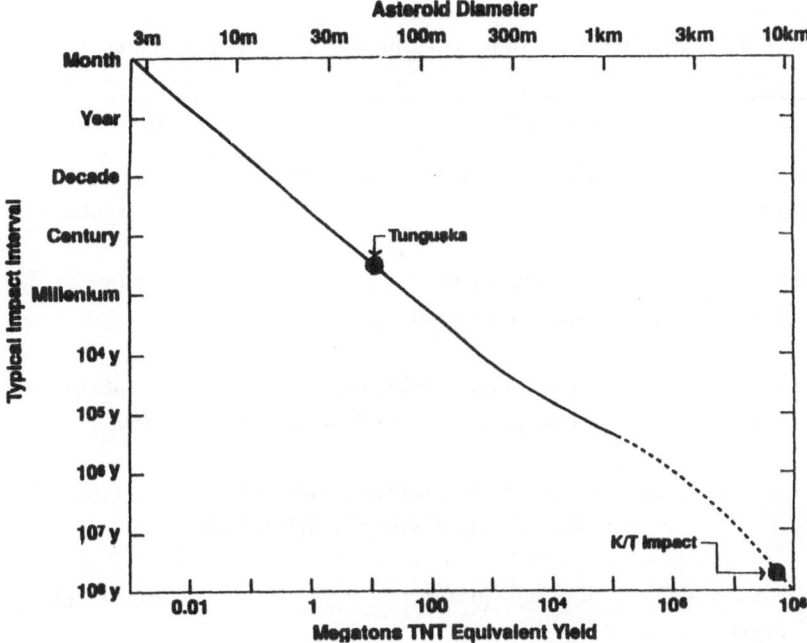

Fig. 2. Estimated frequency curve for impacts on the earth, from astronomical data. The line is a best estimate for the average interval between impacts ≥ the indicated energy yield. Equivalent asteroid diameters are shown (assuming an impact speed of 20 $km\,s^{-1}$ and density of 3 $g\,cm^{-3}$) (after Chapman and Morrison, 1994).

comet and asteroid impacts, as has been suggested by several workers (e.g., Urey, 1973; Alvarez et al., 1980; McLaren and Goodfellow, 1990; Raup, 1990, 1991b; Rampino and Haggerty, 1994). This working hypothesis is testable, in principle, by examining the stratigraphic evidence for impacts at, or near, geologic boundaries that involved mass extinction.

3. Geologic Evidence of Large Impacts

The discovery of the K/T impact layer prompted the search for impact signatures at other geological boundaries (Kyte, 1988; Orth, 1989; Orth et al., 1990). Among the materials considered most diagnostic of impact are shocked mineral grains from the target area (including shocked quartz, stishovite, zircons, etc.), impact glass (microtektites/tektites), microspherules with structures indicating high-temperature origin, and Ni-rich spinels (see Glen, 1994 for a review). These materials have been reported in stratigraphic horizons, ranging from regionally to globally, at, or near, six recorded extinction events in Figure 1, as shown in Table I.

TABLE I

Direct Evidence for Impacts at/or near Extinction Boundaries

Age	Evidence	References
Pliocene (2.3 Myr)	Microcrystites, microtektites?	Kyte, 1988; Margolis et al., 1991
Miocene (~ 15 Myr)	Glass spherules (volcanic? impact?)	Vallier et al., 1978
Late Eocene (~ 36 Myr)	Microtektites, tektites, microspherules, shocked quartz	Montanari et al., 1993
Cretaceous/Tertiary (65 Myr)	Microtektites, tektites, shocked minerals, stishovite, Ni-rich spinels	see Glen, 1994
Triassic/Jurassic (~ 205 Myr)	Shocked quartz	Bice et al., 1992
Frasnian/Famennian (~ 367 Myr)	Microtektites	Wang, 1992; Claeys et al., 1992, 1994

The initial report of evidence of impact was the discovery of enrichments of iridium and other trace metals at the K/T boundary (Alvarez et al., 1980). An iridium anomaly in the parts per billion to tens of parts per billion range is globally well documented at the K/T boundary, and in the search for impact-related iridium anomalies, the K/T boundary has been used as the standard of reference. Criteria for impact signature have thus consisted of high (ppb) levels of iridium, and chondritic or near-chondritic element ratios. Kyte (1988), however, has argued that the relatively high Ir concentrations at the K/T boundary anomaly should not be considered typical of stratigraphic iridium anomalies expected from impacts of comets and asteroids.

The search for iridium has resulted in reports of elevated iridium levels (commonly ≥ 10 times background values) at or close to a number of geologic boundaries associated with mass extinctions (Table II). However, the Ir levels are generally significantly weaker (100s of ppt) than the typical K/T iridium anomaly, and the anomalies are commonly associated with non-chondritic element abundance patterns (Kyte, 1988; Orth et al., 1990; Wang et al., 1991; Wang et al., 1993a,b). This has led to the general conclusion that the Ir peaks are probably unrelated to impact processes (Orth et al., 1990), although Kyte (1988) listed six such sedimentary horizons as "possible" impact horizons (Cenomanian/Turonian [∼ 92 Myr], Callovian/Oxfordian [∼163 Myr], Early/Middle Jurassic [∼ 190 Myr], Permian/Triassic [∼ 250 Myr], Frasnian/Famennian [∼ 367 Myr], and Proterozoic/Cambrian [∼ 540 Myr], see Figure 1).

The meaning of these Ir enrichments is still a subject of debate. Although a number of workers have proposed ways in which Ir could be concentrated at stratigraphic layers by various biological and abiological processes (e.g., Rampino, 1982; Colodner et al., 1992), geological studies suggest that elevated iridium levels may be uncommon in the rock record away from recognized boundaries (Kyte and Wasson, 1986; Alvarez et al., 1990). It should also be noted that times of increased input of extraterrestrial Ir to the oceans (vaporized or dissolved meteoritic material) might also lead to enhanced secondary precipitation in sediments (Kyte, 1988).

If the elevated, but sub-ppb, Ir levels are related to impacts, then other impact debris should be present at the same stratigraphic levels. In this context, it is significant that some biostratigraphic boundaries and transitions marked by relatively small Ir anomalies are now known to be associated with impact-related material -- the Late Eocene interval (∼ 36 − 34 Myr, microtektites, microspherules, shocked quartz), Triassic/Jurassic (∼ 205 Myr, shocked quartz), and Late Devonian (∼ 367 Myr, microtektites). Furthermore, some microtektite layers have no detectable iridium and vice versa, and the widespread Australasian microtektites (0.76 Myr), clearly of impact origin, coincide with an iridium anomaly of only ∼ 160 ppt (Koeberl, 1993).

Moreover, there are several reasons why impact-related Ir anomalies in the geologic record might show lower values than seen at the K/T boundary, and/or non-chondritic element abundance patterns. For one, meteorites differ in iridium content from $> 10^3$ ppb (some irons) to $\sim 10^{-2}$ ppb (eucrites and achondrites), and terrestrial impact melts range from > 30 ppb to background (0.01 ppb), thus impacts of different kinds of extraterrestrial material may produce distributed iridium anomalies of various concentrations (Palme, 1982).

Impacts of comets should produce significantly lower amounts of iridium than asteroid impacts. Because of the rms collision velocity of comets is almost 3 times greater than that of asteroids (~ 60 km s^{-1} versus ~ 20 km s^{-1}), the amount of iridium produced per Joule of impact energy (or for a crater of a given size), for a comet would be only $\sim 1/9$ that of an asteroid. Taking an extreme range, the velocity of a head-on collision with a long-period comet is ~ 6.6 times that of collision with the slowest asteroids, thus the iridium yield per J of such a comet could be $\sim 1/50$ that of an asteroid.

TABLE II

Elevated Iridium Reported at Extinction Boundaries (questionable results marked by ?)

Locality	Ir (ppt)	References
	Pliocene (~ 2.3 Myr)	
Core EL13-3, southeast Pacific	~ 5,000	Kyte, 1988
	Lower/Middle Miocene (~ 11 – 15 Myr)	
DSDP Site 588B, South Pacific	152	Orth, 1989
	Late Eocene (~ 36 Myr)	
Many localities	Up to ~ 4,000	Montanari et al., 1993
	Cretaceous/Tertiary (65 Myr)	
Worldwide	1,000s ppt	see Alvarez et al., 1980; Glen, 1994
	Cenomanian/Turonian (~ 91 Myr)	
Western US, Colombia, S.A., Western Europe	≤560	Orth 1989; Orth et al., 1993
	Jurassic/Cretaceous? (~ 140 Myr)	
Central Siberia	ave. 7,800	Zhakarov et al., 1993
	Callovian/Oxfordian (~ 163 Myr)	
Spain and Poland	1,000-2,400?	Orth, 1989
	Early/Middle Jurassic (~ 190 Myr)	
Italy	~ 3,000?	Rocchia et al., 1986
	Triassic/Jurassic (~ 205 Myr)	
Europe	≤400	McLaren & Goodfellow, 1990
	Permian/Triassic (~ 250 Myr)	
Changxing, China	8,000?	Orth, 1989
Meishan, China	600±400?	Xu et al., 1989
Meishan, China	2,000?	Xu & Yan, 1993
Guangyuan, China	2,480?	Orth 1989; Xu et al., 1989
Nammal, Pakistan	366	Xu et al., 1989
Bolzano, Italy	230	Xu et al., 1989
San Antonio, Italy	3,000?	Brandner, 1988
Tesero, Italy	135	Oddone & Vannucci, 1988
Casera Federata, Italy	100-145	–
Butterloch, Austria	90-95	–
Carnic Alps, Austria	165	Holser et al., 1991
	230	–
Lalung, India	73	Bhandari et al., 1992
	114	–
	Devonian/Carboniferous (~ 360 Myr)	
Texas	380	Orth, 1989; Wang et al. 1993a
	Frasnian/Famennian (~ 367 Myr)	
China	230	Wang et al., 1991
Australia	300	Orth, 1989
Europe	75-160	Orth, 1989
	Ordovician/Silurian (~ 438 Myr)	
Anticosti Island, Canada	58	Orth, 1989
Scotland	≤250	Orth, 1989
China	≤230	Wang et al., 1992, 1993b,c
	Proterozoic/Cambrian (~ 540 Myr)	
China	2,900	Orth, 1989, Xu et al., 1989

Furthermore, the ice-rich composition of comets (≥50% ice?) also means that their Ir content is probably much less than that of asteroids, and could lead to actual comet yields of Ir per J less than 1/100 that of asteroids (Van

Den Bergh, 1994). As an added factor, Vickery and Melosh (1990) calculated that a very energetic impact might send most of its iridium-enriched ejecta into space.

4. Geologic Evidence of Biological Catastrophes

As there is now considerable evidence for an impact at the K/T boundary, McLaren and Goodfellow (1990) argued that other impact-related extinction boundaries might be expected to show a number of geologic features in common with the K/T. These include: (1) a globally synchronous marked negative shift in $\delta^{13}C$ in marine carbonate sediments, indicating a biomass loss and drop in productivity (the Strangelove ocean), coupled with a die-off of marine plankton and proliferation of opportunistic species, followed by recovery and radiation of surviving biota, (2) a marked negative shift in $\delta^{18}O$ in marine sediments, suggesting a brief global warming (e.g., induced by increased greenhouse gases), (3) a positive shift in $\delta^{34}S$ in sediments, suggesting ocean waters with a low dissolved oxygen content, and (4) a reduction in marine biogenic $CaCO_3$ (e.g., Smit, 1990).

Published studies suggest such a common sequence at extinction horizons in the Late Triassic (~ 205 Myr), Late Permian (~ 250 Myr), Frasnian/Famennian (~ 367 Myr), Late Ordovician (~ 438 Myr), and possibly Proterozoic/Cambrian (~ 540 Myr) events (see reviews in Magaritz, 1989; McLaren and Goodfellow, 1990; Rampino and Haggerty, 1994). A number of events show evidence for a sequence of ecological destruction and subsequent recovery that suggests an abrupt and severe ecological disaster (see Rampino and Haggerty, 1994).

5. Extinctions and Evidence of Impacts: Correlations

Of the 24 Phanerozoic extinction peaks in the genus-level data of Sepkoski (1992), six seem to be associated with significant stratigraphic evidence of impacts (large (>ppb) Ir anomalies, shocked quartz, and/or microtektites) and/or large craters (Table I) (Rampino and Haggerty, 1994). At least six more are associated with "possible" evidence of impact consisting of lower (sub-ppb) concentrations of Ir, but still anomalous with respect to background values (Table II). Several other stratigraphic boundaries and faunal transitions (e.g., the Jurassic/Cretaceous [~ 140 Myr], Callovian/Oxfordian [~ 163 Myr] , Devonian/Carboniferous [~ 360 Myr]) may show significant iridium enrichment, but data are scarce (Table II).

Non-productive searches for iridium anomalies and shocked minerals in the geologic record away from boundaries suggest that these features may be uncommon in the geologic record (Kyte and Wasson, 1986; Alvarez et al., 1990; Schmitz et al., 1994). If this is the case, then a simple test of the significance of this correlation might be to consider the chances of hitting a target time series consisting of the 24 extinction events (age error bars taken as the difference in ages given by two recent geological time scales; Palmer, 1983; Harland et al., 1990) with a time series consisting of the six times of diagnostic evidence of impacts. The chance of such a correlation calculated in this way is very low ($\sim 10^{-7}$). The accidental correlation of all 12 "diagnostic" and "possible" indicators of impact with the 24 mass extinctions is of even lower probability. Thus, given the assumption that the two time series are related, the correlation of mass-extinction events and evidence of large impacts is extremely unlikely to be an accidental occurrence.

6. Impact Size and Severity of Extinctions

Natural phenomena (e.g., earthquakes, volcanic eruptions, floods) commonly show a variation in intensity and frequency in which lower intensity events are more common than high intensity events. One can represent this variation by noting the average time interval between events of a given magnitude. Extinctions of life may be viewed this way, and Raup (1990, 1991a,b) constructed a "species kill curve" using extinction data for the last 600 Myr by determining the mean waiting times between extinctions of various severity. If the impact of a large asteroid or comet can produce extinctions, then a relationship should exist between the impactor size and the number of species killed. Raup combined the extinction data with information on the waiting times of terrestrial impact craters of various sizes to construct a possible theoretical relationship between mass extinctions and impact cratering (Fig. 3).

Using the known rate of impacts, and the hypothesized relationship between impacts and mass extinctions, Raup (1990) performed Monte Carlo simulations to determine the appearance of the extinction record supposing that impacts are sufficient to explain the entire record of the past 250 Myr. The results of these simulations compared well with the actual extinction record based on several criteria, including the severity of extinction events, and the general level of background extinction.

Although we realize that severity of mass extinctions is probably related to a number of variables (e.g. ambient climate conditions, susceptibility of fauna and flora, site of impact, etc.), size of impactor may be a first-order cause. The theoretical curve of Raup can be compared with data representing specific large (> 80 km) known impact craters with relatively well-defined ages (Grieve, 1991) that overlap the ages of mass-extinction boundaries (when full dating uncertainties in both are taken into consideration)

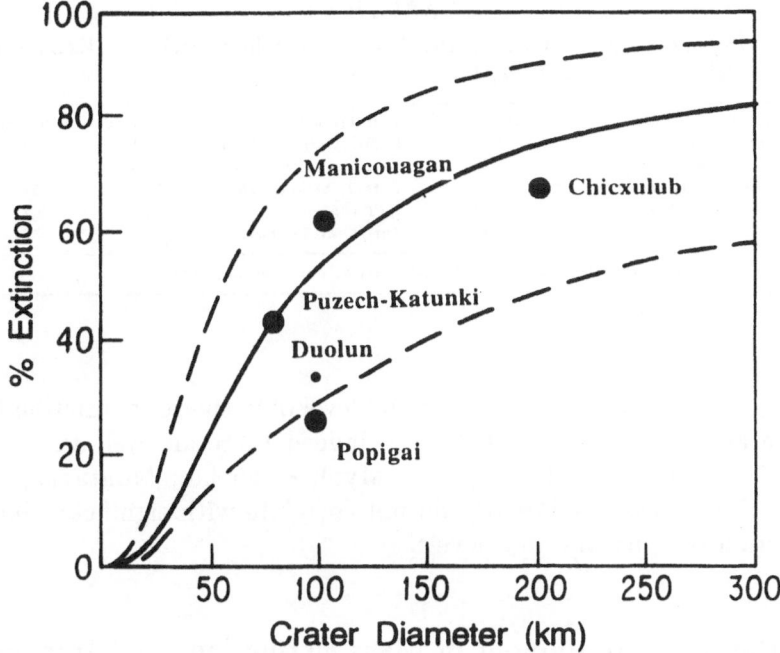

Fig. 3. Kill curve for Phanerozoic marine species (with estimated reasonable error enve-lope) determined from waiting times of large impacts and extinction events (see text and Raup, 1990, 1991a,b). We have plotted the sizes of the largest individual craters with dates overlapping mass extinctions (Table III) against the species kill magnitude of those extinctions (based on culled data). Data points suggest that the "true" kill curve has a step at impacts producing craters ≥100 km diameter (~ 107 Mt events) (see text).

(Table III). No other large, well-dated Phanerozoic craters are known to exist. The Duolun structure in North China is a possible impact structure ~ 100 km in diameter, with an estimated age of ~ 136 Myr (Wu, 1987), which may have occurred near the Jurassic/Cretaceous boundary (~ 140 Myr) (~ 33% extinction of marine species). The observed points agree with the predicted curve within the broad envelope of error permitted by the geologic data, supporting at least a first-order relationship.

The points could be interpreted, however, as indicating a possible step-up in the kill curve at crater sizes of ≥100 km diameter (about 10^7 Mt of energy release), suggesting some kind of a threshold effect. This agrees with the calculations of Toon et al. (1994) showing that the approximate thresholds of dense global dust clouds, and the effect of global-scale surface incineration by heat from re-entering ejecta, occur in the 10^7 to 10^8 Mt range.

The shape of the species kill curves predicts that for craters smaller than ~ 60 − 80 km in diameter there will be no associated extinction pulse that

TABLE III

Large Impact Craters (Grieve, 1991) and Possible Correlative Extinctions (culled data of Sepkoski, 1992, 1994)

Name (Age in Myr)	~Diam. (km)	Extinction (Age)	% Genera	% Species
Puzech-Katunki (220±10)	80	Ladin.-Carn. (~ 230)	20.1	43
Popigai (35±5)*	100	Late Eocene (~ 36)	11.4	26
Manicouagan (212±2)	100	Late Triassic (~ 205)	34.1	62
Chicxulub (65.2±0.4)	200?	K/T (65)	38.5	67
Duolun (~ 136?)	100	Jur./Cret. (~ 140)	12.5	33

* Latest work indicates date of 36 ± 1 Myr (J. Garvin, pers. comm., 1993)

stands above the ~ 20−25% background level of % species extinction (Raup, 1990; see also Jansa et al., 1990), and indeed there are well-dated craters in this size range (e.g., Kara (~ 73 Myr), ~ 65 km; Montagnais (~ 51 Myr), ~ 45 km) that apparently do not correlate with significant increases in extinction over background levels.

7. Periodic Component in Mass Extinctions and Impacts?

The ~ 24 pulses of extinction at the genus level (using the metric of % extinction) that have occurred during the Phanerozoic (approximately the last 540 Myr) give a mean occurrence rate of one every ~ 23 Myr (Figure 1). Raup and Sepkoski (1984) originally identified 12 extinction events at the family level by geologic stages over the last 250 Myr, and reported a statistically significant 26 Myr periodicity in the extinction time series. Later studies found a similar periodicity in extinctions at the genus level (Raup and Sepkoski, 1986). Rampino and Stothers (1986) reported a 29±1 Myr periodic component in the record of vertebrate extinctions. Periods of ~ 26 to 31 Myr have been derived using various subsets of extinction events (family and genus levels), different geologic time scales, and various methods of time-series analysis (e.g., Quinn, 1987; Connor, 1986; Stothers, 1989; Sepkoski, 1989).

The regularity, statistical significance, and reality of the dominant under-lying periodicity have been subjects of intense debate (e.g., Stigler and Wag-ner, 1987, and reply by Raup and Sepkoski, 1988). It is important to note that the detection of a periodic component in the record of marine mass extinctions does not imply either that the record contains a strict periodici-ty, or that every extinction event follow a 26-Myr timescale. The extinction record might be a mixture of periodic and random events.

Part of the problem in acceptance of periodicity as a real manifestation of the geologic record has been the shortness of the record, and the reported

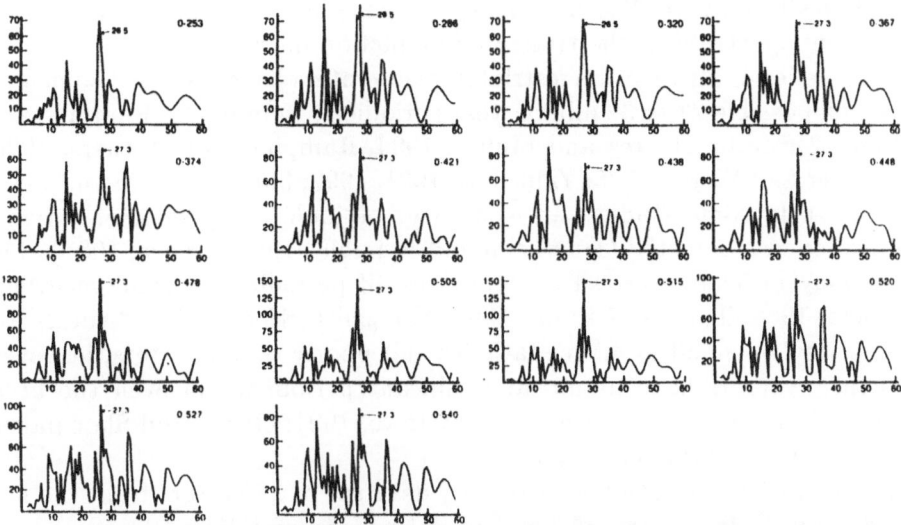

Fig. 4.   Fourier power spectrum for extinction events from Fig. 1 (0-540 Myr), computed as described in Rampino and Caldeira (1992, 1993). X-axis = period in Myr, y-axis = spectral power. The time series were truncated, one extinction event at a time, from 540 Myr to 253 Myr. The highest peak remained in the range 26.5 to 27.3 Myr in all but one of the truncations, where the 27.3 Myr peak was the second highest. The results for the 0-253 Myr sequence gives 26.5 Myr period, similar to that obtained in previous studies (Raup and Sepkoski, 1984, 1986; Rampino and Caldeira, 1992, 1993).

apparent lack of evidence of a statistically significant periodicity in extinction events prior to 250 Myr ago. We performed Fourier analysis (using methods described in Rampino and Caldeira, 1993) on the extended record of 22 extinctions going back $\sim$ 540 Myr, more than doubling the length of the series (culled genera-level data from Sepkoski, 1992, and pers, comm., dated using Palmer, 1983 time scale). For the entire record, we find that the highest peak in the Fourier spectrum is at 27.3 Myr (Fig. 4).

As a test of the significance of the 27.3 Myr peak, we used the 0 to 515 Myr record, and generated 1,000 pseudo-data sets, each containing the same number of randomly dated pseudo-events over the same time interval. Based on this analysis, for example, the probability of generating higher spectral power at 27.3 Myr in the data set is $\sim$ 2%, but the probability of generating higher spectral power at any period between 10 and 65 Myr falls below the 95% confidence level (Rampino and Haggerty, 1994).

As another measure of significance, however, we tested the robustness of the $\sim$ 27 Myr peak, by performing Fourier analysis on the series of truncated extinction time series starting from 0 to 540 Myr, and subtracting one extinction at a time back to 250 Myr (Fig. 4). The robustness of the $\sim$ 27 Myr periodicity in the extinction time series is attested to by the fact that a stable peak between 26.5 and 27.3 Myr remains the dominant feature of the Fourier spectrum in the truncated extinction data sets.

Time-series analyses of terrestrial impact craters have provided some evidence of a possible 28 to 32 million year periodicity in impacts (Rampino and Stothers, 1984a,b; Alvarez and Muller, 1984, Rampino and Stothers, 1986; Shoemaker and Wolfe, 1986; Yabushita 1991, 1992, but see Grieve and Pesonen, 1995). In some studies, an ~ 36 Myr periodic component of variable significance was detected (Durrheim and Reimold, 1987; Grieve, 1991). The record may be characterized as a mixture of periodic and random events (Stothers, 1988; Trefil and Raup, 1987; Yabushita, 1992). These studies are still plagued by small number statistics and poorly dated craters, however, and other workers have found no significant periodicity in even the best-dated craters (e.g., Grieve et al., 1986; Grieve, 1991; Grieve and Shoemaker, 1994; Grieve and Pesonen, 1995).

In the most recent studies, a revised data set of 32 craters from the last 250 Myr (with diameters $\geq$5 km and dating uncertainties $\leq$20 Myr) have been used in time series analysis (see Grieve and Pesonen, 1995). Using a linear time-series analysis technique, Grieve and Shoemaker (1994) found the most significant peak in their time-series analysis of these data at 30 Myr, but the peak failed to meet their rather stringent criterion (3 standard deviations from the mean) by a small margin, and its significance was questioned. We were impressed, however, that the highest peak in the spectrum, was again close to 30 Myr, and we performed a standard Fourier analysis on the newest list of craters. Our results also show the highest peak at 30 Myr, but again the significance level is marginal; clearly more and better crater data are required.

On the other hand, a simple plot of crater frequency with time suggests showers of objects at ~ 0, 35, 65, and 95 Myr ago (Hut et al., 1987), and Stothers (1993) found a significant correlation between impact crater ages and geologic boundaries. The fact that many of the correlative craters seem to be too small (according to the kill curves discussed above) to have produced the associated faunal turnover events is at least consistent with the idea that impacts come in clusters of large and small objects, with many crater still undiscovered (Stothers, 1993).

8. How Well Can the Periodicity in Extinctions and Impact Craters be Determined?

Several investigators have argued that the differences in the periodic components detected in mass extinctions and impact craters may preclude a direct relationship (e.g., Yabushita, 1992, 1994). However, in comparing the periodicities obtained for mass extinctions and impact cratering (and other geological events such as geomagnetic reversals, tectonism, etc., see Rampino and Stothers, 1986; Rampino and Caldeira, 1992, 1993), it is necessary to determine the range of values expected from the data sets in question. This depends on uncertainties in the magnitude of extinction events, completeness of the record of events, uncertainties in radiometric and other dating techniques, differences in geological timescales used, and the ratio of periodic to non-periodic components in the data sets analyzed (Trefil, 1986 unpublished; Trefil and Raup, 1987; Fogg, 1989; Stothers, 1988, 1989; Heisler and Tremaine, 1989).

It is important to note that several studies have demonstrated that the 26 to 33 Myr periods detected in these data sets are not significantly different, and that the range can be readily explained by a combination of dating errors and biases, and signal-to-noise problems. Working with the mass extinction data for the last ∼ 250 Myr, Trefil (1986, unpublished) found that uncertainties in extinction rates introduce an error of ±2 Myr (2σ), and uncertainties in the geologic timescale lead to errors of ±2.5 Myr (2σ) in the detected periodicity. He concluded that there was no significant difference between the 26 Myr period detected in extinctions and 29 to 30 Myr periodic component in impact cratering. Stothers (1989) found that the use of three recently published geological time scales (with differences in ages of extinctions of ±1% to ±5% in the last 250 Myr) gave extinction periods ranging from 25 to 27 Myr, 25 to 30 Myr, and 24 to 33 Myr.

Using numerical simulations, Raup and Trefil (1987) concluded that the observed cratering record was most likely a combination of an ∼ 29 Myr periodic component and random background impacts comprising ∼ 50 to 66% of the total. The range of periods expected from variations in the ratio of cyclic to background impacts was tested by Fogg (1989), who used a computer simulation of impact bombardment of the earth in which the background flux was overlain by a 26 Myr comet-shower cycle. His time-series analysis of impact-related mass extinctions showed periodicities ranging from 24 to 33 Myr in most runs, with the observed period dependent upon the magnitude of the background flux of impactors; < 40% of the runs showed the "true" introduced 26 Myr periodicity. Thus, the impact record might be consistent with a mixture of a random component and an ∼30 Myr periodic component.

9. Periodic Comet Showers and the Galaxy

If the periodic component of ∼ 26 to 30 Myr in the mass extinction and impact cratering pulses is real, then it may be related to the carrousel-like movement of the Solar System through the Milky Way Galaxy. Increased flux of comets (comet showers) might come gravitational perturbations of

Oort comet cloud during the periodic passage of the Solar System through the central plane of the Galaxy (the half-cycle of the oscillation is estimated to be ∼ 26 to 36 Myr depending on galactic models) (Rampino and Stothers, 1984a; Clube and Napier, 1984), although this scenario has been criticised on various grounds (see Weissman, 1986).

Recent calculations by Matese et al. (1994) suggest that time modulation of the flux of new Jupiter-dominated Oort cloud comets could come from gravitational perturbations of the comet cloud by adiabatically varying galactic tides during the in-and-out of plane oscillation. They find that in a galactic model in which half of the disk matter is compact, the peak-to-trough comet flux variation should be ∼ 5 to 1, with a full width of 9 Myr. According to their model, the phase of the nearest cycle peak is 0.6 Myr in the future. For the parameters chosen, Matese et al. (1994) found that the most recent times of peak comet flux were ∼ 30.7, 64.7, and 98.1 Myr ago, and that the cycle interval varied from 29.5 to 34.2 Myr over a 350 Myr run of the model.

In this case, major events in the history of life (and possibly geophysical changes) on Earth may be tied to the dynamics of the Galaxy (Napier and Clube, 1979; Rampino and Stothers, 1986; Rampino and Caldeira, 1992, 1993). Alternative periodic astrophysical models for the Shiva Hypothesis, involving a companion star to the sun (Whitmire and Jackson, 1984; Davis et al., 1984), or a tenth planet (Whitmire and Matese, 1985), seem less likely (e.g., Weissman, 1986; Vandervoort and Sather, 1993).

10. Conclusions

The observed orbital elements and size-frequency distributions of earth-crossing asteroids and comets predict that impactors greater than a few km in diameter ($> 10^7$ Mt events) should collide with the earth on average every few tens of millions of years, with larger ($\sim 10^8$ Mt) events occurring about once every hundred million years. Calculations suggest that the threshold impact size required to cause a detectable mass extinction lies at a few times 10^7 Mt event (with global distribution of fine dust), whereas major mass extinctions seem to require impacts of $\sim 10^8$ Mt (sufficient to cause global scale wildfires). The predicted post-impact environmental effects are expected to lead to mass mortality and subsequent extinction of a large fraction of extant species. Such large climatic perturbations might also destabilize the climatic system, leading to longer term changes in the environment.

A number of extinction events seem to show a generally common pattern in the geologic record, with sharp negative shifts in carbon isotopes in marine sediments suggesting abrupt mass mortality, sudden crash of ocean plankton communities, destruction of terrestrial plant communities, impoverished post-extinction ecosystems, and proliferation of opportunistic survivor species, all suggestive of severe ecological disturbances, mass mortality, and delayed recovery.

Six of the ~ 24 pulses of marine extinction in the last 540 Myr seem to be associated (although not always precisely correlated) with large impact craters and/or stratigraphic evidence of impacts – layers containing high siderophile trace-element anomalies (especially iridium), and/or shocked minerals, tektites and microtektites. An additional six extinction levels are associated with known layers of elevated iridium (and other trace-metal) concentrations above background that might be related to impacts of Ir-depleted objects, possibly comets or non-chondrite-composition asteroids. Alternatively, sediment mixing and trace-metal fractionation may be affecting impact signatures. We believe that although elements of uncertainty remain in many aspects of the problem, and much more work needs to be done, the principle of parsimony suggests that a general relationship between large impacts and mass extinctions is a reasonable working hypothesis.

A periodic component of ~ 26 to 31 Myr has been reported in mass extinction time series for the past 250 Myr, and we report here a robust period of ~ 26.5 to 28 Myr in the record of extinctions of marine organisms going back 540 Myr. A similar period of ~ 28 to 32 Myr has been reported in some sets of dated impact craters, and tests suggests that, if such a periodic component exists in cratering, and considering the uncertainties in dating of geologic events, the periods detected in extinctions and craters are identical within reasonable error. A "Galactic Carrousel" model in which periodic passages of the solar system through the plane of the Milky Way Galaxy lead to gravitational perturbations of the Oort Comet cloud and resulting comet showers might explain the periodic component in cratering and mass extinctions. If supported by further studies, this could provide a connection between critical events here on Earth and the dynamics of the Milky Way Galaxy.

Acknowledgements

We thank M.E. Bailey, S.V.M. Clube, J. Matese, W.M. Napier, J.J. Sepkoski, Jr., E.M. Shoemaker, R.B. Stothers, M. Valtonen, and S. Yabushita for data, discussions, and criticism, and L. Pesonen for a review. M.R.R. was partially supported by NASA Grant NAGW-1697.

References

Alvarez, L.W., Alvarez, W., Asaro, F., and Michel, H.V.: 1980, *Science* **208**, 1095.

Alvarez, W.: 1986, *Eos*, Trans. Amer. Geophys. Union **67**, 649.

Alvarez, W. and Muller, R.A.: 1984, *Nature* **308**, 718.

Alvarez, W., Asaro, F., and Montanari, A.: 1990, *Science* **250**, 1700.

Anders, E., Wolbach, W.S., and Lewis, R.S.: 1986, *Science* **234**, 261.

Benton, M. J.: 1995, *Science* **268**, 52.

Bhandari, N., Shukla, P.N., and Azmi, R.J.: 1992, *Geophys. Res. Lett.* **19**, 1531.

Bice, D.M., Newton, C.R., McCauley, S., Reiners, P.W., and McRoberts, C.A.: 1992, *Science* **259**, 443.

Brandner, R.: 1988, *Ber. Geo. Bundesanst* **15**, 49.

Campbell, J.: 1988, *The Power of Myth*, New York, Doubleday, 224.

Chapman, C.R. and Morrision, D.: 1994, *Nature* **367**, 33.

Claeys, P., Casier, J-G., and Margolis, S.V.: 1992, *Science* **257**, 1102.

Claeys, P., Kyte, F.T., and Casier, J.-G.: 1994, in *Papers Presented to New Developments Regarding the KT Event and Other Catastrophies in Earth History*, LPI Contribution. No. 825, Lunar and Planetary Inst., Houston, 22.

Clube, S.V.M. and Napier, W.M.: 1984, *Mon. Not. Roy. Astron. Soc.* **208**, 575.

Colodner, D.C., Boyle, E.A., Edmond, J.M. N., and Thomson, J.: 1992, *Nature* **358**, 402.

Connor, E.F.: 1986, in Raup, D.M., and Jablonski, D., (eds.) *Patterns and Processes in the History of Life*, Berlin, Springer-Verlag, 119.

Covey, C., Ghan, S.J., Walton, J.J., and Weissman, P.R.: 1990, *Geol. Soc. Amer. Spec. Pap.* **247**, 263.

Croft, S.K.: 1982, *Geol. Soc. Amer. Spec. Pap.* **190**, 143.

Davis, M., Hut, P., and Muller, R.A.: 1984, *Nature* **308**, 715.

Durrheim, R.J. and Reimold, W.U.: 1987, *Proc. Lunar Planet. Sci. Conf.* **XVIII**, 192.

Fogg, M.J.: 1989, *Icarus* **79**, 382.

Gerstl, S.A. and Zardecki, A.: 1982, *Geol. Soc. Amer. Spec. Pap.* **190**, 201.

Glen, W.: 1994, *The Mass Extinction Debates: How Science Works in a Crisis*, Stanford, Stanford Univ. Press.

Goldsmith, D.: 1986, *Griffith Observer* **(Feb.)**, 2.

Gould, S.J.: 1984, *Natural Hist.* **(Aug.)**, 14.

Grieve, R.A.F.: 1991, *Meteoritics* **26**, 175.

Grieve, R.A.F., and Pesonen, L.: 1995, this volume.

Grieve, R.A.F., Sharpton, V.L., Goodacre, A.K., and Garvin, J.B.: 1986, *Earth Planet. Sci. Lett.* **76**, 1.

Grieve, R.A.F. and Shoemaker, E.M.: 1994, in T. Gehrels (ed.), *Hazards due to Asteroids and Comets*, Tucson, Univ. of Arizona Press, 417.

Harland, W.B., Armstrong, R.L., Cox, A.V., Craig, L.E., Smith, A.G., and Smith, D.G.: 1990, *A Geologic Time Scale*, Cambridge Univ. Press.

Heisler, J. and Tremaine, S.: 1989, *Icarus* **77**, 213.

Holser, W.T., Schonlaub, H.P., Boeckelmann, K., and Magaritz, M.: 1991, in W.T. Holser and H.P. Schonlaub (eds.), *The Permian-Triassic Boundary in the Carnic Alps of Austria (Gartnerkofel Region)*, Abhandlung Geol. Bundesanst. **45**, 213.

Hut, P., Alvarez, W., Elder, W.P., Hansen, T., Kauffman, E.G., Keller, G., Shoemaker, E.M., and Weissman, P.R.: 1987, *Nature* **329**, 118.

Jansa, L.F.: 1993, *Palaeogeogr., Palaeoclimatol., Palaeoecol.* **104**, 271.

Koeberl, C.: 1993, *Earth Planet. Sci. Lett.* **119**, 453.

Kyte, F.T.: 1988, *Paleoceanog.* **3**, 235.

Kyte, F.T. and Wasson, J.T.: 1986, *Science* **232**, 1225.

LaBandiera, C.C. and Sepkoski, J.J., Jr.: 1993, *Science* **261**, 310.

Magaritz, M.: 1989, *Geology* **17**, 337.

Margolis, S.V., Claeys, P., and Kyte, F.T.: 1991, *Science* **251**, 1594.

Matese, J.J., Whitman, P.G., Innanen, K.A., and Valtonen, M.J.: 1994, in *Papers Presented to New Developments Regarding the KT Event and Other Catastrophies in Earth History*, LPI Contribution. No. 825, Houston, Lunar and Planetary Inst., 78.

McLaren, D.J. and Goodfellow, W.D.: 1990, *Ann. Rev. Earth Planet. Sci.* **18**, 123.

Melosh, H.J., Schneider, N.M., Zahnle, K.J., and Latham, D.: 1990, *Nature* **343**, 251.
Montanari, A., Asaro, F., Michel. H.V., and Kennett, J.P.: 1993, *Palaios* **8**, 420.
Morrison, D.: 1992, *The Spaceguard Survey, Report of the NASA International Near-Earth Object Detection Workshop*, Pasadena, NASA, Jet Propulsion Lab.
Napier, W.M. and Clube, S.V.M.: 1979, *Nature* **282**, 455.
O'Keefe, J.D. and Ahrens, T.J.: 1982, *Nature* **298**, 123.
O'Keefe, J.D. and Ahrens, T.J.: 1989, *Nature* **338**, 247.
Oddone, W. and Vannucci, R.: 1988, *Mem. Soc. Geol. Ital.* **34**, 121.
Orth, C.J.: 1989, in S.K. Donovan, S.K. (ed.), *Mass Extinctions: Processes and Evidence*, New York, Columbia Univ. Press, 37.
Orth, C.J., Attrep, M., Jr. and Quintana, L.R.: 1990, *Geol. Soc. Amer. Spec. Pap.* **247**, 45.
Orth, C.J., Attrep, M., Jr., Quintana, L.R., Elder, W.P., Kauffman, E.G., Diner, R., and Villamil, T.: 1993, *Earth Planet. Sci. Lett.* **117**, 189.
Palme, H.: 1982, *Geol. Soc. Amer.Spec. Pap.* **190**, 223.
Palmer, A.R.: 1983, *Geology* **11**, 503.
Pollack, J.B., Toon, O.B., Ackerman, T.P., McKay, C.P., and Turco, R.P.: 1983, *Science* **219**, 287.
Quinn, J.F.: 1987, *Paleobiol.* **13**, 465.
Rampino, M.R.: 1982, *Geol. Soc. Amer. Spec. Paper* **190**, 455.
Rampino, M.R.: 1988, *Eos*, Trans. Amer. Geophys. Union **69**, 889.
Rampino, M.R.: 1990, in S.H. Schneider and P.J. Boston (eds.), *Scientists on Gaia*, Cambridge, MIT Press, 382.
Rampino, M.R. and Caldeira, K.: 1992, *Celest. Mech. Dynam. Astron.* **54**, 143.
Rampino, M.R. and Caldeira, K.: 1993, *Earth Planet. Sci. Lett.* **114**, 215.
Rampino, M.R. and Haggerty, B.M.: 1994, in T. Gehrels (ed.), *Hazards due to Asteroids and Comets*, Tucson, Univ. of Arizona Press, 827.
Rampino, M.R. and Stothers, R.B.: 1984a, *Nature* **308**, 709.
Rampino, M.R. and Stothers, R.B.: 1984b, *Science* **226**, 1427.
Rampino, M.R. and Stothers, R.B.: 1986, in R. Smoluchowski, J.N. Bahcall, and M.S. Matthews (eds.), *The Galaxy and the Solar System*, Tucson, Univ. of Arizona Press, 241.
Raup, D.M.: 1989, *Phil. Trans. Roy. Soc. Lond.* **B325**, 421.
Raup, D.M.: 1990, *Geol. Soc. Amer. Spec. Pap.* **247**, 27.
Raup, D.M.: 1991a, *Paleobiol.* **17**, 37.
Raup, D.M.: 1991b, *Extinction: Bad Genes or Bad Luck?* New York, Norton.
Raup, D.M. and Sepkoski, J.J., Jr.: 1984, *Proc. Nat. Acad. Sci. USA* **81**, 801.
Raup, D.M. and Sepkoski, J.J., Jr.: 1986, *Science* **231**, 833.
Raup, D.M. and Sepkoski, J.J., Jr.: 1988, *Science* **241**, 94.
Roccia, R., Boclet, D., Bonte, P., Castellarin, A., and Jehanno, C.: 1986, *Jour. Geophys. Res.* **91**, E259.
Schmitz, B., Jeppsson, L., and Ekvall, J.: 1994, *Geol. Mag* **131**, 361.
Sepkoski, J.J., Jr.: 1982, *Milw. Publ. Mus. Contrib. Biol. Geol.* **51**.
Sepkoski, J.J., Jr.: 1989, *Jour. Geol. Soc. Lond.* **146**, 7.
Sepkoski, J.J., Jr.: 1992, *Milw. Publ. Mus. Contrib. Biol. Geol.* **83**.
Sepkoski, J.J., Jr: 1994, *Geotimes* **March**, 15.
Shoemaker, E.M. and Grieve, R.A.F.: 1994, in T. Gehrels (ed.), *Hazards due to Asteroids and Comets*, Tucson, Univ. of Arizona Press, 417.
Shoemaker, E.M., and Wolfe, R.F.: 1986, in R. Smoluchowski, J.N. Bahcall, and M.S. Matthews (eds.), *The Galaxy and the Solar System*, Tucson, Univ. of Arizona Press, 338.
Shoemaker, E.M., Wolfe, R.F., and Shoemaker, C.S.: 1990, *Geol. Soc. Amer. Spec. Pap.* **247**, 155.
Sigurdsson, H., D'Hondt, S., and Carey, S.: 1992, *Earth Planet. Sci. Lett.* **109**, 543.
Smit, J.: 1990, *Geol. en Mijnbouw* **69**, 187.

Stigler, S.M. and Wagner, M.J.: 1987, *Science* **238**, 940.
Stothers, R.B.: 1988, *Observatory* **108**, 1.
Stothers, R.B.: 1989, *Geophys. Res. Lett.* **16**, 119.
Stothers, R.B.: 1993, *Geophys. Res. Lett.* **20**, 887.
Toon, O.B., Pollack, J.B., Ackerman, T.P., Turco, R.P., McKay, C.P., and Liu, M.S.: 1982, *Geol. Soc. Amer. Spec. Pap.* **190**, 187.
Toon, O.B., Zahnle, K., Turco, R.P., and Covey, C., 1994, in T. Gehrels (ed.), *Hazards due to Asteroids and Comets*, Tucson, Univ. of Arizona Press, 791.
Trefil J. S. and Raup, D.M.: 1987, *Earth Planet. Sci. Lett.* **82**, 159.
Trefil, J.S.: 1986, unpublished ms.
Urey, H.C.: 1973, *Nature* **242**, 32-33.
Vallier, T.L., Bohrer, D., Moreland, G., and McKee, E.H.: 1977, *Geol. Soc. Amer. Bull.* **88**, 787.
Van den Bergh, S., 1994, *Proc. Astron. Soc. Pacific* **106**, 689.
Vandervoort, P.O. and Sather, E.A.: 1993, *Icarus* **105**, 26.
Vickery, A.M., and Melosh, H.J.: 1990, *Geol. Soc. Amer. Spec. Pap.* **247**, 289.
Wang, K.: 1992, *Science* **256**, 1547.
Wang, K., Orth, C.J., Attrep, M., Jr., Chatterton, B.D.E., Hou, H., and Geldsetzer, H.H.J.: 1991, *Geology* **19**, 776.
Wang, K., Chatterton, B.D.E., Attrep, M., Jr., and Orth, C.J.: 1992, *Geology* **20**, 39.
Wang, K., Attrep, M., Jr., Orth, C.J.: 1993a, *Geology* **21**, 1071.
Wang, K., Chatterton, B.D.E., Attrep, M., Jr., and Orth, C.J.: 1993b, *Can. Jour. Earth Sci.* **30**, 1870.
Wang, K., Orth, C.J., Attrep, M., Jr., Chatterton, B.D.E., Wang, X., and Li, J-j.: 1993c, *Palaeogeogr., Palaeoclimatol., Palaeoecol.* **104**, 61.
Weissman, P.R., 1986: in R. Smoluchowski, J.N. Bahcall, and M.S. Matthews (eds.), *The Galaxy and the Solar System*, Tucson, Univ. of Arizona Press, 204.
Whitmire, D. P. and Jackson, A.A., Jr.: 1984, *Nature* **308**, 713-715.
Whitmire, D.P. and Matese, J.J.: 1984, *Nature* **313**, 36.
Wolbach, W.S., Gilmour, I., Anders, E., Orth, C.J., and Brooks, R.R.: 1988, *Nature* **334**, 670.
Wu, S.: 1987, *Abstr. Int. Geol. Correl. Proj. 199 Meeting, Mar. 3-4, 1987*, Beijing, China.
Xu, D., and Yan, Z.: 1993, *Palaeoclimatol., Paleogeogr., Palaeoecol.* **104**, 171.
Xu, D., Yan, Z., Yi-Yin, S., Jin-Wen, H., Qin-Wen, Z., and Zhi-Fang, C.: 1989, *Astrogeological Events in China*, New York, Van Nostrand Reinhold.
Yabushita, S.: 1991, *Mon. Not. Roy. Astr. Soc.* **250**, 481.
Yabushita, S.: 1992, in S.V.M. Clube, S. Yabushita, and J. Henrard (eds.), *Dynamics and Evolution of Minor Bodies with Galactic and Geological Implications*, Dordrecht, Kluwer, 161.
Zhakarov, V.A., Lapukhov, A.S., and Shenfil, O.V.: 1993, *Russian Jour. Geology and Geophysics* **34**, 83.

A JUPITER FRAGMENTED COMET: CAUSE OF THE K/T BOUNDARY RECORD

N.C. WICKRAMASINGHE and MAX K. WALLIS

School of Mathematics, University of Wales College of Cardiff.

Abstract. The extended period of mass extinctions around the K/T boundary correlating with extraterrestrial amino acids in the sediment record constitutes strong evidence of a cometary cause. The input of extraterrestrial matter over 10^5 yr supports the hypothesis of a giant comet, fragmented into subcomets on close encounter with Jupiter, and subsequently perturbed into Earth-crossing orbits. Copious amounts of dust were emitted via this and possibly successive fragmenting encounters, and via normal cometary evaporation. The dynamics of dust from the disintegrating comet fragments favours retention in Earth-crossing orbits of the sub-micron fraction of organic composition. The shroud of dust accreted in the Earth's upper atmosphere varied with time and imposed climatic stresses that caused species extinctions over 10^5 yr. While the iridium peak in the sediments coincides with the Chicxulub crater impactor, other iridium detail suggests that some of the impactor material was reinjected into space and in part re-accreted by Earth from the interplanetary orbits.

1. Introduction

The speculation that mass-extinctions of biological species, such as occurred at the K/T *(Cretaceous/Tertiary)* boundary 65 Myr ago, have an astronomical cause was published as early as 1978 (Hoyle and Wickramasinghe 1978). They estimated that a total mass of 10^{14} g added to the Earth's upper atmosphere in the form of small particles of high albedo for visual wavelengths would produce an inverse greenhouse effect, shielding ground level from sunlight but permitting infrared radiation from the ground to escape into space. They suggested these micron-sized particles might be acquired by the Earth via a close approach to a cometary nucleus and that ice ages and ecodisasters would ensue. A similar idea was used by Napier and Clube (1979) in a wider study of terrestrial catastrophism.

Subsequently Alvarez et al. (1980) made the remarkable discovery that clays in the K/T boundary layer are greatly enriched in the element iridium, a rare element in the Earth's crust but one that is present in meteorites and presumably in comets. This discovery led to their claim that a large asteroidal impactor caused the extinction of the dinosaurs as well as other plant, animal and microbial species some 65 Myr ago. From this datum alone no distinction could be made between the alternatives of an asteroidal or cometary impactor. The discovery of extraterrestrial amino-acids in the boundary clay (Zhao & Bada 1989) – compounds that could not survive high speed impact, stimulated the idea (Zahnle & Grinspoon 1990) that small cometary particles were arriving at Earth over a lengthy 10^5 yr period.

Earth, Moon, and Planets **72**: 461-466, 1996.
© 1996 *Kluwer Academic Publishers*.

The Iridium peak is in the middle of this period and is now recognized to coincide with the sea-bed Chicxulub crater (van den Bergh 1994) and associated tsunami deposits on the shore. If this was a direct cometary strike, the protracted attenuation and extinction of other genera and species, over 100 000 years or more, could have been induced via climatic perturbations resulting from the accretion of cometary dust. There is controversy over whether significant extinctions occurred prior to the impact, as expected on the dust hypothesis. The K/T transition does seem to have been special over the past 250 Myr. Not all extinctions coincide with known craters over 20 km (within dating uncertainties), several craters over 45 km do not coincide with extinction events, and Ir anomalies have been discovered in few cases (Yabushita 1994). However, it seems likely that impacts were responsible for all the major mass extinction events in the geological record, leaving only the relatively minor extinction episodes to be explained by causations other than impacts.

2. Roche fragmentation of a Giant Comet

The scenario we envisage (Wallis & Wickramasinghe 1994; hereafter $W\&W$) is of a giant 10^{22} g comet (300 km diameter) fragmented via close encounter with Jupiter and perturbed into Earth crossing orbits (e.g. Clube and Napier 1990; Steel 1994) some 60 000 yr prior to the Chicxulub impact. Like comet Shoemaker-Levy 9 (SL9) it became a temporary satellite of Jupiter and fragmented, but into Earth-crossing comets of the Jupiter family, rather than impacting Jupiter. Steel (1992) calculates the probability of an Earth impact by such a comet with low inclination i and $q \simeq 1$ AU as about 2×10^{-7} per perihelion passage, i.e. 10^{-8}/yr.

Whether a giant comet fragments like the ordinary-sized Shoemaker-Levy 9, is a matter of keen interest. Dobrovolskis' (1992) solid body analysis indicates that a low strength body fractures from the surface inwards, but does not disassemble until it passes within the Roche limit. Internal friction is supposed to ensure spherical shape is retained. Asphaug & Benz (1994) compute zero-strength assemblages of equal sized fragments as they tend to reassemble under self-gravity after Roche passage; they find the resulting sub-comet numbers to be very sensitive to density. A 300 km comet might divide into 10-100 times more pieces than the 20 they calculate from a 4 km diameter assemblage. Non-spherical shape, rotation and weak attachment forces could substantially change results. The non-uniformities mean that separation speeds are higher and inter-fragment collisions more important. The core of a 300 km comet would be compacted under self-gravity, so fragmentation might result in a 100-200 km core plus many fragile sub-comets. Pending full study of these complex systems, we hypothesise ($W\&W$) that a multiplicity of loosely-bound subcomets results from the Roche break-up, arguing that the 300 km comet might split into 600-2000 subcomets of size $\approx$ 20 km and a hierarchy in sizes down to smallest grains would emerge (Steel 1994), maybe involving multiple close encounters with Jupiter. On our hypothesis, unlike in the case of SL9, the daughter comets followed a subsequent orbit further from Jupiter, though still under that planet's influence. On later orbits, some were perturbed into orbits further from Jupiter, though a fraction came back to suffer fragmenting or impact encounters and create an abundance of all fragment sizes, from comets of the SL9 size ($\sim$ 10 km) down to the smallest grains. Copious amounts of dust are emitted with the fragmentation and via subsequent evaporation of volatiles binding dust grains together. Successive fragmentation of some of the comets will occur and fill the inner solar system with 1000-fold enhanced cometary debris over an extended period. At such raised densities, inter-grain collisions are expected to smash up grains $> 10\,\mu$m on a timescale of 10^4 yr (Bailey et al. 1994).

3. The Extraterrestrial Input

Although many mass-extinction events are found in the geological record, that at the K/T boundary represents perhaps the most dramatic episode in the last 100 million years. It is also remarkable in being the most sharply concentrated event in temporal terms, with the Stevns Klint revealing detail under 10^4 yr resolution (van den Bergh 1994). The observed extinctions of species are spread across a finite but narrow time interval 65 ± 0.05 Myr BP in stepwise fashion (Hut et al. 1987). A similar temporal width is consistent with the distribution of extraterrestrial amino acids out to a metre above and below the K/T boundary layer (Zhao & Bada 1989). Fig. 1 adapted from Zahnle and Grinspoon (1990) adopts a sediment deposition rate of 1.9 cm/kyr and shows the distribution of AIB (alpha-isobutyric acid), as well as the sharply peaked iridium that coincides with the impact event (shown by soot, tsunami deposits, etc.). A second extraterrestrial amino acid, isovaline, is also detected with the AIB. The survival of organics like AIB requires that they reach the Earth not in large cometary lumps, but as small particles of cometary dust that soft-land through deceleration high in the atmosphere. Further work (Bada and Zhao 1992) establishes that the AIB signature continues through the Ir peak, the low point in Fig. 1 being associated with the narrow layer of non-carbonate clay. The material brought in directly with the impactor would possess an enhanced iridium signal but any organic material that was initially present would have been destroyed in the highly energetic impact.

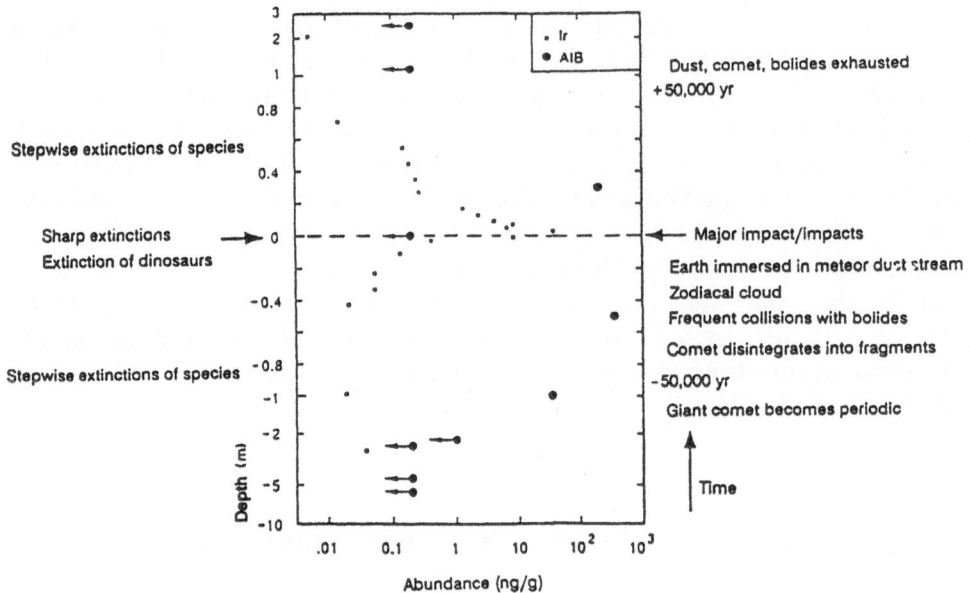

Fig. 1.   Possible interpretation of structures in the K/T boundary (Stevns Klint site)

The main reservation over the Ir and AIB data concerns possible diffusion through the Stevns Klint sediments. The detailed structure of the Ir profile suggests this element diffused little, but the AIB is less clear with sparser data; the low point in the impact-associated non-carbonate clay indicates little diffusion through that layer. Comparative study of data at other K/T sites is required. For AIB alone the surface density deposited through the depth of about 2 metres in the Stevns Klint rocks is 5×10^{-5} g/cm^2 (Zhao and Bada 1989). AIB constitutes perhaps 10^{-4} of cometary dust, implying ($W\&W$) a rate of accumulation of cometary dust of $\sim 10^{-5}$ g/(cm^{2}yr). With a typical residence time in the mesosphere of 2 yr, the average mass loading of the Earth's atmosphere would be 10^{14} g, close to that discussed by Hoyle and Wickramasinghe (1978).

4.   Atmospheric Shroud of Cometary Dust

Micron and submicron-sized grains have been generally ignored on two accounts, first because observational techniques miss them, and second because radiation pressure effects imply they are readily blown out of the solar system. However, the spaceprobes of Halley's comet discovered an abundance of such grains, so we have re-examined ($W\&W$) how radiation pressure affects low density, porous submicron grains. Cometary dust includes a large fraction that is of a complex organic character. Micron-sized cometary grains are in general added to the Earth in a non-destructive manner, slow down at a typical altitude of 100-120 km, and form a dust veil diminishing incident solar radiation. We have calculated the radiative properties of porous grains of *siliceous/organic* composition ($W\&W$) and shown that the 10^{14} g loading induces a temperature drop $\Delta T \simeq -4$ to $-11°C$. Such a drop develops over the apparent rise time of AIB of 10^4 yr (Fig. 1) and would have led to significant climatic change.

Such a dust veil enveloping the Earth fluctuated over the 10^5 year period. A comet like Halley actively evaporates and ejects dust for 100 orbits or more, maybe crusting over and rejuvenating erratically. Submicron particle orbits change over 10^4 yr, so inputs to the Earth would be smoothed over this time interval. There would be short-term peaks on passage of the Earth through the comet trail, but these would average out over the 2 year settling time. The radiative forcing due to dust loading would thus vary primarily on the 10^4 yr timescale, this variation imposing continuing stress on biology throughout the period.

5. Fractionation between Metallic and Organic Material

While cometary organics were destroyed in the impact that produced the Ir peak, how was it that no or little Ir arrived with the cometary AIB? Iron and iridium are products of explosive nucleosynthesis and plausibly condensed in $0.01 - 0.03 \, \mu m$ radius grains that gathered gas condensates and accreted into comets (Hoyle and Wickramasinghe 1970, 1991). Radiation pressure forces, acting with differing efficiencies on different particles, have an important role in separating the various grain types that emerge from comets. Some grains are on unbound orbits and quickly expelled from the solar system. The critical values of β, the ratio of radiation pressure force to gravity, for particle retention depends on the eccentricity of the parent cometary orbit as well as on the position of release (Ishimoto and Mukai 1991). These authors found, counter-intuitively, that porosity can increase the particle β. Our specific calculations ($W\&W$) show that porous organic grains stand a good chance of being retained because of lower β, while iron and graphite particles of the relevant sizes do not satisfy the conditions for retention in bound orbits under any of the three criteria considered.

While the K/T impact coincided with the non-carbonate clay layer and Ir peak, the levels of Ir remained enhanced for several kyr (Fig. 1). The asymmetry in the profile provides evidence that this reflected the real deposition rate rather than an artifact of diffusion in the sediments. If the post-impact tail in Ir was associated with the impactor, one possibility involves dust release from a marginal Roche encounter with Earth, that perturbed the comet and debris into a low-eccentricity orbit, the comet subsequently impacting the Earth. The debris on bound orbits, including iron grains $> 0.3\,\mu$m, will in part be gradually accreted by Earth. But general calculations show that two Earth-encounters are rather improbable ($W\&W$). More probably, the Ir tail arose from material ejected into space from the Chicxulub impact, on the impact theory of Melosh (1988). Ejector speeds reach as much as 85% of the impactor speed, so can readily exceed escape velocity. But much of the escaping debris would still be retained on Earth-crossing orbits that allow gradual re-accretion over 10^4 yr.

References

Alvarez, L.W., Alvarez, W. & Asaro, F.: 1980, *Science* **208**, 1095-1108.

Asphaug, E. & Benz, W.: 1994, this conference; also *Nature* **370**, 120.

Bailey, M.E., Clube, S.V.M., Hahn, G., Napier, W.M & Valsecchi, G.B.: 1994, in *Hazards due to comets and asteroids*, ed. Gehrels, T., U. Ariz. Press.

Bada, J.L. & Zhao, M.: 1992, Symposium F3.1 World Space Congress, Washington, *Adv. Space Res.* in press, 1994.

van den Bergh, S.: 1994, *Pub. Astron. Soc. Pacific* **106**, 689-695

Clube, S.V.M. & Napier, W.M.: 1990, *Cosmic Winter*, Oxford Univ. Press.

Dobrovolskis, A.R.: 1990, *Icarus*, 88, 24-38

Hoyle, F. & Wickramasinghe, C.: 1978, *Astrophys. Space Sci.* **53**, 523-526.

Hoyle, F. & Wickramasinghe, N.C.: 1991, *The Theory of Cosmic Grains*, Kluwer.

Hut, P., Alvarez, W., Elder, W.P., Hausen, T., Kauffman, E.G., Keller, G., Shoemaker, E.M. & Weissman, P.R.: 1987, *Nature* **329**, 118.

Ishimoto, H. & Mukai, T.: 1991, *Proc. 24th ISAS Lunar and Planetary Symp.*, eds. H. Mizutani, H. Oya and M. Shimizu, ISAS Tokyo, p. 148.

Melosh, H. J.: 1988, *Nature* **332**, 687-688.

Napier, W.M. & Clube, S.V.M.: 1979, *Nature* **282**, 455-459.

Steel, D.: 1992, *Origins of Life and Evolution of the Biosphere*, 21, 239-357.

Steel, D.: 1994, this conference.

Wallis, M.K. & Wickramasinghe, N.C.: 1994, *Mon. Not. Roy. Astr. Soc.* **270**, 420-426.

Yabushita, S.: 1994, *Earth Moon Planets* **64**, 207-216.

Zahnle, K. & Grinspoon, D.: 1990, *Nature* **348**, 157-160.

Zhao, M. & Bada, J.L.: 1989, *Nature* **339**, 463-465.

THE TUNGUSKA 1908 EXPLOSION'S REGION AS AN INTERNATIONAL PARK OF STUDIES OF THE ECOLOGICAL CONSEQUENCES OF COLLISIONS OF THE EARTH WITH THE SOLAR SYSTEM SMALL BODIES

G.V. ANDREEV and N.V. VASILYEV

Astronomical Observatory, Box 1106, 634010 Tomsk, Russia; E-mail:
andreev@urania.tomsk.su

Abstract. On the example of studies of the Tunguska 1908 explosion was shown that besides formation of crater and ejection of dust in the atmosphere the many others extraordinary phenomena (including genetic) may be caused by collision the Earth with small asteroid or comet. A Complex Scientific International Program for studies of the Ecological Consequences of a Collision of the Earth with the Solar System Small Bodies is offered by authors taking into account the results of 36th years of investigation of Tunguska Event.

A considerable amount of observational data with respect to Earth-crossing small bodies of the Solar system and the bombardment rate of our planet is available now.

This paper deals with careful studies of all aspects of the ecological consequences caused by collision of the Earth with a small body of the Solar System. It is proposed to use the Tunguska 1908 event for this aim as a recent and powerful one.

On June 30 of 1908 at $00^h 14^m$ UT a catastrophic collision of a small cosmic body with the Earth took place over Central Siberia in the interriver of Podkamennaya and Nizhnyaya Tunguskas (60.9° N; 101.9° E). The Tunguska cosmic body exploded at an altitude of 5–8 km with energy about 20–50 Mtn TNT. This object responsible had a radius of about 40–100 meters. A catastrophic destruction of the region of the fall, climatic changes and the disappearance of species of animals and vegetation caused by changes of conditions of their existence are usually considered as main after-effects of collisional events. On the example of studies of the Tunguska explosion was shown that besides these effects there are also many extraordinary phenomena such as geomagnetic storms,reverse magnetization and thermoluminescent changes of rocks, minerals and soils, isotopic shifts of composition of biotes and abiotes in the catastrophe region and, most unexpectedly, various biological (including genetic) violations of flora, fauna, and man were registered.

These facts have a fundamental meaning since they prove the presence of cosmic factors governing the development of the organic world of the Earth by means of bombardment by small bodies of the Solar System, and evolu-

Earth, Moon, and Planets **72**: 467-468, 1996.

tion, which is caused not only by changes of the environment and climate, but by direct mutations.

Taking into account the above-mentioned material we think that the question of an essential extension and financing investigations of the Tunguska catastrophe of 1908 must be solved and cannot be put off. This requires, first of all, the preservation of the object.of investigation: biocenoses and the landscape of the catastrophe region. A reliable guarantee of safety of the Tunguska region could only be the declaration of it as a cosmic-biospheric reservation, possibly under the aegis of the UNESCO with a corresponding international program of investigation.

The program aim is to study the ecological consequences of the Earth's collisions with the small bodies of the Solar System on the basis of thorough investigations of the Tunguska 1908 Catastrophe; to estimate the probability of such events and to give a prognosis of Earth collisions with known small bodies; to work out the international measures of Earth safety.

The program contains the following complex problems:
1. Elemental and isotope biogeochemistry of the fall region using the example of the Tunguska Meteorite Reserve;
2. Geophysical consequences of the Earth's collisions with the small bodies of the Solar System;
3. Ecological (medical-biological) consequences of collisional events;
4. Prognosis and estimation of the probability of collisions of small bodies of the Solar system with the Earth;
5. Project of international measures of Earth defence against col- lisions with small bodies of the Solar System.

The authors are ready to present a complete text of this proposed program to any institution or to individual scientists.

THE DELIVERY OF ORGANIC MATTER
FROM ASTEROIDS AND COMETS
TO THE EARLY SURFACE OF MARS

GEORGE J. FLYNN

Dept. of Physics, SUNY - Plattsburgh. Plattsburgh, NY 12901 USA

Abstract. Carbon delivered to the Earth by interplanetary dust particles may have been an important source of pre-biotic organic matter (Anders, 1989). Interplanetary dust is shown to deliver an order-of-magnitude higher surface concentration of carbon onto Mars than onto Earth, suggesting interplanetary dust may be an important source of carbon on Mars as well.

1. Introduction

Material from comets and carbonaceous asteroids has been suggested as a major contributor of organic matter to the primitive Earth (Anders and Owen, 1977; Anders, 1989; Chyba, 1987 and Chyba *et al.*, 1990). Chamberlin and Chamberlin (1908) proposed the organic component in planetesimals might have been an important source of pre-biotic organic matter for the early Earth. Oró (1961) suggested a similar role for organic matter from comets. Anders (1989) studied the accretion of objects ranging from dust to large bodies onto the Earth and concluded the major organic contribution is from interplanetary dust small enough to survive atmospheric entry without reaching high temperatures.

2. Accretion rates

Mars has a lower surface gravity than Earth, giving rise to a lower atmospheric entry velocity distribution for interplanetary dust particles at Mars than at Earth. This lower mean entry velocity allows larger particles to decelerate in the atmosphere of Mars without experiencing severe heating. Flynn and McKay (1990) calculate that the fraction of 100 μm diameter particles which survive atmospheric entry at Earth is the same as for 700 μm diameter particles at Mars. This difference is particularly important for the accretion of unmelted meteoritic material onto Mars because the size distribution of interplanetary dust particles is sharply peaked (as shown in Figure 1), with 80% of the continuous, long-term meteoritic mass accreted by the Earth being between 10^{-7} and 10^{-3} grams ($\sim$ 60 to 2700 μm in diameter). Although most particles > 100 μm in diameter melt or vaporize during Earth atmospheric entry, many of these large particles survive atmospheric entry at Mars without melting.

Earth, Moon, and Planets **72**: 469-474, 1996.

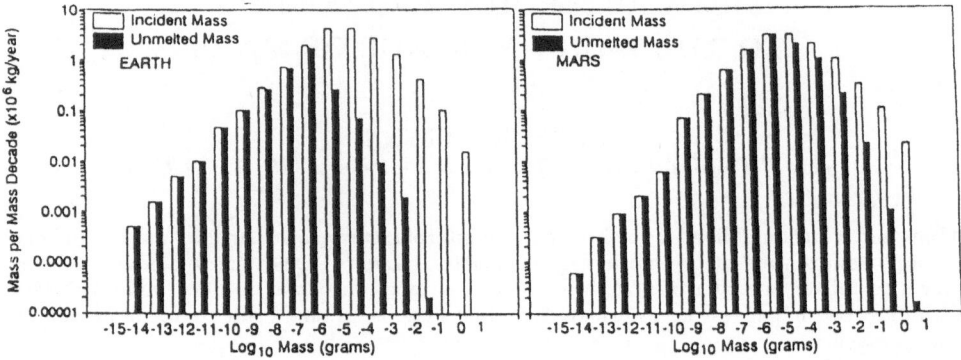

Fig. 1.   The micrometeorite flux measured at Earth (Hughes, 1978) (left) and calculated at Mars (right) in each mass decade are shown along with the mass surviving atmospheric entry without melting or vaporizing.

The present flux of interplanetary dust in each mass decade from 10^1 grams to 10^{-15} grams at Mars, shown in Table I, has been modeled by Flynn and McKay (1990) and Flynn (1991) from satellite, radar and visual meteor measurements of the present flux at Earth (Hughes, 1978) and estimates of the Mars/Earth flux ratio. Flynn and McKay (1990) also calculated the expected velocity distribution of interplanetary dust at Mars using the measured velocity distribution for radar meteors at Earth (Southworth and Sekanina, 1973) and a velocity transformation to Mars, using a method developed by Morgan *et al.* (1988). Using the atmospheric entry heating model developed by Whipple (1950) and extended by Fraundorf (1980), the distribution of temperatures reached by particles in each mass decade during Mars atmospheric entry was calculated. Table I gives the surviving mass fraction, the fraction not heated above 1600K (the melting temperature of the dominant silicate phases in meteorites). The mass of meteoritic material which survives Mars atmospheric entry without melting is estimated to be $8.6 \cdot 10^6$ kg/year (see Table I). Particles in the 100 to 700 μm diameter range, which make only a small contribution to the unmelted meteoritic mass accreted by the Earth because most vaporize on atmospheric entry (see Figure 1), constitute 60% of the long term, continuous meteoritic mass accreted by Mars (see Table I).

The accretion rate for unmelted meteoritic material at Mars is almost three times the $3.2 \cdot 10^6$ kg/year of meteoritic material which Anders (1989) calculates is contributed to the Earth by micrometeorites which survive Earth atmospheric entry without vaporizing. However, since Mars is a smaller planet, the surface density of unmelted meteoritic material on Mars is 12 times that on Earth. Thus the accretion of organic matter in interplanetary dust is potentially much more important on Mars than on Earth.

3. Unaltered organics

Anders (1989) points out some organic matter contributed to Earth by interplanetary dust may be destroyed or altered by exposure to solar ultraviolet radiation or by the temperatures reached on atmospheric entry. Both of these effects are likely to be smaller for Mars than Earth. Ultraviolet radiation has a shallow penetration depth into these black particles. Since larger interplanetary dust particles survive Mars entry than survive Earth entry without melting, the fraction of organic matter located close enough to the surface of a particle to be altered by solar ultraviolet radiation will be smaller in the case of material accreting onto Mars. In addition, since interplanetary dust particles of the same size and space velocity are heated less on Mars entry than on Earth entry, a larger fraction of the organic matter survives Mars entry without thermal alteration.

The amount of organic matter not altered by entry heating can be estimated by considering the fraction of meteoritic material accreted onto Mars without reaching the pyrolysis temperature of the organic matter it contains. Anders (1989) estimates the average pyrolysis temperature of carbonaceous material from meteorites is $\sim 600°C$, and Chyba et al. (1990) calculate that most of the organic material in carbonaceous chondrite meteorites can survive temperatures up to 850K for ~ 1 second. A typical interplanetary dust particle spends only a few seconds within 100K of its peak temperature on atmospheric entry (Flynn 1989). Calculating the fraction of particles not heated above 900K on atmospheric entry, a total of $2.4 \cdot 10^6$ kg/year of meteoritic material, mostly in particles from 60 to 270 μm in diameter, accretes onto Mars with its organic matter intact (see Table I). This is likely to be a lower limit on the number of particles which carry unaltered organic carbon to the surface of Mars since the Whipple (1950) entry heating model only determines the surface temperature of the particle. Particles with surface temperatures over 900K may have lower interior temperatures because of endothermic phase transitions, e.g. alteration of carbonaceous material or dehydration of layer-silicates near the surface (Flynn, 1995), and the organic component may be destroyed in only an outer layer. Textural zoning (Sutton et al., 1992), attributable to temperature gradients, observed in 50 to 100 μm diameter micrometeorites recovered from terrestrial polar ices suggests a mechanism exists to allow the interior to remain cool.

The average carbon content of interplanetary dust particles (5 to 30 μm in diameter) collected from the stratosphere of the Earth has been measured as 10 to 12% (Thomas et al., 1993, and Schramm et al., 1989), which is

TABLE I

Meteoritic Contribution to Carbon on Mars

Mass Range (grams)	Size Range (μm)	Mass Flux (kg/year)	Unmelted Fraction T<1600K	Unmelted Mass (kg/year)	Fraction T<900K	Unaltered Carbon (kg/year)
$10^1 - 10^0$	12400–26800	$0.02 \cdot 10^6$	0	0	0	0
$10^0 - 10^{-1}$	5760–12400	$0.1 \cdot 10^6$	0.01	$0.01 \cdot 10^5$	0	0
$10^{-1} - 10^{-2}$	2680–5760	$0.3 \cdot 10^6$	0.05	$0.2 \cdot 10^5$	0	0
$10^{-2} - 10^{-3}$	1240–2680	$1 \cdot 10^6$	0.2	$2 \cdot 10^5$	0	0
$10^{-3} - 10^{-4}$	580–1240	$2 \cdot 10^6$	0.5	$10 \cdot 10^5$	0.01	$0.2 \cdot 10^5$
$10^{-4} - 10^{-5}$	270–580	$3 \cdot 10^6$	0.7	$20 \cdot 10^5$	0.07	$2.1 \cdot 10^5$
$10^{-5} - 10^{-6}$	120–270	$3 \cdot 10^6$	0.9	$30 \cdot 10^5$	0.24	$7.2 \cdot 10^5$
$10^{-6} - 10^{-7}$	60–120	$1.5 \cdot 10^6$	0.98	$15 \cdot 10^5$	0.46	$6.9 \cdot 10^5$
$10^{-7} - 10^{-8}$	27–60	$0.6 \cdot 10^6$	1	$6 \cdot 10^5$	0.75	$4.5 \cdot 10^5$
$10^{-8} - 10^{-9}$	12–27	$0.2 \cdot 10^6$	1	$2 \cdot 10^5$	0.88	$1.8 \cdot 10^5$
$10^{-9} - 10^{-10}$	6–12	$0.07 \cdot 10^6$	1	$0.7 \cdot 10^5$	0.95	$0.7 \cdot 10^5$
$10^{-10} - 10^{-11}$	3–6	$0.006 \cdot 10^6$	1	$0.06 \cdot 10^5$	1	$0.06 \cdot 10^5$
$10^{-11} - 10^{-12}$	1–3	$0.002 \cdot 10^6$	1	$0.02 \cdot 10^5$	1	$0.02 \cdot 10^5$
$10^{-12} - 10^{-13}$	0.6–1	$0.0009 \cdot 10^6$	1	$0.009 \cdot 10^5$	1	$0.009 \cdot 10^5$
$10^{-13} - 10^{-14}$	0.3–0.6	$0.0003 \cdot 10^6$	1	$0.003 \cdot 10^5$	1	$0.003 \cdot 10^5$
$10^{-14} - 10^{-15}$	0.1–0.3	$0.00006 \cdot 10^6$	1	$0.0006 \cdot 10^5$	1	$0.0006 \cdot 10^5$
TOTALS		$12 \cdot 10^6$		$8.6 \cdot 10^6$		$2.4 \cdot 10^6$

about 2.5 to 3 times the carbon content of the most carbon rich meteorites, the CI carbonaceous chondrites. A carbon content of 45% was reported for one interplanetary dust particle (Thomas et al., 1993). The small masses of interplanetary dust particles ($\sim 10^{-8}$ grams) have, thus far, precluded quantitative determination of the fraction of this carbon that is present in organic molecules. However Clemett et al. (1993) detected polycyclic aromatic hydrocarbons in interplanetary dust particles, confirming the presence of an organic component. Radicati di Brozolo et al. (1986) detected carbon clusters (C_2-C_{15}) and protonated species, and Rietmeijer (1990) has identified turbostatic carbon which he suggests may have been formed by dehydrogenation of organic compounds originally present in the interplanetary dust. Gibson (1992) reviewed the identifications of carbon compounds in interplanetary dust particles.

The carbon content of the larger micrometeorites (> 50 μm in diameter) is not well established since most of these particles melt or vaporize on Earth atmospheric entry. However, Yates et al. (1991) have extracted carbon from melted meteoritic spherules recovered from the terrestrial polar ices. The carbon isotopic composition is consistent with that of the macromolecular organic material from carbonaceous chondrite meteorites (Yates et al., 1991). Carbon abundances from 200 ppm to 5000 ppm were measured in the spherules (Yates *et al.*, 1991). However, much of the pre-atmospheric carbon may have been lost when these particles melted on atmospheric entry.

Assuming, following Anders (1989), that the average carbon content of interplanetary dust is 10%, then the present accretion rate of unaltered meteoritic carbon onto Mars is $2.4 \cdot 10^5$ kg/year. If this accretion rate were constant over the age of the Solar System, meteoritic material would have contributed a total of $1 \cdot 10^{15}$ kg of carbon to the surface of Mars, an amount comparable to estimates of the current terrestrial biomass ($6 \cdot 10^{14}$ kg; Chyba *et al.*, 1990).

The total accretion of meteoritic carbon onto the surface of Mars must be even higher, since the meteoritic flux was as much as 10^3 times the present flux during the first half-billion years of the Solar System (Anders, 1989). The accretion of meteoritic matter onto the Moon is fit by a two component model: a rapidly decaying component ($t_{1/2} = 40$ million years) exceeding the present flux by an order-of-magnitude or more 4.0 billion years ago, and a relatively constant flux near the present value over the past 3.6 billion years (Wasson *et al.*, 1975). Thus, during its early history, the accretion of meteoritic material onto Mars is likely to have substantially exceeded the value calculated using current flux estimates.

4. Martian biogenesis

The failure of the Viking gas chromatograph experiments to detect organic carbon (Biemann *et al.*, 1977) at the two landing sites sampled on Mars places an upper limit on the surviving organic component in the present surface soils. However, Banin (1988) has suggested that the high redox potential on the present surface of Mars may cause the decomposition of organic matter contributed by meteoritic infall. In addition, Stoker *et al.* (1989) have shown in laboratory simulations that organic matter would decompose due to solar ultraviolet radiation on the present surface of Mars more rapidly than the calculated current accretion rate.

The conditions on the surface of Mars early in its history are likely, however, to have been considerably different from the current conditions. McKay and Stoker (1989) suggest that about 3.5 billion years ago conditions on the surface of Mars may have been suitable for the origin of life. This is just at the end of the era when the accretion of organic material from interplanetary dust particles is expected to have been at its highest. During this period the atmospheric density on Mars may have been much higher than at present, and the atmospheric composition may have differed considerably from the present composition. These factors are likely to have allowed the survival of organic matter for considerably longer during that earlier era. In addition, Anders (1989) suggests, for the terrestrial case, that moderate ultraviolet exposure could make the organic material more rather than less interesting. The presence of liquid water on the surface of Mars (Carr, 1979) during this era, coupled with the continuous, planet-wide infall of organic matter from the interplanetary dust may have provided conditions appropriate for the development of life on Mars. Boston *et al.* (1992) have suggested that if microorganisms developed in antiquity on the surface of Mars, they might later have migrated to deeper levels where they would be protected from the present harsh surface conditions.

References

Anders, E.: 1989, *Nature* **342**, 255-257.

Anders, E. and Owen, T.: 1977, *Science* **198**, 453-465.

Banin, A.: 1988, in *Workshop on Mars Sample Return Science*, LPI **TR 88-07**, 35-36.

Biemann, K., Oró, J., Toulmin III, P., Orgel, L.E., Nier, A.O., Anderson, D.M., Simmonds, P.G., Flory, D., Diaz, A.V., Rusneck, D.R., Biller, J.E. and Lafleur, A.L.: 1977, *Journ. Geophys. Res.* **82**, 4641-4658.

Boston, P.J., Ivanov, M.V. and McKay, C.P.: 1992, *Icarus* **95**, 300-308.

Carr, M.H.: 1979, *Journ. Geophys. Res.* **84**, 2995-3007.

Chamberlin T.C. and Chamberlin, R.T.: 1908, *Science* **28**, 897.

Chyba, C.F.: 1987, *Nature* **330**, 632-635.

Chyba, C.F., P.J Thomas: 1990, *Science* **249**, 366-373.

Clemett, S.J., Maechling, C.R., Zare, R.N., Swan, P.D. and Walker, R.M.: 1993, *Science* **262**, 721-725.

Flynn, G.J.: 1989, *Proc. 19th Lunar Planet. Sci. Conf.*, 673-683.

Flynn, G.J.: 1991, in *The Environmental Model of Mars* (ed. K. Szegö) 121-124 (Pergamon Press PLC, London) pp.121-124.

Flynn, G.J.: 1995, *Lunar Planet. Sci. XXVI*, 405-406.

Flynn, G.J. and McKay, D.S.: 1990, *Journ. Geophys. Res.* **95**, No. B9, 14,497-14,509.

Fraundorf, P.: 1980, *Geophys. Res. Lett.* **10**, 765-768.

Gibson, E.K.: 1992, *Journ. Geophys. Res.* **97**, No. E3, 3865-3875.

Hughes D.W.: 1978, in *Cosmic Dust* (ed. J.A.M. McDonnell), John Wiley, New York 123-185.

McKay, C.P. and Stoker C.R.: 1989, *Rev. Geophysics* **27**, 189-214.

Morgan, T.H., Zook, H.A. and Potter, A.E.: 1988, *Icarus* **75**, 156-170.

Oró, J.: 1961, *Nature* **190**, 389-390.

Radicati de Brozolo, F., Bunch, T.E., Chang, S. and Brownlee, D.E.: 1986, in *Trajectory determinations and collection of micrometeoroids on the space station*, LPI **TR 86-05**, 77-79.

Rietmeijer, F.J.M.: 1990, *Lunar Planet. Sci. XXI*, 1014.

Schramm, L.S., Brownlee, D.E and Wheelock, M.M: 1989, *Meteoritics* **124**, 99-112.

Southworth, S.A. and Sekanina, Z.: 1973, *NASA Conf. Report* **CR-2316**.

Stoker, C.R., Mancinelli, R., Tsay, F.D., Kim, S.S. and Sculley, J.: 1989, *Lunar Planet. Sci. XX*, 1065-1066.

Sutton, S.R., Prinz, M., Maurette, M., Nehru, C.E., Weisberg, M.K. and Bajt, S.: 1992, *Lunar Planet. Sci. XXIII*, 1391-1392.

Thomas, K.L., Blanford, G.E., Keller, L.P., Klock, W. and McKay, D.S.: 1993, *Geochim. Cosmochim. Acta* **57**, 1551-1566.

Wasson, J.T., Boynton, W.V., Chou, C.-L. and Baedecker, P.A.: 1975, *The Moon* **13**, 121-141.

Whipple, F.L.: 1950, *Proc. Nat. Acad. Sci. U.S.A.* **36**, 687-695.

Yates, P.D., Wright, P.I., Pillinger, C.T. and Hutchison, R.: 1991, *Meteoritics* **26**, 412.

EXPERIMENT "TSAREV" AND DIFFERENTIATION OF CHONDRITIC BODIES

J.I. ZETZER
*Institute of Dynamics of Geospheres, Russian
Academy of Sciences, 117979 Moscow, Russia*

and

A.V. VITYAZEV
*Institute of Planetary Geophysics, United Institute of Earth
Physics, Russian Academy of Sciences, 123810 Moscow, Russia*

Abstract. As a check on the differentiation processes of planetesimals the experiment "Tsarev" was carried out. The main problems of the experiment were: investigation of melting and liquation processes in the melt of primitive meteorite substance; analysis of the composition of metallic and silicate phases, including the distribution of rare-earth elements and comparison of the obtained phase with iron and differentiated meteorite composition. The heating of a sample of the L-chondrite "Tsarev" (volume 15 cm^3) was produced by intensive microwave radiation because this method of heating provides the possibility of uncontact entry of energy into a sufficiently large sample. Upon the heating of the meteorite "Tsarev" (with the maximum temperature $1500 \pm 50°$K) this sample was melted and two phases appeared and were separated in the gravity field: the composition of Fe-Ni-S phase includes 15 rare-earth elements and the silicate phase is composed of the main oxides. There is also clear evidence for the presence of liquation processes of second order, e.g. separation of Fe-Ni from Fe-S component.

1. Introduction

According to the modern "moderate hot" models of the origin of the terrestrial planets, the melting and differentiation of their interiors began long before they attained their current dimensions. The high velocity collisions of the planetesimals lead to heating and melting of the major portion of the matter. It results in the process of differentation: metal sinks to the center, displacing the lighter silicates which form the mantle. A second order differentation processes leads to formation of the observed variations in the differentiated bodies (Vityazev et al., 1990).

To check the differentation processes the experiment "Tsarev" was carried out. The main problems of the experiment were:
– investigation of melting and liquation processes in the melt of primitive meteorite substance;
– analysis of the composition of metallic and silicate phases, including the distribution of trace elements;
– comparison of the obtained phase with iron and differentiated meteorite composition.

Earth, Moon, and Planets **72**: 475-480, 1996.

2. Experiment

The realization of this experiment has to obey several conditions: 1) uncontact heating; 2) sufficient volume of the sample; 3) vacuum or inert atmosphere; and 4) the controlling of the rate of heating and cooling.

The first and second conditions define the temperature and hydrodynamics parameters, which are necessary to realize the differentiation process. The third condition prevents oxidation of the surface layers of the sample. The fourth condition makes it possible to fix the stages of the differentiation process in the various inner sections. The sample is derived from the L-chondrite "Tsarev", which fell on 6 Dec. 1922 in the Volgograd region and had a total recovered mass of about 1 ton.

The heating of the sample was produced by intensive microwave radiation. Most dielectrics are transparent for radiation of this wavelength. The use of microwave radiation not only provides the possibility of uncontact entry of energy into a sufficiently large sample. It also provides an intensive and at the same time "gentle" regime of heating. The "gentle" regime means that the physico-chemical transformations take place only as a result of the heating.

Fig. 1 shows the skeleton diagram of the experiment. The sample size was about 15 cm^3; it was placed in the waveguide. The end of the waveguide was placed in a vacuum chamber. The waveguide had holes, which provided degassing of the sample during its heating. The power was conveyed to the sample from a magnetron generator. The generator produced microwave radiation with 12 cm wavelength and a power of about 5 kW. Preliminary estimates showed that there was compensation of the heating losses in the sample (temperature about 2000°K for a microwave power of several kilowatt. Some efforts were made to increase the performance of microwave heating:

– to increase the absorption efficiency the sample was placed in a maximum of the standing wave; for this purpose a metallic plate was placed on the end of the waveguide; this plate had small holes for degassing of the sample during its heating;

– to decrease the heating losses the sample was placed in quartz glass suspended in the waveguide by a quartz bar.

The measurement of processes in the sample during its interaction with intensive microwave radiation had some difficulties: first, every sensor placed in the sample interacted with the radiation, distorting the phenomena of interaction; secondly, intense microwave radiation was a source of hard electromagnetic interferences, which hampered the measurements.

This is why we had to work out special methods of measurement, without contact and immune to interferences. We have used various independent methods of measuring the parameters for increasing the reliability of results

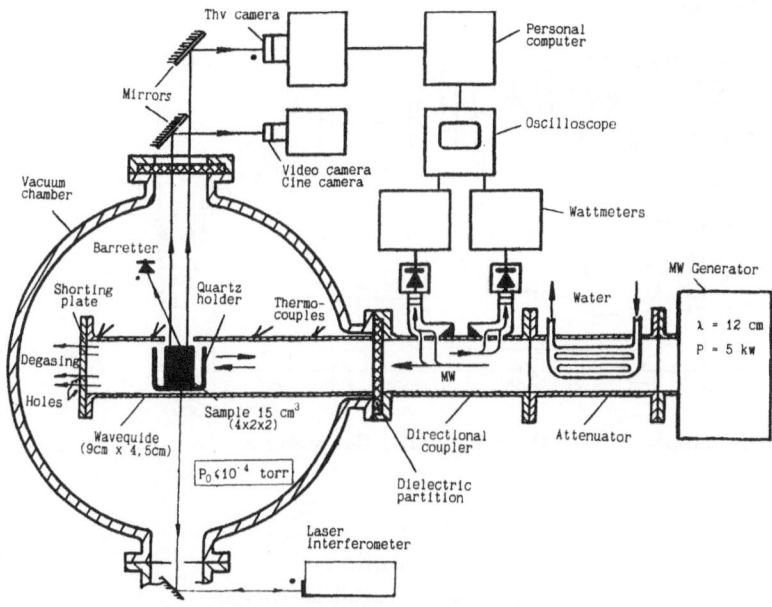

Fig. 1. Skeleton diagram of the experiment

in each experiment. The temperature distribution was measured by thermo-vision, laser interferometer, barrette and thermochemical indicators, placed on the surfaces of the sample (Fig. 1).

Fig. 2 shows the temperature of the sample of the meteorite "Tsarev" and the power of the generator as functions of time. The value of the temperature is an average of all sensors. As shown in this figure the temperature of the sample was about 1400°K during ~ 15 minutes. The speed of heating was ~ 4.5 deg/s, and the speed of cooling was ~ 1.5 deg/s. The sample was analyzed after heating, giving the following main results:

1. The sample of the meteorite "Tsarev" was melted by microwave radia-tion. The temperature maximum was 1500 ± 50°K.

2. The results of controlling (by EPMA and SIMS) of the microchemical and phase composition of the samples showed that after the melting of the L-chondrite "Tsarev" two phases appeared and were separated in the gravity field:

    – the composition of Fe-Ni-S phase includes 15 trace elements, in agree-ment with the average composition of iron meteorites;

    – the silicate phase in composition of main oxides (Table 1) is close to the eucrite-howardite-diogenite association, reminding of some ultraba-sic terrestrial rocks and practically coinciding with the Dreibus-Wänke (1990) model of SPB;

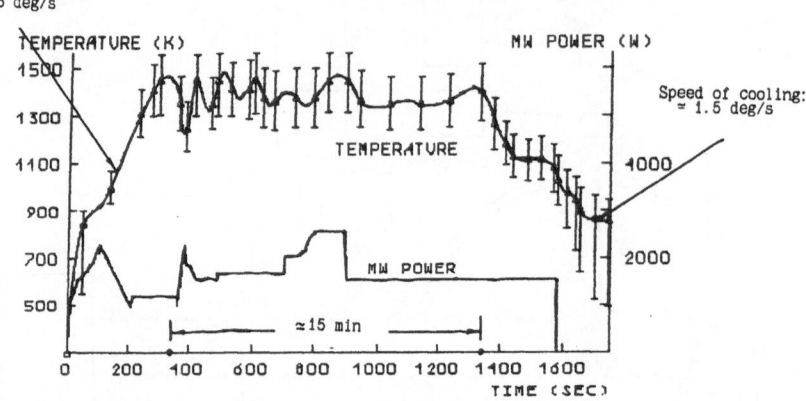

Fig. 2.   The temperature of the sample and the power of the generator as functions of
time

TABLE I

Comparison of chemical composition between the silicate phase
from the chondrite and SPB.

	1	2	3
SiO_2	47.25	46.35	44.40
TiO_2	0.04	1.11	0.14
Al_2O_3	3.04	3.22	3.02
Cr_2O_3	0.81	0.56	0.76
FeO	19.39	19.62	17.90
MnO	0.35	0.44	0.46
MgO	25.08	25.23	30.20
CaO	2.65	2.53	2.45
Na_2O	0.59	0.35	0.50
K_2O	0.09	0.00	—
P_2O_5	0.01	0.13	0.16

1. Chemical composition of the silicate phase extracted from the Tsarev 1 chondrite;

2. Chemical composition of the silicate phase extracted from the Tsarev 2 chondrite;

3. Chemical composition of Shergottite parent body (SPB) by Dreibus and Wänke.

– there is clear evidence for the presence of liquation processes of second
order, e.g. separation of Fe-Ni and Fe-S components.

3. Discussion

We discuss here three main questions connected with early differentiation of chondritic bodies:
– the closeness of characteristic times of formation of meteoritic parent bodies and the beginning of differentiation of interiors of the terrestrial planets;
– the problem of difference in the bulk compositions of the Moon and the Earth;
– the problem of difference between compositions of Martian and Earth cores.

1.   The well known isochronism of formation of parent bodies of main types of meteorites (4.6 ± 0.1 Gyr) and oldest lunar rocks testifies that during the planet formation some part of the preplanetary bodies could avoid the high degree of heating (C1-meteorites). The other bodies were subject to strong heating leading to significant melting and had a chance before their cooling to pass certain stages of differentiation. Leaving aside the possible role of ^{26}Al, we emphasize the leading role of high-velocity collisions of planetesimals with each other and with growing planets.

2.   For interpreting the known distinctions of the bulk compositions of the Moon and the Earth today, two alternate dynamic models are attractive: the coaccretion model of lunar origin in a swarm of fragments captured by the Earth into a prelunar disk, and the model of a unique giant (mega) impact of a Mars-like body during the final stage of the Earth growth. Note that there exists a more consecutive model (Vityazev et al., 1990) which takes account of the existence of a numerous population of 100–1000 km sized bodies during the main stage of planetary formation. During their falls to the Earth an ejection of some portion of the substance ($< 10\%$ of projectile mass) into geocentric and heliocentric orbits had to occur. This model naturally explains the volatile depletion in the lunar material. It can explain also the similarity of bulk compositions of the Moon and the primitive Earth mantle. The latter was depleted in iron in comparison with ordinary chondrites, but enriched with respect to the modern mantle. Moreover, part of the primitive Al-Ca-Ti rich Earth crust was predominantly ejected in the course of macroimpacts.

3.   Experiments similar to the above-described show that at moderate pressure we can expect a content of the Fe-Ni-S component in the metal phase, that explains naturally the composition of iron meteorites and the assumed compositions of the Mars and Mercury cores. For explanation of the massive Earth core one can assume a large quantity of reduced iron in the planetesimals which formed the Earth. A more interesting proposal by Ringwood (1984) and confirmed in experiments is that at pressure > 70 GPa, FeO goes into the metal phase. For the Earth such pressure was achieved

when the size was about 0.4 of the present value. So, in this model we have a good explanation for the lower content of FeO in the modern Earth mantle in comparison with the Moon.

References

Vityazev, A.V., Pechernikova, G.V. and Safronov, V.S.: 1990, *Terrestrial Planets: Their Origin and Early Evolution.* Nauka, Moscow.

Dreibus, G. and Wänke, H.: 1990, Comparison of the chemistry of Moon and Mars. *Adv. Space Res.* **10**, (3)7 – (3)16.

Ringwood, A.E.: 1984, The earth's core: its composition, formation and bearing upon the origin of the Earth. *Proc. Roy. Soc. Lond.* **A395**, 1–64.

FIELD-ALIGNED CURRENT GENERATION AT PLASMA CLOUDS OR BODIES WITH PLASMA SHELLS MOVING IN MAGNETIC FIELDS

B.G. GAVRILOV, I.M. PODGORNY and J.I. ZETZER

Institute for Dynamics of Geospheres of RAS, Moscow, 117334, Russia

Abstract. Field-aligned currents (FAC) can be generated during a plasma jet or the motion of a body with a plasma shell in a transverse magnetic field. They can be investigated in an active space experiment. For the proper choice of diagnostics and for evaluation of expected results a laboratory simulation is carried out. The preliminary results show FAC generation in conditions correlated with conditions in space.

1. Introduction

The origin of field-aligned currents (FAC) and their influence on the state and dynamics of environmental media is one of the fundamental problems of space physics. FAC can transfer momentum to the ionosphere from plasma flow at high altitudes and heat the ionosphere by Joule losses. They exist in the Earth magnetosphere and magnetospheres of other planets and some astrophysical objects. FAC could cause changes in their state and dynamics. They are also apparently responsible for solar flares [Podgorny and Podgorny, 1993]. At the present time there are several ideas about the mechanism of FAC generation. Here we restrict ourselves to consideration of one of them.

FAC can be generated at plasma clouds or bodies with a plasma shell moving perpendicular to magnetic field lines [Gavrilov et al., 1994a]. The plasma flow in the magnetosphere acts like a MHD generator with the ionosphere as a load. The plasma is polarized under the influence of the Lorentz force. At rather high altitude the parallel magnetospheric conductivity $\sigma_\parallel$ is much greater than the cross conductivity $\sigma_\perp$. Hence, the current should propagate along magnetic field lines with the Alfvén velocity until it reaches the lower ionosphere and closes by Pedersen currents. FAC generation should occur also in induced magnetospheres of celestial bodies without intrinsic magnetic field (Venus, comets). Magnetic field lines of induced magnetospheres are connected with the interplanetary magnetic field. The electric field $\mathbf{E} = -\mathbf{V} \times \mathbf{B}/c$ produced by the solar wind can be projected along magnetic field lines inside the magnetosphere. Between the magnetic field lines, which penetrate deeply into the plasma shell, the Pedersen current should close the oppositely directed FAC. The solar wind electric field can reach ~ 10 mV/m. The magnetic field of the induced magnetosphere can be of several tens of the interplanetary magnetic field. In that case the electric field in

Earth, Moon, and Planets **72**: 481-486, 1996.
© 1996 *Kluwer Academic Publishers*.

the plasma shell of different celestial bodies may be $\sim 10 - 100$ mV/m. The value is comparable with the electric field in the Earth auroral region. This estimation shows that FAC and aurora generation are expectable around Venus and comets.

Fast plasma jet creation and FAC generation because of the electric field $-\mathbf{V} \times \mathbf{B}/c$ were expected in the Jupiter magnetosphere after discovery of a gas torus along the Io orbit. The gas ionization due to solar ultraviolet radiation and effect of critical velocity ionization should produce a plasma jet with a velocity of the order of the Io orbital velocity, and FAC generation should occur. A rather successful attempt was made to detect the FAC system during the Ulysses mission [Dougherty et al., 1993]. They observed a FAC system with $J = 20$ mA/m, which apparently may be connected with the Io gas torus.

2. FAC generation in active experiments

The complex solving of the problem of FAC generation is too complicated to be analyzed in a purely theoretical way. The observations and measurements of electric and magnetic fields in situ are also very incomplete and have irregular character. Even the artificial satellite and ground-based observations do not provide enough information, because usually it is difficult to separate different phenomena and to investigate them in controlled conditions. Such a possibility can be supplied only in controlled active high altitude experiments or in a model laboratory experiment.

The active experiments for investigation of the geophysical phenomena exited by artificial plasma clouds or jets, which were carried out up to now, have demonstrated some interesting physical effects (the fine-scale stratification of the plasma cloud, plasma instabilities, the precipitation of the fast particles, the appearance of the polarization electric field and others). But one of the most important problems of the magnetosphere-ionosphere interaction—the manner in which field-aligned currents can be generated—was not solved. Under such circumstances only a specially worked out active magnetosphere experiment may give some important information.

The main goal of the international Active Geophysical Rocket Experiment (AGRE) [Adushkin et al., 1994; Gavrilov et al., 1994b] is to excite some part of the ionosphere and magnetosphere by a calibrated impulsive high-energy plasma jet at a height above $500 - 600$ km. The polarization field in the vicinity of a plasma jet is $E = VB/(1 + \Sigma_p/\sigma_{pl}L)$, where V is the plasma velocity, B is the magnetic field at the height of the injection, Σ_p is the height-integrated Pedersen conductivity of the ionosphere, σ_{pl} is the plasma conductivity, and L is the specified size of the jet. For the conditions of our experiment we may adopt $V = 20$ km/s, $B \simeq 0.4$ Gauss, and $\Sigma_p \simeq 10^{12}$ esu.

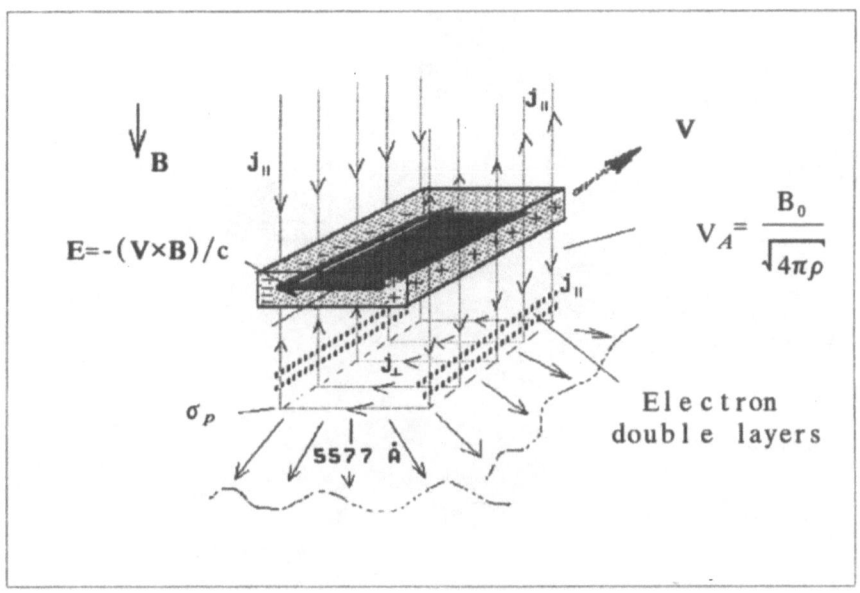

Fig. 1.   The diagram of FAC generation at plasma moving in a magnetic field

So the strength of the polarization field is $\mathbf{E} = [\mathbf{V} \times \mathbf{B}] \simeq 0.8$ V/m. At the plasma cloud diameter about 10 km the potential drop is $U = EL \simeq 10^4$ V. The current density in the whole circuit is determined by the value Σ_p and should be $J = E\Sigma_p \simeq 10^{-2}$ A/cm. Furthermore, if the corresponding FAC density is large enough and the electron drift velocity in the FAC is of the order of magnitude of the thermal electron velocity, electron double layers or anomalous resistivity regions may be created. So a potential drop may appear along field lines and cause acceleration of electrons downward. As a consequence the fast electrons precipitate into the upper atmosphere and produce aurora and additional ionization (Fig. 1).

For investigation of phenomena caused by FAC (Joule heating of the iono-sphere, increase of electron density, fast electron precipitation and aurora) a lot of measurements should be conducted at the experiment. But some important physical phenomena are difficult to predict exactly. Because of this, it is necessary to check theoretical results in laboratory experiments. To simulate the set of conditions which are expected in space experiments the principle of limited simulation [Podgorny, 1978] was used.

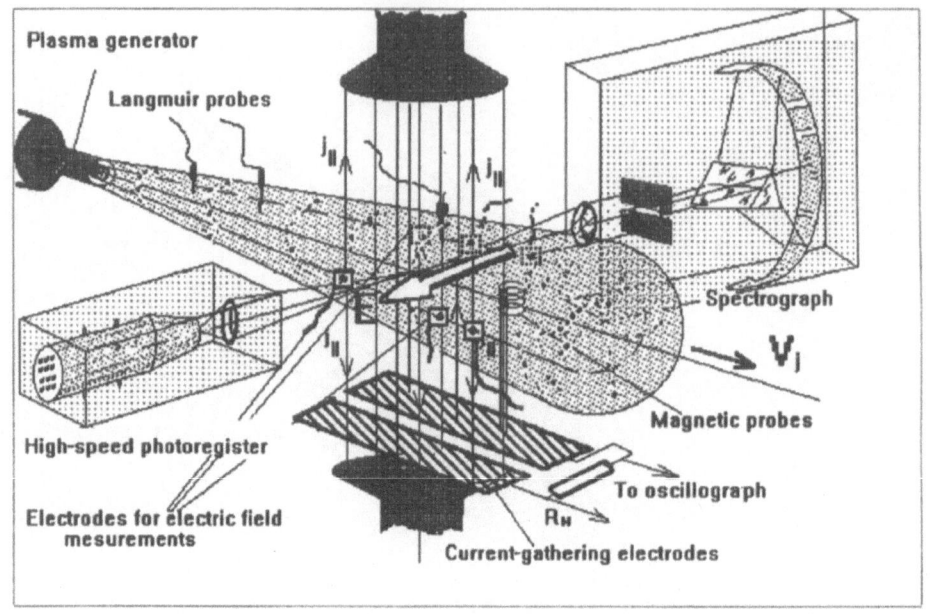

Fig. 2.   The scheme of the laboratory simulation experiment

3. Laboratory Simulation

Experimental set FACEL (Field-Aligned Current Experiment in Laboratory) includes a high-vacuum chamber, a plasma generator, a pulsed magnet and a multipurpose diagnosis equipment (Fig 2).

The laboratory injector is a coaxial type electrodynamic plasma accelerator. For the simulation of the cross-conductivity of the lower region of the ionosphere current-assemble electrodes were used. The plasma jet velocity in the model experiment is several tens of km/s. The characteristic size of the jet is $L = 5 - 10$ cm, the plasma density is $n \simeq 10^{14} - 10^{15}$ cm^{-3}. The background neutral atom density ($n_0 = 10^{14}$ cm^{-3}) was determined from the condition in space: $n_0 \leq 1/(\sigma_{ce} L)$, where σ_{ce} is the charge-exchange cross-section.

The multipurpose diagnosis equipment includes sensors for measurements of ion and electron density and temperature, electric and magnetic field disturbance, velocity and pressure of plasma flow. The preliminary results show that the plasma jet induces an electric field which is determined by the Ohm law $\mathbf{j}/\sigma = \mathbf{E} - \mathbf{V} \times \mathbf{B}/c$. The electric field produces a FAC system which is closed in the special current receivers with controlled conductance.

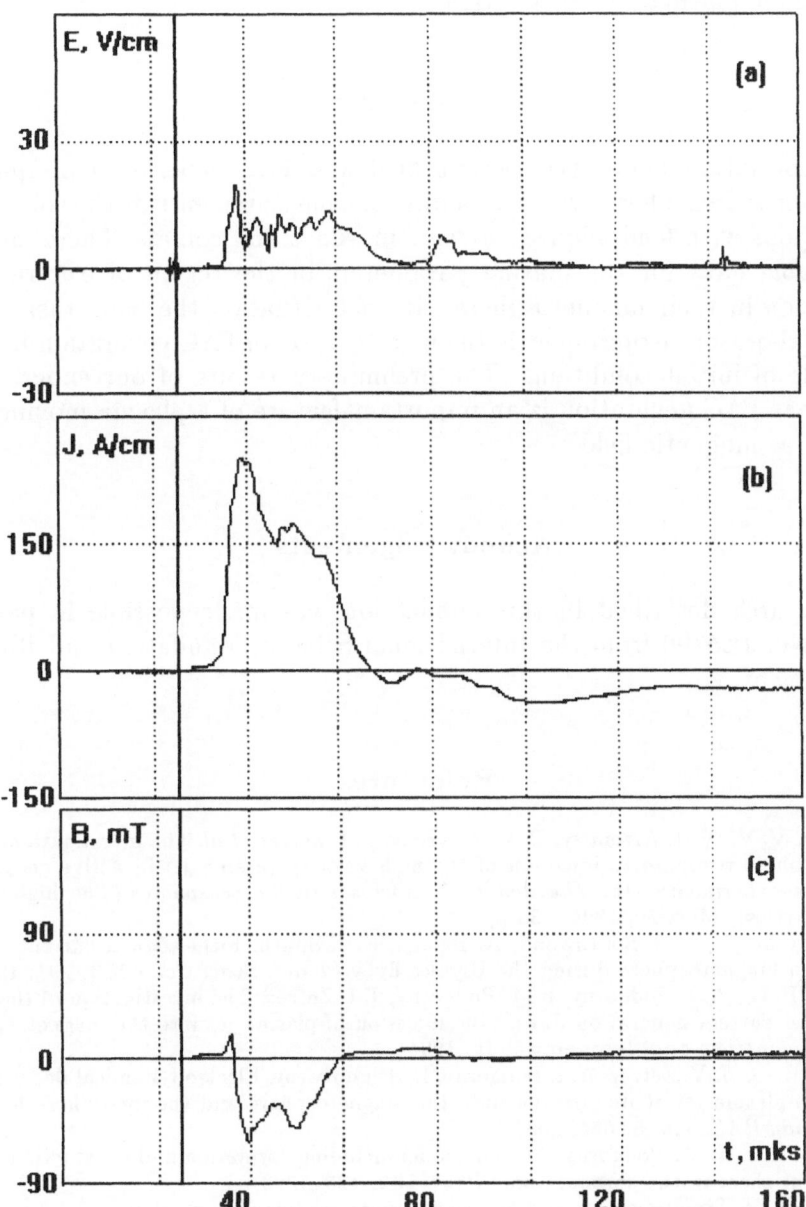

Fig. 3. Oscillograms of (a) the electric field $-\mathbf{V} \times \mathbf{B}/c$, (b) the diamagnetic signal and (c) the field-aligned current at the plasma jet moving in a transverse magnetic field

They are located near the surface of magnet's poles. The oscillograms of some measured parameters are shown in Fig. 3.

The typical FAC value was about 100 A/cm. This current produces electrodynamic plasma jet deceleration.

4. Conclusion

The information about the electric field and FAC generation in space is very poor today. There are no simultaneous measurements in the solar wind and regions with field-aligned currents in Venus and comets. There are also no reliable data for the plasma parameters in the region of electric field generation in their magnetospheres. In such situation the main task of the future laboratory experiment is the investigation of FAC generation in wide intervals of initial conditions. The preliminary results of our experiments show that FAC generation is an important feature of a plasma moving in a transverse magnetic field.

Acknowledgements

The research described in this publication was made possible in part by Grant No. JE5100 from the International Science Foundation and Russian Government.

References

Adushkin V. V., V. I. Artem'ev, B. G. Gavrilov, J. I. Zetzer *et al.* The investigation of the ionosphere response to injection of the high velocity plasma jet in active geophysics rocket experiments. In: *The dynamics processes in the geospheres* (The high power geophysics), Moscow, 1994, 335 pp.

Dougherty M. K., D. J. Southwood, A. Balogh, E. J. Smith. Field-aligned currents in the Jovian magnetosphere during the Ulysses flyby, *Planet. Space Sci.* **41**, 4, 291, 1993.

Gavrilov B. G., A. I. Podgorny, I. M. Podgorny, J. I. Zetzer. The investigation of the field-aligned current generation during the injection of plasma jet into the magnetosphere, *Geomagnetism and Aeronomy* **5**, 16, 1994.

Gavrilov B. G., J. Y. Zetzer, A. I. Podgorny, I. M. Podgorny. Electrodynamical deceleration of the plasma jet at its injection into the magnetosphere and the ionosphere heating, *Doklady RAS.* **336**, 5, 684, 1994.

Podgorny A. I., I. M. Podgorny. A solar model including formation and destruction of the current sheet in the corona, *Solar Phys.* **139**, 125, 1993.

Podgorny I. M. Simulation studies of space, *Fundamentals Cosmic Phys.* **1**, 1, 1978.

CONFERENCE SUMMARY: COMETS

FRED L. WHIPPLE

Smithsonian Institution, Astrophysical Observatory, USA

After apologies to authors whose papers are omitted because of my personal bias, of posters, which I did not have energy to study, and of highly theoretical papers because my math is rusty, my comments on comet papers in this highly interesting meeting follow.

Innanen opened the meeting with a fine discussion of galactic tides upsetting the Oort Cloud of comets, and then, with Matese, Whitman and Valtonen, calculated a 5 to 1 variation of cometary impact to the solar system with a 32 million-year period. They then would produce the distribution of inclinations for short-period (S-P) comets by splitting.

The contribution of passing stars in producing the observed comets was discussed by several authors, with Mylläri and Orlov noting that among stars within 25 pc. of the Sun, there have been almost no close approaches in the near past, but many more in the future. Zheng, Valtonen, Korpi, Mikkola and Rickman describe how the (S-P) comets can be derived from the Oort Cloud without recourse to a "Kuiper Belt". A few comet splittings help quantitatively.

After Bailey's beautiful description of the evolution of comets, I derived comet "ages" from the extraordinary calculations by Levison and Duncan, correlating "ages" with nongravitational forces, comet brightening rates and dust/gas ratios, young comets being more active with more gas, but no more erratic than old comets.

Tancredi showed that comet orbits can "remember" no more than one close approach to Jupiter. He, with Fernández and Rickman, found that comets need to be about 1 km in diameter to be found. However, Brandt, with A'Hearn, Randall, Schleicher, Shoemaker and Stewart, think there are many more very small ones within 2 AU of the Sun and propose to search for them. I wish them the best of luck as the search should clear up the question of the final demise of comet fragments. Very faint new or (L-P) comets, however, should be rare.

Storrs finds the usual less–and–one dust–to–gas ratio for 35 comets, while Keller would have the dust overwhelm the gas. He, with Kührt, would make the inactive crusts of comets much deeper and more rigid than usually ascribed to the icy conglomerate model. This model is now subject to a much more precise definition physically than ever before because of more and improved observations. Knollenberg, Kührt and Keller demonstrate this trend in their study of the dust–gas output of Halley's Comet. Wallis, for

Earth, Moon, and Planets **72**: 487-488, 1996.
© 1996 *Kluwer Academic Publishers.*

example, discusses several chemical and physical processes undergone by grains on comet surfaces, leaving a very complex situation theoretically.

Williams would have comets eject grains at much higher velocities than by outgassing, to account for the large spreading in meteoroid streams. This long–standing problem needs much more attention.

We came to Mariehamn to appraise, not to mourn the passing of Comet Shoemaker-Levy 9. The great god Jupiter tore out the heart of this comet. The wake lasted several months as we pathologists watched the path of the comet's bones and end–trails streaming across the sky until they were finally cremated in a grand display. This display was beautifully reproduced by A'Hearn and a number of other observers, to the pleasure of all the participants at the meeting. The phenomena provided new clues about the structure of comet nuclei.

Particularly significant were the simulations of tidal breakup made by Asphaug and Benz. They could not produce the 20+ resultant fragments if the original comet (assumed density 0.5 gm cm^{-3} and radius 4 km) broke up by brittle fracture. It had to be weaker than uncompacted snow made up of many uncemented units packed together under gravity like a bag of marbles, followed by considerable gravitational reassembly. During the final plunge Jupiter's tidal force only distorted the fragments.

Nakamura and Yoshikawa, by orbital integrations make collision of (S-P) comets with Jupiter a once–in–a–thousand–year event with a lesser chance for longer period comets in agreement with Obrubov.

The pressing problem of dead comets among the NEA's deserves much more attention and seems to be getting it by Steel, Babadzhanov and several others.

Another pressing problem is the significance of the Comet Encke, Taurid–Meteor complex. Steel, Clube and Napier support the idea that it is a huge collection of debris from a truly gigantic comet providing a cosmically short–term hazard to the inner Solar–System. Morbidelli deals with the transient character of such objects and shows how comet and asteroid type orbits can be interchanged. Having originally pointed out the reality of the Taurid–Encke relation half a century ago, I take no further responsibility for its effects.

CONFERENCE SUMMARY: ASTEROIDS

A.W. HARRIS

Jet Propulsion Laboratory, California Institute of Technology,
MS 183-501, Pasadena, CA 91109 U.S.A.

The title of this paper is misleading on two counts. First, I do not intend to give a uniform summary of the conference, since that can be had by reading the abstracts of the papers in this volume. Instead I will give a personal perspective of the highlights and, to me, the "lowlights" of the conference. Secondly, because the boundaries between asteroids, comets, and meteors are diminishing to the point of vanishing, I shall feel free to comment on papers that you may regard as dealing with "comets" or "meteors", as well as those relating primarily to "asteroids".

Certainly the star of the show has been P/Shoemaker-Levy 9. I am somewhat distressed by the preoccupation with the question of whether S-L 9 is a comet or an asteroid. Some might even claim that it is a meteor stream! I am reminded of the fable of the blind men feeling different parts of an elephant and trying to name the creature (See figure). It is not useful to carry on a debate over nomenclature

Earth, Moon, and Planets **72**: 489-492, 1996.

for the purpose of claiming the object as "one of our own." Rather we should be looking for physical properties that will lead to real insights as to what is in the outer solar system and how it got there. I note in particular that although there were a variety of seemingly testable predictions made before the impacts, and certainly there has been a wealth of observations of the impacts, I have not yet seen many recantations. Thus it seems that everyone, in their own view, was right! This calls into question whether any of the theories were really testable by observation, and if not, why bother to formulate them? But I am being too cynical. I really do believe that the theories are testable, and I suspect many of the constraining observations are successfully made. It will take more time to digest the results and see who is right among the theorists. I look forward to seeing these results in the months ahead.

Regarding specific presentations, I was impressed by the work of Asphaug and Benz on the tidal breakup of the S-L 9 parent, and implications for its strength and density. With an apology for lack of modesty, I note my own paper on the effect of non-spherical shape and spin of the nucleus. I look forward to seeing in the future a calculation marrying the two approaches, that is hydrocode numerical models of the tidal breakup of spinning, non-spherical bodies. The numerical works of modeling the impact dynamics of the nuclei into the atmosphere of Jupiter are truly impressive. By matching the predictions from various models to the observed events, we can hope to gain some certainty in the correct way to model and scale these events, which can have important implications for the currently popular question of giant impacts on the Earth. The spectral signatures of non-volatile elements detected indicate that silicate compounds existed in the nucleus. This is no surprise. Unfortunately, it will be very difficult to assess whether the apparent near-absence of more volatile species has any meaning at all. Most such elements could have been dredged up from the Jovian atmosphere in much greater quantities than would be expected from even a very volatile-rich comet nucleus, so even if detected it would be hard to interpret the meaning.

Turning now to other "asteroid" presentations, I found the discussions of the "sun impact" track of orbital evolution, mentioned by Farinella, Froeschle, Morbidelli and others to be an interesting new result of fundamental importance. This casts the degree of uniqueness of the Kruetz group of sun-grazing comets into a new light, and indicates a significant sink for the population of Earth-crossing asteroids. I see good progress in the derivation of proper orbital elements, and it now seems possible to extract believable "structures" of families (e.g. Zappalà's work on the Dora family). Likewise, more sophisticated proper element extraction is allowing progress in defining families of objects in and near resonances.

In the field of observations, the widespread availability of CCD systems has resulted in serious photometric work becoming within reach of very small observatories, including those of amateurs, and in spectacular progress obtaining photometry and spectrophotometry of extremely small bodies with larger telescopes. Thus we are seeing an explosion in the numbers of objects surveyed, and extension of observations to distant objects, small family members, and NEAs observed at large distance, rather than just at very infrequent close passes by the Earth. I have been impressed by the spectacular IR images of the S-L 9 impacts on Jupiter. I am not sure how much application there is of this technology for asteroid observations, but it bears watching.

The recent results from Spacewatch, reported by Jedicke, are impressive. At present, Spacewatch is discovering more than half of all NEAs worldwide. This explosion in discovery rate is leading to provocative and important new questions, such as the population of NEAs in the size range of tens of meters, and whether there is a special population of objects in very Earth-similar orbits. On the local front, I enjoyed very much the tour of the Lumparn structure. It is nice to see progress in identifying impact structures on the Earth, which is necessary for correlating the observations in the sky with events on the Earth.

Now I must turn, with some regret and trepidation, to "roast" a few papers that I found unconvincing. I am skeptical of the reality of the claimed population of small NEAs in very Earth-similar orbits. The claimed distribution, of relatively low eccentricity but substantial inclination, seems implausible and most likely the result of discovery biases. The results of Bottke *et al.* are suggestive of this conclusion. I applaud Jedicke's stated intention to revisit the question of the reality of the claimed population.

I am also skeptical of the existence of "micro-comets" suggested by Brandt *et al.* I believe they dismiss too casually the failure of Spacewatch to discover any such small comets. It is true that the Spacewatch camera is not optimized for such discoveries, but even allowing for that, an estimate of the number of small comets expected to have been discovered is ~10. Instead they have seen none. (Spacewatch has discovered two comets, but not especially small ones.) Thus preliminary evidence is that the comet population in the inner solar system "peters out" at a size range of the order of 1 km diameter. Jewitt, in his excellent review, gave us good physical reasons to expect that this should be so: smaller objects become devolatilized to the core in a short time. Thus "micro-comets" are probably there, they are hiding as small NEAs. Nevertheless, Jack Brandt's advice is well worth heeding: never trust a theory without bothering to *look*.

Finally, I remain skeptical of "comet showers" and "coherent catastrophism" (*e.g.*, Clube). I am particularly bothered by the proponents of these theories accusing the rest of us of blind prejudice in rejecting aspects of the hypothesis. For every Galileo or Copernicus whose theories were resisted, the scientific landscape is littered with Pons's and Fleischmans whose ideas were perhaps cruelly rejected, but justly so. Thus ideas must be tested in a cold light of skepticism, and not promoted by name calling.

Likewise, the comparison with Darwin being "wrong" is irrelevant. The philosophical importance of evolution *vs*. creationism vastly overshadows the minor quibble of gradualism *vs*. catastrophism. Furthermore, the profound aspect of "cosmic catastrophism" is that biologic evolution may be largely driven by cosmic impacts. Whether these events are quasi-periodic or truly random is only a minor sidelight.

A warning sign of a non-productive, if not outright wrong, avenue in science is a lack of progress in proving a hypothesis. Consider the example of the K-T boundary impact hypothesis. We see good and continual progress from the Alvarezes' first suggestion that the boundary clay is of cosmic origin: mapping out the layer and iridium anomaly worldwide, the discovery of microtektites in the layer, the gradient in thickness and grain size of the layer leading to the Caribbean as the impact site, and culminating with the identification of Chixcalub as the "smoking gun." More recently still, we have seen the "test" posed at the latest "Snowbird Conference" (curiously held in Houston, TX), which most observers conclude clearly shows that the extinctions associated with the K-T boundary were essentially instantaneous, within the resolution achievable in the geologic record. I see no such pattern of progress in the "Taurid complex" or "Comet shower" hypotheses. To the contrary, there appears to be mounting evidence (*e.g.* the paper by Morbidelli) that the clustering interpreted as the "Taurid complex" is largely coincidental or the result of selection effects. Likewise, the above mentioned result that K-T extinctions were instantaneous removes any indication of a "comet shower" at that time.

I would like to conclude by thanking the hosts of this meeting for a timely and stimulating conference. Even, or perhaps especially, the most controversial papers served a good purpose in triggering thought and discussion on a wide variety of topics.

Acknowledgment

My participation in this conference was supported by the Jet Propulsion Laboratory, California Institute of Technology, under contract with NASA.

Rebuttal of the comments by A.W. Harris

D.I. Steel

Anglo-Australian Observatory, Private Bag, Coonabarabran, NSW 2357, Australia; and Department of Physics and Mathematical Physics, University of Adelaide, South Australia

EDITORIAL NOTE: *The following comments were submitted by Dr. Steel as part of his paper (p. 279). Our editorial decision was to print them as a separate rebuttal of parts of A.W. Harris' conference summary. This way we hope to record, with the consent of both authors, one of the important discussions of the conference.*

First, Harris confuses the concept of coherent catastrophism with the idea of comet showers (note he uses 'hypothesis' – singular). These phenomena, if they exist, are quite distinct. Second, in my oral review I clearly stated that Darwin was incorrect in that he favoured the *process* of evolution occurring through a series of small steps under environmental pressures caused by factors that one sees in action continuously. This is informally termed 'gradualism', or more correctly 'substantive uniformitarianism'. Darwin was wrong, in that this is *not* the main process through which evolution occurs. It is important here to recognize that there is a gulf of a difference between a *fact*, and an *understanding*. For example, there is no mistaking the fact of gravity. If one releases a stone, it falls to the floor. Both Newton and Einstein realized that, but one of them developed a much better understanding of *why* it falls. As John Herschel wrote in 1830, "We must never forget that it is principles, and not phenomena – the interpretation, not mere knowledge of facts – which are the objects of inquiry to the natural philosopher." Darwin got the phemonenon right – the fact of evolution – but his underlying principle was incorrect. Neither could it be argued that Darwin was wrong only insofar as Newton was 'wrong': Newton's theory fitted the facts as they were then known, whereas Darwin had the data to hand to contradict his hypothesis, for example the work of Cuvier, but rejected them. In my view, the issue of evolution *vs.* creationism is merely a clash between science and non-science, whilst the gradualism *vs.* catastrophism debate involves an attempt to understand how the universe works, a truly scientific pursuit.

Third, Harris's claim that the random, or otherwise, nature of impacts is "only a minor sidelight" ignores a large literature on the subject. Obviously many have considered this to be a matter worthy of study. Without knowing the temporal distribution of impacts it is not possible to understand their effect upon the evolution of life. In any case, the temporal (and physical) nature of impacts is the province of astronomical – as opposed to biological or geological – science, and thus rightly part of the subject matter of this conference.

Fourth, Harris states that there is "mounting evidence" that the claimed Taurid Complex is either coincidental or due to selection effects. The only example that

Earth, Moon, and Planets **72**: 493-494, 1996.

he gives is a paper presented at Mariehamn by A. Morbidelli; that paper is now submitted to *Icarus* with a title 'The dynamics of objects in orbits resembling that of P/Encke', with G.B. Valsecchi being the first author. Valsecchi *et al.* set out to investigate the objects that I, along with Clube, Napier and Asher, have suggested to be aligned with P/Encke and the Taurid meteoroid stream. Valsecchi *et al.* concluded that "...the groupings are either due to chance, or to some other selection effects worth investigating; a third possibility is that there *has been* the hierarchical fragmentation of one or two larger bodies that has given origin to the Taurid complex asteroids...". Harris has therefore misrepresented that paper, which does *not* explain the clustering in apsidal orientation. Regarding selection effects, the only paper that I know of on this topic is that by J. Klačka (see my paper herein), which has yet to be published.

Fifth, Harris's analysis of progress since 1980 with regard to the K/T boundary event ignores historical fact. The Alvarez *et al.* paper was an epoch-making advance, but it did not come out of a vacuum. Halley suggested cometary impacts as being significant in 1694, being joined later by many others; in the modern era, asteroid impacts as agents capable of causing mass extinctions/geological upheavals were mooted by Nininger in 1942; De Laubenfels invoked impacts as potential dinosaur-killers in 1956; in 1958 Öpik suggested that all discontinuities in the geological record are due to massive impacts; Urey pointed out in 1973 that tektites (presumably of impact origin) have ages similar to known boundary events; and in 1979 Clube and Napier suggested that various geological and biological events could be linked to impacts (for example through coincidences between dated large craters and boundary ages). This is, at least in part, the background for the later remarkable development in understanding post-1980. With regard to "proving a hypothesis" (does Harris mean 'substantiating'?), note also the example of plate tectonics, from Wegener's suggestion in the 1920s through to its eventual acceptance in the 1960s/70s. Not all paradoxical theories are correct; but the fact that any specific hypothesis does not become regarded as orthodox in short order is not an argument that it is wrong. Proponents of paradoxical hypotheses in general do not so much object to blind prejudice on the part of critics as to misrepresentation of the facts.

As regards the 'Snowbird test', all that this implies is that the foraminiferal extinctions were abrupt (within a few $\times 10^4$ yr). These seem to be temporally-linked to the large iridium anomaly, and therefore the largest impact around that time, but it does not provide evidence contradicting the comet shower hypothesis. There may have been many impacts around that time (within a few Myr), but none sufficiently energetic to leave a palaeontological/physical signal strong enough to have been recognized to date. This is not the case with other boundary events, however; for example there are (i) Multiple microtektite strewn fields, multiple iridium anomalies, and multiple craters, associated with the Eocene/Oligocene transition; (ii) Similar evidence from the Frasnian/Famennian boundary; and (iii) At least two distinct events contributing to the Late Permian extinctions.

Conference Summary: Meteoroids

Z. Ceplecha
No. 253, 28167 Stříbrná Skalice, Czech Republic

Summarizing our meeting for its meteoroid part is mostly the task of definition, what meteoroids are. Some are certainly small pieces of comets as we know from meteor-stream comet associations. Jewitt in his invited talk draw our attention to definition of terms: asteroid and comet. It is certainly used as phenomenalistic description of the bodies, but it is also used in variance as denoting composition of the bodies. I heard the term "fireball" at least in three completely different meanings at this Conference. For me it is primarily a dazzling bright meteor in the Earth's atmosphere: meteoric fireball. We should somehow redefine our terminology. Enormous progress in observing technics also brought asteroids to meteors and vice versa (I mean the phenomenalistic definition of these words). A new population of bodies in the 10-m size range was revealed. Some tell they are small extinct comets, may be building blocks of them, may come once a while in a form of giant comets into the inner solar system, some prefer origin by impacts of asteroids onto the terrestrial planets.

Nevertheless, there could be no reasonable doubt that these bodies are the most important interplanetary contributors to the Earth environment during time spans of years. Almost half of the influx of interplanetary bodies to the Earth comes from 10- and 100-m size bodies. We got the first hint about it since sixtieth just from the first results of fireball networks, from the observations of somewhat smaller bodies (meteoroids) mostly smaller than 1-m size. Recently the discovery and direct observations of 10-m size bodies (asteroids) put the extrapolation – done from fireball networks – to firm ground. In their invited talk, Jedicke and Gehrels denoted these bodies as an "unheralded" population: certainly from the point of view of extrapolating from our knowledge of asteroids of larger sizes, they were surprising, but from the point of view of extrapolating from small sizes, from knowledge of meteoroids producing bright fireballs we observe, they were expected.

Duncan Steel in his invited talk presented here a thorough evidence of very short time supply of giant comets into the inner parts of the solar system, something as one giant comet each 10^5 years; why not a source for the abundant bodies of 10 m sizes? Clube, Wickramasinghe and Wallis proposed Chiron-like bodies as potential progenitors of smaller bodies in the inner solar system on the same time scale. They may cause a temporary excess of cometary over hard-rock material. We need still more observational evidence, but all this starts to form a well defined scenario. As Brandt et al. posed it, we should try to learn more about small comets with radii less than 1 km: their guess of 10-meter-size comets in the shell from 1 to 2 AU is not far from the observed impacts of these bodies onto Earth. Bottke et al. prefer only asteroidal source for these bodies and came out with Earth-Moon system or Venus impact ejecta as source of 10-m size bodies. I would like to change sign of dt values in their model and denote these "sources" rather sinks of the 10-m-size and larger bodies. We observe them coming continuously into the Earth's atmosphere, enough for sure.

From meteoroid observations we know that bodies of 1- to 10-m sizes contain two basic populations: about 50% of carbonaceous bodies, about 40% of soft cometary bodies, composed of material similar to meteoroids from comet Giacobini-Zinner, and only few percent of hard rock bodies. Napier found two strong dynamical groups of Near Earth Objects connected with two comets: Encke and Giacobini-Zinner. Morbidelli presented here a dynamical study of the Encke's comet system together with the Taurid-meteoroid system. His conclusion that Taurids are both of asteroidal and cometary nature corresponds to what we observe for larger meteoroids of Taurid membership: it is something we were puzzled with Dick McCrosky many years ago, when we wanted to use Taurids as a good cometary signature to calibrate our atmospheric trajectory observations.

These 10-m size bodies collide with the Earth several times in a year. We should also use the Earth atmosphere the same way as we use it for getting data on smaller meteoroids. Observations of the whole globe with the aim of following the atmospheric trajectories of 10-m size bodies from satellites may be the solution of this problem. Recent publication (Flagstaff centennial meeting) by Spalding et al. of data on Feb 1, 1994 huge event (10 to 20-m size) over Pacific close to Marshall Islands is an example what could be done by sensors onboard satellites. Systematic work on observing meteoric fireballs originating from 10-m size bodies with such global systems is urgently needed.

I will also mention the Shoemaker-Levy 9 impact on Jupiter. We heard a preliminary survey by A'Hearn. I will pay attention to impact models of bodies of different sizes in connection to Earth impacts. Our knowledge of impacts of meteoroids into the Earth atmosphere is a very detailed and covers bodies up to several meters in size. Our observations, namely photographic, can yield a set of accurately timed and precise data of the atmospheric flight for each individual body. We can check our theoretical models by such data. We are working with something like ±20 m precision of distances and heights for individual observed points on meteoroid atmospheric trajectory. Our theories are checked by these observations, and moreover calibrated by photographically observed meteorite falls.

Contrary to the regime of atmospheric interaction of meteoroids of several meters in size, most of the theoretical models of very large bodies, say 10 m and larger, contain ablation mostly in the form of explosive fragmentation. After the Shoemaker-Levy 9 impacts on Jupiter, we can say that the pancake model, represented e.g. by the most cited paper of Chyba et al. in Nature and used by Sekanina to predict the maximum brightness of the Jupiter impacts, and also another simple model presented at this conference by Kruchinenko, these models simply did not work. The reason I see in highly idealized simple theoretical approach. On the other hand more sophisticated models by Nemchinov et al. and by Svetsov presented also at this conference are closer to what was observed on Jupiter. The most sophisticated model was computed by Bouslough and Crawford at Sandia National Laboratories; it will soon appear in Geophysical Research Letters and in Transactions of the American Geophysical Union. This model had predicted what then really happened: a back-thrown plume on Jupiter with high velocities in opposite direction, emitting light mainly in IR.

Data on the Peekskill meteorite fall from videorecords of its fireball trajectory from several locations were presented at this conference by myself, Brown et al. Its almost horizontal trajectory also well demonstrated that fragmentation of meteoroids in the Earth atmosphere starts much sooner (7 to 10 Mdyn/cm^2 in this case) before the material strength is reached (may be some 300 Mdyn/cm^2 for the recovered meteorite). This holds for any fragmentation of bodies we photographed as fireballs so far, i.e up to several meters in size.

Another equally well recognized observational evidence is known from spectral records of meteors. Spectral data on fainter meteors were presented at this conference by Borovička and Boček. Based also on Borovička's recent work on fireball spectra, they were able to study three different components in visible and near infrared regions, i.e. cool meteoric, hot meteoric and hot atmospheric components. The fainter the meteor, the more abundant are the atmospheric lines. At this occasion I want to stress a well known observational result for fireballs. The most important regime of atmospheric motion of larger meteoroids up to several meters in size is the ablation with final stage of hot vapor from ablated material. Very detailed spectral records of such meteoroids in the visual pass-band are existing for bodies up to several meters in size (down to -21 absolute maximum magnitude), down to a height of 16 km and for various velocities and they all show an overwhelming radiation of rather low excited metalic atoms (several eV; temperatures 3000 to 5000 K) originating from the ablated meteoric material. Fritzsimmons et al. got spectra of plumes of the Shoemaker-Levy-9 impacts on Jupiter: their spectra are low excited metallic spectra very close to spectra of meteor wakes.

I left the classical meteor topics to the end of my survey. There were many papers on meteor showers presented at this conference. This is quite well understood: there were Perseids coming so soon. Maximum of Perseids happens perhaps just now, almost perfectly at the same time, I am presenting this summary here. Williams predicted richer shower this year than the last one. Perseids destroyed Olympus platform last year and caused some problems on some other spacecrafts. Williams' prediction depends on initial conditions, i.e. on ejection velocities. For well established meteor showers (as Perseids are) we should be able to get these velocities from observations of the shower for many consecutive years and from modeling the stream. Ejection velocities derived using this method by Williams tend to be higher than velocities derived from commonly used simple formulae. Ziolkowski studied Taurid complex by modeling its asteroid component and came to conclusion of common origin of the whole complex. Babadzhanov proposed asteroid (2329) Orthos as progenitor of 8 showers: four of them have been observed. Gustafson and Adolfsson computed influence of radiation pressure on orbits of small meteoroids. Pittich and Klačka presented even more general view of electromagnetic fields changing orbits of small meteoroids. Šidlichovský and Nesvorný presented a theoretical modeling of temporary dust grain capture in exterior resonances with individual planets, solving three body problem with Poynting-Robertson effect included. Quist, Grün et al. presented experimental results for impacts of 10^{-11} to 10^{-22} g, and derived crater scaling law and vapor production.

A survey paper by Elford was presented as a poster and filled up, at least partly, the space of *non-invited paper* on meteors. He attacked general problems of all information we can get from atmospheric trajectories of smaller meteoroids and he paid a special attention to radar observations of meteors. He stressed the fact that composition and structure of these bodies can be obtained from observations. New HF radars have made possible to detect short-lived weak meteor showers extending thus the possibility of establishing more stream-cometary associations.

I'll finish this survey on problems of meteoroids by stressing the importance of impacts for the present development of the solar system. Fresh craters on surfaces of all atmosphereless bodies tell this. Grieve in his invited talk pointed out that in addition to about 140 known impact structures on the Earth surface, knowledge of some 3 to 5 new ones are added each year. Times when community of geologists took everything for volcanoes is gone. Impacts can cause each few hundred thousand years some "nuclear winter" effect, influencing all living world on the globe. Impact structures are one of the important ground truths. The other ground truth we find in observations of motion, radiation and other effects of meteoroids penetrating through the atmosphere and in meteorite bodies and falls. We came now to something I will call Jovian truth: observations of impacts of giant bodies into the atmosphere of Jupiter. There is certainly an analogy and we could learn a lot for giant impacts on Earth. Evaluation and interpretation of immense number of observations of Jupiter phenomena and their modeling will last rather a long time. However, the next time we all researchers of the Small Bodies in the Solar System and their Interactions with the Planets will come together again, we may know more about the bodies, more about giant impacts.

INDEX

504

THE KLUWER LATEX STYLE FILE

Kluwer Academic Publishers has developed a special style file for authors who want to submit LATEX articles. KLUWER.STY is a general LATEX style file which is used for all Kluwer journals, irrespective of the publication's size or layout. (The specific journal characteristics are added later during the production process.) Authors are kindly requested always to use KLUWER.STY when creating a LATEX article for a Kluwer journal.

Instruction File
Although KLUWER.STY is very similar to the ARTICLE.STY and uses many of the standard LATEX commands, there are some differences. These are explained in the accompanying instruction file - KAPINS[number].TEX

Getting the Kluwer Style File

KLUWER.STY is offered at a number of servers around the world. Unfortunately, those are unauthorized copies and authors are strongly advised not to use them. Kluwer can only guarantee the integrity of files obtained directly from Kluwer.

Gopher-server, E-mail or Air Mail
Authors can obtain KLUWER.STY and the instruction file from Kluwer Academic Publishers' GOPHER-server. This free service is available at:

GOPHER.WKAP.NL
IP-Number: 192.87.90.1
WWW URL: gopher://gopher.wkap.nl/

The stylefile may be requested from:

Kluwer Academic Publishers, Editorial Department,
P.O.Box 17, 3300 AA Dordrecht, The Netherlands.
Telephone: (0)78-6392392, Fax: (0)78-6392254
E-mail: EDITDEPT@WKAP.NL

Earth, Moon and Planets **72**: 507–508, 1996.
© 1996 *Kluwer Academic Publishers.*

The files can be sent either by e-mail or on diskette. Don't forget to mention the journal's name, your e-mail number, and postal address.

Submitting Manuscripts

Please send your completed LaTeX article on diskette, together with the appropriate number of hard copies to the address listed in the *Instructions to Authors* of your journal.

Via E-mail

Experience has shown that sending articles via e-mail can sometimes result in lacunas appearing in the article. That's why Kluwer prefers to receive LaTeX articles on diskettes, accompanied by the hard copies. If you must send the LaTeX article via e-mail, don't forget to send the requisite number of hard copies by air mail.

Questions

Should you have any questions or encounter problems using KLUWER.STY, please contact Kluwer Academic Publishers for assistance.